FIELD THEORY, DISORDER AND SIMULATIONS

World Scientific Lecture Notes in Physics

ISSN: 1793-1436

---

*Published titles**

*For the complete list of published titles, please visit
http://www.worldscientific.com/series/wslnp

FIELD THEORY, DISORDER AND SIMULATIONS

GIORGIO PARISI
II Università di Roma Tor Vergata

World Scientific
Singapore • New Jersey • London • Hong Kong

Published by

World Scientific Publishing Co. Pte. Ltd.
5 Toh Tuck Link, Singapore 596224
USA office: 27 Warren Street, Suite 401-402, Hackensack, NJ 07601
UK office: 57 Shelton Street, Covent Garden, London WC2H 9HE

British Library Cataloguing-in-Publication Data
A catalogue record for this book is available from the British Library.

World Scientific Lecture Notes in Physics — Vol. 49
FIELD THEORY, DISORDER AND SIMULATIONS

ISBN-13 978-981-02-0964-3
ISBN-10 981-02-0964-9
ISBN-13 978-981-02-1356-5 (pbk)
ISBN-10 981-02-1356-5 (pbk)

CONTENTS

Part III. Simulations

INTRODUCTION

This volume contains a number of my lectures, given over the years at several schools and conferences, together with some research papers that are meant to complement them. The lectures are naturally grouped into three subjects that have been the main themes of my research: *Field Theory, Disordered Systems and Computer Simulations.*

Field Theory

The status of quantum field theory is now quite different from what it was when I started to work in physics. At that time, relativistic quantum field theory, with the exception of perturbative quantum electrodynamics, was considered to be an ill-defined and "dangerous" field. In the early seventies, the situation changed dramatically: theoretical and experimental progress demonstrated that quantum field theory is a necessary description for strong interactions. Moreover, the work of Wilson showed that the theory of strongly interacting fields can be controlled in the nonperturbative region, and that one can predict the critical exponents for second-order phase transitions.

The first lecture of the section devoted to field theory (originally written for Cargese 1973), "Field Theoretic Approach to Second-Order Phase Transitions in Two- and Three-Dimensional Systems," contains a full discussion of the method used to compute critical exponents using renormalization group ideas and the Callan–Symanzik equation. This paper is particularly interesting because it contains the first field theoretical computation of the critical exponents at fixed dimensions, without using the epsilon expansion in $4-\varepsilon$ dimensions. In the second paper, "On Nonrenormalizable Interactions," the methods described in the first are applied to the nonperturbative construction of nonrenormalizable interactions.

In "An Introduction to Scaling Violations," one finds a simple introduction to the theory of quantum chromodynamics and to the perturbative computations (based on the renormalization group) which can be done in the short-distance region.

"The Physical Basis of the Asymptotic Estimates in Perturbation Theory" is a presentation of some of the main results obtained in the study of the large-order behaviour of perturbation theory and related problems (e.g. Borel resummability). Perturbation theory is our main tool for studying field theory, and it is crucial to understand its limits and its convergence. "Critical Exponents and Large-Order Behaviour of Perturbation Theory" is an application of the techniques described in the previous paper to the computation of critical exponents using the renormalized perturbative expansion.

"The Borel Transform and the Renormalization Group," "Singularities of the Borel Transform in Gauge Theories" and "On Infrared Divergences" are, respectively, a short lecture and two papers devoted to the study of the large-order behaviour of perturbation theory (or equivalently of the singularities of the Borel transform) in renormalizable theories, where new phenomena are present. The case of asymptotically-free field theories is studied in detail.

"Quartic Oscillator" is related to the previous papers: it deals with the convergence problems of an asymptotic expansion in a quite different context: one studies the properties of the WKB expansion for a one-dimensional Hamiltonian with a potential equal to x^4. Finally "Trace Identities for the Schroedinger Operator and the WKB Method" is a short related paper in which one derives the trace identities for the same Hamiltonian.

Disordered Systems

My best contribution to physics is possibly the theory of broken replica symmetry that was first constructed in order to solve the spin glass model with long range interactions. This theory is rather complex and contains some points that are still not fully understood. The theoretical approach based on broken replica symmetry is quite general, and has been successfully applied to many other systems, like neural networks and interfaces in random media.

In "An Introduction to the Statistical Mechanics of Amorphous Systems," one finds a full discussion of the replica method as applied to spin glasses, to stochastic differential equations and to localization in disordered systems. The theory of broken replica symmetry is described in detail.

Stochastic differential equations are also the subject of the next paper, "Supersymmetric Field Theories and Stochastic Differential Equations." Here it is shown how some stochastic differential equations yield to relativistically invariant probability distributions.

"Spin Glasses and Optimization Problems Without Replicas" contains the reformulation of the broken replica theory in probabilistic terms using a self-consistent method (called cavity approach). The techniques used here are much more standard from a mathematical point of view: the replica method is not used and the dimensions of matrices is never smaller than 1.

While in "Spin Glass Theory" one finds a short technical review of spin glasses, in "On the Emergence of Tree-Like Structures in Complex Systems," the main results of the theory of spin glasses are presented using plain language; indeed this contribution was aimed at a nonspecialized audience.

The paper "On the Multifractal Nature of Fully Developed Turbulence and Chaotic Systems" has nothing to do with spin glasses. Rather, it contains a different, more geometrical, approach to some disordered systems, based on the new concept of multifractals.

Simulations

Computer simulations in physics play a role that is intermediate between experiments and theory. Indeed, when one wants to explain a new phenomenon in a complicated setting, one first proposes a simplified model which one tries to analyze theoretically. Computer simulations, when possible, are extremely useful for understanding the properties of a new model for many reasons.

• They allow a comparison between the predictions of the model and the experiments. In this way, it is possible to find out whether the proposed model is adequate in describing the experimental data.

• Theoretical predictions for the result of the computer simulations can be done in a rather simple setting where there are no doubts on the correctness of the chosen Hamiltonian. In computer simulations, every quantity is directly accessible to measurements, and this feature allows a more severe test of the theory.

• Sometimes the system is so complicated that no detailed theoretical predictions can be done for some quantities. This is the case of quantum chromodynamics (QCD), where quantitative predictions on the hadronic spectrum may be obtained only via numerical simulations.

The main subject of the remaining lectures is numerical simulations for QCD. This subject is studied from different points of view: general theory, real simulations and construction of dedicated hardware.

"Recent Progresses in Gauge Theories" is a simple introduction to gauge theory and numerical simulations, written at the time of the first numerical computations for lattice gauge theories.

"The Strategy for Computing the Hadronic Mass Spectrum" and "Prolegomena to Any Future Computer Evaluation of the QCD Mass Spectrum" are two lectures written when first generation simulations were already done and the people involved were examining, with a critical eye, all possible sources of systematic errors and trying to find the best strategy for reducing the statistical error.

"A Short Introduction to Numerical Simulations of Lattice Gauge Theories" and "Principles of Numerical Simulations" are two lectures in which the general principles of numerical simulations are exposed; some of the more recent results for lattice gauge theories are also presented.

In "The APE Computer: An Array Processor Optimized for Lattice Gauge Theory Simulations," one finds a description of the APE computer (hardware and software). The results that have been obtained using this very powerful computer (10^9 floating point operations per second) have been presented in some of the previous papers.

"The APE-100 Computer: (I) The Architecture," contains the description of the architecture and the general principles of software for the new generation APE-100 computer that should reach the speed of 10^{11} floating point operations per second. At this time, a prototype of a section of APE-100 has been constructed; it reaches the speed of 6×10^9 floating point operations per second.

The list of people to whom I have been scientifically indebted in this twenty-year period is too long to be inserted here. I thank all of them warmly. I also thank the organizers of the congresses and schools where I have presented the material collected here. In particular, I am grateful to the organizers of Cargese and Les Houches, who have been so kind as to invite me so many times.

Part I. Field Theory

Journal of Statistical Physics, Vol. 23, No. 1, 1980

Field-Theoretic Approach to Second-Order Phase Transitions in Two- and Three-Dimensional Systems

Giorgio Parisi[1]

Received November 15, 1979

We review the physical principles which are at the basis of recent field-theoretic computations of the critical exponents in two- and three-dimensional systems. We concentrate on those points that do not show up explicitly in the more standard ϵ-expansion: they must be discussed with care if one uses a perturbative approach at fixed space dimensions (the loop expansion). We present in detail simple computations of the critical exponents, while we summarize the results of longer and more accurate computations.

KEY WORDS: Field theory; second-order transitions; critical exponents; ϵ-expansion; loop expansion; renormalization group; Callan–Symanzik equation.

1. INTRODUCTION

There have been many recent approaches toward a deeper understanding of second-order phase transitions. The phenomenological Kadanoff–Widom scaling laws[1,2] are in quite good agreement with the experimental data and the high-temperature expansions (for a review see Ref. 3). Also, the universality hypothesis[1] (independence of the critical exponents from the detailed structure of the interaction) seems to be satisfied. The intensive use of the renormalization group[4–7] has produced a neat derivation of the scaling laws for static and dynamic phenomena and has clarified the deep reasons for the validity of the universality hypothesis. Very simple approximate computations have been done for many systems, ranging from spin-glasses[8] to Reggeon field theory.[9]

In the framework of a field-theoretic approach high-precision estimates of the critical exponents of the three-dimensional Ising model have been done.[10,11] These estimates involve an explicit evaluation of all diagrams up

[1] Laboratori Nazionali INFN, Frascati, Italy.

to six loops and approximate predictions for the size of the neglected higher loop diagrams.[12]

The aim of this paper is to describe in detail the foundations of the method used in these high-precision estimates; indeed, this paper is a shortened, revised version of unpublished 1973 Cargese Lecture Notes,[13] which were the basis of the computations of Refs. 10 and 11.

To present a microscopic derivation of the scaling laws is rather difficult. Many subtle questions must be settled: thermodynamic quantities must be computed just at the point where their dependence on the temperature is not analytic; any sort of high- or low-temperature expansion is divergent. The theory of second-order phase transitions is dominated by the quest for functions which are regular near the transition.

In this paper we concentrate on the derivation of the static scaling laws for a magnetic system above the transition at zero external field. Our treatment has many points in common with the standard approach[14–18]; the main difference is that we always work in the massive (finite correlation length) theory also at the critical temperature. This approach bypasses the problems connected with the infrared divergences present in a perturbative approach; the theory admits a closed formulation in a system of arbitrary dimensions.

This paper is divided into nine sections. In Section 2 we present the model which we will consider as the prototype of a system undergoing a second-order phase transition and which we will study in the other sections. In Section 3 we explain why a straightforward approach does not work near the critical temperature, and how infrared divergences arise. In analogy with quantum electrodynamics, we conjecture that the introduction of renormalized quantities will avoid the problems. All the results derived in the rest of the paper are based on this conjecture: a rigorous proof is lacking, although there are results, based on the Lebowitz inequality, which go in this direction.[19] In Section 4 we derive the Kadanoff–Widom scaling laws and prove that the critical exponents are connected to renormalized correlation functions computed at some finite value of the external momenta. In the next section we study the behavior of the correlation functions at the critical temperature. An exact expression for these correlation functions has been found. If it is expanded in powers of the bare coupling constant, infrared divergences appear; however, if a different expansion is used, finite results are obtained also at the critical temperature. Infrared divergences show up in a non-analytical dependence of the correlations functions on the bare coupling constant. In Section 6 we present simple applications of the formalism; critical exponents are computed using very simple approximations; the results of much more lengthy and accurate computations are reported. In the next section we show that the range of allowed values for the renormalized coupling constant is a closed interval. In Section 8 we show how the concept of a universal critical behavior arises naturally in this framework; we derive a universal equation for the correlation functions at the critical temperature. In the last section we present our final conclusions and comments.

In the appendix we explain some notations used in this paper.

2. THE MODEL

In this paper we concentrate on a special kind of continuous Ising model in the limit of zero lattice spacing. In presence of an external magnetic field H the partition function is

$$Z(H, \beta) \propto \int d[\phi] \exp\left[- \int d^D x \, \mathscr{L}(x) \right] \tag{2.1}$$

where

$$\begin{aligned} \mathscr{L}(x) &= \tfrac{1}{2}\partial_\mu \phi \, \partial_\mu \phi + \tfrac{1}{2} M^2(\beta)\phi^2 + (u/4!)\phi^4 \\ M^2(\beta) &= M^2(0) + \beta; \qquad \beta = 1/KT \end{aligned} \tag{2.2}$$

D is the dimension of the space and $\int d[\phi]$ stands for functional integration. In the two- and three-dimensional cases, there are rigorous proofs of the existence of the model and a precise mathematical meaning can be given to the words "functional integration." The only ultraviolet divergences present can be absorbed by assuming that $M^2(0)$ is a polynomial in g with infinite coefficient; i.e., in the language of field theory, the interaction is super-renormalizable for $D < 4$.[19]

In the limit $u \to 0$ we obtain the Gaussian model and the perturbative expansion in powers of u can be easily obtained. It is identical to the Feynman graph expansion for a relativistic ϕ^4 theory in $D - 1$ space, one time dimension. It has been rigorously proved that the correlation functions of the two theories are connected through a Wick rotation.[20,21] The correlation functions of statistical mechanics are the Schwinger functions[22] (Wightman functions[23] at imaginary time in relativistic theory). In Fourier space they are the analytic continuation of the time-ordered functions in the Euclidean region.

This model can be generalized by introducing a multiplet of N fields which form a representation of the $O(N)$ group. The partition function is[6,24]

$$Z(H_i, \beta) \propto \int \prod_1^N{}_i \, d[\phi_i] \exp\left[- \int d^D \mathscr{L}(x) \right] \tag{2.3}$$

where

$$\mathscr{L}(x) = \tfrac{1}{2}\partial_\mu \phi_i \, \partial_\mu \phi_i + \tfrac{1}{2} M^2(\beta)\phi_i\phi_i + (1/4!)(\phi_i\phi_i)^2 \tag{2.4}$$

If we take $N = 1$, this model reduces to the first one. For $N = 2$ we recover both the Ginzburg–Landau[25] partition function of a superconductor and the Landau[26,27] partition function of superfluid helium near the λ transition. A special kind of continuous Heisenberg model is obtained for $N = 3$. In the limit N goes to infinity we get the spherical model.[28]

At each order in perturbation theory the N dependence of the correlation functions comes from multiplicity factors which multiply each Feynman diagram. These factors are polynomial in N and they can be analytically continued to noninteger N. It has been suggested that the $N = 0$ correlation functions are connected to some properties of the self-avoiding walk problem.[29] It may be of interest to note that for $N = -2$ the two-point correlation function above the transition coincides with that of the Gaussian model.[30]

The partition function of a "realistic" spin-1/2 model can be represented in a similar way: even higher powers of ϕ are present and the Lagrangian is no longer a polynomial.[31] According to the conventional wisdom, higher order couplings are irrelevant as far as the critical behavior is concerned; the model (2.2) and the Ising model on a real lattice belong to the same universality class. The arguments which lead to this belief are briefly discussed in Section 8.

The following simple remark will be quite useful in the rest of the paper: the argument of the exponential is a pure number and all the quantities have the dimension of a length to some power. Defining the length dimension to be -1, we find

$$\begin{gathered}
[x] = -1, \qquad [d/dx] = 1, \qquad [\phi] = (D-2)/2, \qquad [M] = 1, \\
[u] = 4 - D, \qquad [\phi^2] = D - 2, \qquad [G_2] = -2, \qquad [\Gamma_2] = 2 \\
[G_N] = D - N(\tfrac{1}{2}D + 1), \qquad [\Gamma_N] = D - N[-1 + \tfrac{1}{2}D], \\
[G_{N\phi^2}] = [G_N] - 2, \qquad [\Gamma_{N\phi^2}] = [\Gamma_N] - 2, \qquad [D_{\phi^2\phi^2}] = D - 4
\end{gathered} \tag{2.5}$$

where the square brackets stand for "dimension of," and G_N and Γ_N are respectively the Fourier transforms of the connected N-point correlation function and of the amputated, one-particle, irreducible N-point correlation function. The notation is defined in the appendix.

Dimensional analysis may be used extensively; e.g.,

$$G_2(P, u, M) = \frac{1}{M^2} A\left[\frac{u}{M^{4-D}}, \frac{P}{M}\right] = \frac{1}{P^2} A^1\left[\frac{u}{P^{4-D}}, \frac{P}{M}\right] = \cdots \tag{2.6}$$

3. THE ASSUMPTIONS

The simplest approximation we can make is to take $u = 0$, or neglect terms proportional to high powers of the field. Naive arguments suggest that this approximation may be valid only for dimensions greater than four. However, in lower dimensions also it gives a first semiqualitative description of a second-order phase transition.[1]

In this situation, the model can be easily solved.[32] The two-point correlation function is

$$G_2(K, T) = 1/(K^2 + M^2) \tag{3.1}$$

All other connected correlation functions are zero. The transition temperature is at $M^2 = 0$. The transition is characterized by the fact that the correlation functions become singular at zero external momenta. These singularities are produced from the nonexponential decrease of correlation functions in the configuration space at large distances.

Introducing the reduced temperature $\tau \propto (T_c - T)/T_c$, we can write $G_2(K, \tau)$ for small τ as

$$G_2(K, \tau) = \tau^{-\gamma} f(K/\tau^{\nu}) \tag{3.2}$$

where

$$\gamma = 1, \qquad \nu = \tfrac{1}{2}, \qquad f(x) = 1/(1 + x^2) \tag{3.3}$$

The scaling laws[1] state that a formula similar to (3.2) is valid near the transition for not too large K and also for u different from zero. The critical exponents γ and ν and the function h may, however, be different from (3.3). It is also assumed that (3.2) has a finite, nonzero limit when $\tau \to 0$ at K different from zero:

$$G(K, 0) = 1/K^{2-\eta}, \qquad 2 - \eta = \gamma/\nu \tag{3.4}$$

The main problem of the theory of second-order phase transitions is to prove this hypothesis and to compute ν, γ, and h.

If u is different from zero, we can develop the partition function and the correlation function in powers of u, expanding the exponential in (2.10) (see the appendix)

$$\begin{aligned} Z \propto \sum_{k=0}^{\infty} \frac{u^k}{k!} \int d[\phi] \int \prod_{i=1}^{k} d^D y_i\, \phi^4(y_i) \\ \times \exp \int d^D x \left[-\tfrac{1}{2}\partial_u \phi(x)\, \partial^u \phi(x) - M^2 \phi^2(x) \right] \end{aligned} \tag{3.5}$$

Unfortunately, this expansion is not convergent and it may be regarded only as an asymptotic expansion. It has been proved[33] that (3.5) is not convergent, but its Borel sum exists and is equal to the functional integral (2.10). These problems are connected with the nonexistence of the theory for negative u.

Near the phase transition new pathologies arise. The first effect of introducing a nonzero coupling constant is a shift in the transition temperature. This difficulty may be bypassed by making a perturbation in u not at fixed M^2, but at fixed $M^2 - M_c^2 = \bar{M}^2$. Then $M_c^2(u)$ is the point at which the phase transition occurs. Near the transition, the most singular part of the dependence of the correlation function on the temperature comes from $\bar{M}^2$ and not from u. We can consistently assume that $\bar{M}^2$ is proportional to τ and neglect the temperature dependence of u. The error involved in this approximation affects only terms that are not singular near the critical temperature.

Although the introduction of $\bar{M}$ improves the situation, it remains hopeless. The dimensionless coupling constant in which we are making our expansion is $u/\bar{M}^{4-D}$; it goes to infinity when $\bar{M}$ goes to zero in any dimension less than four. The perturbative expansion is useless if $\bar{M}$ is small, i.e., near the transition. It is well known that it is very hard to reconstruct the behavior of a function $f(x)$ when x goes to infinity from the knowledge of the first few terms of its Taylor expansion around $x = 0$.

The situation is still worse if we start directly from $M = 0$ and limit ourselves to the study of correlation functions at nonzero external momenta. In this case the perturbative expansion does not exist, because of infrared divergences.

Let us study a simple example: the diagram

(3.6)

in the massless case is proportional to K^{-4+D} if $2 < D < 4$. Its contribution to Γ_4 is proportional to the integral

$$\int d^D p \frac{1}{p^2(p+K)^2} \tag{3.7}$$

This integral is infrared-convergent at K different from zero if $D > 2$ and ultraviolet-convergent if $D < 4$; in this interval of dimension, power counting implies a simple power behavior in K.

The chain of N bubbles

(3.8)

factorizes into a product of N integrals and is proportional to $K^{(D-4)N}$. If $(D - 4)N < -D$, the Fourier transform with respect to K cannot be performed, and the vertex is no longer an L^1 function. Any diagram having this chain as subgraph is divergent, e.g.,

(3.9)

We will see later that these divergences are connected to the fact that correlation functions are not C^∞ functions of u around $u = 0$ in the zero-mass case. This effect is a subtle one; it is present only at an order $N > D/(4 - D)$. If we work in a dimension near enough to four, it appears only at a very high order in the coupling constant and it disappears completely in dimensions infinitesimally near to four. We stress that any computation of the critical exponents done in $4 - \epsilon$ dimensions[5] using the renormalization group in the zero-mass theory does not contain any information on the critical behavior of a system in 3.99 dimensions without an additional hypothesis on the resummation of these infrared singularities.

The conclusion is that the limit $\bar{M} \to 0$ at fixed u is strictly connected with the limit $u \to \infty$ at fixed $\bar{M}$ and cannot be studied by treating u as a small perturbation.

Similar problems are also present in renormalizable theories such as quantum electrodynamics (QED) in four dimensions: at each finite order in perturbation theory, the presence of the ultraviolet divergence destroys the possibility of performing an expansion in powers of the bare coupling constant.[34] After more than fifteen years of theoretical work, this difficulty has been overcome by introducing renormalized parameters such as the physical charge and mass of the electron and by performing an expansion of correlation functions in powers of the renormalized parameters.[35,36] The perturbative expansion for correlation functions at low external momenta is perfectly well defined and regular, although the bare parameters are infinite at each order in perturbation theory. The drawback is that in the large-momentum region the effective coupling constant is of order $\alpha \lg K^2$, and clearly blows up when the momenta go to infinity.[36] Renormalized perturbation expansion is useless in this region because the large-momentum region is controlled by the bare coupling constant and not by the renormalized one. The same phenomenon happens also in our case if $D = 4$. If $D < 4$, problems in the critical region arise from the fact that the dimensionless bare coupling constant $u/\bar{M}^{(4-D)/2}$ goes to infinity, as in QED. We hope that the introduction of the renormalized integral equation will be of some help. The perturbative solution of these equations produces the renormalized perturbation expansion (RPE). The analogy with QED suggests that we can trust RPE only for the results concerning correlation functions computed at low external momenta, but we should not believe in any direct computation of the bare quantities or of the behavior of correlation functions in the large-momentum region, although all these quantities have a well-defined perturbative expansion in powers of the renormalized coupling constant.

We are led to the following conjecture: All renormalized correlation functions (to be defined later) at fixed external momenta have a finite limit when the bare coupling constant goes to infinity at fixed m. The whole theory of second-order phase transitions is explicitly or implicitly founded on this hypothesis, whose intuitive justification comes from the idea that a too strong repulsive interaction shields itself, producing a finite result. An explicit realization of this phenomenon may be found in the large-N limit[37–39] at all orders in $1/N$ or in the $D = 1$ case (anharmonic quantum oscillator). We note that the bare coupling constant is the limit of the four-point functions when the external momenta become very large. Our conjecture states that we can reach a situation where the four-point function goes to infinity in the large-momentum region, but no infinity is present in the correlation functions in the finite-momentum region. In a very rough sense, the low- and the high-momentum correlation functions are decoupled in the integral equations. The large-momentum behavior of correlation functions comes from the internal region of integration where all the momenta are large, and the main contribution to the correlation functions in the low-momentum region comes from the region of integration where all the momenta are low.[40] We suppose that we can find solutions to these equations whose high-energy behavior is singular with a perfectly regular low-energy behavior. This may be possible if the large-momentum behavior of the four-point function is such that no new ultraviolet divergences are created when the bare coupling constant goes to infinity. If this condition is not violated, the low-momentum behavior is quite insensitive to the high-momentum behavior. The construction of RPE is straightforward and it is discussed in detail in many books (see, e.g., Refs. 36). We present here only a sketch of the fundamental steps. We introduce renormalized ϕ and ϕ^2 fields which are proportional to the bare ones:

$$\phi_R = Z_1^{-1/2}\phi, \qquad \phi_R{}^2 = Z_2^{-1}\phi^2 \tag{3.10}$$

The constants of proportionality are fixed from the conditions

$$\frac{d}{dK^2}[\Gamma_2{}^R(K^2)]|_{K^2=0} = 1, \qquad \Gamma^R_{2,\phi^2}(0,0) = 1 \tag{3.11}$$

The renormalized mass and coupling constant are defined as

$$m^2 = \Gamma_2{}^R(0) = [G_2{}^R(0)]^{-1}, \qquad g = \Gamma_4{}^R(0,0,0)m^{D-4} \tag{3.12}$$

The Z_1, Z_2, and g are clearly dimensionless. Note that in general $\phi_R{}^2$ is only proportional but not equal to $(\phi_R)^2$. It is easy to check that

$$\begin{aligned} m^2 &= M^2[1 + O(u/M^{4-D})] \\ Z_1 &= 1 + O[u^2/M^{2(4-D)}] \\ Z_2 &= 1 + O(u/M^{4-D}) \\ g &= u/m^{4-D} + O[u^2/M^{2(4-D)}] \end{aligned} \tag{3.13}$$

The physical meaning of these parameters follows from the relation

$$G_2(K) = Z_1/[m^2 + K^2 + O(K^4)] \tag{3.14}$$

m^{-1} is the correlation length ξ; $\bar{m}^{-2} = \bar{m}^{-2} \cdot Z_1$ is the magnetic susceptibility χ; Z_2 is $Z_1(d/d\tau)\bar{m}^2 = Z_1(d/dM^2)\bar{m}^2$; and $D_{\phi^2,\phi^2}(0)$ is the specific heat, i.e., the second derivative with respect to the temperature of the logarithm of the partition function. All the correlation functions have an expansion in powers of g at fixed m, where no divergences appear in four dimensions also. However, in this framework we cannot compute the bare coupling constant without studying the large-momentum behavior of the theory, going outside of the range of validity of RPE. The problem that we have to solve is to compute the bare coupling constant using as input only correlation functions in the low-momentum region, or to compute the large-momentum behavior from the low-momentum behavior. This problem seems very hard because we know that the two momentum regions are decoupled in the integral equations. The answer to such problems is contained in the following sections and is the main result of this work.

We recall that the following has been rigorously proved, using the Lebowitz inequality[19]:

$$0 \leqslant g \leqslant A \tag{3.15}$$

where A is a computable constant of order 1. If $g(u)$ is a monotonically increasing function (or has only a finite number of oscillations), as is reasonable, then $\lim_{u \to \infty} g(u)$ exists and is finite. Although a crucial inequality is missing to extend this argument to all Green's functions, the rigorous result (3.15) strongly supports the hypothesis of finiteness of the renormalized Green's functions in the infinite coupling limit.

The reader may observe that the whole procedure seems terribly complicated. Why does one not use the integral equation directly in the large-momentum region, which decouples from the low momenta, and solve the equations in the large-momentum region? The answer is that this alternative approach has been tried in the past.[40–44] Solving integral equations is not an easy job and it is hard to make good approximations.

An explicit computation of the state equations seems to be nearly impossible. As far as the critical exponents are concerned, safe results are obtained in the $1/N$ expansion[44] and they coincide with those obtained in the conventional approach. Rather good results for the critical exponents have been obtained also in the case $N = 1$ by truncating in an appropriate way the nonlinear integral equations for the massless theory and by making full use of the constraints dictated by conformal invariance.[45]

4. THE SCALING LAWS

The problem we will study in this section is how to compute the bare coupling constant and the renormalization constants as functions of the renormalized coupling constant.

It is possible to prove by direct inspection of the diagrams in perturbation theory that

$$Z_1^{-1} = \lim_{\lambda \to \infty} \lambda^2 K^2 G_2{}^R(\lambda K), \qquad |K| \neq 0$$

$$(Z_2 Z_1)^{-1} = \lim_{\lambda \to \infty} \lambda^2 K^2 G^R_{2\phi^2}(\lambda K, \lambda p), |K| \neq 0, \quad |p| \neq 0, \quad |K+p| \neq 0$$

$$Z_1{}^2 u = \lim_{\lambda \to \infty} \Gamma_4{}^R(\lambda p_1, \lambda p_2, \lambda p_3), \qquad |p_1| \neq 0, \quad |p_2| \neq 0, \quad |p_3| \neq 0$$

$$|p_1 + p_2| \neq 0, \quad |p_1 + p_3| \neq 0$$

$$|p_3 + p_2| \neq 0, \quad |p_1 + p_2 + p_3| \neq 0 \tag{4.1}$$

The interaction does not change the large-momentum behavior of the correlation functions.

However, (4.1) is of no practical interest insofar as it involves correlation functions computed in the very large-momentum region, where we cannot trust RPE. Our goal is to find formulas which are equivalent to (4.1), but involve only correlation functions computed at fixed external momenta.

We introduce the differential operator $\bar{\Delta}$ defined by[46,47]

$$\bar{\Delta} = \bar{m} \frac{\partial}{\partial \bar{m}^2}\bigg|_u = \frac{\partial}{\partial \lg \bar{m}^2}\bigg|_u \tag{4.2}$$

Using definitions (3.10) and (3.11), we prove the following chain of identities:

$$\bar{m}^2 \frac{\partial G_N}{\partial \bar{m}^2}\bigg|_u = \bar{m}^2 \frac{\partial M^2}{\partial \bar{m}^2}\bigg|_u \frac{\partial G_N}{\partial M^2}\bigg|_u = \bar{m}^2 \frac{\partial M^2}{\partial \bar{m}^2}\bigg|_u G_{N\phi^2}$$

$$= \bar{m}^2 \frac{\partial M^2}{\partial \bar{m}^2}\bigg|_u Z_2 G_{N\phi_R^2} = \bar{m}^2 G_{N\phi_R^2} \tag{4.3}$$

The action of the $\bar{\Delta}$ operator on correlation functions can be computed in RPE; finite results are obtained also if the theory is renormalizable, like QED, in four dimensions. Instead of $\bar{\Delta}$ we will use the operator Δ, which is proportional to $\bar{\Delta}$; the proportionality constant is fixed from the condition $\Delta m^2 = m^2$. We introduce Δ only to be closer to the standard notation.[46] The following functions can be defined.

$$Z_1^{\ 2}\Delta\Gamma_4(0,0,0) = m^{4-D}h(g)\Delta\frac{\partial}{\partial K^2}\Gamma_2^R(K^2) = -c_1(g);$$
$$\Delta\Gamma^R_{2,\phi_R^2}(0,0) = c_2(g) + c_1(g) \tag{4.4}$$

Using definitions (3.10)–(3.12), we find

$$\Delta g = -\tfrac{1}{2}(4-D)g + h(g) - 2gc_1(g) \equiv b(g) = -\tfrac{1}{2}(4-D)g + O(g^2)$$
$$\Delta Z_1 = c_1(g)Z_1, \qquad \Delta Z_2 = c_2(g)Z_2 \tag{4.5}$$

We introduce the notation

$$u = m^{4-D}N(g) \tag{4.6}$$

We apply the Δ operator to both sides of (4.6):

$$\Delta u = 0 = m^{4-D}[\tfrac{1}{2}(4-D) + b(g)\,\partial/\partial g]N[g] \tag{4.7}$$

The solution of the differential equation (4.2) is uniquely fixed by the condition (3.13)

$$N(g) = g\exp\int_0^g\left(\frac{D-4}{2b(g')} - \frac{1}{g'}\right)dg' \tag{4.8}$$

$b(g)$ is negative for small g, $N(g)$ is monotonically increasing in the region between 0 and the first zero of $b(g)$. In this region we can define an inverse function ρ such that

$$g = \rho[u/m^{4-D}] \tag{4.9}$$

The same technique can be used to obtain

$$Z_1(g) = \exp\int_0^g \frac{c_1(g')}{b(g')}dg', \qquad Z_2(g) = \exp\int_0^g \frac{c_2(g')}{b(g')}dg' \tag{4.10}$$

Let us now try to use these formulas to study the critical behavior of the partition function.

In the limit $u/m^{4-D} \to \infty$ the integral defining $N(g)$ must diverge, and this is possible only if the $b(g)$ function has a zero at a positive point g_c. If we suppose that the function has a simple zero with slope b', then

$$b' = \frac{d}{dg}b(g)\bigg|_{g=g_c} > 0 \tag{4.11}$$

Equations (4.8)–(4.10) simplify in the limit $u \to \infty$:

$$\frac{u}{m^{4-D}} = H\left(\frac{g_c-g}{g_c}\right)^{-(4-D)/2b'}, \qquad g = g_c\left[1-\left(\frac{u}{Hm^{4-D}}\right)^{-2b'/(4-D)}\right] \tag{4.12}$$

where

$$H = g_c \exp \int_0^{g_c} \left[-\frac{4-D}{2b(g')} - \frac{1}{g} - \frac{4-D}{2b'(g_c - g)} \right] dg' \tag{4.13}$$

If $b' = 0$, but $b'' \neq 0$,

$$g = g_c - \frac{1}{\lg(u/m^{4-D})} + \cdots \tag{4.14}$$

In any case g goes to g_c when $u^{1/4-D}/m$ goes to infinity. Note that the bare coupling constant has a negative power in (4.12).

Putting (4.12) into (4.10), we obtain

$$Z_1 = m^{2c_1^c}\left\{1 + O\left[\left(\frac{m}{u^{1/4-D}}\right)^{2b'}\right]\right\}, \qquad Z_2 = m^{2c_2^c}\left\{1 + O\left[\left(\frac{m}{u^{1/4-D}}\right)^{2b'}\right]\right\} \tag{4.15}$$

where $c_1^c = c_1(g_c)$ and $c_2^c = c_2(g_c)$. Using the definitions of Z_1 and Z_2, we find

$$dm^2/dM^2 = dm^2/dT = m^{2(1/2\nu - 1)} \tag{4.16}$$

Relation (4.16) implies that, if $\nu > 0$,

$$m = \tau^{\nu}, \qquad Z_1 = m^{\eta} \tag{4.17}$$

where

$$\eta = 2c_1^c, \qquad \nu = 1/(2 - 2c_2^c) \tag{4.18}$$

The relations (2.13) and (3.14) imply the scaling law for the two-point correlation function:

$$G_2(K, \tau) = (1/\tau^{\gamma})\{f_0[K/\tau^{\nu}] + \tau^{\nu\omega} f_1[K/\tau^{\nu}] + \cdots\} \tag{4.19}$$

where

$$\gamma = \nu[2 - \eta], \qquad \omega = 2b' \tag{4.20}$$

The limit $\tau \to 0$ is done at K/τ^{ν} constant; the study of the behavior of correlation functions in the limit $\tau \Rightarrow 0$ at fixed K requires a different and more complicated analysis, which will be the object of the next section.

The corrections to scaling laws have a simple power behavior, which is connected to b'. Similar scaling laws can be derived for high-order correlation functions. The same technique can be used to study the behavior of the specific heat $C = D_{\phi^2\phi^2}(0)$. We introduce the function

$$\Delta[D_{\phi_R{}^2\phi_R{}^2}(0)] = (1/m^{4-D})l(g) \tag{4.21}$$

Using the definition of the renormalized field, we find

$$\frac{\partial}{\partial \lg m^2} D_{\phi^2\phi^2}(0) = \Delta D_{\phi^2\phi^2}(0) = (Z_2)^2 \Delta D_{\phi_R{}^2\phi_R{}^2}(0) + 2c_2(g) D_{\phi^2\phi^2}(0) \tag{4.22}$$

The solution of this equation is

$$C(g) \equiv D_{\phi^2\phi^2}(0) = \frac{1}{u}\int_0^g \frac{dg'}{b(g')} N(g')[Z_2(g')]^2 l(g') \tag{4.23}$$

Near the transition it simplifies to

$$C \simeq C_1 m^{\alpha/\nu}/\alpha + C_0 = \tilde{C}_1 \tau^{\alpha}/\alpha + C_0 \tag{4.24}$$

where

$$\alpha = 2 - D\nu \tag{4.25}$$

and C_0 and C_1 are computable constants. For $\alpha = 0$ the singularity of the specific heat is logarithmic, and for α negative we find a constant term plus an irregular term. The presence of the regular term C_0 allows a positive specific heat also for negative α.

We stress that all our results rely on the assumption that correlation functions have a limit when the bare coupling constant goes to infinity at fixed mass, i.e., they have a limit when g goes to g_c. If we relinquish this hypothesis, no conclusion can be drawn, e.g., if $c_2(g) = \sin[1/(g_c - g)]$ we find an oscillatory behavior. If $l(g)$ is singular at $g = g_c$, we obtain a violation of the scaling law for the specific heat.

5. THE CORRELATION FUNCTION AT THE CRITICAL TEMPERATURE

In this section we study the behavior of correlation functions at the critical temperature, i.e., in the zero-mass theory. These correlation functions are connected by an infinite scale transformation with the correlation function of the massive theory at the infinite bare coupling constant. Our aim is to express the correlation functions in both situations as an integral over correlation functions of the massive theory. Such an integral should be dominated by the region of integration where the external momenta are comparable with the mass.

For the sake of simplicity we study only the correlation function of the ϕ field. A similar analysis can be performed on the correlation function of ϕ^2. From (3.10) and (4.5) we derive the identity[46–48,14]

$$\begin{aligned}\Delta G_N^R(P|m^2, g) &= \left[\Delta m^2 \frac{\delta}{\delta m^2} + \Delta g \frac{\delta}{\delta g} + \Delta Z \frac{\delta}{\delta Z}\right] G_N^R(P|m^2, g) \\ &= \left[m^2 \frac{\delta}{\delta m^2} + b(g)\frac{\delta}{\delta g} + \frac{N}{2} c_1(g)\right] G_N^R(P|m^2, g)\end{aligned} \tag{5.1}$$

Using dimensional analysis, it can be written as

$$\Delta G_N^R(\lambda P|m^2, g) = \left[-\frac{1}{2}\lambda\frac{\partial}{\partial\lambda} + b(g)\frac{\delta}{\delta g} + \frac{1}{2}d_N + \frac{N}{2}c_1(g)\right] G_N(\lambda P|m^2, g)$$
$$d_N = [G_N] \tag{5.2}$$

A similar equation can be derived for the Γ_N functions

$$\Delta\Gamma_N^R(\lambda P|m^2, g) = \left[-\frac{1}{2}\lambda\frac{\partial}{\partial\lambda} + b(g)\frac{\delta}{\delta g} + \frac{1}{2}\delta_N - \frac{N}{2}c_1(g)\right]\Gamma_N^R(\lambda P|m^2, g)$$
$$\delta_N = [\Gamma_N] \tag{5.3}$$

Equations (5.3) are identities which are valid at each order in perturbation theory; however, we can also consider them as partial differential equations for the function G_N (Γ_N), assuming ΔG_N ($\Delta\Gamma_N$) is known. This linear, inhomogeneous differential equation has an infinite number of solutions. However, as shown in Ref. 49, there is only one solution which satisfies the physical requirement of being regular in λ and g around the lines $\lambda = 0, g = 0$. This solution is

$$G_N^R(\lambda P|m^2, g)$$
$$= \int_0^g \frac{dg'}{b(g')}\left[\frac{R(g)}{R(g')}\right]^{d_N}\left[\frac{Z_1(g)}{Z_1(g')}\right]^{-N/2}\Delta G_N^R\left(\lambda\frac{R(g)}{R(g')}P|m^2, g'\right)$$
$$= -2\lambda^{d_N}\int_0^\lambda \frac{dx}{x}x^{-d_N}\left[\frac{Z_1(g)}{Z_1[\bar g(g, \lambda/x)]}\right]^{-N/2}\Delta G_N^R\left(xP|m^2, \bar g\left(g, \frac{\lambda}{x}\right)\right) \tag{5.4}$$

where we have introduced the new functions

$$R(g) = [N(g)]^{-1/(4-D)}, \qquad R^{-1}(x) = \rho[x^{-(4-D)}],$$
$$\bar g(g, x) = R^{-1}[xR(g)] \tag{5.5}$$

The functions N, Z, and ρ are defined in Eqs. (4.8)–(4.10). These functions have the following limits:

$$g \to 0:\quad R(g) \to g^{-1/(4-D)}, \qquad g \to g_c:\quad R(g) \to A(g_c - g)^{1/2b'}$$
$$A = g_c^{-1/2b'}H^{-1/(4-D)}$$
$$x \to \infty:\quad R^{-1}(x) \to x^{-(4-D)}, \qquad x \to 0:\quad R^{-1}(x) \to g_c - (x/A)^{2b'}$$
$$\bar g(g, 1) = g$$
$$x \to \infty:\quad \bar g(g, x) \to N(g)x^{-(4-D)} \tag{5.6}$$

Note that the function ΔG can be computed from the same loop integral as G; the only difference is that one propagator $1/(K^2 + m^2)$ is changed to

$-1/(K^2+m^2)^2$. This substitution suppresses the high-momentum region in the loop integration. At each finite order in perturbation theory, we find for generical momenta P,[48–50]

$$\Delta G_N^R(\lambda P)/G_N^R(\lambda P) \to 0 \tag{5.7}$$

However, perturbation theory is useless in the large-momentum region for very large coupling constant. Nevertheless, we conjecture that (5.7) is true. The whole analysis of the behavior of correlation functions at the critical temperature depends heavily on this assumption. It is interesting to note that (5.7) is satisfied at all orders in perturbation theory both for renormalizable and superrenormalizable theories. Nonperturbative arguments may only suggest the consistency of (5.7). For example, one can try to compute $G_{N\phi^2}(P,0)$ from $\Delta G_{N\phi^2}(P,0)$. If one assumes that $\Delta G_{N\phi^2}/G_{N\phi^2} \to 0$, one finds that $\Delta G/G \to 0$. The analysis is not so simple, because of the exceptional momentum configuration. However, the problem can be studied in great detail.[50] 48

The main consequence of (5.7) is that the relevant region of integration in x remains bounded when λ goes to infinity also. Three different limits can be studied: (I) λ goes to infinity at fixed $g \neq g_c$; (II) λ goes to infinity at $g = g_c$; (III) $\bar{M}^2$ goes to zero at fixed P and u, or equivalently $\lambda \to \infty$ and $g \to g_c$ together. Let us study, in order, these three limits.

When $\lambda \to \infty$, $x/\lambda \to 0$. In this limit we can freely substitute in $g(\lambda/x, g)$ its asymptotic limit $(x/\lambda)^{4-D}N(g)$ and extend the integral to infinity. The leading term comes from the region where g is very small. If the first nonzero contribution to G is of order k

$$\begin{aligned} G_N^R(P|m^2,g) &= g^k G_N^{R(k)}(P|m^2) + O(g^{k+1}) \\ \Delta G_N^R(P|m^2,g) &= g^k\, \Delta G_N^{R(k)}(P|m^2) + O(g^{k+1}) \end{aligned} \tag{5.8}$$

we find

$$\begin{aligned} G_N^R(\lambda P|m^2,g) &\underset{\lambda\to\infty}{\longrightarrow} Z_1^{-N/2}(g)\left[\frac{u}{m^{4-D}}\right]^k \lambda^{-[k(4-D)-d_N]} \\ &\quad\times \int_0^\infty \frac{dx}{x}\, x^{-d_N+k(4-D)}\, \Delta G_N^{R(k)}(xP|m^2) \\ &\sim Z_1^{N/2}(g)\left[\frac{u}{m^{4-D}}\right]^k G_N^{R(k)}(\lambda P|m^2) \end{aligned} \tag{5.9}$$

We have just recovered the expansion in powers of the bare coupling constant, which is known to give the dominant term in the large-momentum region. It is easy to see that (4.1) is identically satisfied.

If we start directly from $g = g_c$, the situation changes dramatically. The bare parameters are infinite; however,

$$\bar{g}(g_c, x) \equiv g_c, \qquad \lim_{g \to g_c} Z_1[\bar{g}(g, x)]/Z_1(g) = x^{2c_1'} \tag{5.10}$$

Equation (5.5) simplies to

$$G_N{}^R(\lambda P|m^2, g_c) = -2\lambda^{d_N + Nc_1'} \int_0^\lambda \frac{dx}{x} x^{-d_N - Nc_1'} \Delta G_N{}^R(xP|m^2, g_c) \tag{5.11}$$

Our assumption implies that the integral is convergent when $\lambda \to \infty$. The upper limit of integration can be taken equal to infinity; neglecting terms that vanish in this limit, we obtain

$$G_N{}^R(\lambda P|m^2, g) \underset{\lambda \to \infty}{\longrightarrow} -2\lambda^{d_N - Nc_1'} \int_0^\infty \frac{dx}{x} x^{-d_N - Nc_1'} \Delta G_N{}^R(xP|m^2, g_c) \tag{5.12}$$

The integral no longer has any λ dependence. The correlation function in the large-momentum region satisfies a simple scaling law. The functional form of the correlation function is given by an integral on the correlation function computed in the low-momentum region. We have solved the problem of computing functions in the large-momentum region, where renormalized perturbation expansion is not convergent, using as input only correlation functions computed in the low-momentum region, where the renormalized perturbation expansion is more reliable.

We are now ready to study the limit $m \to 0$ at fixed u of the correlation function of the unrenormalized field.

Using dimensional arguments and the definition of renormalized correlation functions, we can derive the following chain of identities:

$$\begin{aligned} G_N{}^R(\mu P|m^2/\lambda^2, u) &= \lambda^{-d_N} G_N{}^B(\lambda\mu P|m^2, u\lambda^{4-D}) \\ &= Z_1^{N/2}(g)\lambda^{-d_N} G_N{}^R(\lambda\mu P|m^2, g) \end{aligned} \tag{5.13}$$

where g is a function of λ, u, and m

$$g = \rho[\lambda^{4-D}u/m^{4-D}] = R^{-1}[m/\lambda u^{1/(4-D)}] \tag{5.14}$$

Using (5.4) and (5.5), we obtain

$$G_N{}^B\left(\mu P|\frac{m^2}{\lambda^2}, u\right) = -2\mu^{d_N} \int_0^{\lambda\mu} \frac{dx}{x} x^{-d_N} Z_1^{N/2}\left[\bar{g}\left(g, \frac{\mu\lambda}{x}\right)\right] \Delta G_N{}^R\left[xP|m^2, \bar{g}\left(g, \frac{\mu\lambda}{x}\right)\right] \tag{5.15}$$

From the definition of $\bar{g}$, Eq. (5.5), it follows that

$$\bar{g}\left(g, \frac{\mu\lambda}{x}\right) = R^{-1}\left[\frac{\mu\lambda}{x} R(g)\right] = R^{-1}\left(\frac{\mu}{x}\frac{m}{u^{1/(4-D)}}\right) \tag{5.16}$$

Without loss of generality we can take $m^2 = u^{2/(4-D)}$; any other value of m^2 can be reached by changing λ

$$G_N^B\left(\mu P \middle| \frac{u^{2/(4-D)}}{\lambda^2}, u\right)$$
$$= -2\mu^{d_N} \int_0^{\lambda\mu} \frac{dx}{x}\, x^{-d_N} Z_1^{N/2}\left[R^{-1}\left(\frac{\mu}{x}\right)\right] \Delta G_N^R\left(xP|u^{2/(4-D)}, R^{-1}\left[\frac{\mu}{x}\right]\right) \tag{5.17}$$

We stress that (5.17) is an identity which can be derived without assumptions. If we send λ to infinity, the integral may diverge or converge. The possible region of divergence comes from x very large. In this situation $R^{-1}[\mu/x] \to g_c$. If the high-momentum behavior of ΔG_N at $g = g_c$ is good [see Eq. (5.7)], the integral is convergent. However, the limit $\lambda \to \infty$ does not exist in the perturbation expansion. $R^{-1}(1/x)$ is a finite-order polynomial of $x^{(4-D)}$ and clearly diverges when $x \to \infty$. If we include enough higher orders in the bare coupling constant, the decrease of $\Delta G_N(xP)$ when $x \to \infty$ can no longer compensate for the faster and faster increase of $R^{-1}(1/x)$, and infrared divergences appear. However, if we use a function R which satisfies (5.6) and ΔG_N computed at any order in perturbation theory, infrared divergences disappear.

If the zero-mass theory holds, or our two assumptions on the finiteness of the renormalized correlation functions in the limit $u \to \infty$ at fixed m and on the large-x behavior of $\Delta G_N^R(xP)$ are valid, the upper limit of integration can be shifted to infinity

$$G_N^B(\mu P|0, u) = -2\mu^{d_N} \int_0^{\infty} \frac{dx}{x}\, x^{-d_N} Z^{N/2}\left[R^{-1}\left(\frac{\mu}{x}\right)\right]$$
$$\times\, \Delta G_N^R\left[xP|u^{2/(4-D)}, R^{-1}\left(\frac{\mu}{x}\right)\right] \tag{5.18}$$

Let us study Eq. (5.18) in the two different limits. When $\mu \to \infty$, the situation is very similar to the first case we have studied. An expansion in powers of u/μ^{4-D} can be obtained. This expansion breaks down at the same order in u at which infrared divergences appear in perturbation theory. The coefficient of the next power in μ is no longer analytic in u. When $\mu \to 0$,

x/μ goes to infinity, and at the leading order in μ, barring corrections of order $\mu^{2b'}$, we find

$$G_N{}^B(\mu P|0, u) \underset{\mu \to 0}{\longrightarrow} -2\mu^{d_N + Nc_1'} \int_0^\infty \frac{dx}{x} x^{-d_N - N\,2c_1'} \Delta G_N(xP|u^{2/(4-D)}, g_c) \tag{5.19}$$

The coefficient of $\mu^{d_N + Nc_1'}$ is strongly nonanalytic in the coupling constant: e.g., if we consider the two-point correlation function, we obtain

$$G_2{}^B(P) \underset{P \to 0}{\longrightarrow} a u^{-2c_1'/(4-D)} P^{2-2c_1'} \tag{5.20}$$

where a is a pure number:

$$a = -2 \int_0^\infty dx\, x^{1} \; m^{c_1'} \; \Delta G_2{}^R(x|m^2, g_c) \tag{5.21}$$

Correlation functions are no longer C^0 in the coupling constant in the small-momentum region. It is not a surprise that the attempt to compute the Taylor expansion of a discontinuous function produces infrared divergences.

Comparing (5.11) with (5.18), we verify that the low-momentum behavior of the zero-mass theory and the large-momentum behavior of the theory computed at $g = g_c$ are the same.

It is clear that in this approach infrared divergences are completely washed out: the only welcome trace of their presence comes from the nonanalyticity in the coupling constant. The problem of computing the zero-mass behavior of correlation functions is reduced to the computation of correlation functions at finite values of the external momenta. The scaling law (4.14) is satisfied also at $K \neq 0$, $\tau \to 0$. The corrections to the scaling law are ruled by the same exponent ω

$$G_2(0, \tau) = \tau^{-\gamma}[1 + O(\tau^{\nu\omega})], \qquad G_2(K, 0) = (1/K^{2-\eta})[1 + O(K^{\omega})] \tag{5.22}$$

6. SIMPLE EXAMPLES

The general analysis of the correlation function near the transition for the partition function (2.1) is now completed. The critical exponent and correlation function of the zero-mass theory can be computed as integrals over the correlation functions of the massive theory. Before looking for the possible generalizations of the model, we want to study a few concrete examples.

We compute the b, c_1, c_2, and l functions in the one-loop approximation. The relevant Feynman diagrams are shown in Fig. 1. Diagrams (a), (b), and (c) contribute respectively to the functions b, c_2, and l. No diagram contributes to c_1.

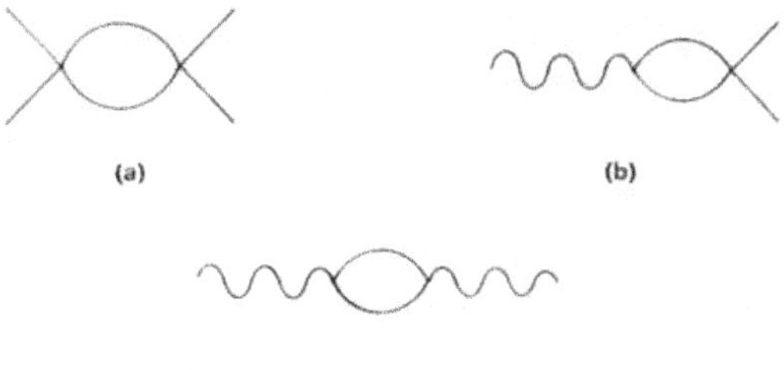

Fig. 1. Diagrams involved in the computation of the b, c_2, and l functions in the one-loop approximation.

The final result is[51]

$$\Gamma_4(0,0,0) = u - u^2 \frac{N+8}{6} I(m^2) + O(u^3)$$

$$\Delta\Gamma_4 = \frac{N+8}{6} u^2 m^2 \frac{\partial}{\partial m^2} I(m^2) + O(u^3)$$

$$\frac{N+8}{6} m^{2(4-D)} g^2 m^2 \frac{\partial}{\partial m^2} I(m^2) \tag{6.1}$$

$$I(m^2) = \frac{1}{(4\pi)^{D/2}} \Gamma\left(2 - \frac{D}{2}\right) m^{2(2-D/2)}$$

$$m^2 \frac{d}{dm^2} I(m^2) = -\frac{1}{(4\pi)^{D/2}} \Gamma\left(3 - \frac{D}{2}\right) m^{2(2-D/2)}$$

Note that when $D \to 4$, Γ_4 diverges (a renormalization of the coupling constant is needed) but $\Delta\Gamma_4$ remains finite. It is very important to realize that $\Delta\Gamma_4$ is nonzero in four dimensions only because Γ_4 is divergent.

In the same way we find

$$\Delta Z_2 = \frac{1}{2} \frac{g}{(4\pi)^2} \frac{N+2}{6} \Gamma\left(3 - \frac{D}{2}\right), \qquad \Delta G = -\frac{1}{2} \frac{N(m)^{D-4}}{(4\pi)^2} \Gamma\left(3 - \frac{D}{2}\right) \tag{6.2}$$

These formulas imply

$$b(g) = -\frac{4-D}{2} g\left[1 - \frac{g}{g_c}\right] + O(g^3)$$

$$c_1(g) = 0 + O(g^2), \qquad c_2(g) = c_2^c g/g_c + O(g^2)$$

$$l(g) = \frac{N}{2(4\pi)^2} \Gamma\left(3 - \frac{D}{2}\right) + O(g) \tag{6.3}$$

$$c_2^c = \frac{4-D}{2} \frac{N+2}{N+8}, \qquad g_c = \frac{(4\pi)^{D/2} 3(4-D)}{(N+8)\Gamma(3-D/2)}$$

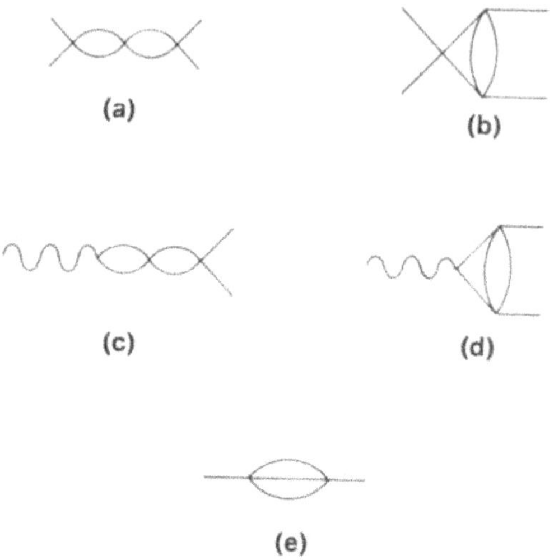

Fig. 2. Diagrams involved in the computation of the b, c_1, and c_2 functions in the two-loop approximation.

Putting (6.3) into (4.8) and (4.10), we obtain

$$
\begin{aligned}
\frac{u}{m^{4-D}} &= \frac{g}{1-g/g_c} \\
\frac{dm^2}{dT} &= \left[1+\frac{u}{g_c m^{4-D}}\right]^{2c_2'/(4-D)} \\
C &= \frac{N}{2(4\pi)^2}\,\Gamma\!\left(3-\frac{D}{2}\right)\int_{-\infty}^{m^2}\frac{d\tilde{m}^2}{\tilde{m}^{(2+4-D)}}\left[1+\frac{u}{g_c\tilde{m}^{4-D}}\right]^{-4c_2'/(4-D)} \\
G_2(K, m^2) &= \frac{1}{K^2+m^2}
\end{aligned}
\qquad (6.4)
$$

Near the transition, these formulas simplify to

$$
\begin{aligned}
g &= g_c - \frac{m^{4-D} g_c^{2}}{u} \\
\frac{\partial m^2}{\partial T} &= \left[\frac{u}{g_c}\right]^{2c_2'/(4-D)} m^{-2c_2'} \\
C &= \frac{N}{2(4\pi)^2}\,\Gamma\!\left(3-\frac{D}{2}\right)\left[\frac{g_c}{u}\right]^{+4c_2'/(4-D)} m^{D-4+4c_2'}
\end{aligned}
\qquad (6.5)
$$

The critical exponents are

$$\nu = \frac{1}{2 + [(N+2)/(N+8)](D-4)} \tag{6.6}$$
$$\gamma = 2\nu, \qquad \eta = 0, \qquad \alpha = 2 - D\nu.$$

For the three-dimensional Ising model we obtain

$$\gamma = 1.2, \qquad \nu = 0.6, \qquad \eta = 0, \qquad \lambda_c \equiv g_c\Gamma(3 - D/2)3/(4\pi)^{D/2} = 1 \tag{6.7}$$

For D sufficiently small we find $\nu < 0$. A negative value of ν implies that the system has no transition at finite temperature. The reason is simple. The integrated form of the scaling law $dM^2/dm = m^{(1/\nu)-1}$ is

$$M^2 - M_0^2 \propto (1/\nu)m^{1/\nu} \tag{6.8}$$

where M_0^2 is an integration constant. If $1/\nu$ is positive, M_c^2 (the critical value of the temperature) is simply M_0^2; however, if $1/\nu$ is negative or zero, M_c^2 is located at $-\infty$ and cannot reach finite temperature. It depends on the details of the model whether $M_c^2 \to -\infty$ corresponds to zero temperature.

A transition is also present in the one-dimensional Ising model $(1/\nu = 1 > 0)$, while no transition is correctly found in the spherical model if $D \leqslant 2$ $(\nu < 0)$. It is of interest to note that at this simple order the exponents for the spherical model are the exact ones. In the case of the Ising model the corrections to the classical exponents seem to be in the right direction. How can the approximation be improved?

A first possibility, advocated by Wilson,[5] is based on the observation that g_c is of the order $4 - D$ when D is near to 4. This property is not destroyed in high orders in perturbation theory. we can compute both g_c and the critical exponents as a formal power of $\epsilon = 4 - D$. This is possible because the linear term in $b(g)$ is proportional to $-\frac{1}{2}\epsilon g$ and in the limit $\epsilon \to 0$, $b(g)$ is a finite, nontrivial function of g. The finiteness of $b(g)$ in dimension 4 is far from being a trivial statement: it is connected to the existence of the renormalized four-dimensional ϕ^4 theory in perturbation theory.[36]

In this situation we need to compute only the first k-loop diagrams to get the critical exponents up to the order ϵ^k. In the case of the N-component model, the following results have been found:

$$\gamma = 1 + \frac{N+2}{2(N+8)}\epsilon + \frac{(N+2)(N^2+22N+52)}{4(N+8)^3}\epsilon^2 + O(\epsilon^3)$$
$$\eta = \frac{\epsilon^2(N+2)}{2(N+8)^2}\left\{1 + \left[\frac{6(3N+14)}{(N+8)^2} - \frac{1}{4}\right]\epsilon\right\} + O(\epsilon^4) \tag{6.9}$$

If we use Eq. (6.9) in the case $N = 1$, $D = 3$, we get

$$\gamma = 1.244, \qquad \eta = 0.037, \qquad \nu = 0.628 \tag{6.10}$$

Higher order corrections up to ϵ^4 have been computed.[12,51] The expansion is not convergent; the generical terms increase like $\epsilon^k k!$.[52] However, there are good arguments which suggest that the correct results are obtained by using the Borel summation technique. Indeed one finds, for $N = 1$, $D = 3$,[51]

$$\gamma = 1.235 \pm 0.004, \qquad \eta = 0.0333 \pm 0.0001, \qquad \nu = 0.628 \pm 0.002 \tag{6.11}$$

A different approach, which avoids the use of noninteger dimensions at intermediate steps, consists in computing directly the k-loop contributions at fixed dimensions and in a first approximation neglecting higher orders in g.[13,53] The two methods produce the same results up to the order ϵ^k but are different at finite ϵ. The final answer is

$$\begin{aligned}
2b(\lambda) &= -(4-D)\lambda + \frac{N+8}{9}\lambda^2 - \left[\frac{10N+44}{81}f(D) - \frac{N+2}{81}h(D)\right]\lambda^3 + O(\lambda^4)\\
C_2(\lambda) &= \frac{N+2}{18}\lambda - [6f(D) - h(D)]\frac{N+2}{324}\lambda^2 + O(\lambda^3)\\
C_1(\lambda) &= -\frac{N+2}{324}h(D)\lambda^2 + O(\lambda^3)\\
\lambda &= \frac{3\Gamma(3-D/2)}{(4\pi)^{D/2}}g
\end{aligned} \tag{6.12}$$

where

$$\begin{array}{lll}
f(4) = 1.0, & f(3) = 0.2/3, & f(2) \simeq 0.56\\
h(4) = 1.0, & h(3) \simeq 0.59, & h(2) \simeq 0.46
\end{array} \tag{6.13}$$

Unfortunately, at this approximation the function $b(\lambda)$ in Eq. (6.12) has no zero in dimensions less than 3.5 if $N = 1$. We know a priori that the loop expansion is not convergent and that resummation techniques should be used; although the Borel technique is the best suited to cope with this divergent series, we use for simplicity a Padé approximant. The presence of a zero is restored and the critical exponents can be evaluated.

In Fig. 3 we show the functions $\nu(D)$ computed to the first and the second order in ϵ, computed in the one- and two-loop approximations, using the Taylor expansion, and that in the two-loop approximation using the Padé approximant. The upper part of HP is the analytic continuation of the lower part and it has no physical meaning as far as phase transitions are concerned. No transition is found in the one-dimensional Ising model, i.e., $1/\nu$ slightly negative.

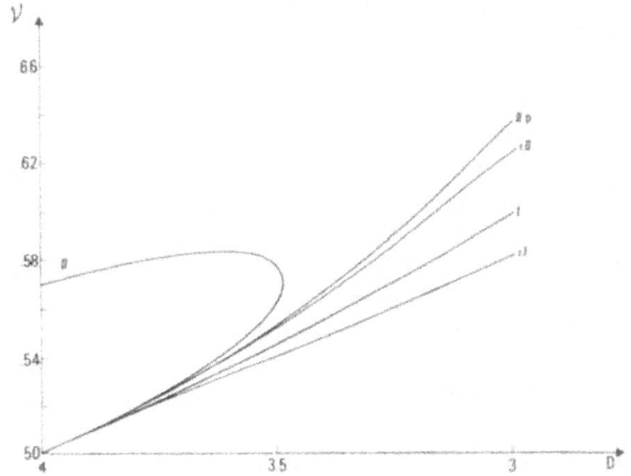

Fig. 3. Plot of the critical exponent ν of the Ising model in different approximations: ϵI and ϵII are, respectively, the first and the second order in ϵ; I and II are the critical exponents produced respectively by the one-loop and the two-loop approximations using the g_c given by the Taylor expansion, while in IIP we use the g_c computed from the [2, 1] Padé approximant.

In the three-dimensional Ising case we find

$$\gamma = 1.256, \qquad \eta = 0.033, \qquad \nu = 0.638, \qquad \lambda_c \simeq 1.73 \tag{6.14}$$

Always at the two-loop level, but using the Padé Borel resummation technique, we get[10]

$$\gamma = 1.247, \qquad \eta = 0.028, \qquad \nu = 0.633, \qquad \lambda_c = 1.60 \tag{6.15}$$

The success of this computation has led people to use more industrious techniques: all the diagrams up to six loops have been computed (1042 diagrams),[54] and asymptotic estimates for higher loop diagrams have been found.[12]

$$b(\lambda) = \sum_k b_k \lambda^k, \qquad b_k \xrightarrow[k\to\infty]{} C(-A)^k \Gamma(k + 9/2)$$
$$C = 0.03996, \qquad A = 0.147742 \tag{6.16}$$

The final result is[10,11]

$$\gamma = 1.241 \pm 0.002, \qquad \eta = 0.031 \pm 0.004$$
$$\nu = 0.630 \pm 0.0015, \qquad \lambda_c = 1.416 \pm 0.005 \tag{6.17}$$

For the two-dimensional Ising model a similar procedure, starting from the knowledge of only four-loop diagrams, gives[11]

$$\begin{aligned} \gamma &= 1.79 \pm 0.09, & \eta &= 0.13 \pm 0.07 \\ \nu &= 0.97 \pm 0.08, & \lambda_c &= 1.85 \pm 0.10 \end{aligned} \tag{6.18}$$

The results in the two-dimensional case are worse than those in the three-dimensional case. There are two reasons for this: the effective value of g is larger and the precision is decreased; the result of any computation of this kind is automatically an analytic function of N; however, we know that the critical exponents are discontinuous around $N = 2$: the critical exponent jumps from 1/4 to 0 for N greater than 2.[55,56] If we do not consider the contribution of very high orders in g, poor results should be obtained for N too near to 2.

7. THE RANGE OF THE RENORMALIZED COUPLING CONSTANT

An interesting problem is the determination of the possible values that the renormalized coupling constant may assume. It is clear that when the bare coupling constant u is in the 0 to ∞ range, the renormalized coupling ranges from 0 to g_c. A theory with g greater than g_c cannot be obtained as limit of theories where the bare coupling constant is finite, but it could be obtained as an analytic continuation in the renormalized coupling constant from the good region $g \leqslant g_c$. We shall see below that there is a cut, starting from g_c, which forbids the analytic continuation to greater values of the renormalized coupling constant, and that g_c is the greatest allowed value. This result implies that, if ultraviolet divergences are eliminated using dimensional regularization, the four-dimensional theory is always free.[6]

Let us assume that all the correlation functions are C^∞ functions of the coupling constant and that $b(g)$ has a simple zero at $g = g_c$. This hypothesis is not consistent with the differential equation (5.23) unless an infinite number of independent sum rules is satisfied; this last possibility is rather unrealistic; moreover, in the framework of the $1/N$ expansion it is possible to check that the sum rules are not satisfied and a cut is present at g_c. To be specific, let us consider the case of the Γ^6 function computed at zero external momentum. Then Eq. (5.4) reads

$$\begin{aligned} \Gamma_6(g) &= \int_0^g \frac{dg'}{b(g')} \left[\frac{R(g)}{R(g')}\right]^{2D-6} \left[\frac{Z_1(g)}{Z_1(g')}\right]^3 \Delta\Gamma_6(g) \\ &= \int_0^g dg' \frac{g_c - g'}{b(g')} \frac{f(g)}{f(g')} \frac{(g_c - g)^{B_6}}{(g_c - g')^{B_6+1}} \Delta\Gamma_6(g') \end{aligned} \tag{7.1}$$

where $B_6 = (D - 3 + 3c_1^c)/b'$ and the function $f(g)$ is C^∞ around g_c. If B_6 were negative, Eq. (7.1) could be written as

$$\Gamma_6(g) = f(g)(g_c - g)^{B_6} \int_0^{g_c} dg' \frac{(g_c - g')^{B_6}}{b(g')f(g')} \Delta\Gamma_6(g')$$

$$- \int_g^{g_c} dg' \frac{(g_c - g)^{B_6}}{b(g')(g_c - g')^{B_6}} \frac{f(g)}{f(g')} \Delta\Gamma_6(g') \tag{7.2}$$

If B_6 is positive, but is not an integer, Eq. (7.2) is still true provided that the integrals are defined as analytic continuations in B_6. In both cases, the second integral is C^∞ around g_c; the first term, if it is not identically zero, has a cut of the type $(g_c - g)^{B_6}$.[49]

Equation (7.2) implies that the renormalized Green's functions must be singular at g_c; this equation, however, cannot be correct as it stands: if $D = 3$ the value of B_6 is only slightly positive and becomes negative for D slightly less than 3. Indeed there is a loophole in the argument; if $\Gamma_6(g)$ has a cut at $g = g_c$, then $\Delta\Gamma(g)$ also must have a cut of the same strength as $\Gamma_6(g)$.

It has been shown in Ref. 57 that under a technically simplifying hypothesis, we can write the following equation (we have neglected operator mixing):

$$[b(g)\, \partial/\partial g + 3 - D - C_6(g)]\Gamma^6(g) = \tilde{\Delta}\Gamma_6(g) \tag{7.3}$$

where the singularity of $\tilde{\Delta}\Gamma_6(g)$ at $g = g_c$ is less strong than that of $\Gamma_6(g)$, the function $C_6(g)$ being the "anomalous dimension" of the operator ϕ^6:

$$C_6(g) \sim \frac{12 + 3N}{8 + N} \lambda + O(\lambda^2) \tag{7.4}$$

The quantity $A_6 = 3(D - 2) + C_6(g_c)$ is the effective dimension of the renormalized ϕ^6 operator at the critical point:

$$\langle \phi^6(x)\phi^6(0)\rangle \underset{x \to 0}{\to} \frac{1}{x^{2A_6}}, \qquad A_6 = 6 + 2\frac{2 + N}{8 + N}\epsilon + O(\epsilon^2) \tag{7.5}$$

An equation similar to (7.2) holds, where now $B_6 = (A_6 - D)/2b'$. The power of the singularity is controlled by the dimension of the ϕ^6 operator, as it should be.[58,59] The disaster $B_6 < 0$ happens only if $A_6 < D$, which would be a definite signal of serious problems (see next section).

The physical origin of these singularities is quite clear; an a priori argument can be given for their existence; the only difficulty is to see their appearance in the present formalism. In the limit $\omega \to \infty$ the following expansion for the renormalized Green's functions is expected to hold:

$$\Gamma^R(u) \to \Gamma^R(\infty) + \sum_n R_n u^{(D-A_n)/\epsilon}, \qquad D = 4 - \epsilon \tag{7.6}$$

where the A_n are the effective dimensions of all the possible local operators. Let us consider for simplicity only the ϕ^n operators with even n ($n \neq 2$). Equation (.) clearly shows why we need the condition $A_n > D$, and comparing it with Eq. (4.12), we get the identification

$$2b' = A_4 - D \tag{7.7}$$

It is easy to see that Eq. (7.6) implies the presence of a cut in the renormalized constant complex plane of the form

$$(g_c - g)^{(A_6 - D)/(A_4 - D)} \tag{7.8}$$

As a last remark, we notice that in $4 - \epsilon$ dimensions we have a relation between the singularities of the Green's functions at $g = g_c = O(\epsilon)$ and the singularities due to ultraviolet divergences of the Borel transform of the Green's functions.[57] We only quote the result:

$$\Gamma_6(\lambda) \sim (\lambda_c - \lambda)^{B_6}[\pi/\sin(\pi B_6)]\,\tilde{\Delta}C_6(2) \tag{7.9}$$

which can be written also as

$$\operatorname{Im}\Gamma_6(\lambda) \sim (\lambda - \lambda_c)^{B_6}\theta(\lambda - \lambda_c)\pi\tilde{\Delta}C_6(2) \tag{7.10}$$

where $\Delta\tilde{C}_6(t)$ is the Borel transform of $\Gamma_6(\lambda)$ with respect to λ [see Eq. (6.12)]:

$$\tilde{\Delta}\Gamma_6(\lambda) = \sum_n^{\infty} t_n\lambda^n, \qquad \tilde{\Delta}C_6(Z) = \sum_1^{\infty}{}_n \frac{t_n Z^n}{(n-1)!} \tag{7.11}$$

8. ON UNIVERSALITY

Until now we have discussed only a $\lambda\phi^4$ interaction. It is clear that the whole approach is of little interest if the results cannot be extended to much more general interactions.

We consider a more complicated model

$$Z \propto \int d[\phi] \exp \int d^D x\, \mathscr{L}(x) \tag{8.1}$$

where

$$-\mathcal{L}(x) = \frac{1}{2}\partial_\mu\phi(x)\,\partial^\mu\phi(x) + \frac{1}{2}M^2\phi^2 + \frac{u_0}{4!}\phi(x^4)$$
$$+ u_1\phi^6(x) + u_2[\Delta\phi(x)]^2 + \cdots \qquad (8.2)$$

The field ϕ is still defined on a continuous space M, and the u's are arbitrary functions of the temperature. The number of parameters may be arbitrary.

For simplicity we assume that all interaction constants u, M_c^2, and $\bar{M}^2 = M^2 - M_c^2(u)$ are regular functions of the temperature [$M_c^2(u)$ is the value of the bare mass at the critical temperature]. If in the limit $\bar{M}^2 \to 0$ thermodynamic quantities are regular functions of the other parameters, we can neglect their temperature dependence near the transition. This is the general case, although more complicated situations can occur near a tricritical point.[50] In conclusion, without any loss of generality we assume that near a simple critical point the partition function depends on the temperature only through the $\bar{M}^2$ term: $\bar{M}^2 \propto T - T_c$. The techniques used in the other sections can be easily extended. Renormalized fields, mass, and coupling constants are defined in such a way that the only dimensional parameter is the mass. For example, in a theory described by the Lagrangian (8.2), a possible choice of coupling constants could be

$$g_0 = \Gamma^4(0, 0, 0)m^{D-4}, \qquad g_1 = \Gamma^6(0, 0, 0, 0, 0)m^{2D-6}$$
$$g_2 = (d/dK^2)^2\Gamma^2(K^2)|_{K^2=0}m^2 \qquad (8.3)$$

It may be convenient to define the coupling constants in such a way that if the interaction contains only a ϕ^4 term, all coupling constants except g_0 are zero.

We introduce the Λ operator

$$\Delta = m^2 \left.\frac{\partial}{\partial m^2}\right|_{u_i} = \left.\frac{\partial}{\partial \lg m^2}\right|_{u_i} \qquad (8.4)$$

and we define a set of b functions

$$\Delta g_i = b_i(g_0, \ldots, g_j) \qquad (8.5)$$

In compact notation

$$\left.\frac{\partial}{\partial \lg m^2}\right|_{\mathbf{u}} \mathbf{g} = \mathbf{b}(\mathbf{g}) \qquad (8.6)$$

The integration of the system (8.6) yields the renormalized coupling constants as functions of the bare ones and the renormalized mass

$$\mathbf{g} = \rho(m, \mathbf{u}) \tag{8.7}$$

The bare coupling constants play the role of integration constants or boundary conditions to the solutions of the differential equations in the limit $m \to \infty$.

Approaching the transition, m goes to zero and $\log m$ goes to $-\infty$. If the limit

$$\mathbf{g}_c = \lim_{m \to 0} \rho(m, \mathbf{u}) \tag{8.8}$$

exists and is finite, then

$$\mathbf{b}(\mathbf{g}_c) = 0 \tag{8.9}$$

We say that g_c is a fixed point if (8.9) is satisfied. However, if $g_c^{(1)}$ is a particular fixed point, the actual limit in (8.8) may be different from $g_c^{(1)}$ unless this fixed point is attractive and the set of bare coupling constants is inside its domain of attraction.[5] The first condition implies that the real part of all the eigenvalues of the matrix

$$B_{ik} = \left.\frac{\partial b_i}{\partial g_k}\right|_{g = g_c} \tag{8.10}$$

are positive. Its eigenvalues control the deviation from scaling near the transition. The exponent ω defined in (4.20) is $1/2B_{\min}$, where $B_{\min}$ is the minimum eigenvalue of B. If we recall Eq. (7.6), we find that the eigenvalues B_n of the matrix B are connected to the effective dimensions of the renormalized operator by

$$B_n = A_n - D \tag{8.11}$$

The condition $A_n > D$ (i.e., absence of relevant operators) imposed in the previous section is equivalent to the stability of the fixed point.

Both the critical exponent and the detailed form of the correlation functions in the critical region depend only on the fixed point. They are the same in all the theories which belong to the domain of abstraction of the same fixed point. The dependence of the critical exponent on the detailed form of the interaction is absent; only discontinuous behavior can be found. [We have assumed that the system of equations $b(g) = 0$ has only a discrete set of solutions. If lines of zeros are present, the critical exponent can have a continuous dependence on the parameters of the bare interaction.]

We stress that by increasing the complexity of the interaction, new fixed points can be created, but the old ones cannot disappear. However, they can become unstable. The results of the $\lambda\phi^4$ model can be applied to other models only if the matrix A does not develop negative eigenvalues when new interactions are introduced, and the bare coupling constants of the new interactions are not too large.

In dimensions near enough to 4 all local interactions which involve more derivatives or higher powers of the fields have positive eigenvalues, and the smallest one is b'. If $N \neq 1$, the only possible relevant operators must be quadrilinear or bilinear with different dependence on the internal degree of symmetry; their presence accounts for the different behavior of the isotropic and anisotropic Heisenberg models.

We have just arrived at the conclusion that a wide class of interactions have the same renormalized correlation functions at the critical point. It would be nice to compute these correlation functions without committing ourselves to any particular interaction. This can be done by deriving integral nonlinear equations for the renormalized correlation functions which are valid only at the critical point.

Comparing (5.4) with (5.11), we find that the fixed point is characterized by the fact that the correlation functions $G_N(\Gamma_N)$ can be computed as integrals over the correlation functions $\Delta G_N(\Delta\Gamma_N)$ evaluated at the same value of the coupling constant. If we are not at the fixed point, (5.4) also involves $\Delta G_N(\Delta\Gamma_N)$ functions computed at all possible values of the coupling constant. Moreover, the $\Delta G_N(\Delta\Gamma_N)$ can be easily computed from the knowledge of the first $N + 2$ of the $G(\Gamma)$ functions. The correlation functions of $\phi_R{}^2$ with the other N of the ϕ_R fields can be written as an integral over some nonlinear combination of the correlation function of the $N + 2$ field, using the relation $\phi_R{}^2 \propto (\phi_R)^2$.

The generating function of the connected (amputated one-line irreducible) correlation functions satisfies a nonlinear functional equation

$$G = F(G) \qquad [\Gamma = \tilde{F}(\Gamma)] \tag{8.12}$$

This equation has the same structure as the Schwinger equation[31] for the generating function of a polynomial interaction. The main difference is that no possible small parameter exists in (8.12): it is not clear how to construct a simple algorithm which would allow the iterative solution of this equation.

9. CONCLUSIONS AND OUTLOOK

In this paper we have formulated a field-theoretic approach to the theory of second-order phase transitions, based on the renormalization group; in this approach, spaces of noninteger dimensions play no role and everything can be done without leaving the three (two) dimensional space.

Simple computations of the critical exponents of the three-dimensional Ising model have been shown in detail; reasonable results are obtained. If longer computations are done, better results are obtained: the critical exponents are estimated with an error less than 10^{-3}.

We cannot claim to have proved the validity of the scaling laws: we have only derived them, starting "reasonable" hypotheses. However, it is gratifying that these hypotheses can be explicitly checked in the $1/N$ expansion.

The extension of this method to other systems undergoing a second-order phase transition is straightforward: indeed, it has been used to compute the critical exponents of the Reggeon field theory with good accuracy.[60]

APPENDIX

In this appendix we define the notations used in the text.

We consider the partition function (2.10) in the presence of a point-dependent magnetic field H

$$Z[H] \propto \int d[\phi] \exp\left[-\int d^D x\, \mathscr{L}(x, H)\right] \tag{A.1}$$

where

$$\mathscr{L}(x, H) = -\frac{1}{2}\partial_\mu \phi(x)\, \partial^\mu \phi(x) + \frac{1}{2} M^2 \phi^2(x) + \frac{u}{4!}\phi^4(x) + H(x)\phi(x) \tag{A.2}$$

is a functional of the magnetic field H. We define the free energy functional

$$G[H] = \ln\{Z[H]\} \tag{A.3}$$

Z is the generating functional of the correlation function of the field ϕ, while G is the generating functional of the connected (truncated) correlation functions

$$\left.\frac{\delta^N Z}{\delta H(x_1) \cdots \delta H(x_N)}\right|_{H=0} \equiv \langle \phi(x_1) \cdots \phi(x_N)\rangle$$

$$= \int d[\phi]\, \phi(x_1) \cdots \phi(x_N) \exp\left[-\int d^D x\, \mathscr{L}(x, 0)\right]$$

$$\frac{\delta^N G}{H(x_1) \cdots H(x_N)} = \langle \phi(x_1) \cdots \phi(x_N)\rangle_c = G_N(x_1 \cdots x_N) \tag{A.4}$$

or

$$G[H] = \sum_{0}^{\infty} \frac{1}{n!} \int \prod_{1}^{n} H(x_i)\, d^D x_i \prod_{1}^{n} \phi(x_i) \tag{A.5}$$

The magnetization of the system is

$$\mu(x, H] = \partial G / \delta H(x) \tag{A.6}$$

$\mu(x, H]$ is a function of x and a functional of H.

The functional (A.6) can be inverted, producing the magnetic field as a functional of the magnetization $H(x, \mu]$.

The free energy at constant magnetization can be introduced by performing a Legendre transformation

$$\Gamma[\mu] = \int d^D x\, \mu(x) H(x, \mu] - G[H[\mu]] \tag{A.7}$$

$\Gamma[\mu]$ is the generating functional of the connected one-line (one-particle) irreducible correlation functions

$$\left. \frac{\delta^N \Gamma}{\delta\mu(x_1) \cdots \delta\mu(x_N)} \right|_{\mu = 0} = \Gamma_N(x_1 \cdots x_N) \tag{A.8}$$

If $u = 0$, the partition function Z can be explicitly computed

$$Z_0[H] \propto \exp - \int dx\, dy\, \tfrac{1}{2} D(x - y, M) H(x) H(y) \tag{A.9}$$

where $D(x - y, M)$ satisfies the differential equation

$$(-\Delta_x + M^2) D(x - y, M) = (-\Delta_y + M^2) D(x - y, M) = \delta^D(x - y) \tag{A.10}$$

If u is different from zero, (A.1) can be formally written in compact notation as

$$Z[H] \propto \exp\left\{ -\frac{u}{4!} \int \left[\frac{\delta}{\delta H(x)} \right]^4 d^D x \right\} Z_0[H] \tag{A.11}$$

The expansion of the first exponential in (A.10) in powers of u reproduces the standard perturbation expansion, Eq. (3.5).

The functional Z satisfies the Schwinger functional differential equation:

$$\left[(\Delta_x + M^2) \frac{\delta}{H(x)} + \frac{u}{3!} \left(\frac{\delta}{\delta H(x)} \right)^3 + H(x) \right] Z[H] = 0 \tag{A.12}$$

Equation (A.12) is very similar to a Kirkwood–Salisbury equation; it yields the N-point correlation functions as an integral over the $N + 2$ correlation functions

$$\frac{\delta Z}{\delta H(x)} = \int d^D y\, D(x-y, M)\left[H(y) + \frac{u^3}{3!}\left(\frac{\delta}{\delta H(y)}\right)^3\right] Z[H] \quad \text{(A.13)}$$

The ϕ field is defined as the square of ϕ^2. Its correlation functions are defined by

$$Z_{N,\phi^2}(x_1 \cdots x_N, y) = \langle \phi(x_1) \cdots \phi(x_N)\phi(y)\phi(y)\rangle \quad \text{(A.14)}$$

From (A.14) one can obtain the connected and the amputated correlation functions of ϕ^2.

In this paper one normally considers the Fourier transforms of the correlation functions G_N and Γ_N. They are defined as follows:

$$G_N(P_1 \cdots P_{N-1}) = \int d^D x_1 \cdots d^D x_{N-1} \exp\{i[P_1 x_1 + \cdots + P_{N-1} x_{N-1}]\}\, G_N(x_1 \cdots x_{N-1}, 0)$$

$$G_{N\phi^2}(P_1 \cdots P_{N-1}, K) = \int d^D x_1 \cdots d^D x_{N-1}\, d^D y \quad \text{(A.15)}$$

$$\times \exp\{i[P_1 x_1 + \cdots + P_{N-1} x_{N-1} + Ky]\}\, G_{N+2}(x_1 \cdots x_{N-1}, 0, y, y)$$

$$D_{\phi^2\phi^2}(P) = \int d^D x\, e^{iPx} \langle \phi^2(x)\phi^2(0)\rangle_c$$

Similar definitions are valid for the Γ_N functions. Sometimes the vector K is omitted from the argument of $G_{N\phi^2}$; in such a case it is supposed to be zero, e.g.,

$$G_{2\phi^2}(P) = \int d^D x\, d^D y\, e^{iPx} G_{2\phi^2}(x, 0, y) \quad \text{(A.16)}$$

The correlation functions of the renormalized theory field are denoted by $G_N{}^R(\Gamma_N{}^R)$. To simplify the notation the subscript R is omitted in Section 8. In Section 5 the correlation function of the bare field acquires a subscript B to prevent any confusion.

REFERENCES

1. L. P. Kadanoff *et al.*, *Rev. Mod. Phys.* **39**:395 (1967).
2. B. Widom, *J. Chem. Phys.* **43**:3893 (1965).
3. C. Domb and H. S. Green, eds., *Phase Transitions and Critical Phenomena* (Academic Press, 1975), Vol. III.

4. C. Di Castro and G. Jona-Lasinio, *Phys. Lett.* **29A**:332 (1969).
5. K. G. Wilson and M. E. Fisher, *Phys. Rev. Lett.* **28**:234 (1972); K. G. Wilson, *Phys. Rev. Lett.* **28**:548 (1972); K. G. Wilson and J. Kogut, *Phys. Rep.* **12C**:77 (1974).
6. E. Brezin, J. C. Le Guillou, B. G. Nikel, and J. Zinn Justin, *Phys. Lett.* **44A**:227 (1973).
7. C. De Dominicis and L. Peliti, *Phys. Rev. B* **18**:353 (1978), and references therein.
8. A. B. Harris, T. C. Lubensky, and J. H. Chen, *Phys. Rev. Lett.* **36**:415 (1976).
9. M. Moshe, *Phys. Rep.* **43C**:197 (1977).
10. G. A. Baker, B. G. Nikel, M. S. Green, and D. I. Meiron, *Phys. Rev. Lett.* **36**:1351 (1976); *Phys. Rev. B* **17**:1365 (1978).
11. J. C. Le Guillou and J. Zinn Justin, *Phys. Rev. Lett.* **39**:95 (1977); Saclay Preprint DPh-T/79/94.
12. E. Brézin and G. Parisi, *J. Stat. Phys.* **19**:269 (1978), and references therein.
13. G. Parisi, Cargese Lectures Notés, Columbia University preprint (1973), unpublished.
14. C. Di Castro, *Nuovo Cimento Lett.* **5**:69 (1972).
15. K. Symanzik, unpublished, quoted in G. Mack, *Lecture Notes in Physics*, Vol. 17 (Springer Verlag, 1972).
16. W. F. Jegerlehrer and B. Schoer, *Acta Physica Austriaca Suppl.* (1973); D. K. Mitter, *Phys. Rev. D* **7**:2927 (1973); E. Brézin, J. Le Guillou, and J. Zinn Justin, *Phys. Rev. D* **8**:2418 (1973); F. De Pasquale and P. Tombesi, *Nuovo Cimento* **12B**:43 (1972).
17. E. Brézin, J. C. Le Guillou, and J. Zinn Justin, in *Phase Transitions and Critical Phenomena*, C. Domb and M. S. Green, eds. (Academic Press, 1976), Vol. VI.
18. D. J. Amit, *Field Theory, the Renormalization Group and Critical Phenomena* (McGraw-Hill, 1978).
19. J. Glimm and A. Jaffe, Cargese Lectures Notes (1976).
20. K. Symanzik, *J. Math. Phys.* **7**:510 (1966).
21. K. Osterwalder and J. Schrader, *Comm. Math. Phys.* **31**:83 (1973).
22. J. Schwinger, *Proc. Nat. Acad. Sci. (U.S.)* **44**:956 (1958).
23. R. Streater and A. S. Wightman, *PCT, Statistics and All That* (Benjamin, 1964).
24. A. A. Migdal, *Sov. Phys. JEPT* **32**:552 (1971).
25. V. L. Ginzburg and L. D. Landau, *Sov. Phys. JEPT* **20**:1060 (1950).
26. L. D. Landau and M. E. Lifshitz, *Statistical Physics* (Addison-Wesley, 1969).
27. A. A. Migdal, *Sov. Phys. -JEPT* **28**:1036 (1969).
28. A. Stanley, *Phys. Rev.* **176**:718 (1968).
29. P. G. De Gennes, *Phys. Lett.* **38A**:339 (1972).
30. R. Balian and G. Toulouse, *Phys. Rev. Lett.* **30**:544 (1973).
31. A. M. Poliakov, *Sov. Phys. JEPT* **28**:533 (1969).
32. T. H. Berlin and M. Kac, *Phys. Rev.* **86**:821 (1952).
33. J. P. Eckman, J. Magnen, and R. Sénéor, *Comm. Math. Phys.* **39**:251 (1976).
34. P. A. M. Dirac, *The Principles of Quantum Mechanics* (Oxford, The Clarendon Press, 1930).
35. J. S. Schwinger, *Selected Papers on Quantum Electrodynamics* (New York, Dover, 1958).
36. J. D. Bjorken and S. D. Drell, *Relativistic Quantum Field Theory* (New York, McGraw-Hill, 1965); C. Itzykson and J.-B. Zuber, *An Introduction to Field Theory* (McGraw-Hill, in press).
37. E. Brézin and D. J. Wallace, *Phys. Rev. B* **7**:1967 (1973).
38. K. Wilson, *Phys. Rev. D* **4**:2911 (1973).
39. G. Parisi, in Proc. Colloquium on Lagrangian Field Theory (Marseille 1974), *Nucl. Phys. B* **100**:368 (1975); *Cargese Lecture Notes* (1976).
40. A. M. Poliakov, *Sov. Phys. -JEPT* **30**:151 (1970).
41. A. M. Poliakov, *JEPT Lett.* **12**:381 (1970).
42. A. A. Migdal, *Phys. Lett.* **37B**:98, 386 (1971).
43. G. Parisi and L. Peliti, *Lett. Nuovo Cim.* **2**:627 (1971).

44. G. Parisi and L. Peliti, *Phys. Lett.* **41A**:331 (1972).
45. E. S. Fradkin, M. Y. Palchik, and V. N. Zaikin, Saclay preprint (1977).
46. K. Symanzik, *Comm. Math. Phys.* **18**:222 (1970).
47. J. Callan, Jr., *Phys. Rev. D* 1:1541 (1971).
48. K. Symanzik, *Comm. Math. Phys.* **23**:61 (1971).
49. G. Parisi, *Nuovo Cimento* **21A**:179 (1974).
50. S. Weinberg, *Phys. Rev.* **118**:838 (1960).
51. D. I. Kazakov, O. V. Tarasov, and A. A. Vladimirov, Dubna Preprint E2-12249 (1979).
52. E. Brézin, J. C. Le Guillou, and J. Zinn Justin, *Phys. Rev. D* **15**:1544 (1977).
53. G. Parisi, Rome University Preprint No. 426 (1973); and in *Proceedings of the Temple Conference on Field Theory and Critical Phenomena* (1973).
54. B. G. Nikel, D. I. Meiron, and G. A. Baker, Jr., University of Guelph Report (1977).
55. J. M. Kosterliz and D. J. Thouless, *J. Phys. C* **6**:1181 (1973); for a review see D. R. Nelson, *Phys. Rep.* **49C**:255 (1979).
56. A. M. Poliakov, *Phys. Lett.* **59B**:79 (1975); A. A. Migdal, *Sov. Phys.—JEPT* **42**:743 (1976); E. Brézin and J. Zinn Justin, *Phys. Rev. Lett.* **36**:961 (1976); *Phys. Rev. B* **14**:3110 (1976).
57. G. Parisi, *Phys. Lett.* **76B**:65 (1978); and *Phys. Rep.* **49C**:215 (1979).
58. K. Symanzik, Desy Preprint (1977).
59. Y. Araf'eva, *Theor. Math. Phys.* **31**:279 (1977).
60. J. L. Cardy, *Phys. Lett.* **67B**:97 (1977).

ON NON-RENORMALIZABLE INTERACTIONS

G. PARISI

(I.H.E.S. - BURES-sur-YVETTE)

I. Introduction

Nonrenormalizable interactions have always been the black sheep of field theory. Long time ago (1) it was supposed that non-renormalizable interactions are characterized by having Green functions which are not C^{∞} in the coupling constant : if this interpretation is correct, the ultraviolet divergences found in the perturbative expansion arise from the non existence of the quantities which are computed in the standard approach (i.e., the coefficients of the Taylor expansion for zero coupling constant).

The first attempts in this direction were done using or the ξ-limiting procedure (2) or the peratization technique (3). However they were mainly inconclusive ; the full understanding of the problem required a better non perturbative knowledge of quantum field theory which is now given by the modern theory of second order phase transitions (4).

The purpose of these lectures is to study the existence and the properties of non-renormalizable interactions at the light of the knowledge gathered in the study of critical phenomena. We do not prove any rigorous theorem and we are able to control our results only in favorable cases where a divergence-free perturbative expansion can be used. We also do serious attempts to understand the general case which is outside the range of perturbation theory. Our interest is more concentrated on general structural properties than on specific numerical computations.

These lectures are divided in nine sections :

In section II we explain the relation between critical phenomena and non-renormalizable interactions and we present a general method for computing the Green functions of a non-renormalizable theory.

In section III we study two different concrete applications of the methods described in the previous section.

In section IV we discuss an explicit example of a non-renormalizable interaction for which a new divergence-free perturbative expansion has been constructed.

In section V we show that the Green functions constructed in the previous section are not C^∞ in the coupling constant at the origin. The general structure of the irregular terms is found.

In section VI we write the C.S. (Callan Symanzik) equation and we use the properties of its solutions to recover the general structure of zero coupling singularities found in the previous section.

In section VII we see how the C.S. equation can be used as a self consistency requirement, to compute the values of the counter-terms which are arbitrary in the standard perturbative approach.

In section VIII we show how the C.S. equation may be used as a substitute of the equations of motion.

Finally in section IX we present our conclusions.

II. The Relation with Critical Phenomena

It is not well known that non-renormalizable interactions can be regarded as superrenormalizable interactions with infinite bare coupling constant (5). Although this statement may sound very deep it is trivial ; its only utility is due to the fact that changing the language we transform an unsolved problem (the infinite cutoff limit in non-renormalizable interaction) in a solved problem (the infinite bare coupling limit in superrenormalizable interactions). The reader must realise that contrary to the appearence there are many theories in which the infinite coupling limit can be controlled. Indeed the celebrated Kadanoff scaling law for critical phenomena (6) (which it is considered to be well understood) is equivalent to the existence of the renormalized Green functions of a superrenormalizable interaction in the infinite coupling limit (7).

The connection between critical phenomena (massless theory) and infinite coupling theories is quite simple : if the coupling constant has positive mass dimension, the dimensionless coupling constant (on which dimensionless quantities do depend) goes to infinity when the mass goes to zero.

Let us now see some example of NRI (non-renormalized interactions) which are obtained as infinite coupling limit of SRI (superrenormalizable interactions), some of them are well known. Let us consider a vector field of mass m interacting with a conserved current with a coupling constant, λ , the space time dimension D being less than 4 . When we send the mass and the coupling constant to infinity at fixed λ^2/m^2 , we recover the local Fermi current-current interaction. Similarly the $\lambda\Phi^4$ interaction in the infinite coupling limit becomes the non linear σ model (8) which is characterized by the measure $\delta(\Phi^2-1)$. This last result can be formally proved using the identity

(2.1) $$\lim_{\lambda\to\infty} \exp-\lambda(\Phi^2-1)^2 \alpha \delta(\Phi^2-1)$$

A perturbative check of the identity of the two theories can be found in ref.(8) and in Zinn-Justin's lectures at this school.

These results are not surprising ; indeed the coupling constant of the SRI plays the same rôle of the cutoff of the NRI. The familiar divergences in the infinite cutoff limit of NRI are traded with the more familiar divergences of the standard perturbative expansion when $\lambda \to \infty$. (The value at infinity of a non trivial polynomial is always infinity!) However, the modern theory of second order phase transitions teaches us that these divergences are spurious : the renormalized coupling constant (g) acquires a finite value g_c when the bare coupling becomes infinite ; the renormalized Green functions can be computed in this limit : one uses their expansion in powers of the renormalized coupling constant to evaluate them at $g = g_c$ (7), (the so called infrared stable fixed point).

Using these definitions the statement at the beginning of this chapter reads "non-renormalized interactions are superrenormalizable interactions computed at the infrared stable fixed point". It is now clear which is the general procedure which we can follow to construct a NRI : we put a cutoff in the interaction, we interpret the cutoffed theory as a SRI and we find its infrared stable fixed point : this can be done looking for the zeros of a well defined function (9) $(\beta(g_c)=0)$. The renormalized Green functions of the SRI at the infrared unstable fixed point are also the renormalized Green functions of the NRI.

The weak point of this approach is that the infrared stable fixed point may not exist (e.g. the system undergoes always a first order and never a second order transition) or if it exists, it cannot be easyly found . A general classification of NRI can be done using as a criterion the control that we have on the position and on the existence of the zero. There are essentially three different cases:

a) g_c is very small ; it may be of order ε in $4-\varepsilon$ (4) or $2+\varepsilon$ (8) dimensions or it may be of order $1/N$ in the theories invariant under the $O(N)$ group (10). In this case the value of g_c is quite well known : there are asymptotic expansions in $1/N$ or ε for small values of $1/N$ of ε which allow the evaluation of g_c with very great accuracy.

b) g_c is of order one ; however the function $\beta(g)$ has a zero also if one includes only the contributions coming from one loop diagrams. Taking care of higher order diagrams the zero does not disappear and its position can be extimated with improved accuracy. This situation is realized in the three dimensional $\lambda\phi^4$ theory (7), (11). Also in this case very precise results may be obtained.

c) The function $\beta(g)$ does not have any non trivial zero in the one loop approximation and no one has computed the contribution from a number of loops to high enough to see if improving the approximation, a stable zero appears.

In this last case no conclusions can be drawn ; unfortunately this is the situation for a local current-current interaction in 4 dimensions ; consequently the problem of constructing a finite non renormalizable realistic model for weak interaction will not be solved in these lectures. (My personal opinion is that such a construction can hardly be obtained without giving up the unification among weak and electromagnetic interactions, which may be a too high price to pay).

At this stage the reader which is not too familiar with the theory of critical phenomena will be amazed by the existence of case a . He would think that the author must have done a terrific trick to transform an infinite coupling theory in a theory with infinitesimal coupling and maybe he will doubt on the ability of the author to cope with such difficult problems. Consequently it may be useful to spend some words to explain the origin of case a ; the existence of a computable expansion for the critical exponents in powers of ϵ has a similar origin.

Let us consider a $\lambda\Phi^4$ interaction in $4-\epsilon$ dimension, ϵ being a small but non zero number. Given an expansion for a quantity in powers of the bare or the renormalized coupling, we will call it "good" or "bad" if the coefficients of the Mc Laurin expansion have respectively a finite or an infinite limit when $\epsilon\to 0$ It is clear that although the differente between a "good" and a "bad" expansion is defined only in the limit $\epsilon\to 0$ this distinction will play a crucial rôle also for small ϵ . The renormalizability of the theory in $4-\epsilon$ dimensions implies that the expansion of the Green functions in powers of the bare coupling constant is "bad", while the expansion of the renormalized Green functions in powers of the renormalized coupling constant is "good". The function $\lambda(g)$ which gives the bare coupling λ as function of the renormalized one (g) has also a "bad" expansion. It is a crucial observation to note that the function $\lambda(g)$ has the representation (7), (9).

$$\lambda(g) = g \exp \int_0^g [-\epsilon/\beta(g')-1/g']dg' \tag{2.2}$$

where the $\beta(g)$ has a "good" expansion! Moreover the point $\lambda\to\infty$ corresponds to the point g_c such that $\beta(g_c) = 0$. At the one loop level we find

(2.3) $\beta(g) = -\epsilon+g^2+O(g^3)$.

For small ϵ the high order terms are negligible when $g = O(\epsilon)$ (the expansion for β is "good"). We finally get $g_c=\epsilon+O(\epsilon^2)$ which is the first term of a systematical expansion of g_c in powers of ϵ .

Roughly speaking the physical picture is that for small ϵ the ultraviolet divergences nearly shield the interaction so that a theory with a large bare coupling constant is always reduced to a theory with a renormalized coupling constant of order ϵ .

Having explained the general ideas laying behind our approach to the study of NRI , we think that it is better to consider some specific applications of them. This will be the subject of the next section.

III. Two Examples

In the first part of this section we study a current-current interaction among Fermions in dimension $D(2 < D < 4)$; the Lagrangian density is :

(3.1) $$\mathcal{L} = \sum_{1}^{N} {}_i\bar{\Psi}_i(\not\partial+m)\Psi_i+\frac{1}{2}GJ_\rho J^\rho \; ; \; J_\rho = \sum_{1}^{N} {}_i\bar{\Psi}_i\gamma_\rho\Psi_i$$

where Ψ_i is a N component spin 1/2 field.

This interaction is non-renormalizable : the expansion in powers of G can be defined only after the introduction of a cut-off and at any finite order in G , divergent results are obtained in the infinite cutoff limit. Our goal is to show that these divergences disappear using non pertubative techniques (12).

We will use the following cutoffed Lagrangian density :

(3.2) $$\mathcal{L}_c = \sum_{1}^{N} {}_i\bar{\Psi}_i(\not\partial+m)\Psi_i+uA_\rho J^\rho+\frac{1}{2} A_\rho(\mu^2-Z\Box)A^\rho$$.

Where $G = u^2/\mu^2$ and Z plays the rôle of the inverse of the cutoff ; in the limit $Z \to 0$ we recover the local Lagrangian (3.1) : this fact can be easily proved integrating over the A_ρ field. The dimensionless bare parameters which are invariant under a redefinition of the A_ρ field are :

$$(3.3) \qquad g_B = u/[\mu^{(4-D)/2}.Z^{(D-2)/2}] \; ; \; r = u^2 m^{(D-2)}/\mu^2 \quad .$$

When Z goest to zero, the coupling constant g_B goes to infinity if $D > 2$; consequently the limit $Z \to 0$ can be controlled directly only when the interaction is renormalizable.

Renormalized fields, masses and coupling constant can be defined in the usual way (e.g. the propagator of the A field is $1/(p^2+\mu^2)+O(p^4)$. The bare quantities can be computed as functions of the renormalized ones ; using the CS equation and the Ward identities, we find :

$$(3.4) \quad \mu_R = \mu \; ; \; u_R = u \; ; \; Z(g) = \exp \int_0^g [(D-4)/(2\beta(g')) - 1/g']dg'$$

where $\beta(g) = (D-4).g/2+Ag^3+O(g^5)$ $(A>0)$ and g is the dimensionless renormalized coupling constant $(g=u/\mu^{(4-D)/2})$. When D is near to 4 , the function β has a zero at $g_c = [(4-D)/2A]^{1/2}+O(4-D)$. At $g = g_c$; $Z = 0$; the Green functions of the NRI (3.1) concede with those of the SRI (3.2) computed at a particular value of the coupling constant, which is small and where perturbation theory can be successfully applied. The non-renormalizable Fermi interaction turns out to be an intermediate boson theory in which the value of the dimensionless renormalized coupling constant is fixed by the condition $\beta(g_c)=0$; the Fermi coupling is proportional to the inverse of the mass (μ) of the Bose field : the weak coupling limit is obtained when μ goes to infinity.

If 4-D is not small (e.g. D = 3) , $g_c = O(1)$: it is not clear if we can get sensible results using the perturbative expansion in powers of g . We do not know the answer in the case of the Lagrangian (3.1), however we will show that in a different case using a similar procedure one can estimate the Green functions of the NRI with a few per cent accuracy also when g_c is not small.

As we have discussed in the previous section, the non linear σ-model can be regarded as an infinite coupling $\lambda\Phi^4$ theory ; consequently its Green functions coincide with those of the $\lambda\Phi^4$ theory computed at the infrared stable fixed point ; in particular the off shell scattering amplitude at zero momenta in the single phase region $(\langle\Phi\rangle=0)$ is equal to the renormalized coupling constant (g_c) satisfying the condition $\beta(g_c) = 0$. One can try to estimate the position of the zero of the function $\beta(g)$ using perturbation theory in g (the renormalized coupling constant of the $\lambda\Phi^4$ interaction). An explicit computation shows that (7), (11) :

$$\beta(g) = -g+g^2-.42g^3+.35g^4-.38g^5+.50g^6+O(g^7) \quad . \tag{3.5}$$

We know from general theorems that the expansion of β in powers of g has zero radius of convergence (13) and it is an asymptotic expansion (14). If the series is summed using the powerful Padé-Borel technique (15), taking as input only the first 2,3,4,5,6 terms we find (11) respectively g_c=1,1.60,1.42,1.43,1.42 . There are few doubts that the true value of g_c is 1.42 with an error of a few per cent (the precise amount of the error may be a matter of debate).

This example shows that, if an enough high number of diagrams is computed, the approach we propose may be able to produce accurate results for a NRI also when the coupling constant is not small. Having succeeded to construst non trivial NRI's, we would like to understand some structural properties (e.g. the nature of the singularities at zero coupling constant, the validity of the standard divergent pertubative expansion). In principle this can be done without changing the technique, however for simplicity we prefer to study these problems using an explicit renormalizable expansion in which general properties can be verified order by orber. The construction of such an expansion will be the subject of the next section.

IV. The Large N Expansion

We are now going to see that in some non-renormalizable interactions the $1/N$ expansion is renormalizable (16). More precisely there are NRI's, symmetric under the action of the $O(N)$ group, in which the Green functions can be exactly computed in the limit $N\to\infty$ and a systematic expansion in powers of $1/N$ can be constructed. At each order in $1/N$ the only divergences present can be absorbed by mass, wave function and coupling constant renormalization. The $1/N$ expansion may not be suited for pratical computations, however each order can be written in a closed form ; this technique is therefore quite useful to derive general properties.

For example let us consider the quadrilinear interaction of a scalar field in a D dimensional space $(4 < D < 6)$; similar considerations can be extended to a quadrilinear interaction of Fermions for $2 < D < 4$. The theory is non-renormalizable ; its Lagrangian density is :

$$(4.1)\qquad \mathcal{L} = 1/2 \sum_1^N{}_i (\partial_\mu \Phi_i) + 1/2(m^2 + (g/N)^{1/2}\sigma) \sum_1^N{}_i \Phi_i^2 + 1/2\sigma^2$$

where Φ_i is an N component scalar field ; σ is an auxiliary field which can be eliminated reproducing the usual $\lambda\Phi^4$; the Lagrangian is invariant under the group $O(N)$. Notice that our definition of coupling constant has the opposite sign of the conventional one : in our notations, if the coupling constant (g) is positive, the hamiltonian is formally unbounded from below ; the theory is well defined only for negative g , however the 1/N expansion can be constructed only for positive g . This contradiction is peculiar of this model and it is absent in the slightly more complicated 1/N expansion for a current-current interaction or for the nonlinear σ-model. For simplicity we disregard the problem and we consider here only the Lagrangian (4.1) ; we think that the study of the essentially selfadjointness of the hamiltonian and the construction of a formal perturbative expansion are disconnected at this level of sophistication.

In the limit $N\to\infty$ the Green functions are those of the free field theory, the only exception being the renormalized propagator of the σ field, which is equal to :

(4.2)

$$D(p^2) = 1/(1-g\Pi(p^2))$$

$$\Pi(p^2) = \int [q^2+m^2)^{-1}((q+p)^2+m^2)^{-1}-(q^2+m^2)^{-2}]d^Dq \quad .$$

Standard dimensional arguments imply that $\Pi(p^2)$ is finite for $D < 6$. The lack of convergence of the integral (4.2) forbids the application of the 1/N expansion for $D \geq 6$. In the large p region one finds that $\Pi(p^2) \simeq Ap^{(D-4)}$ $(A < 0)$. For negative g the $D(p^2)$ propagator has a pole in the Euclidean region $(p^2 > 0)$; to avoid the presence of this unwanted singularity we are bound to take g positive.

At the first order in 1/N the elastic scattering amplitude for the process i+ → j+m || i ,ℓ, j, and m are indexes which refer to the internal degrees of freedom) is

(4.3) $$A_{i,\ell;jm}(s,t,u)=(\delta_{i\ell}\delta_{jm}D(s)+\delta_{i},\delta_{\ell m}D(t)+\delta_{im}\delta_{j\ell}D(u))/N$$

s,t and u being the Mandestam variables. Elastic unitarity is satisfied at the order $1/N$. At high energies s-wave amplitude goes to a constant and the differential cross section $d\sigma/dt$- at wide angles scales like $s^{(D-2)/2}F(\theta)$, as suggested by naive scale invariance (17). There is a striking difference among these results and those obtained from the first order in pertubation theory, where the unitarity bounds and asymptotic scaling invariance are both violated. At the first non trivial order in $1/N$ all the Green functions are asymptotically scale and conformal invariant. The field ϕ has canonical dimension $(D-2)/2$ while the dimension of the field σ has jumped from its canonical value $(D-2)$ to 2. The behaviour in the large momentum region is g independent, the only dependence from the coupling constant is in the point where asymptotics set in.

The diagrammatic rules for constructing the $1/N$ expansion are very simple and they will not described here (10). The main difference with standard perturbation theory consists in the systematic use of the renormalized σ-propagator $D(p^2)$. Using simple power counting arguments one can show that the $1/N$ expansion is renormalizable in the usual sense (16) : only a finite number of superficial divergent diagrams are present. All the divergences disappear after mass, wave function and coupling constant renormalization. As far as ultraviolet divergences are concerned we find the same situation as in quantum electrodynamics.

This result is not unaspected. Using general arguments it can be shown that in an asymptotically scale invariant theory, if the dimensions of the fields are not too small or too high, the number of superficially divergent diagrams is finite and the theory can be consequently renormalizable (18). Notice that the shift in the dimensions of the field σ makes the effective dimensions of the interaction Lagrangian equal to the space time dimensions

D . The effective coupling constant in the large momentum region is dimensionless as in the standard renormalizable theories, its value is not arbitraty but fixed (it is g independent). Each order of the $1/N$ expansion is no more a polynomial in the coupling constant, indeed one finds a very complicated structure at $g = 0$. The study of this structure will be the main subject of the remaining part of these lectures.

V.- The Structure around $g = 0$

The presence of ultraviolet divergences in the standard perturbative approach and the absence of these divergences in the $1/N$ expansion suggest that the Green functions are not C^{∞} in the coupling constant. We will prove that at each order in $1/N$, terms proportional to non-integer powers of g are present ; their origin is clear : as suggested by unitarity something like an effective cutoff is present at momenta of order $g^{1/(D-4)}$. A term which in the cutoffed perturbation theory is proportional to $g^n \Lambda^\nu$, becomes $g^{n+\nu/(D-4)}$; the conjecture of T.D . Lee (2) done in the framework of the ξ-limiting procedure finds here its explicit realization.

A greater insight on the structure of the irregular terms maybe given by studying in detail a simple example ; we consider the contribution to the six points function at zero momentum coming from only one diagram (i.e. the one looking like a hexagon) (16) ; it can be written as :

$$(5.1)\quad \Gamma^6(g)=g^3 I(g)=g^3\int d^Dk(k^2+m^2)^{-3}[1-g\Pi(k^2)]^{-3} \quad .$$

This integral exists for any positive value of g , however a divergent integral is obtained if we perform enough derivatives respect to g at the point $g=0$. In order to understand the precise nature of the singularities of the function $I(g)$ at $g=0$, we study the positions of the pôles of its Mellin transform (16, 19-21) :

$$(5.2)\quad \begin{aligned} M(s) &= \int_0^\infty dg\; g^{s-1} I(g) \\ I(g) &= (2\pi i)^{-1} \int_{i\infty}^{+i\infty} g^s M(s) ds \quad . \end{aligned}$$

One easily finds :

$$(5.3)\quad M(s)=\Gamma(D/2)\Gamma(s)\Gamma(3-s)/2. \int_0^\infty \Pi(X)^{-s} X^{(D-2)/2}/(X^2+m^2)^3 dX \quad .$$

The pôles of $M(s)$ have two different origines : some arise from the pôles of the Γ functions in front of the integral, others are produced by divergences in the integral itself. If the theory is superrenormalizable $(D < 4)$, the integral is a regular function in the negative s plane ; there are pôles on the positive s plane but they are not interesting for small g : the integration path can always be shifted to the left. When the theory becomes non-renormalizable $(D > 4)$, these pôles migrate via the point at infinity from the positive to the negative s plane and they become relevant in determining the expansion of $I(g)$ in broken powers of g. These extra pôles are located at $s = 2i/(4-D)-j \; (i,j \in Z^+)$. For irrational dimensions none of these pôles collides and one obtains the following double expansion :

$$(5.4)\quad I(g)=\Sigma_i f_i(D)g^i + \Sigma_k\Sigma_i h_{i,k}(D)g^{i+k/(D-4)} \quad .$$

This analysis can be extended to any Feynman diagram of the $1/N$ expansion ; a distinction must be done between the pôles that come form the last integration and pôles of the integrand itself ; the difference between the two cases is similar to that between only superficial divergent diagrams and divergent subdiagrams. In the general case one obtains a double expansion similar to (5.4).

The functions f and h have pôles when two or more of them multiply the same power of g (D must be rational) ; finite results are obtained : the divergences cancel out, and integer powers of ln(g) appear. In other words when two or more pôles of the Mellin transform collide, a higher order pôle is produced.

If D is irrational, dimensional regularization maybe used to define the divergent integrals of the standard perturbative expansion ; it is not difficult to check (16) that this procedure gives correctly the functions f(D) (the functions h(D) are obviously missing). This fact implies that the functions f(D) have a closed representation for fixed N ; such a representation does not exist for the functions h(D) , which must be considered as the substitutes of the divergent counterterms of the old perturbative expansion. Information on their behaviour at finite N can be obtained imposing the cancellation of their singularities with those of the functions f(D) .(This situation has many points in common with the one described by Symanzik) (22).

If we want to use the expansion (5.4) for small values of the coupling constant and not small 1/N (e.g. N=1) , it is imperative to get information on the irregular terms independently from the 1/N expansion. This cannot be done unless we have under control the large momenta behaviour of the theory : the functions h(D) comes from the integration region where the momenta are of order $g^{1/(4-D)}$. Although we can cope with this problem using the techniques described in sections II and III, it would be nice to get direct estimates in the infinite cutoff limit. The rest of these lectures will be devoted to the study of this problem.

VI. - The Solution of the Callan Symanzik Equation

The proof of the validity of CS (Callan Symanzik) equation (9), (23) can be extended to NRI, provided that the renormalized Green functions depend only on the renormalized mass and coupling constant. For example in the case of the $\lambda\phi^4$ interaction for $D < 6$ we find :

$$(6.1)\quad [-\omega\frac{\partial}{\partial\omega} + \beta(g)\frac{\partial}{\partial g} + d_n - n\gamma(g)]\Gamma_n(\omega p,g) = \Delta\Gamma_n(\omega p,g) = m^2\Gamma_{n,\phi_R^2}(\omega p,g)$$

$$\beta(g)=(D-4)g-(g^2+O(g^3);C=(N+8)/N\Gamma(3-D/2),\gamma(g)=O(g^2)\;;$$

$$d_n=2D-n(D-2)\quad ,$$

where g is the renormalized coupling constant (we use the same sign convention as in section IV), Γ_n are the one particle irreducible Green functions and Γ_{n,ϕ_R^2} is the zero momentum insertion of the renormalized ϕ^2 field. If the higher order terms are neglected $\beta(g)$ has a zero at $g_u = (D-4)/C$; the corrections to the position of the zero are small if $1/N$ or $D-6$ are small.

Let us consider eq.(6.1) as a differential equation in Γ ($\Delta\Gamma$ is supposed to be known) ; a family of solutions can be found using the method of the characteristic curves (24) (the solution of a first order differential equation is not unique if we do not specific the boundary conditions). However, if $\exists g_u$ such that $\beta(g_u)=0, \beta'(g_u) < 0$ (g_u is the ultraviolet stable fixed point), there is only one solution which remains finite at $p = 0$ when $g \to g_u$; it is given by (25) :

$$(6.2)\quad \Gamma_n(p,g) = \int_{g_u}^{g} dg'/\beta(g').F_n(g)/F_n(g')\Delta\Gamma_n(R(g)/R(g')p,g')$$

$$R(g) = g^{1/(D-4)}\exp\int_0^g [1/\beta(g')-(D-4)^{-1}g'^{-1}]dg'$$

$$Z(g) = \exp\int_0^g \gamma(g')/\beta(g')dg'\;;\; F_n(g) = R(g)^{-d_n}Z(g)^n\quad .$$

If the interaction is superrenormalizable or renormalizable but asymptotically free (26), $g_u=0$; if the interaction is non renormalizable or renormalizable but not asymptotically free, $g_u \neq 0$. There are no doubts that in the first case the Green functions are regular near g_u , however the condition of regularity near g_u seems very plausible also in the second case (24). The integrated form (eq. 6.2) of the CS equation may be used to investigate the large momentum behaviour of the Green functions : if $\lim_{\lambda\to\infty} \Delta\Gamma(\lambda p,g)/\Gamma(\lambda p,g) = 0$ as it happens at any order of perturbation theory, after some manipulations which are described in ref.(25), we find that asymptotic scale invariance is satisfied :

$$(6.3)\quad \Gamma_n(\lambda p,g) \xrightarrow[\lambda\to\infty]{} a^n(g)\lambda^{-[d_n-n\gamma(g_u)]}\int_0^\infty dx\, x^{[d_n-n\gamma(g_u)-1]}\, \Delta\Gamma_n(xp,g_u).$$

An interesting feature of eq.6.2 in the case of NRI is that Γ_n is not C^∞ at $g=0$ also if the functions $\Delta\Gamma_n$, β and γ are analytic around $g=0$. For sake of simplicity let us condiser the case $p=0$: eq.(6.2) can be written

$$(6.4)\quad \Gamma_n(g) = \int_0^g dg'/\beta(g').F_n(g)/F_n(g')\Delta\Gamma_n(g') +$$
$$+ F(g)\int_{g_u}^0 dg'/[\beta(g')F_n(g')\Delta\Gamma_n(g') \quad .$$

After the split both integrals in eq. 6.4 are divergent, however, if D is irrational, they can be defined using analytic regularization ; the first term is C^∞ in g (always for irrational dimension) while the second term is proportional to $g^{-d_n/(D-4)}$ (16), (25) . Comparing eq. 6.4 with eq. 5.5 we see that we have obtained an explicit representation for the functions $h(D)$ as integrals from 0 to g_u . This fact explains why the computation of $h(D)$ is not simple (it involves the knowledge of the Green functions up to $g = g_u$) , however it may be used to give rough estimates of $h(D)$. If the same argument is used to estimate the singularities at $g=0$ of the function Γ_{n,Φ_R^2}, assuming that $\Delta\Gamma_{n,\Phi_R^2}$ is regular, we find that $\Delta\Gamma_n = g^{(-d_n+2)/D-4)}$ + regular terms, i.e. the power of the irregular term of $\Delta\Gamma$ is greater than the power of the irregular term of Γ.

In conclusions, if the functions β, γ and $\Delta\Gamma$ are reasonable, eq. 6.2 has the virtue to implement automatically asymptotic scale invariance and to generate functions which are not C^∞ starting from a C^∞ input.

VII. - The Counterterms

In the standard perturbative approach to NRI finite results are obtained adding counterterms to the Lagrangian, whose number increase with the order of perturbation theory ; the infinite part of the counterterms is fixed but their finite part is arbitrary : for a given value of the coupling constant we obtain an infinite class of theories. We will argue that the true NRI (which is uniquely defined by the $1/N$ expansion) belongs to this class and it is the only one in which the CS equation is satisfied and the Green functions are finite at the ultraviolet stable fixed point.

As discussed in Symanzik's lectures the introduction of counterterms corresponds to the use of the Lagrangian :

$$\mathcal{L}_E = \mathcal{L}_0 + \sum F_i(g,m,D) O_i \tag{7.1}$$

where $\mathcal{L}_0$ is the Lagrangian without counterterms for NRI the sum runs over all possible local operators O_i, the functions F have singularities at rational dimensions which cancel out with those coming from the perturbative expansion of $\mathcal{L}_0$.

Improving the analysis of section V on the structure of singular terms in the framework of the $1/N$ expansion, we find that the irregular terms h(D) can be generated by a suitable form of the counterterms. Indeed the singularities of the Mellin transform of the Green functions arise either from the superficial divergence of a diagram, or from the divergence of a subdiagram in the large momenta region ; the divergent term is a polynomial in the external momenta and corresponds to the insertion of a local operator. The final result is :

$$F_i(g,m,D) = g^{-d_i/(D-4)} f_i(gm^{(D-4)}, m^2 g^{2/(D-4)}, D) \quad ,$$

where d_i is the mass dimension of the operator O_i and for irrational dimension the functions f_i are C^∞ in both variables ; they can explicitly be computed in the $1/N$ expansion.

Let us see now how we can fix the functions f_i without using the $1/N$ expansion. If extra counterterms are introduced, we can compute in perturbation theory the Green function of the renormalized ϕ and ϕ^2 fields. A CS equation can be written for these Green functions : it will be satisfied in the correct theory but it will not be satisfied for an arbitrary choice of the counterterms. At a finite order (k) in g only a finite number (ℓ) of counterterms is needed. The finite part of the ℓ counterterms can be reconstructed from the knowledge of the value of ℓ different Green functions (usually computed at zero external momenta). Let us concentrate our attention on these ℓ Green functions. In the conventional approach their values are arbitrary and can be treated as ℓ independent parameters ; however they can be computed using the integrated form of the CS equation (6.2). As far as we are not using the exact form of the Green functions, but only an approximated one, we cannot pretend that eq.(6.2) is exactly satisfied ; a reasonable requirement is that the discrepance among the l.h.s. and the r.h.s. must be of order g^{k+1}. It easy to check that this will not happen for an arbitrary choice of the counterterms, imposing such a requirement we find a set of

ℓ coupled non linear equation, whose solution fix the value of all the ℓ Green functions and consequently of the counterterms. The simplest case is realized when no counterterms are needed to define $\Delta\Gamma$ and the counterterms appear only in Γ : eq. 6.4 gives the correct values of the counterms and there is no system of non linear equations to be solved. In the general case also the function $\Delta\Gamma$ will depend on the counterterms.

Unfortunately this technique to find the counterterms is not stable respect to the increase in the order in perturbation theory, in the sense that the value we obtain for a fixed counterterm depends on the order k of perturbation and it is not at all clear what happens when k goes to infinity. (We are also unable to prove the existence of a solution for large k). We think however that these results are interesting because the counterterms are computed using a procedure which remain internal to the theory : it does not involve the introduction of any specific cutoff (we can use dimensional regularization) ; moreover asymptotically ASI is satisfied at any stage : we pick that particular solution of the CS equation (eq. 6.2) which remains finite at the ultraviolet stable fixed point.

VIII. - The Callan Symanzik Equation as a Substitute of the Equations of Motion

It is well known that the equations of motion for the field ϕ (let us restrict to the case of a scalar interaction) are equivalent to an infinite system of coupled integral non-linear equations for all the connected Green functions. Performing formal manipulations one obtains a system of a finite number of equations for only the low degree Green functions (27) (the so-called Dyson equations). The standard perturbative expansion can be generated by the iterative solution of these equations. Although there are solutions of the Dyson equations which are asymptotically scale invariant, (28) (29), ASI (asymptotic scale invariance) is not automatically implemented : if we try to solve the Dyson equations by iteration and we start from a zero order approximation which violates ASI, the amount of violations of ASI increases after each iteration up to the point where ultraviolet divergences appear. This is the origin of the divergence present in the standard treatment of NRI interactions ; it also explains the lack of convergence of the perturbative expansion in the large momenta region for a renormalizable interaction.

Longtime ago it was noticed that in the case of SRI the CS equation can be used as a substitute of the equations of motion : (24), (30), $\Delta\Gamma_n$ can be written as an integral over the first n+2 Γ_n functions ; if both sides of eq. 6.1 are expanded in powers of g, we obtain the k-order of perturbative theory for Γ_n as function of the first (k-1)-orders of perturbation theory for $\Gamma_i(i=2,N+2)$. If we write the CS equation at the infrared stable fixed point $g_c(\beta(g_c)=0$, i.e. the bare coupling constant is infinite), we obtain a system of integral equations for the Green functions (6) which is the equivalent of Wilson's fixed point condition for the Hamiltonian (4).

In the same spirit we can consider eq. 6.2 as an infinite set of non-linear equations for the Γ_n functions ; the requirement that in the limit $g\to 0$ we recover the first non zero order of perturbative theory must be imposed as boundary condition. This boundary condition is automatically satisfied if we solve eq. 6.2 by iterations using as a zero order approximation the first order of perturbation theory. This approach becomes particulary interesting when the interaction is no more superrenormalizable, because ASI is built in the formalism. If we apply the same procedure to NRI we discover that no ultraviolet divergence is present and that terms no C^∞ in the coupling constant appear in the Green functions (eq. 6.4).

Although also in the case of SRI it is not known if the solution of this kind of infinite coupled non linear equations can be found by iteration (i.e. if the iterative solution converges)we think that the methods described in this section maybe used to perform approximated computations of the Green functions and, perhaps, to obtain non formal results on non-renormalizable and renormalizable interactions (the case of asymptotically free interactions seems to be the most promising).

In this discussion we have been cavalier with two major points; the definition of the renormalized Φ^2 field (Φ_R^2) and the existence of a zero of the $\beta(g)$ functions.

The first problem is due to the divergence of the Green functions of the renormalized field (Φ_R) at two coinciding points, the remedy is well known and consistes in the introduction of the field $\Phi_R^2 = Z_2(\Phi_R)^2$ (Z_2 is an infinite constant). Divergence free equations can written (23), (24), at the price of introducing the whole machinery of Beth-Salpeter kernels, two particle irreducible Green functions... However this is only a technical complication. The second problem may give us serious troubles : eq. 6.2 makes sense only if g_u exists. If the interaction is asymptotically free, $g_u=0$, however if the interaction is not asymptotically free or it is not renormalizable, g_u is the first zero of the $\beta(g)$ function. If we start from a zero order approximation in which the $\beta(g)$ does not have a zero, eq. 6.2 cannot be written for absence of a candidate for g_u . If the $\beta(g)$ does not have a zero at the one loop level, it is not clear which should be the starting points of our iterative procedure (maybe the methods described in the previous section can be useful in this case). This difficulty is much more serious and goes to the heart of the construction of NRI. In this framework there is practically no difference between renormalizable non asymptotically free and non-renormalizable interactions:

if we would be able to construct in this way a $\lambda\phi^4$ interaction in 4 dimensions with the good sign of the coupling constant, we should have no serious problems to construct the nearby NRI in $4+\epsilon$ dimensions). We note en passant that the conformal invariant self consistency conditions for the propagator and the vertex (29) do not show any pathology (31) when we go from a renormalizable to a non-renormalizable interaction provided that a non-trivial solution exists in the renormalizable case : the values of the coupling constant and of the anomalous dimensions seem to be regular functions of the dimension of the space (5).

IX.- Conclusions

We hope to have convinced the hypothetical reader of all the previous sections that the concept of a finite non-renormalizable interaction is not contradictory and that using an appropriate perturbative expansion accurate results maybe obtained. We consider quite gratifying the validity of asymptotic scale invariance. We stress that our results are valid only for a limited class of interactions and there are many interesting cases which have not been the object of systematic investigations (e.g. a non abelian current-current interaction and the Einstein theory of gravity), however we think that we have settled a general scheme in which the properties of a particular non-renormalizable interaction can be investigated.

References

1) P.J. Redmond and J.L. Uretski, Phys. Rev. Letters 1, 145 (1958), Ann. of Phys. 9, 106 (1960).

2) T.D. Lee, Phys. Rev. 128, 899 (1962).

3) G. Feinberg and A. Pais, Phys. Rev. 132, 2724 (1963).

4) K. Wilson and J. Kogut, Phys. Reports 12, 75 (1974).

5) G. Parisi, "Some considerations on nonrenormalisable interactions", on the Proceedings of the "Colloquium on Lagrangian Field Theory", Marseille 1974.

6) L.P. Kadanoff, et al. Rev. Mod. Phys. 39, 395 (1967).

7) G. Parisi, Lectures given at the Cargese Summer School, July 1973, Columbia University, Preprint.

8) E. Brezin and J. Zinn-Justin, Phys. Rev. Letters 36, 639 (1976) and Phys. Rev. B (October 1976).

9) K. Symanzik, Comm. Math. Phys. 18, 227 (1970).

10) E. Brezin and D.J. Wallace, Phys. Rev. B7, 1967 (1973).

11) G.A. Baker, B.G. Nickel, M.S. Green and D.I. Meiron, Phys. Rev. Letters 36, 1351 (1976).

12) G. Parisi (in preparation).

13) A. Jaffe, Comm. Math. Phys. 1, 127 (1965).

14) J.T. Eckan, J. Magnen and R. Sénéor, Comm. Math. Phys. 39, 251 (1975).

15) S. Graffi, V. Grecchi and B. Simon, Phys. Letters 32B, 631 (1970).

16) G. Parisi, Nucl. Phys. B100, 368 (1975).

17) K. Wilson, Phys. Rev. D4, 2911 (1973).

18) G. Mack and L. Todorov, Phys. Rev. D8, 1764 (1973).

19) V. de Alfaro and E. Predazzi, Nuovo Cimento 39, 235 (1965).

20) G. Parisi, Nuovo Cimento Lettere 6, 450 (1973).

21) K. Symanzik, Nuovo Cimento Lett. 8, 771 (1973), Cargese Lectures 1973, DESY preprint 73158 .

22) K. Symanzik, Comm. Math. Phys. 45, 79 (1975).

23) J. Callan, Phys. Rev. D1, 1541 (1971).

24) K. Symanzik, Comm. Math. Phys. 23, 61 (1971).

25) G. Parisi, Nuovo Cimento 21A, 179 (1974).

26) K. Symanzik, Nuovo Cimento Lettere 6, 420 (1973).

27) J.D. Bjorken and S.D. Drell, Relativistic Quantum Field, Theory (New York, McGraw-Hill, 1965).

28) A.M. Poliakov, Soviet Physics JEPT 28, 533 (1969).

29) G. Parisi and L. Peliti, Nuovo Cimento Lettere 2, 623 (1971).

30) G. Parisi, Phys. Letters 39B, 643 (1972).

31) K. Symanzik, Nuovo Cimento Lettere 3, 734 (1972).

AN INTRODUCTION TO SCALING VIOLATIONS

G. PARISI
Laboratori Nazionali di Frascati
Frascati (Italy)

Abstract: The theory of scaling violations in deep inelastic scattering is presented using the parton model language ; intuitive physical arguments are used as far as possible. In the comparison between theory and experiments particular attention is payed to the consequences of the opening of the threshold for charm production.

Resumé : On utilise ici le language du modèle a partons pour exposer la théorie de la violation de la loi d'échelle dans la diffusion très inélastique, en employant autant que possible des arguments intuitifs. On compare ensuite théorie et donnés expérimentales en étudiant avec attention particulière les conséquences de l'overture du seuil pour produire du charm.

1. - INTRODUCTION[x]

ὅσων ὄψις ἀκοὴ μάθησις, ταῦτα ἐγὼ προτιμέω [o]
(Heracleitus)

I think that deep inelastic scattering is one of the best processes which can be used to test our theoretical understanding of strong interactions. The success of the Bjorken scaling law and the ability of the parton model to explain the experimental data are the main historical motivations for our present belief in the quark model.

It has now been realized that the naive quark-parton model is inconsistent and that small violations of the scaling law must be present: more accurate data seem to agree with this conclusion. The standard theoretical arguments which are used to study scaling violations are mainly based on sophisticated field theory techniques such as Wilson expansion at short distances and on the light cone, anomalous dimensions, bilocal operators.... All this theoretical machinery has been essential to derive unambigous and correct results, however we have departed from the physically intuitive approach which makes the standard parton model so appealing.

In this introduction to the violations of the scaling law, we try to recover the physical interpretation of the theory; to this end the language of the parton model will be used to derive and interpret the theoretical results. We hope that this paper will partially fill the gap between the conclusions of the parton model (which are physically motivated but incorrect) and the conclusions of a field theoretical analysis (which are correct but whose intuitive interpretation has been lost somewhere[+].

2. - THE PARTON MODEL

Let us briefly review the main ideas which are behind the parton model[6] in order to understand how they must be modified to account for the violations of the Bjorken scaling law.

(x) - Part of the results presented here have been obtained by the author in collaboration with G. Altarelli and R. Petronzio[1-3].

(o) - The things of which there is seeing and hearing and parception, these do I prefer.

(+) - This point of view is not new: a similar approach has been advocated by Polyakov[4] and by Kogut and Susskind[5].

In a deep inelastic process an highly virtual photon of mass Q^2 interacts with the pointlike constituents (partons) of the hadron. In the Breit frame the photon carries no energy and the proton has a momentum P proportional to $(Q^2)^{1/2}$. For high Q^2, P is large and the proton looks like a highly Lorentz contracted pancake; the time (τ) of interaction is proportional to $(Q^2)^{-1/2}$. For small τ we can safely suppose that the photon scatters incoherently on each parton; the cross section for deep inelastic scattering depends on the parton distribution seen when we look inside the hadron with a resolution time τ.

The cross section for longitudinaly (σ_L) and transverse (σ_T) polarized photons can be written using two independent structure functions[7] : $F_1(x, Q^2)$, $F_2(x, Q^2)$, x being equal to $2M\nu/Q^2$.

For spin 1/2 partons:

(2.1) $$F_2(x, Q^2) = \Sigma_i e_i^2 \, x N_i(x, \tau) , \qquad \tau = (Q^2)^{-1/2} ,$$

where $N_i(x, \tau)$ is the number of partons of the i-th type, having charge e_i and carrying longitudinal momentum xP; σ_L/σ_T is proportional to $\langle p_\perp^2 \rangle / Q^2$, where $\langle p_\perp^2 \rangle$ is the mean squared value of the transverse momentum carried by the partons.

This is quite general: we have only assumed that the electromagnetic current couples to point-like constituents and that the final state interaction does not change total cross sections at very high energies: after the interaction with the photon the system evolves in time with its own hamiltonian.

The Bjorken scaling law follows from the assumption that:

(2.2) $$\lim_{\tau \to 0} N(x, \tau) = N(x) \neq 0 .$$

In very short times partons cannot modify their distribution inside the hadron: they move slowly and they can be considered free on a short time scale.

Two main assumptions are thus involved in the derivation of the Bjorken scaling law:

a) The hadron interacts with an highly virtual photon via some point-like constituents (partons). Final state interactions can be neglected.

b) The constituents cannot change their momentum too fast: their interactions can be neglected in the limit $\tau \to 0$.

However what is the rationale for these assumptions? In any reasonable quantum field theory in 4 dimensional space-time the first one is valid, the second one is false[(8-9)].

For example in quantum electrodynamics the validity of both assumptions would imply that the radiative corrections scale with the energy and are the same both for $e\bar{e}$ and $\mu\bar{\mu}$ scattering. Anyone working in high energy physics knows that this is not the case and that radiative corrections show a logaritmic dependence on E/m.

If the first hypothesis is true, even in presence of scaling violations the parton model inequalities in deep inelastic scattering (e. g. $1/4 \leq F_2^n(x, Q^2)/F_2^p(x, Q^2) \leq 4$) are unchanged. The failure of the second hypothesis implies that the Bjorken scaling law is no more valid and that more complicated scaling laws are satisfied. These new scaling laws depend on the detailed dynamics of the strong interactions and their verification would be quite important.

Before discussing what happens in the strong interaction case, I want to clear up the situation in a more familiar case, i. e. quantum electrodynamics. This will be done in sections 3 and 4. In section 5 I will present the theoretical results based on a coloured gauge theory of strong interactions. In sections 6 and 7 I will compare the theoretical results with the experimental data on electron and neutrino scattering.

3. - THE COSTITUENTS OF THE ELECTRON

*ἐν δὲ μέρει κρατέουσυ περυπλομένοιο κύκλοιο, καὶ
φθίνει εἰς ἄλληλα καὶ αὔξεται ἐν μέρει αἴσης*[(x)]

(Empedocles)

Pure quantum electrodynamics is a good place to study the violations of the Bjorken scaling law. They show up in very simple and familiar formulae: the equivalent number of photons in an electron on energy E (momentum P = E) is:

$$N_\gamma(x, P) \simeq \frac{\alpha}{2\pi}\frac{4}{x}\ln(P/m_e) + O(\alpha^2) , \tag{3. 1}$$

(x) - In turn they (elements) get the upper hand in the revolving cycle, and perish into one another and increase in the turn appointed by their fate.

where x is the fraction of longitudinal momentum carried by the photon. If we interpret $1/P \simeq (1/Q^2)^{1/2}$ as the resolution time τ, we obtain that the equivalent number of photons in the electron is:

$$N_\gamma(x,\tau) = \frac{\alpha}{2\pi}\frac{4}{x}\ln(1/m_e\tau) + O(\alpha^2)\,, \qquad \tau m_e << 1\,. \tag{3.2}$$

This quantity goes to infinity when $\tau \to 0$ and the assumption b) (eq. 2.2) of the last section is violated. Moreover for each photon of momentum xP there must be an electron of momentum (1-x)P; the momentum distribution of the electrons inside the electron is:

$$N_e(x,\tau) = \delta(x-1) + \frac{\alpha}{2\pi}\left[\frac{4}{1-x} - 2C\,\delta(x-1)\right]\ln(1/m_e\tau)\,. \tag{3.3}$$

The constant C is fixed by the condition that the total number of electrons is not changed by the interaction:

$$\int_0^1 N_e(x,\tau)\,dx = 1\,. \tag{3.4}$$

Stricly speaking C is logaritmically divergent ($C = 2\int_0^1 \frac{dx}{1-x}$). The two divergences in eq. (3.4) cancel each other.

However eqs. (3.1) and (3.3) cannot be used directly to study the limit $\tau \to 0$: the neglected higher order terms become important when $\alpha \ln \tau \simeq 1$. Let us first study the effect of multiple photon emission (see Fig. 1). The key step is to concentrate one's attention on the time derivative of the number of electrons; the variable $L = -2\ln(m_e\tau)$ is introduced for convenience.

From eq. (3.3) we find:

$$\frac{dN_e(x,L)}{dL} = -\frac{1}{2}\frac{dN_e(x,\tau)}{d\ln\tau} = \frac{\alpha}{2\pi}\frac{2}{1-x} - C\,\delta(x-1) = \frac{\alpha}{2\pi}P_{ee}(x)\,. \tag{3.5}$$

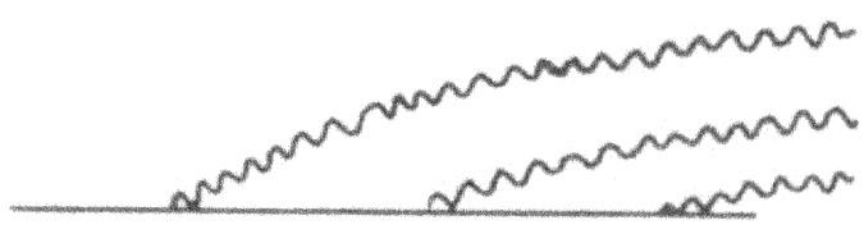

fig.1

FIG. 1 - A typical diagram contributing to multiple photon emission.

Eq. (3.5) suggests that the transition probability for electron bremsstrahlung is independent of L. However the electron distribution is L dependent: the change in time of the electron distribution must be the product of the transition probability p and the actual electron distribution at "time" L. One is led to the following "master" equation:

(3.6) $$\frac{dN_e(x,L)}{dL} = \frac{\alpha}{2\pi}\int_x^1 \frac{dy}{y} N_e(y,L)\, p_{ee}(x/y) =$$

$$= \frac{\alpha}{2\pi}\left[-C N_e(x,L) + \int_x^1 N_e(y,L)/(y-x)\, dy\right].$$

The first term arises from the decrease of $N_e(x, L)$ due to the bremsstrahlung of electrons staying at the point x: it is naturally proportional to $N_e(x, L)$. The second term represents the increase in the number of electrons at the point x due to bremsstrahlung of electrons carrying momentum $y > x$, the relative loss of electron momentum being x/y.

Eq. (3.6) can be easily solved by computer; qualitative statements can be made studying the L dependence of the moments:

(3.7) $$M_e^N(L) = \int_0^1 \frac{dx}{x} x^N N_e(x,L).$$

Substituting eq. (3.6) in the derivative of eq. (3.7) we obtain:

(3.8) $$\frac{dM_e^N(L)}{dL} = \frac{\alpha}{2\pi}\int_0^1 \frac{dx}{x} x^N N_e(x,L) \int_0^1 \frac{dy}{y} y^N p_{ee}(y) =$$

$$= -\frac{\alpha}{2\pi} M_e^N A_{ee}^N, \qquad A_{ee}^1 = 0 \qquad A_{ee}^N > 0 \ (N > 1),$$

whose solution is:

(3.9) $$M_e^N(L) = M_e^N(L_o) \exp\left[-\frac{\alpha}{2\pi} A_{ee}^N (L-L_o)\right].$$

M^1 is the total number of electrons in the system and it is a constant. M^2 is the total momentum in P units carried by the electrons and it goes exponentially to zero: the whole momentum is transferred from the electron to the photon system. Increasing L, $N_e(x, L)$ shifts towards $x = 0$ and asymptotically it is concentrated at this point.

Eqs. (3.6-3.9) are valid in the so called leading logaritm approximation (terms proportional to $(\alpha L)^n$ are retained and terms proportional to $\alpha(\alpha L)^n$ are neglected).

The transition probabilities p_{ee} contain higher orders in α; however these new terms are not L dependent and no qualitative conclusion is changed; to neglect them is a good approximation for all values of L if α is not too large.

A similar equation can be written for the photons:

$$\frac{dN_\gamma(x,L)}{dL} = \frac{\alpha}{2\pi}\int_x^1 \frac{dy}{y} N_e(y,L)\, p_{\gamma e}(x/y)\,. \tag{3.10}$$

The following relation holds:

$$p_{\gamma e}(x) = p_{ee}(1-x)\,. \tag{3.11}$$

However the situation is not so simple: the photon itself may split in a $e\bar{e}$ pair, each of the new born e or $\bar{e}$ may emit a photon and so on. The whole process is quite similar to the evolution of an electromagnetic shower in lead. A typical diagram is shown in Fig. 2.

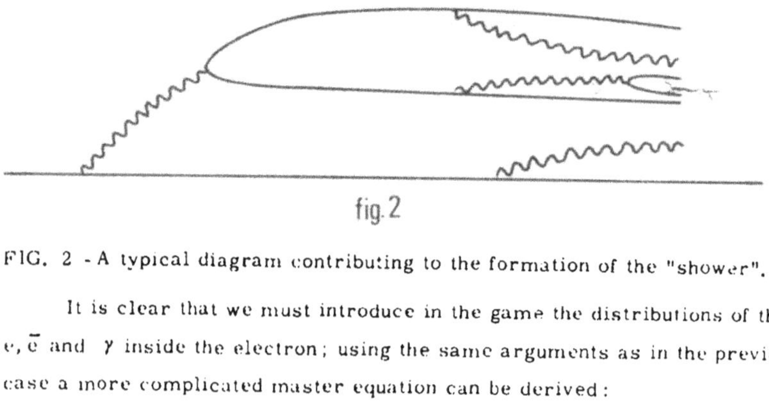

fig. 2

FIG. 2 - A typical diagram contributing to the formation of the "shower".

It is clear that we must introduce in the game the distributions of the e, $\bar{e}$ and γ inside the electron; using the same arguments as in the previous case a more complicated master equation can be derived:

$$\frac{dN_e(x,L)}{dL} = \frac{\alpha}{2\pi}\int_x^1 \frac{dy}{y}\left[N_e(y,L)\, p_{ee}(x/y) + N_\gamma(y,L)\, p_{e\gamma}(x/y)\right],$$

$$\frac{dN_{\bar{e}}(x,L)}{dL} = \frac{\alpha}{2\pi}\int_x^1 \frac{dy}{y}\left[N_{\bar{e}}(y,L)\, p_{\bar{e}\bar{e}}(x/y) + N_\gamma(y,L)\, p_{\bar{e}\gamma}(x/y)\right], \tag{3.12}$$

$$\frac{dN_\gamma(x,L)}{dL} = \frac{\alpha}{2\pi}\int_x^1 \frac{dy}{y}\Big\{N_\gamma(y,L)p_{\gamma\gamma}(x/y) + \left[N_e(y,L) + N_{\bar{e}}(y,L)\right] p_{\gamma e}(x/y)\Big\} ,$$

where:

(3. 13)
$$p_{ee}(y) = p_{\bar{e}\bar{e}}(y) \quad p_{\gamma e}(1-y) = p_{\gamma\bar{e}}(1-y) ,$$

$$p_{\bar{e}\gamma}(y) = p_{e\gamma}(y) = p_{e\gamma}(1-y) = \frac{1}{2}\left[y^2 + (1-y)^2\right] ,$$

$$p_{\gamma\gamma}(y) = - C_\gamma \delta(y-1) ,$$

$$C_\gamma = \frac{1}{2}\int dy \left[p_{e\gamma}(y) + p_{\bar{e}\gamma}(y)\right] = \frac{1}{3} .$$

The meaning of these equations is quite clear. The last equation implies that the number of photons which disappear at the point x it is equal to the number of new born $e\bar{e}$ pairs carrying total momentum x. The functions $p_{e\gamma}$ and $p_{\gamma e}$ are related to the longitudinal distributions of bremsstrahlung photons and of Dalitz pair electrons(x).

It is interesting to note that the derivative of the difference of the number of electrons and positrons does not depend on the γ distribution:

(3. 14)
$$\Delta N(x,L) = N_e(x,L) - N_{\bar{e}}(x,L) ,$$

$$\frac{d\Delta N(x,L)}{dL} = \int_x^1 \frac{dy}{y} \Delta N(y,L)\, p_{ee}(x/y) .$$

The L evolution of this difference decouples from that of the other functions. Also this coupled set of equations can be easily solved with a computer: the knowledge of the three functions N_e, $N_{\bar{e}}$ and N_γ at a particular value of L in the region $1 > x > x_m$ allows us to compute them at any value of L, in the same x region.

It is possible to study the behaviour of the moments of the distributions, if one defines a three component vector

---

(x) - The possibility of using these formulae to compute higher order processes in quantum electrodynamics has been suggested by Cabibbo. This technique has been applied to the study of the reactions $e^+e^- \to e^+e^-\rho$ (10) and $e^+e^- \to e^+e^-e^+e^-$ (11).

$$(3.15) \qquad M_i^N(L) = \int_0^1 \frac{dx}{x} x^N N_i(x, L) \qquad \begin{array}{l} i = 1 \leftrightarrow e \\ i = 2 \leftrightarrow \bar{e} \\ i = 3 \leftrightarrow \gamma \end{array}$$

one finds :

$$(3.16) \qquad \frac{dM_i^N(L)}{dL} = - \frac{\alpha}{2\pi} A_{iK}^N M_K^N(L) ,$$

where A is a three by three matrix. If we denote by λ_a^N and $\vec{u}_a^N$ the three eigenvalues and eigenvectors of A^N, the solution of (3.16) can be written using the vectorial notations as :

$$(3.17) \qquad \vec{M}^N(L) = \sum_1^3 {}_a M_a^N \vec{u}_a^N \exp - \frac{\alpha}{2\pi}(L-L_o) \lambda_a^N .$$

The quantities M_a^N are fixed by the boundary condition $\vec{M}^N(L)\big|_{L=L_o} = \vec{M}^N(L_o)$.

For N = 2 one of the eigenvalues is 0, reflecting the conservation of the total momentum carried by the constituents. When $L \to \infty$ the distributions of both electrons and photons shifts towards 0, the ratio of the momentum carried by the electrons and the positrons goes to one and the total momentum carried by the "valence" electron goes to zero, while the momentum carried by the sea of $e\bar{e}$ pairs and by the photons goes to a constant. In the limit $L \to \infty$ an equilibrium situation is reached : the momentum lost by the electrons via bremsstrahlung is equal to the momentum refilling due to the creation of Dalitz pairs. The mean value of the momentum carried by each constituent goes to zero and this degradation of momentum is the origin of the progressive concentration of the functions N(x, L) near x = 0. Up to now, we have considered only the distributions in longitudinal momentum. The transverse momentum distribution can be studied using similar techniques ; one finds[12] :

$$\frac{\sigma_L}{\sigma_T} \propto \frac{\langle p_\perp^2 \rangle}{p^2} = O(\alpha) .$$

Unfortunately the situation is not so simple : we have neglected the possibility that an $e\bar{e}$ pair annihilate in a photon which subsequently splits in an $e\bar{e}$ pair and so on. A typical diagram is shown in Fig. 3.

To study this phenomenum a new concept must be introduced : vacuum polarization. The effect of these new diagrams can be accounted for, by the introduction of an effective L dependent coupling constant.

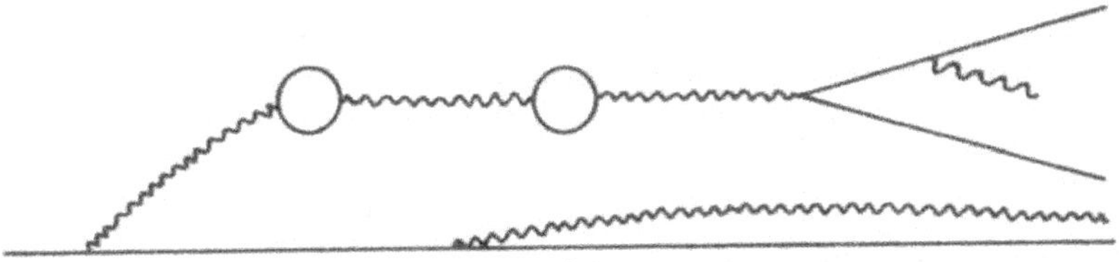

fig.3

FIG. 3 - A typical diagram contributing to vacuum polarization.

We prefer to discuss the consequences of vacuum polarization and to write the final formulae in this section; we postpone to the next section the discussion on the rationale and on the physical meaning for the introduction of an effective time dependent coupling constant. The correct formulae are obtained by substituting α by $\alpha(L)$ in eqs. 3. 12 and 3. 16. The function $\alpha(L)$ satisfies the differential equation:

$$(3.18) \qquad \frac{d\alpha(L)}{dL} = \beta\alpha^2(L) + O\left[\alpha^3(L)\right] , \qquad L \gg 1 ,$$

whose solution is:

$$(3.19) \qquad \alpha(L) = \frac{\alpha(L_0)}{1 - \beta(L-L_0)\alpha(L_0)} , \qquad \alpha(L),\ \alpha(L_0) \ll 1 .$$

Two different possibilities are open(13):

a) $\beta > 0$,

b) $\beta < 0$.

In case a) the effective coupling increases with L, also if we start from a small value of α, increasing L we are projected in the strong coupling regime where we cannot justify our approximation of neglecting higher order in α in the transition probabilities p. What will finally happen in this case is still an open problem: no general consensus has been reached on this point.

Case a) is realized in pure QED; the energies at which the perturbative expansion become useless are gigantic: they are of the order of the mass of the universe.

Case b) is better understood: increasing L the effective coupling constant decreases; also if we start from a relative large value of α we finally end up with a small value of $\alpha(L)$ ($\alpha(L) \to -1/\beta L$ when $L \to \infty$). In this kind of theory the large L limit can be controlled using a perturba tive estimate of the transition probabilities, whatever the value of the coupling constant in the low momentum region.

There is no problem to solve the modified eq. (3. 12) by computer. Eq. (3. 16) becomes now:

$$\frac{dM_i^N(L)}{dL} = -\frac{\alpha(L)}{2\pi} A_{iK}^N M_K^N(L) \,, \tag{3.20}$$

whose solution is:

$$\vec{M}^N(L) = \sum_1^3 {}_a M_a^N \vec{u}_a^N \left[1 - \beta(L-L_0)\alpha(L_0)\right]^{-\frac{\lambda_a^N}{2\pi\beta}} \,. \tag{3.21}$$

We now have in our hands the tools which are needed to study the violations of the scaling law in deep inelastic scattering. We are able to compute how the distribution of the pointlike constituents depends on the resolution time. We have seen that when the resolution time goes to zero ($L \to \infty$) a continuous process of interchange of momentum among the bare constituents is present, the laws which regulate this phenomenum can be summarized in the "master" equation (3. 12).

4. - VACUUM POLARIZATION

It is a common day experience that salt can be easily dissolved in water but not in oil. This fact is due to the high value of the static dielectric constant $\varepsilon_s = 80$ ($\varepsilon_s = 1$ in vaccum). The force between two charges is:

$$F = \frac{q_1 q_2}{\varepsilon_s} \frac{1}{r^2} \,, \qquad r \to \infty \,, \tag{4.1}$$

at large distances. However, at distances smaller than the radius of the water molecule (a), one recovers the more familiar:

$$F = q_1 q_2 \frac{1}{r^2} \,, \qquad r \ll d \,. \tag{4.2}$$

It is possible to define a function $\varepsilon(r)$ such that:

$$F = \frac{q_1 q_2}{\varepsilon(r)} \frac{1}{r^2} \,, \qquad \varepsilon(0) = 1 \,, \quad \varepsilon(\infty) = \varepsilon_s \,. \tag{4.3}$$

This effect arises from the orientation of the water dipoles in presence of an electric field. The scale of the phenomenum is naturally given by d.

Equivalently one would define an r dependent effective charge and write:

$$(4.4) \qquad F = q_1(r)\, q_2(r)\, \frac{1}{r^2} \;, \qquad\qquad q_1(r) = q_1/\sqrt{\varepsilon(r)} \;.$$

A typical plot of $q(r)$ as function of L is shown in Fig. 4.

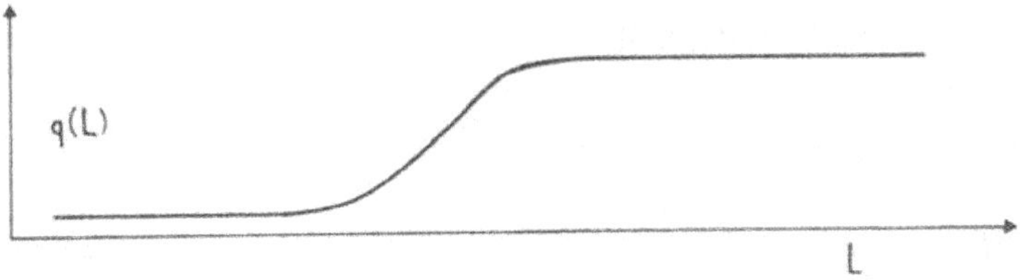

FIG. 4 -The effective charge in water as function of $L = -\ln(r/d)$, r being the distance and d being the radius of water molecules.

The polarization of water decreases the force among Na^+ and Cl^- ions and allows the solution of salt in water : charged ions in water are nearly asymptotically free at large distances while they have a strong interaction at short distances.

A more dramatic effect can be found in metals : here $\varepsilon_s = \infty$ and the effective charge goes to zero exponentially at large distances : the charge is completely shielded.

In quantum electrodynamics the role of water is played by the virtual $e\bar{e}$ pairs which fill the vacuum. The presence of a charge modifies their distribution and produces a polarization of the vacuum which alters the value of the effective charge seen at large distances. The inverse of the mass of the virtual pair corresponds to the radius of water molecules : the shielding effect reaches a constant at distances larger than $1/2\, m_e$. However there is no upper bound to the mass of a virtual pair so that the effective charge changes its value also at very short distances. At distances of order 10^{-100} cm the effective coupling constant becomes of order 1 and non linear phenomena in the electric field are quite important. It is not clear what happens at so short distances, however this problem is not relevant here.

We hope we have clarified why the effective coupling constant in Quantum electrodynamics depends on the distance r and by relativistic invariance also on the resolution time τ. The fact that the force among different (equal) sign charges is attractive (repulsive) implies that in all possible materials, vacuum included, $\varepsilon_s > 1$ and the effective charge at large distances is smaller than the bare charge: $q(\infty) < q(0)$. We can conceive a world in which the force among charges of the same sign is attractive and among charges of opposite sign is repulsive. We will call the matter of which this world is made up "enantion". The static polarizability of the enantion is always less than 1. Also in this case we can introduce a distance-dependent effective coupling constant: the effective charge seen at large distances is always greater that the bare one:

$$q(\infty) = \frac{q(0)}{\sqrt{\varepsilon_s}} > q(0) \ . \tag{4.5}$$

Let us chose a particular kind of enantion in which $\varepsilon_s = 0$ and let us suppose that the radius of the molecules has a continous distribution which ranges from zero up to a maximum length d. In this case $q(\infty)/q(0) = \infty$. If the effective charge seen at large distance is finite, the effective charge at very small distance must be equal to zero (see Fig. 5). Two ions in enantion behave as free at short distances while the interaction remains strong at large distances.

Why are we interested in such a devious system? The reason is simple: there are models of strong interactions in which the polarizability properties of vacuum are just the same as those of enantion. These model belong to case b) of section 3 and have a coupling constant which is asymptotically zero at short distance. I think that it is interesting to have a concrete example of a system in which the interaction among pointlike particles fades at short distances.

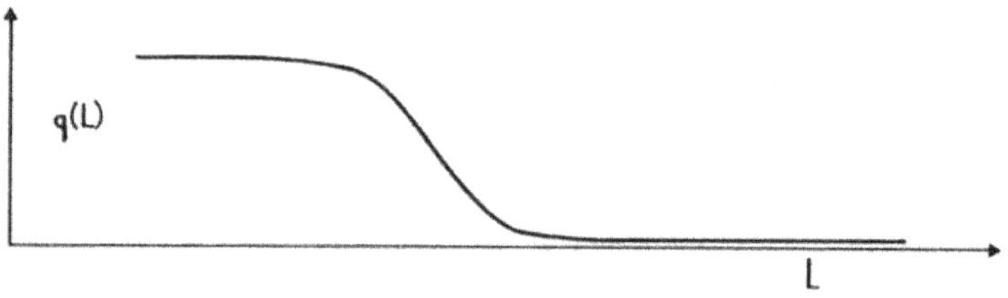

FIG. 5 -The effective charge in enantion as function of $L = -\ln(r/d)$, r being the distance and d being the maximum radius of enantion molecules.

5. - THE STRUCTURE OF STRONG INTERACTIONS

τέσσαρα γὰρ πάντων ῥιζώματα πρῶτον ἄκουε(x)

(Empedocles)

In the most popular model of strong interactions the hadrons are composed of 4 quarks (p, n, λ and p')[14]; the three different colours of quarks interact via the exchange of an octet of coloured gluons. Electromagnetic and week currents are colour singlets; the theory is invariant under the group SU(3) colour.

The effective coupling constant of the theory satisfies the equation[15-17]:

$$\text{(5.1)} \qquad \frac{d\alpha(L)}{dL} = -\frac{25}{12\pi}\alpha^2(L) + O(\alpha^3) \qquad (L = \ln Q^2),$$

whose solution is:

$$\text{(5.2)} \qquad \alpha(L) = \frac{\alpha(L_o)}{1 + \frac{25}{12\pi}(L - L_o)\alpha(L_o)} .$$

The situation is the same as in enantion. Although the coupling constant of strong interactions is large at distances of order $1/m_\pi$, it is possible that at rather shorter distances it becomes smaller and smaller and that a perturbative approach can be used in the deep inelastic region. If this is the case, it is possible to obtain sharp predictions for the breaking of the Bjorken scaling law for very high Q^2.

We denote by $N_{q_i}(x, L)$ i - 1, 4, $N_{\bar{q}_i}(x, L)$ i = 5, 8 and $N_g(x, L)$ respectively, the longitudinal momentum distributions of quarks, antiquarks and gluons inside an hadron. The L dependence of these distribution functions can be computed using the transition probabilities for the processes: $q \to q + g$, $g \to q + \bar{q}$ and $g \to g + g$. The first two are present also in quantum electrodynamics, while the third is peculiar to non abelian gauge theories.

The following master equation holds:

$$\text{(5.3)} \qquad \frac{dN_{q_i}(x, L)}{dL} = \frac{\alpha}{4\pi}\int_x^1 \frac{dy}{y}\left[P_{qq}(x/y)N_{q_i}(y, L) + P_{qg}(x/y)N_g(y, L)\right],$$

(x) - Hear first the four roots of all things.

$$\frac{dN_g(x,L)}{dL} = \frac{\alpha}{4\pi}\int_x^1 \frac{dy}{y}\left[p_{gg}(x/y)N_g(y,L) + p_{gq}(x/y)\sum_{1}^{8}{}_i N_{q_i}(y,L)\right],$$

where

(5.4)

$$p_{qq}(y) = \frac{4}{3}\left[\frac{4}{(1-y)_+} - \delta(y-1) - 2 - 2y\right],$$

$$p_{gq}(y) = \frac{4}{3}\left[\frac{2(1-y)^2+2}{y}\right],$$

$$p_{qg}(y) = \frac{3}{16}\left[2(1-y)^2 + 2y^2\right],$$

$$p_{gg}(y) = 3\left[\frac{4}{(1-y)_+} + \frac{4}{y} + 4y(1-y)\right] - 2\delta(y-1),$$

$\frac{1}{(1-y)_+}$ is a distribution defined by:

(5.5)
$$\int_x^1 \frac{dy}{y}\frac{1}{(1-\frac{x}{y})_+}N(y) = \ln(1-x)N(x) + \int_x^1 \frac{dy}{y}\frac{1}{1-\frac{x}{y}}\left[N(y)-N(x)\right].$$

The following consistency conditions are satisfied:

(5.6)

$$p_{qq}(y) = p_{gq}(1-y),$$

$$p_{qg}(y) = p_{qg}(1-y),$$

$$p_{gg}(y) = p_{gg}(1-y).$$

Eqs. (5.3-5.5) can be directly derived from the standard results of ref. (18--20) using the technique employed in ref. (21).

Higher orders in α have been neglected. σ_L/σ_T is of order α and is therefore asymptotically zero.

Let us try to use these formulae to compute the violations of the scaling law in deep inelastic scattering on nucleons.

Electron and neutrino deep inelastic scattering gives us very good information on the x distribution of quarks inside the nucleon, however no in formation is available on the gluon distribution; we only know that gluons must be present in the nucleon: they carry about 0.48 of the total momentum. Unfortunately the theoretical predictions for scaling violations depend on the form of the gluon distribution. Two phenomena contribute to the scaling violations: firstly the shift of the quark and antiquark distributions due to gluon bremsstrahlung, secondly the creation of quark-antiquark pairs. Only the second process depends on the distribution of gluons. However it is quite reasonable that the sea will be negligible for x near to one ($x > 0.5$). Model independent conclusions can be reached only in this region.

If we want to be more quantitative we can try to put upper and lower bounds on the scaling violations using two extreme models of gluon distributions.

The first unreasonable possibility is that the gluons are concentrated at $x = 0$; $N_g(x, L) = 0.48\,\delta(x)/x$.

In this case the L derivative of the structure function is[21]:

(5.7)
$$\frac{dF_2(x,Q^2)}{d\ln q^2} = \frac{\alpha(Q^2)}{3\pi}\left\{\left[3 + 4\ln(1-x)\right]F_2(x,Q^2) + {} + x\int_x^1 dy\left[\left(-2\left(1+\frac{x}{y}\right) + \frac{4}{1-\frac{x}{y}}\right)F_2(y,Q^2) - \frac{4F_2(x,Q^2)}{1-\frac{x}{y}}\right]\right\}.$$

The value of the effective coupling constant $\alpha(Q^2)$ appears as a factor. Using as input[22]

(5.8)
$$F_2^p(x) = (1-x)^3\left[1.274 + 0.5989(1-x) - 1.675(1-x)^2\right]$$

we obtain curve I of Fig. 6 for $\alpha(Q^2) = 0.4$. Notice that for such an high value of α corrections coming from the higher order terms may not be completely negligible. In this case we have neglected the gluon contribution which is positive: curve I is a lower bound on the derivative.

A physical motivated upper bound can be obtained supposing that the gluon distribution is proportional to the quark distribution in the region x near to one: for example we can assume that the x distribution is exactly $1.92(1-x)^3/x$. In this case one obtains the curve III of Fig. 6. In the region

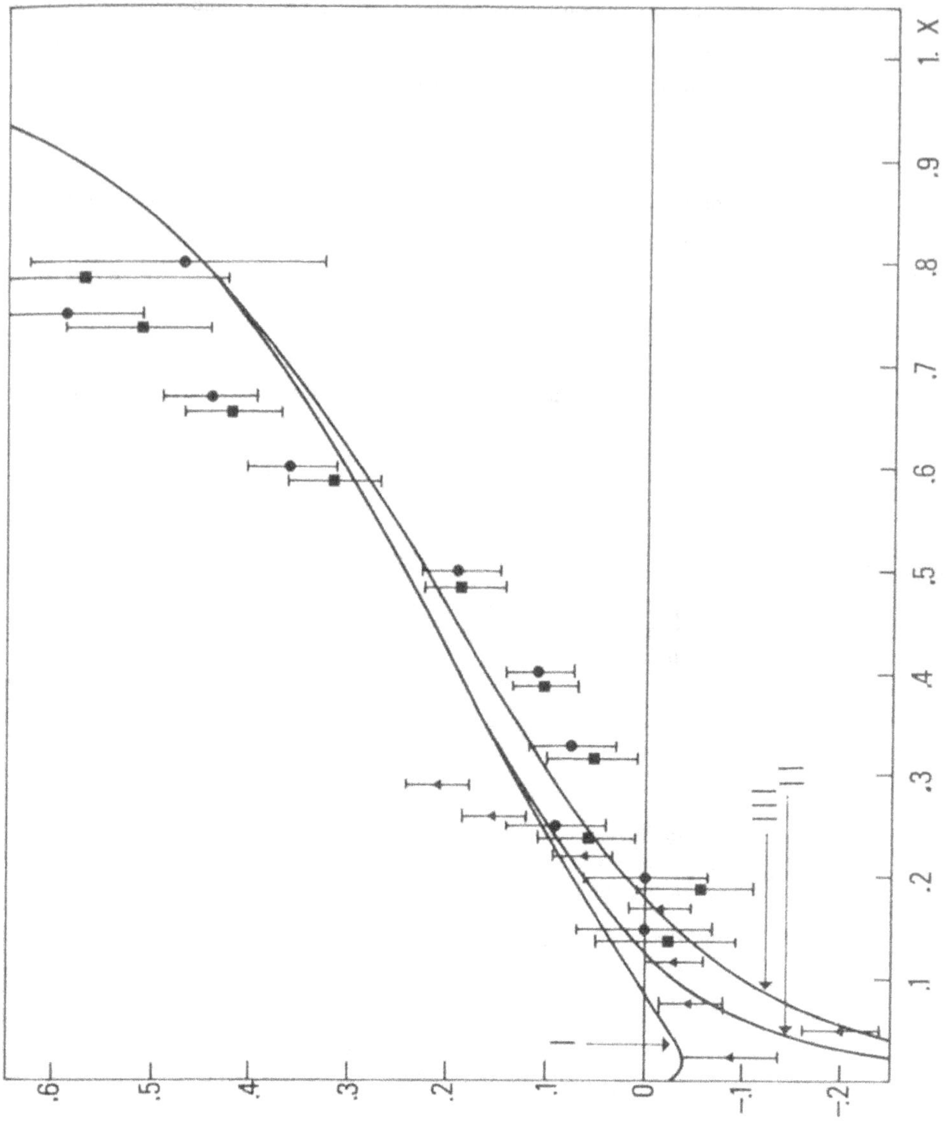

FIG. 6 - Curve I, II and III are respectively the predictions for $-\partial \ln F_2^p(x,Q^2)/\partial(\ln Q^2)$ assuming respectively, I the concentration of all gluons at $x = 0$, II an educated guess for the gluon distribution, and III a distribution $(1-x)^3$ for the gluons; the same predictions are obtained for the neutron with an accuracy of 0.02; α = 0.4 has been assumed. (♦) and (●) are respectively the experimental points for proton and deuterium[27]; (▲) are the experimental points for iron[28].

of large x there is no significant difference between the two curves for the two extreme choices of the gluon distribution. The difference is concentrated in the region of low x and it is due to the increase of the sea.

An educated guess for the gluon distribution can be obtained as follows: suppose that at a low value of L only p and n quarks are present in the proton. Using the master equation (5.3) one can compute the quark, antiquark and gluon distributions for all values of L. If we impose the constraint that, at a particular value of L, the structure functions coincide with eq. (5.8) we are able to fix the quark and gluon distributions at that particular L. Without entering into the details of how it can be done, we show directly the results: the predictions for the derivative of the structure functions are represented by curve II of Fig. 6.

A consistency check[2] of this model can be done comparing the predicted quark and antiquark distributions with the experimental data coming from neutrino and antineutrino scattering at Gargamelle. The agreement is not bad (see Fig. 7): notice that we have no free parameter and that we have used as input only data coming from deep inelastic electron scattering.

It seems to me that the predicted antiquark distribution is too concentrated near $x = 0$ (better data are needed to prove this conclusion); it is reasonable to suppose that the predicted gluon distribution has the same defect and that we are understimating the number of gluons in the large x region. My personal conclusion is that the correct prediction is between curve II and III. The ambiguity due to our ignorance of the gluon distribution is not large and sharp predictions can be made in the real asymptotic region.

Similar results can be obtained for the neutron structure functions. The difference among curves I, II and III for the neutron and the proton would hardly be observable in Fig. 6. It is always less than 0.02.

These predictions are done in the region of very high Q^2 where $\alpha(Q^2)$ and M^2/Q^2 are small numbers. In the next section we shall see that in the intermediate Q^2 region where actual experiments are done, extra ambiguities are present which make the comparison between theory and experiments less straightforward.

6. - THE COMPARISON WITH EXPERIMENTS

ἁμαρτίης αἰτίη ἡ ἀμαθίη τοῦ κρέσσονος[x]

(Democritus)

When precise data on deep inelastic e-p scattering appeared in 1970 it was clear that violations of the Bjorken scaling were present[24]. These violations disappeared when the variable x' was used[25]; x and x' are asymptotically equal; the difference is only relevant at "low" values of Q^2. The amount and the very existence of scaling violations depends on the choice of the "correct" variable.

Up to now no strong theoretical argument has been found which allows a choice between x or x' or any other similar variable. However the choice of the "best" variable can be done using the experimental data plus a theoretical criterion of what we mean by the "best" variable.

In 1970 an experimental proof of Bjorken scaling was strongly desirable and the "best" variable was the one for which Bjorken scaling was better satisfied. In 1975 it was discovered that it is impossible to find a variable for which the Bjorken scaling law is satisfied both for proton and neutron deep inelastic scattering[26]. The experimental observation of scaling violations in the proton at fixed x' ($0.5 \leq x' < 0.7$) (see Fig. 8) suggests the use of a variable different from x', on the contrary the lack of scaling violations in the neutron at fixed x' would imply that x' is the "best" variable (see Fig. 9).

It is possible to use a new scaling variable x_{1975} for the proton and the old x' for the neutron, and this may be a simple phenomenological way to summarize the data. I think that it would be quite hard to find a theoretical justification in the framework of the parton model for the use of two different scaling variables: the criterion that the "best" variable must minimize the violations of Bjorken scaling, has led us to a dead end.

A new criterion is needed: we propose that the best variable should be such that scaling violations are the same for the neutron and the proton, at least in the large x region. If we use the experimental data[27,28] to com-

---

(x) - The cause of errors is ignorance of better.

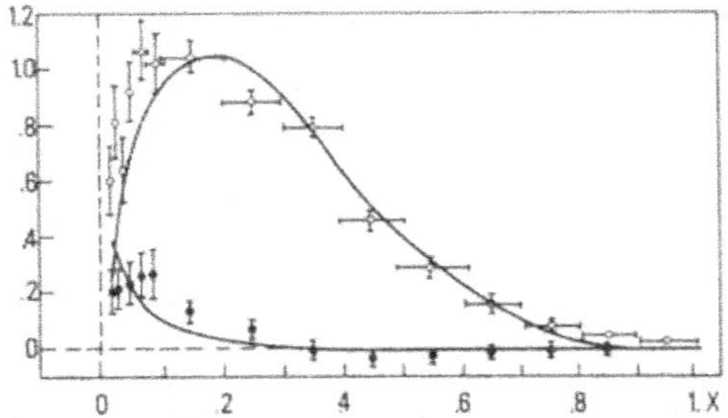

FIG. 7 - Our predictions for the amount of quarks (ϕ) and antiquarks (ϕ) in an isospin zero target are presented together with the experimental values extracted from neutrino and antineutrino scattering[23].

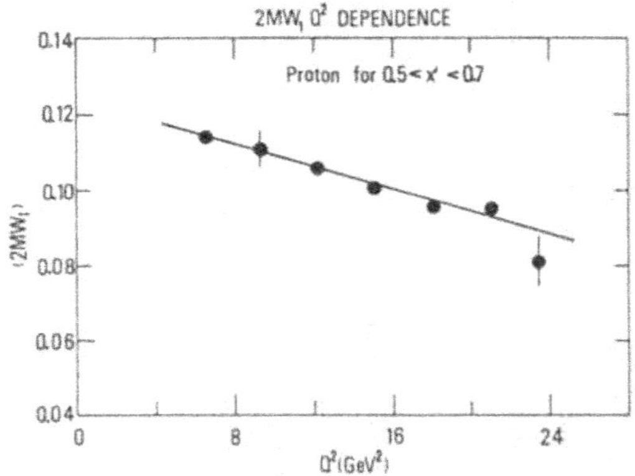

FIG. 8 - The experimental data for the mean value of F_1^p in the interval $0.5 \le x' \le 0.7$ plotted against Q^2.

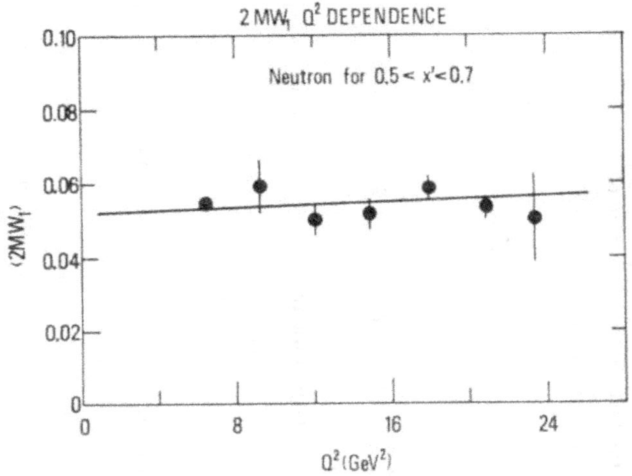

FIG. 9 - The experimental data for the mean value of F_1^n in the interval $0.5 \le x' \le 0.7$ plotted against Q^2.

pute the logaritmic derivative of the proton and of the neutron structure functions at fixed x, we find that they are roughly equal (see Fig. 6).

How is it possible that the two logaritmic derivatives at fixed x are equal and those at fixed x' are different? The answer can be easily found using the identity:

(6.1)
$$\left.\frac{\partial \ln F}{\partial Q^2}\right|_{x'} = \left.\frac{\partial \ln F}{\partial Q^2}\right|_{x} + \left.\frac{\partial \ln F}{\partial x}\right|_{Q^2} \cdot \left.\frac{\partial x}{\partial Q^2}\right|_{x'} .$$

In the x region we are interested in, one find that:

(6.2)
$$\frac{\partial \ln F_2^n}{\partial x} \simeq 1.2 \frac{\partial \ln F_2^p}{\partial x} .$$

Any change of variable modifies the Q^2 derivative of the neutron data more strongly than the proton data.

The variable x (and not x') satisfies the new criterion we have proposed, and we are going to use it in the rest of the paper (see Fig. 6). We stress that, if our intuition is wrong and if the predicted scaling violations must be compared with the derivative of the experimental data at fixed x', the present experimental evidence excludes that the observed scaling violations come from the mechanism described in this paper. However the data are not accurate enough to fix unambigously which is the best variable: any variable not too far from x would also satisfy our criterion within the experimental errors. The problem of the best variable arises from the existence of scaling violations due to the finite mass of the nucleons and of the quarks; these violations disappear asymptotically, however in the low Q^2 region it is impossible to disentangle the scaling violations which die as Q^2 is increased, from those which survive in the limit $Q^2 \to \infty$. The theory of these mass dependent scaling violations is practically lacking: the situation can be clarified in the framework of the so called covariant parton model of Landshoff and Polkingorne[29], unfortunately the analysis has not been carried out in detail.

Another problem is present: our asymptotic predictions do not distinguish among F_1 and F_2 (σ_L is asymptotically zero), however at present energies the logaritmic derivative of F_1 is systematically larger than that of F_2[27,30]. Now it is not clear which function should be compared with the theoretical predictions: the chosen function must satisfy the requirement of minimizing the scaling violations due to finite mass effects.

The observed Q^2 dependence of the function F_2 can be well fitted using $\alpha = 0.4$ (see Fig. 6), a similar agreement between theory and experiment would be obtained using F_1 instead of F_2: in this case we would get $\alpha = 0.5$.

I would like to conclude that the observed scaling violations can be accounted for by interactions among partons with a coupling constant of order 0.4 - 0.5 in the few GeV^2 range. However there is still another effect which increases the error on the value of α: the large value of the coupling constant changes rather drastically with Q^2. The data in the central x region have $\langle Q^2 \rangle$ 3-6 GeV^2, while the data at x near to 1 have $\langle Q^2 \rangle$ 8-12 GeV^2.

In principle changing Q^2, we should also change the value of the effective coupling constant; in this particular instance this is not true because we are changing both Q^2 and x together. The effect we are talking about, is of the same order of magnitude as the neglected terms proportional to α^2 in the transition probabilities p (eq. (5.4)). We must realize that eq. (5.3) is asymptotically correct also if we substitute $\alpha(Q^2)$ by $\alpha(Q^2/(1-x))$; eq. (5.1) implies:

$$\alpha(Q^2/(1-x)) \simeq \alpha(Q^2) - \frac{25}{12\pi} \ln(1-x)\, \alpha^2(Q^2) \; . \tag{6.3}$$

The difference is of order α^2. Notice that $\langle (1-x) Q^2 \rangle$ is roughly constant in a wide x region in the SLAC sample.

The effect of the neglected second order terms has not been computed at the present moment; it can be easily be of order of 30%, expecially in the region $x \sim 1$ where higher order contributions are expected to be enhanced. Terms proportional to α^2 are not negligible because our preferred value for the coupling constant is not small; they will distort the theoretical predictions in the region $x \sim 1$ and they will also change the Q^2 dependence of the moments of the structure function for N very large.

In our theoretical predictions we have also neglected the effect of the Q^2 dependence of the r.h.s. of eq. (5.7); the error we have introduced is rather small and can be easily corrected using the data themselves and not their scaling fit (5. 8) in the r. h. s. of eq. (5. 7).

If I take care of all these ambiguities, I would estimate:

(6. 4) $$0.25 \leq \alpha(6\ \mathrm{GeV}^2) \leq 0.5\ .$$

Correspondly:

(6. 5) $$0.4 \lesssim \alpha(1\ \mathrm{GeV}^2) \leq 1.2\ .$$

The determination of the value of α is based mainly on the SLAC data. If high quality data coming from an high energy μ beam becomes available in the future for a large interval of Q^2, the determination of α can be improved. I hope that at that time the theoretical ambiguities will be solved: the transition probabilities will be computed at order α^2 and the scaling violations due to the finite mass of the proton will be understood.

7. - SCALING VIOLATIONS AND THE SEARCH FOR CHARM

The parton model gives rather interesting predictions when it is applied to neutrino and antineutrino induced reactions. In this paper we concentrate our analysis on the charged current processes; a similar analysis can be done for the case of neutral currents. If only V-A currents are present, we find:

(7. 1) $$\sigma_\nu = \frac{2G^2 M E_\nu}{\pi}\left[M_q^\nu + \frac{1}{3} M_{\bar q}^\nu\right], \quad \sigma_{\bar\nu} = \frac{2G^2 M E_{\bar\nu}}{\pi}\left[M_{\bar q}^{\bar\nu} + \frac{1}{3} M_q^{\bar\nu}\right]$$

$$\langle y\rangle_\nu = \frac{1}{2}\,\frac{6M_q^\nu + M_{\bar q}^\nu}{6M_q^\nu + 2M_{\bar q}^\nu}, \quad \langle y\rangle_{\bar\nu} = \frac{1}{4}\,\frac{M_q^{\bar\nu} + 6M_{\bar q}^{\bar\nu}}{M_q^{\nu} + 3M_{\bar q}^{\bar\nu}},$$

where σ denotes total cross section and y is the ratio between the neutrino (the antineutrino) energy and the energy given to the hadron system E_h: $y = E_h/E^\nu$. M_q^ν and $M_{\bar q}^\nu$ ($M_q^{\bar\nu}$ and $M_{\bar q}^{\bar\nu}$) are respectively the effective momentum carried by the quarks and the antiquarks which interact with the neutrino (with the antineutrino).

In the 4 quark model different results hold below and above the threshold for creation of charmed particles in the final state; below threshold we find:

$$(7.2)\qquad M_q^{\nu} = \cos^2\theta_c M_n^2 + \sin^2\theta_c M_\lambda^2 , \qquad M_q^{\bar\nu} = M_p^2 ,$$

$$M_{\bar q}^{\bar\nu} = \cos^2\theta_c M_{\bar n}^2 + \sin^2\theta_c M_{\bar\lambda}^2 , \qquad M_{\bar q}^{\nu} = M_{\bar p}^2 .$$

Above threshold, transitions involving the p' quark are switched on:

$$(7.3)\qquad M_q^{\nu} = M_n^2 + M_\lambda^2 , \qquad M_q^{\bar\nu} = M_p^2 + M_{p'}^2 ,$$

$$M_{\bar q}^{\bar\nu} = M_{\bar n}^2 + M_{\bar\lambda}^2 , \qquad M_{\bar q}^{\nu} = M_{\bar p}^2 + M_{\bar p'}^2 .$$

It is commonly assumed that the quark distributions inside the nucleon can be divided into a valence contribution, an SU(3) symmetric sea of quarks and antiquarks and a charmed sea. If the target has isospin zero we get:

$$(7.4)\qquad M_p^2 = \frac{V^2}{2} + S^2 , \quad M_n^2 = \frac{V^2}{2} + S^2 , \quad M_\lambda^2 = S^2 , \quad M_{p'}^2 = C^2 ,$$

$$M_{\bar p}^2 = S^2 , \qquad M_{\bar n}^2 = S^2 , \qquad M_{\bar\lambda}^2 = S^2 , \quad M_{\bar p'}^2 = C^2 .$$

If we neglect the sea, no antiquarks are present in the nucleon: the antineutrino over neutrino total cross section ratio is below threshold:

$$(7.5)\qquad R \equiv \sigma_{\bar\nu}/\sigma_\nu = 1/3 \cos^2\theta_c \simeq 0.35 .$$

At Gargamelle energies $R = 0.39$[23]; only a small contamination of antiquarks is present in the nucleon at low energy. The x distributions of quarks and antiquarks are shown in Fig. 7: the mean value of x of antiquarks ($\langle x_S \rangle$) is much smaller than that of the valence quarks ($\langle x_V \rangle$). The data suggests that at $Q^2 = 1\ \mathrm{GeV}^2$ (the mean value of Q^2 in the Gargamelle experiment is about 1 GeV^2) the following relations hold:

$$(7.6)\qquad V^2 = 0.46 , \quad S^2 = 0.01 , \quad C^2 = 0 , \quad G^2 = 0.48 .$$

Obviously the data give no information about the amount of charmed quarks present in the proton; for simplicity I have assumed that the charmed component of the proton can be neglected in the low Q^2 region. The conservation of the total momentum implies the sum rule

$$V^2 + 6\,S^2 + 2\,C^2 + G^2 = 1\,, \tag{7.7}$$

which has been used to fix the momentum carried by the gluons (G^2).

Violations of the Bjorken scaling law are due to the presence of a threshold for charm production and to the Q^2 dependence of the quark distributions. The first effect is characteristic of neutrino scattering. It will be shown here that both effects are needed to explain the observed violations of the scaling law in neutrino deep inelastic scattering: in the framework of the 4 quark model it is not simple to fit the experimental data neglecting the Q^2 dependence of the parton distributions.

The Q^2 dependence of the momentum carried by each component of the proton can be easily computed: proceeding as in section 3 we can derive from eqs. (5.3-4) an equation having the same form as eq. (3.20); its solution is[3]:

$$\begin{aligned} V^2(Q^2) &= B_8\, G^{-32/75}\,, \\ S^2(Q^2) &= \frac{3}{56} + \frac{1}{14} B_o\, G^{-56/75} + \left(\frac{1}{24} B_{15} - \frac{1}{6} B_8\right) G^{-32/75}\,, \\ C^2(Q^2) &= \frac{3}{56} + \frac{1}{14} B_o\, G^{-56/75} - \frac{1}{8} B_{15}\, G^{-32/75}\,, \\ G^2(Q^2) &= \frac{4}{7} - \frac{4}{7} B_o\, G^{-56/75}\,, \end{aligned} \tag{7.8}$$

where

$$G(Q^2) = 1 + \frac{25}{12\pi}\,\alpha(\mu^2)\ln Q^2/\mu^2\,. \tag{7.9}$$

The constants B_o, B_8 and B_{15} can be fixed by requiring that eq. (7.6) be satisfied at $Q^2 = 1\ \mathrm{GeV}^2$.

In Fig. 10 the results have been plotted for $\alpha(1) = 0.5$, as functions of Q^2. Since G is a slowly varying function of Q^2 we can compute G from an effective Q^2 value

$$Q^2_{\mathrm{Eff}} = 2\,M\,E\,\langle xy\rangle\,, \tag{7.10}$$

where $\langle xy\rangle$ is the average value of xy, which is different for neutrino and antineutrino. At fixed energy E neutrino data involve larger value of

Q^2_{Eff} than antineutrino data.

Our predictions for the momentum carried by the charmed quarks must be taken cum grano salis : the effects of the large mass of the charmed quarks has not been taken into account. A more precise analysis would be needed to study effects that depend crucially on the amount of charmed quarks in the proton.

The cross sections and the y distribution below and much above the threshold for charm production can be easily computed.

The effects of the threshold may be simulated by a simple θ function in the mass W of the produced hadronic system :

$$\sigma_T(x,y) \simeq \sigma_B(x,y) + \sigma_C(x,y)\,\theta(W - W_T) \quad , \tag{7.11}$$

where σ_B is the cross section below the threshold for charm production, σ_C is the asymptotic cross section for producing a charmed final state and W_T is an effective threshold mass. Simple kinematical arguments, due to Barnett, suggest that :

$$W_T^2 = m_{p'}^2 / \langle x \rangle \quad , \tag{7.12}$$

where $m_{p'}$ is the mass of the charmed quark and $\langle x \rangle$ is the mean value x of the quarks from which the p' is produced. Charm is produced by neutrinos mainly out of valence quarks, by antineutrinos out of sea quarks. Using $m_{p'} = 2$ GeV, $\langle x_V \rangle = 0.25$ and $\langle x_S \rangle = 0.13$ we estimate :

$$W_T^{\nu} \simeq 4 \text{ GeV} \quad , \qquad W_T^{\bar\nu} \simeq 5.5 \text{ GeV} \quad . \tag{7.13}$$

The higher value of the effective threshold for antineutrino is caused by the exoticity of the hadronic final state (B = 1, C = -1).

In Figs. 11 and 12 we show our predictions for $\langle y \rangle_{\nu}$ and R respectively for various values of W_T and α. The data for $\langle y \rangle_{\bar\nu}$ come from the HPWF collaboration. For simplicity the same threshold has been used for neutrino and antineutrino.

When $\alpha = 0$ the scaling violations due to the strong interactions are absent and when $W_T = \infty$ the charm threshold never opens. It is apparent that both $\alpha \neq 0$ and $W_T < \infty$ are needed to fit the data for $\langle y \rangle_{\bar\nu}$; in this case R is predicted to rise with E. If the momentum carried by each quark were Q^2 independent, R would stay almost constant and be insensitive to the charm threshold; in fact the increased proportion of momentum carried by the sea makes R to behave as in Fig. 12. While this prediction is not supported by the published data of the Caltech group[32] (although not excluded within quoted errors) a sharp rise of R has been reported by the HPWF group[33, 34].

In the infinite energy limit very simple predictions are obtained:

$$\sigma_\nu = \sigma_{\bar\nu} = \frac{G^2 M E}{\pi} \frac{2}{7} , \qquad \langle y \rangle_\nu = \langle y \rangle_{\bar\nu} = \frac{7}{16} . \tag{7.14}$$

If scaling violations were absent, eq. (7.6) implies that the fraction $\Delta\sigma/\sigma$ of charmed final states would not exceed 10% even at infinite energy. The predictions with scaling violations included are shown in Fig. 13. If charmed particle have an average branching ratio into muons of the order of 5% to 10%, the observed yield of events with muons of opposite charge is obtained[35, 36].

Dimuons with equal charge[35] may come from the production of a charmed quark-antiquark pair in an event with $\Delta C = 0$. A very rough estimate of the order of magnitude of the cross section for the creation of two charmed particles is :

$$\sigma_{C\bar{C}} \simeq C^2 \sigma_T . \tag{7.15}$$

A careful study of the effects due to the high p' mass would be needed to understand if this mechanism may explain the observed yield of equal sign dimuons. It is also possible that the equal sign dimuons come from the decay of a massive b quark[37] produced out of a p' quark.

A distinctive feature of the scaling violations due to the strong interaction is that the x distributions of quarks and antiquarks shift toward zero with increasing Q^2. This effect has been observed in electroproduction and should also be observed in neutrino production. Predictions for the behaviour of the structure functions at fixed x can be made using the same techniques as in section 5. However it may be convenient to concentrate on global quantities such as $\langle x \rangle$.

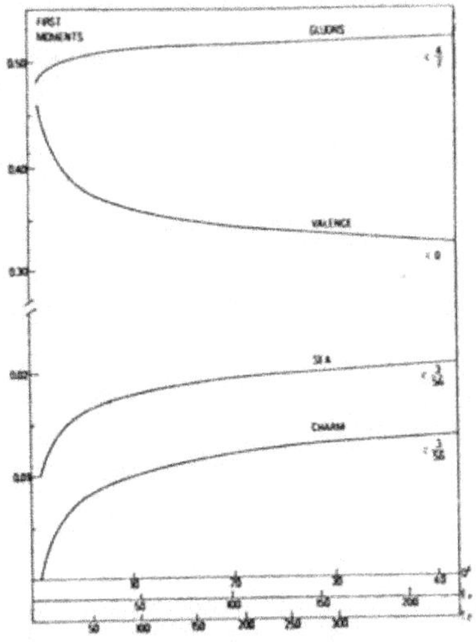

FIG. 10 - Momenta carried by the gluons, the valence quarks, the SU(3) symmetric sea and the charmed sea. The arrows indicate the asymptotic values. $G + 2V + 6G + 2C = 1$ is identically satisfied. The curves have been computed using $\alpha(1\ \mathrm{GeV}^2) = \alpha = 0.5$.

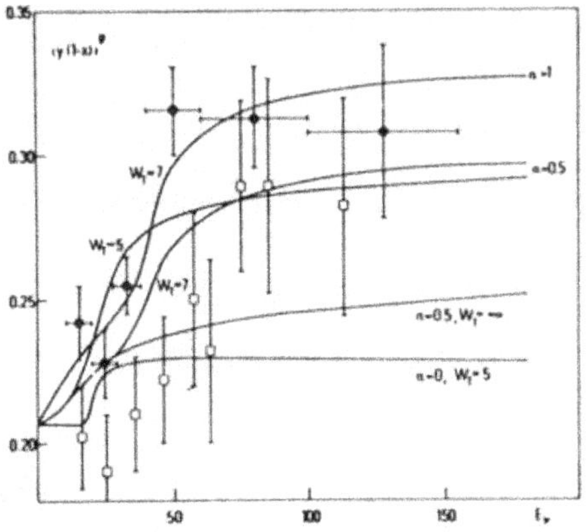

FIG. 11 - Average value of $y_{\bar{\nu}}$ for different values of α and W_T, the effective invariant mass for charm threshold. $\alpha = 0$ corresponds to Q^2 independent parton distributions. $W_T \to \infty$ corresponds to neglecting effects for charm production. Both effects seem to be needed to reproduce the data. α is the coupling constant at $Q^2 = 1\ \mathrm{GeV}^2$. The experimental points are taken from ref. (31).

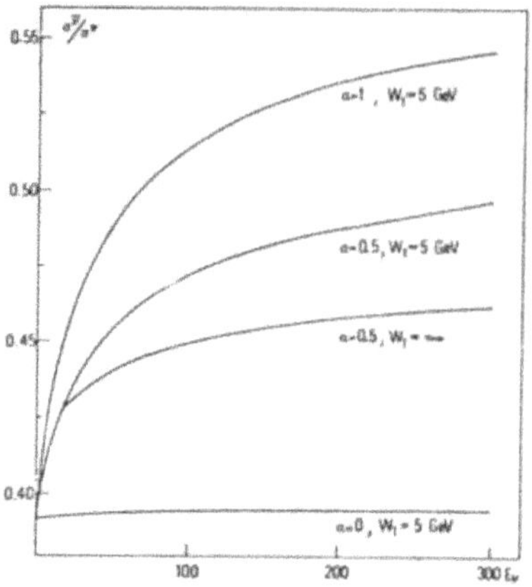

FIG. 12 - The ratio $\sigma^{\bar{\nu}}/\sigma_{\nu}$ for different values of α(1 GeV^2) and W_T, the effective invariant mass for charm threshold.

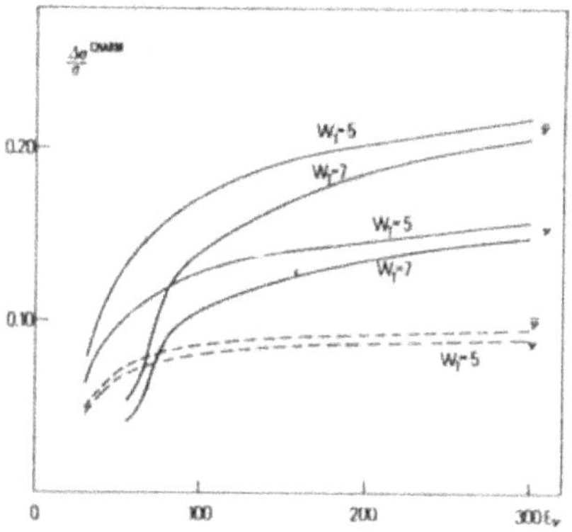

FIG. 13 - The prediction for the fraction $\Delta\sigma/\sigma$ of charmed final states for neutrino and antineutrino. The dashed line is obtained with Q^2 independent parton distributions, the full lines are obtained assuming $\alpha = 0.5$ at $Q^2 = 1$ GeV^2; W_T is the effective invariant mass for charm production.

Evolution equations like eq. (7. 8) can be written also in this case[3]. This problem will not be studied here : the interested reader can find a careful treatment of this and of many other phenomena concerning scaling violations in neutrino scattering in the paper of Altarelli presented at this Rencontre de Moriond[38].

8. - CONCLUSIONS

In this paper we have shown that in relativistic quantum field theory the breaking of the Bjorken scaling law can be understood in terms of successive fragmentations of the partons. The parton model relations among electron, neutrino and antineutrino scattering are preserved, provided we use Q^2 dependent parton distributions.

The scaling violations observed in deep inelastic electron scattering can be understood using a strong interaction coupling constant of the order 0. 4 in the few GeV range. Using this value for the strong interaction coupling constant we compute the scaling violations in neutrino and antineutrino scattering. The most interesting prediction is a large increase in the momentum carried by the sea of antiquarks with increasing energy. The mean value of y in antineutrino scattering and the ratio of the total antineutrino and neutrino cross sections are consequently affected. Without this effect it is hard to understand the present experimental data in the framework of the 4 quark model. It appears that a correct treatment of the violations of the Bjorken scaling law is a necessary ingredient in any successfully analysis of neutrino scattering at present energies.

The author is really grateful to R. K. Ellis for a critical reading of the manuscript.

REFERENCES

(1) - G. Parisi and R. Petronzio, A dynamical approach to deep inelastic scattering, Rome preprint 617 (1975); to be published on Nuovo Cimento.

(2) - G. Parisi and R. Petronzio, On the breaking of Bjorken scaling, Rome preprint; submitted to Physics Letters B.

(3) - G. Altarelli, G. Parisi and A. Petronzio, Charmed quarks and asymptotic freedom in neutrino scattering, Rome preprint, submitted to Physics Letters B.

(4) - A. M. Poliakov, Sov. Phys. -JEPT 32, 296 (1971).

(5) - J. Kogut and L. Susskind, Phys. Rev. D9, 697 (1971).

(6) - S. D. Drell, D. J. Levy and T. M. Yan, Phys. Rev. 187, 2159 (1970).

(7) - We use the same notation as G. Altarelli, Riv. Nuovo Cimento 4, 335 (1974).

(8) - G. Parisi, Phys. Letters 42B, 114 (1972).

(9) - G. Parisi, Serious difficulties with canonical dimensions, Frascati report LNF-72/94 (1972).

(10) - N. Cabibbo and M. Rocca, CERN preprint Th 1974.

(11) - N. Cabibbo and G. Parisi (unpublished).

(12) - D. Bailin, A. Love and D. V. Nanopoulos, Lett. Nuovo Cimento 9, 501 (1974).

(13) - K. Symanzik, Lett. Nuovo Cimento 6, 420 (1973).

(14) - S. L. Glashow, J. Iliopoulos and L. Maiani, Phys. Rev. D2, 1285 (1970).

(15) - G. 't Hooft, Marseille Conference 1972 (unpublished).

(16) - H. D. Politzer, Phys. Rev. Letters 30, 1346 (1973).

(17) - D. J. Gross and W. Wilzek, Phys. Rev. Letters 30, 1343 (1973).

(18) - H. Georgi and H. D. Politzer, Phys. Rev. D9, 416 (1974).

(19) - D. J. Gross and W. Wilczek, Phys. Rev. D9, 980 (1974).

(20) - A. Zee, F. Wilzeck and S. L. Treiman, Phys. Rev. D10, 2881 (1974).

(21) - G. Parisi, Phys. Letters 43B, 207 (1973); 50B, 367 (1974).

(22) - G. Miller et al., Phys. Rev. D5, 528 (1972).

(23) - CERN-Gargamelle collaboration, Phys. Letters 46B, 274 (1973).

(24) - D. Wilson, Review talk presented at the Kiev Conference (1970).

(25) - E. D. Bloom and F. J. Gilman, Phys. Rev. Letters 25, 1140 (1970).

(26) - R. Taylor, Review talk presented at the Stanford Conference (1975).

(27) - A. Bodek et al., SLAC Publ. 1445 (1975).

(28) - C. Chang et al., Phys. Rev. Letters 35, 901 (1975).

(29) - N. Cabibbo, G. Parisi, M. Testa and A. Verganelakis, Lett. Nuovo Cimento 4, 569 (1970).

(30) - E. M. Riordan, Phys. Letters 52B, 249 (1973).

(31) - A. Benvenuti et al., Further data on the high-y anomaly in inelastic antineutrino scattering, Preprint HPWF 76/1 (1976).

(32) - B. C. Barish et al., Phys. Rev. Letters 35, 1316 (1975).

(33) - A. K. Mann, Talk presented at this Rencontre.

(34) - C. Rubbia, Talk presented at this Rencontre.

(35) - A. Benvenuti et al., Phys. Rev. Letters 35, 1199, 1203 and 1249 (1975).

(36) - B. G. Barish et al., Cal. Tech. Report 58-485 and 68-510 (1975).

(37) - H. Harari, SLAC-Pub-1589 (1975).

(38) - G. Altarelli, Talk presented at this Rencontre.

THE PHYSICAL BASIS OF THE ASYMPTOTIC ESTIMATES IN PERTURBATION THEORY

G. PARISI +

Laboratoire de Physique Théorique de l'Ecole Normale Supérieure ++, 24 rue Lhomond 75231 Paris Cedex 05 France

I. INTRODUCTION

During last year, the asymptotic behaviour in K of the Kth order of the perturbative expansion has been widely investigated with remarkable success[1-3]. If the interaction is superrenormalizable, the problem has been solved for both interacting bosons[4,5] and fermions[6-8] : of course technical details may change from theory to theory. When the interaction is renormalizable new difficulties arise[9,10]; they are not completely understood at the present moment.

The aim of these lectures is to present a general review of the subject and to discuss the motivations of these investigations. The asymptotic estimates obtained are quite interesting : they are an essential step toward the use of the perturbative expansion outside the weak coupling regime. Indeed, if the perturbative expansions were convergent, by computing a high enough number of Feynman diagrams it would be possible to obtain numerical results with arbitrarily small accuracy also in the strong coupling region. Unfortunately, the perturbative series have a zero radius of convergence. We can obtain reasonable answers for strong coupling only if we use resummation techniques which cope with non convergent series. The success of these techniques is not guaranteed : it depends on the properties of the asymptotic estimates.

+ On leave of absence from INFN (Frascati-Roma) Italy.

++ Laboratoire Propre du C.N.R.S. associé à l'Ecole Normale Supérieure et à l'Université de Paris Sud.

To be concrete, let us consider a typical example. The function A(g) (g being the coupling constant) has a formal expansion in powers of g

$$A(g) = \sum_{0}^{\infty}{}_K A_K g^K \tag{I.1}$$

Our aim is to get an expression of A(g) in terms of the A_K. The solution is not unique; however, if the function A(g) is analytic in the domain $\mathcal{D}$ ($|g| < R^{-1}, |\arg(g)| < \delta + \frac{\pi}{2}$, $\delta > 0$) and the bounds $\left| \frac{d^K}{dg^K} A(g) \right| < (K!)^2 R^K$ are valid inside $\mathcal{D}^2$, (this happens in many field theories[11-13]; e.g. super-renormalizable $g\phi^4$ and Yukawa theories) A(g) can be unequivocally written as :

$$A(g) = \int_0^\infty B(bg)\exp(-b)db = g^{-1}\int_0^\infty \exp(-b/g)B(b)db \tag{I.2}$$

The function B(b) (the Borel transform of A(g)) is defined by the convergent expansion :

$$B(b) = \sum_{0}^{\infty}{}_K A_K b^K/K! \tag{I.3}$$

inside its radius of convergence and by analytic continuation on the positive real axis; the conditions we have imposed on the function A(g) imply that B(b) is analytic in a sector containing the positive real axis ($|\arg b| < \delta$); the following representation holds

$$B(b) = (2\pi i)^{-1}\oint_\varphi G(g)/g \exp(b/g)dg \tag{I.4}$$

φ being the border of the domain .

The Borel transform is a quite naturel concept in quantum field theory : in a rough sense it has the meaning of the contribution to A(g) of those field configurations having an action equal to b/g, as it can be seen from the functional integral representation.

The asymptotic estimates for the A_K are connected to the singularities of the Borel transform : if

$$A_K \underset{K\to\infty}{\to} K!\, R^K \cos(\theta K)\, K^\alpha \left[C + O(K^{-1})\right] \tag{1.5}$$

the nearest singularity to the origine is at $b = R^{-1}\exp(\pm i\theta)$.

Although the Borel sum (eqs. (I.2)-(I.3)) gives a unique definition of the function A(g), at first sight, it seems that the knowledge of the first few A_K is not sufficient to obtain convergent sequence of approximants to A(g) : if we restrict the sum in the definition of B(b) to a finite number of terms, eqs. (I.2)-(I.3) reproduce the standard non convergent perturbative expansion. However, we can take advantage of the analyticity properties of the Borel transform to map conformally the real positive line on the interval 0-1 [14-16]

$$A(g) = g^{-1} \int_0^1 \exp(-b(z)/g)\, F(z)dz$$

$$F(z) = B(b(z)) \frac{db}{dz} \tag{I.6}$$

$$b(0) = 0; \quad b(1) = \infty$$

The conformal map must be chosen such as not to map the singularities of the Borel transform inside the unity circule. A convergent sequence of approximants is given by :

$$A^N(g) = g^{-1} \int_0^1 \exp\left[-b(z)/g\right] F^N(z)dz$$

$$F(z) = \sum_{0}^{\infty}{}_K F_K z_K \tag{I.7}$$

$$F^N(z) = \sum_{0}^{N}{}_K F_K z_K$$

where the F_K are linear combinations of the A_K .

In theories in which representations like (I.6) can be established, we have an algorithm based on the Feynman diagrammatical expansion which allows us to compute physically interesting quantities for strong coupling constants with arbitrarily small accuracy. This technique has been succesfully applied to superrenormalizable $g\phi^4$ interactions, also for very large values of the coupling constant[14-16].

In the general case, a disaster may happen which ruins our beautiful dreams of computing physical quantities using the perturbative expansion in the strong coupling region. If the angle θ in eq. (I.5) is equal to zero, the Borel transform has a singularity on the real axis and the perturbation expansion cannot be

Borel summed. However, a disaster of such a magnitude (we do not know how to define the theory for strong coupling) must be a punishment for a wrong doing. Indeed the presence of singularities on the real line in the Borel transform is the signal of existence of non perturbative effects which are ignored in the perturbative expansion. The study of the singularities of the Borel transform (i.e. the asymptotic estimates) aims to find and to compute these non perturbative effects. Let us discuss a few examples.

1) The interacting Hamiltonian is unbounded from below : no ground state exists and the perturbative vacuum is a metastable state (a resonance). Its energy has an imaginary part which is exponentially small for a small coupling constant. This phenomenon is typical of theories in which the coupling constant has the wrong sign.

2) The interacting Hamiltonian is bounded from below. However there are more than one classical vacua and the tunneling between these vacua is exponentially small but non zero. There is a non perturbative effect due to the mixing between the different vacua. This usually happens if we study the perturbation expansion around a classical vacuum which breaks an unbreakable symmetry (i.e. a discrete symmetry in one dimension or any symmetry in a finite volume).

3) The short (or the large) momentum behavior of the Green functions is different in the free and in the interacting theories. If P_c (the cross over momentum between the free and the interacting behaviour) is a power of the coupling constant (as when the interaction is non renormalizable or super-renormalizable but massless[17]) the perturbative expansion is undefined because of ultraviolet or infrared divergences. If P_c is exponentially small or large as in a renormalizable theory, the perturbative expansion is well defined, however, a remnant of this phenomenon appears as a singularity of the Borel transform.

In the following sections of these lectures, we will see how to compute explicitly these non perturbative effects and consequently how to derive asymptotic estimates for the high orders of perturbation theory. Semi-classical methods will be used to study cases 1 and 2 while renormalization group arguments will be the main tools in the discussion of case 3. Unfortunately, we are not able to reach definite conclusions in this last case. Finally, in the last section we will discuss our understanding of Yang-Mills theories.

II. THE UNSTABLE VACUUM

In this section we study the asymptotic estimates connected to the metastability of the vacuum (case 1). To be definite let us consider the $g\phi^4$ interaction in dimension $D < 4$. The Lagrangian density is

$$\mathcal{L}(x) = \frac{1}{2} (\partial_\mu \phi)^2 + \frac{1}{2} \phi^2 + \frac{1}{4} g\phi^4 \tag{II.1}$$

We concentrate our attention on the "vacuum energy" $E(g)$ i.e. the sum of all the vacuum to vacuum connected diagrams. We suppose that $E(g)$ is analytic on the cut complex g plane and the discontinuity on the negative real axis is related to the mean life of the vacuum for negative values of g. The interaction among particles is repulsive for positive g, but attractive for negative g; an attractive interaction among bosons leads to a collapse. Let us see why. Denoting by $E_N(g)$ the energy of a state of N particles with overlapping wave functions, we roughly find :

$$E_N(g) \simeq Nm + gN^2 \tag{II.2}$$

For negative g there are collapsed states of negative energy. The vacuum decays in one or more of these objects with probability roughly proportional to $\exp(-E_B)$; E_B being the potential barrier, i.e. the maximum of $E_N(g)$ which is proportional to $1/g$. The imaginary part of the energy is proportional to the mean life and it is of order $\exp((-R|g|)^{-1})$; R being an appropriate constant.

The asymptotic estimate

$$E_K \sim K!(-R)^K$$
$$E(g) = \sum_0^\infty{}_K E_K g^K \tag{II.3}$$

can be easily derived from the result

$$\mathrm{Im}\, E(g) = \exp(-|Rg|^{-\alpha}) \qquad g < 0 \qquad \Longrightarrow$$
$$E_K \sim \Gamma(\alpha^{-1}K)(-R)^K \tag{II.4}$$

The particles in the collapsed state will be in a coherent state. They will be described by a classical field satisfying the classical equation of motion. Semi-classical methods may be applied to compute the constant R. For technical reasons it is convenient to study the problem in the Euclidean version of field theory. We have to find the non trivial solution φ_c of the classical equations of motion having a minimal action. In this case, the classical equations of motion have non trivial solutions only for negative g.

It has been shown that if $\mathcal{A}_c/|g|$ is the action corresponding to φ_c, the imaginary part of E(g) is $\exp(-\mathcal{A}_c/|g|)$ and the nearest singularity of the Borel transform of E(g) is located at $b = -\mathcal{A}_c$. The existence of this singularity of the Borel transform for negative b forbids us to define the theory for negative g. These results can be obtained[18] starting from the functional integral representation for E(g)

$$E(g) = \lim_{V\to\infty} -V^{-1} \ln \int d|\varphi|_V \exp - \mathcal{A}(\varphi)$$

$$\mathcal{A}(\varphi) = \int d^D x \mathcal{L}(x) \tag{II.5}$$

and integrating on those field near to φ_c. The final result is

$$\mathrm{Im}\, E(g) \simeq |\det M|^{-1/2} \exp -\mathcal{A}_c/|g| \quad \Longrightarrow$$

$$E_K \sim |\det M|^{-1/2} (K-1)!\ (-\mathcal{A}_c)^{-K} \tag{II.6}$$

M being defined as

$$\mathcal{A}(\varphi_c + \tilde{\varphi}) = \mathcal{A}(\varphi_c) + \frac{1}{2} (\tilde{\varphi}, M\tilde{\varphi}) + O(\tilde{\varphi}^3) \tag{II.7}$$

Notice that any solution of the classical equation of motion (in particular φ_c) is a stationary point of the integral and, in the limit of small g, the saddle point method of integration yields the correct result. Using this technique and taking care of the first corrections to the saddle point method, the terms of $O(K^{-1})$ in eq. (II.3) can also be computed.

It is believed that using the same method all the singularities of the Borel transform can be computed; however, there is no numerical evidence of this conjecture.

For simplicity, in this sketchy derivation of the asymptotic estimates, we have neglected all problems connected to the symmetries of the Lagrangian, in particular the need of collective coordinates to take care of translational invariance[18] has not been discussed.

If we substitute bosons by fermions there is only a crucial difference[6] : fermions satisfy the Pauli exclusion principle and they cannot be in a coherent state : a description based on a classical field is consequently forbidden. The correct results can be obtained bu using the same intuitive argument as before. The energy of a system of N interacting fermions with overlapping wave functions is

$$E_N(g) \sim N^{1+1/d} + gN^2 \tag{II.8}$$

$d = D-1$ being the number of spatial dimensions. Fermi statistics implies that the kinetical energy term is no more proportional to N : at high density at each point of the space we must fill the Fermi sphere up to a momentum proportional to $\rho(x)^{1/d}$, $\rho(x)$ being the local density.

Using the same arguments as before we find that

$$E_B \sim g^{-\frac{d+1}{d-1}} \quad \text{Im}\, E(g) \simeq \exp\left[-(Rg)^{-D/(D-2)}\right] \tag{II.9}$$

The reader will recognize in this argument the familiar results of the Thomas Fermi approximation[19]. In order to obtain quantitative results we must extend the Thomas Fermi approximation to field theory. In the standard derivation of the TF approximation (e.g. in atomic physics), one obtains an approximate form for the fermionic energy in terms of the field in which the fermions are imbedded. This expression is valid for strong external field and it is obtained without having to discuss the fermionic wave functions. The same strategy can also be applied to field theory. The generating functional for a Yukawa theory can be written as

$$Z_V(g) = \int d[\psi] d[\bar\psi] d[\sigma] \exp\left[-\int_V \mathcal{L}(x)\, d^D x\right]$$

$$\mathcal{L}(x) = 1/2\, ((\partial_\nu \sigma)^2 + \mu^2\sigma^2) + \bar\psi(\not\partial + m)\psi + \lambda\bar\psi\psi\sigma \tag{II.10}$$

$$g = \lambda^2$$

Integrating over the fermionic fields, we find

$$Z_V(g) = \int d[\sigma] \exp\left[-\int_V \mathcal{L}_F(\sigma) d^D x\right] D(\lambda\sigma)$$

$$\mathcal{L}_F(\sigma) = \frac{1}{2}\,(\partial_\nu\sigma)^2 + \mu^2\sigma^2 \;; \qquad D(\lambda\sigma) = \det(\not\partial + m + \lambda\sigma) \tag{II.11}$$

The determinant which is an entire function of λ , appears at a positive power because of Fermi statistics. In the bosonic case, it appears to a negative power and its zeros are responsible for the leading part of the discontinuity along the cut in the complex g plane. In the fermionic case non analytic terms in the coupling constant may only arise from the integration in the region where $\lambda\sigma$ is large. In this region, the determinant is asymptotically proportional to

$$\det\left[\not\partial + m + \lambda\sigma\right] \sim \exp - S_D \left[\int (\lambda\sigma(x))^D d^D x\right]$$

$$S_D = 2\,\Gamma(-D/2)(4\pi)^{-D/2} \tag{II.12}$$

This result on the large field behaviour of the determinant is obtained[7] by using the same arguments as those needed to derive the standard TF approximation[19]. It is now a simple job to derive the asymptotic estimates

$$Z_K \sim \Gamma\left[\left(1 - \frac{2}{D}\right)K\right] R^K \cos(2\pi K/D)$$

$$Z(g) = \sum_0 {}_K Z_K g^K \tag{II.13}$$

The Borel transform is now an entire function : it may be convenient to introduce the Borel transform of order $\alpha = 1 - 2/D$ defined by

$$Z(g) = \int db \exp(-b^{1/\alpha}) B_\alpha(bg) \tag{II.14}$$

The asymptotic estimates give informations on the nearest singularity to the origine of $B_\alpha(b)$.

The introduction of fermions does not change the basic principles of the approach, although rather delicate estimates[8] are needed to control the large field behaviour of det $(\not\partial + eA)$ which is relevant in the study of the asymptotic estimates in QED.

III. THE WRONG VACUUM

In this section we study the asymptotic estimates for a theory in which the perturbation expansion is done around the wrong vacuum, (case 2 of the introduction). This happens in a one dimensional scalar theory (quantum mechanics) in which the potential has two degenerate minima. The simplest example is given by the Lagrangian

$$\mathcal{L}(\phi) = \frac{1}{2}(\partial_\mu\phi)^2 + \frac{1}{2}\phi^2(1-\lambda\phi)^2$$

$$\lambda^2 = g \tag{III.1}$$

The sum of all the vacuum to vacuum connected diagrams yields formally the ground state energy of the Hamiltonian

$$H = \frac{1}{2}p^2 + \frac{1}{2}(1-\lambda x)^2 x^2$$

$$E(g) = \sum_0^\infty {}_K E_K g^K = -\lim_{T\to\infty} T^{-1} \ln \int d[p] \exp\left[-\int_0^T \mathcal{L}(\phi)dx\right] \tag{III.2}$$

For small coupling constant, the Hamiltonian (III.2) has two nearly degenerate levels, whose wave functions are the symmetric and anti-symmetric combinations of those of two harmonic oscillators centered respectively at 0 and λ^{-1}. The splitting between the levels can be computed by semi classical methods[20] and it is proportional to $\exp(-1/(6g))$. Of course, the standard perturbative expansion is obtained by perturbing around the classical vacuum $\phi = 0$ (or equivalently $\phi = \lambda^{-1}$) . The asymptotic estimates can be obtained by a simple modification of the approach described in the previous sections.
The Kth order of the perturbative expansion can be formally written as

$$E_K = (2\pi i)^{-1} \oint dg/g^{K+1} E(g) \tag{III.3}$$

In the limit K going to infinity, E_K is evaluated by inserting eq. (III.2) into eq. (III.3) and considering a simultaneous saddle point in the variables ϕ and g.

The saddle point equations for ϕ are the classical equations of motion. The existence of solutions for real values of the coupling constant, (the so called instantons[21-23]), having an action equal to $+\mathcal{A}_c/g$, implies the asymptotic estimates

$$E_K \sim K!\,(2\mathcal{A}_c)^{-K} \tag{III.4}$$

Consequently, the Borel transform B(b) of E(g) has a singularity on the real line at $b = 2\,\mathcal{A}_c = \frac{1}{3}$.

The derivation of the asymptotic estimates would be straigthforward except for the following technical point which is the origine of the factor 2^{-K} in eq. (III.4). The solution to the classical Euclidean equation of motion which governs the large order behaviour has to satisfy periodic boundary conditions. In the case of degenerate minima such a solution does not exist. In particular, the well known instanton which interpolates between the two minima (its action being 1/(6g)) and corresponds to the quantum mechanical tunneling, obviously does not satisfy periodic boundary conditions. The correct result can be obtained by expanding around an interacting instanton-anti-instanton path configuration and integrating over a parameter describing the separation of the instanton-anti-instanton pair[24]. The instanton anti-instanton path configurations have, asymptotically, twice the action of a single instanton 1/3g; they are not solutions of the equation of motion, but the derivative of the action with respect to the path vanishes exponentially when the separation between the instantons goes to infinity. They form a family of quasi solutions of the equation of motion.

Using this technique, the correct asymptotic estimates have been obtained, and they agree with the computed first 73 orders of the perturbative expansion.

Unfortunately, the perturbation theory is no more Borel summable, the Borel transform being singular on the real axis, so that the problem remains : how to extract the true ground state energy from the perturbation series ?

In order to discuss this question we must sharpen our concepts and introduce a few definitions.

Given a function A(g) having the asymptotic expansion

$$A(g) = \sum_{0}^{\infty}{}_K A_K g^K \qquad |A_K| < K! \quad R^K \tag{III.5}$$

we define the perturbative Borel transform by

$$B^P(b) = \sum_{0}^{\infty}{}_K A_K b^K/K! \tag{III.6}$$

for $|bR| < 1$ and by analytic continuation elsewhere.

The true Borel transform is defined by the representation

$$A(g) = \frac{1}{g}\int_0^\infty \exp(-b/g)\, B(b)\, db \tag{III.7}$$

where B(b) may also be a distribution. The asymptotic expansion

is Borel summable if $B^p(b) = B(b)$ and it is analytic on the real positive axis.
We can also introduce the concept of asymptotic Borel transform if

$$A(g) = \frac{1}{g}\int_0^T \exp(-b/g)B(b)db + O(\exp(-T/g)) \qquad (III.8)$$

and we do not require the convergence of the integral in the limit $T \to \infty$.

The perturbative expansion fixes $B^p(b)$; how can we compute $B(b)$? The situation is not simple : the presence of exponentially small terms due to level splitting implies that $B(b)$ contains a non analytic term proportional to $\theta(b - 1/6)$, while the nearest singularity of $B^p(b)$ is located at $b = 1/3$. Moreover, both the symmetric level and the anti-symmetric level have the same perturbation expansion and consequently, the same perturbative Borel transform, their true Borel transform being obviously different. The best that can be hoped is a relation of the kind

$$\bar{B}(b) = \mathrm{Re}\, B^p(b)$$

$\bar{B}(b)$ being the half sum of the Borel transform for the symmetric and anti-symmetric levels. Eq. (III.9) is a suggestive possibility which is quite likely wrong.

In principle, the function $B(b)$ may be fixed studying the perturbative expansion around the instanton and the instanton-anti-instanton path configurations, however, this problem remains unsolved, especially for what concerns the many instanton contribution.

There is still an additional difficulty. The representation (III.7) implies that $E(g)$ is analytic at least in the sector

$$\mathrm{Re}(g^{-1}) > \delta \qquad (III.10)$$

The WKB approximation strongly suggests that for complex g in the right plane $E(g)$ is singular, due to level crossing[25]. This phenomenon produces a sequence of cuts in the right plane accumulating toward the origine. Only the sector $|g| < r$, $|\arg(g)| < \theta$, $\theta \sim 10^{-2}$ is singularity free. The analytic properties of $E(g)$ clash with those implied by eq. (III.7) : $B(b)$ may be the Borel transform of $E(g)$ only in the asymptotic sense (III.8).

How to reconstruct the function $E(g)$ from the knowledge of its asymptotic Borel transform ? The solution is not unique. The simplest possibility is given by a two step Borel transform.

In the previous chapter we have seen that we can introduce the Borel transform of order α. The validity of such a representation implies that the function $E(g)$ is analytic in the sector $Re(g^{-\alpha}) > \delta$. We can pick an α compatible with the analyticity properties of $E(g)$ and write it as

$$E(g) = \int_0^\infty \exp(-\beta^{1/\alpha}) B_\alpha(g\beta)\, \beta^{(1/\alpha - 1)}\, d\beta \qquad \text{(III.11)}$$

The function $B_\alpha(\beta)$ must be computed not from the non-convergent Taylor expansion around the origine, but as an integral over the asymptotic Borel transform $B(b)$. A simple example is given if $\alpha = 1/2$.

$$E(g) = \int_0^\infty d\beta \exp(-\beta^2)\, \beta B_{1/2}(g\beta)$$

$$B_{1/2}(\beta) = 4/\pi^{1/2} \int_0^\infty db \exp(-b^2) B(2b\beta) \qquad \text{(III.12)}$$

In the case in which there is no sector free of singularities we can generalize the two step procedure by introducing a "Borel transform of infinitesimal order"

$$E(g) = \frac{1}{g} \int_0^\infty F(s \exp(-1/g)) \exp(s) ds \qquad \text{(III.13)}$$

The function $F(s)$ can be formally written as

$$F(s) = \int_0^\infty B(b)/\Gamma(b+1)\, s^b\, db \qquad \text{(III.14)}$$

This last construction introduced by 't Hooft[9] may be quite useful if the function $B(b)$ increases as $\Gamma(b)$ at infinity and has strong oscillations.

IV. THE LANDAU GHOST

At least in the massless case, there is no difficulty in extending the arguments of the previous section to a renormalizable interaction. In the massive case, the Euclidean equation of motion may have no solutions. However, a family of quasi solutions (in the sense of the previous section) can be constructed. Ultraviolet divergences are present in the intermediate steps of the calculations; however, they cancel at the end giving finite estimates for the high orders of the perturbative expansion in the renormalized coupling constant. In other words, the formation of collapsed states does not show any new physically interesting feature when the interaction becomes renormalizable.

Unfortunately, semi-classical methods are well adapted to control the configuration to the functional integral coming from one field configuration (a finite dimensional manifold of field configurations) which is essentially different from zero, but the study of the contributions of an infinite dimensional manifold of field configurations (each of them being infinitesimally near to zero) is outside the range of application of semi-classical methods. This last region of integration becomes important when the theory is renormalizable and it is the origine of the ultraviolet divergences. The estimates for the singularities of the Borel transform obtained in the previous sections are undoubtedly correct, however, there may be other singularities which correspond to non perturbative effects connected to the renormalization procedure.

For simplicity, let us consider the massive $g\,\phi^4$ interaction in 4 dimensions. It is well known that in the leading logarithmic approximation one finds a Landau ghost[26] in the large momentum region, i.e. a pole of the 4 point amputated Green function $\Gamma_4(P|g)$, in the Euclidean region at momentum $P_G \sim \exp(1/\beta' g)$; β' is defined by

$$\beta(g) = \beta' g^2 + O(g^3)$$

$\beta(g)$ being the Callan-Symanzik beta function. The theory we are considering is not asymptotically free and β' is positive.

Let us assume for the time being that this feature of the leading logarithmic approximation is shared by the exact theory. The Landau ghost is present in any theory in which there are no ultraviolet stable fixed point, i.e. the function $\beta(g)$ has no zero for positive g and $\int_\varepsilon^\infty \beta^{-1}(g)dg < \infty$. If we insert a 4 point function with a Landau ghost in the D-S (Dyson-Schwinger) renormalized equations of motion, the region of integration near the ghost will give an imaginary contribution proportional to $P_G^{-\alpha}$. Consequently, the Borel transform has a singularity on the positive line at $b^{-1} = (\alpha/\beta')^{-1}$, corresponding to an additional contribution to the asymptotic estimates proportional to

$$G_K \sim K!(\beta'/\alpha)^{-K} \qquad \alpha > 0 \qquad \text{(IV.2)}$$

This new singularity is quite interesting and it cannot be studied with semi-classical methods. If the interaction is super-renormalizable, each diagram contributing to the Kth order is bounded by $Cnst^K$. The K! behaviour is due to the increase in the number of diagrams. The asymptotic estimates may be obtained by computing the contribution of the generical diagram using statistical methods[27,28]. If the interaction is renormalizable only a tiny minority of the diagrams of the Kth order has a number n of subdivergent diagrams of the same order as K. At K-n fixed, their number increases as $Cnst^K$. Contrary to the first appearance they cannot be neglected in the asymptotic estimates. Each of them gives a contribution proportional to K! [9,10]. These diagrams are at the origine of the asymptotic estimates (IV.2) : they have a structure rather different from those contributing to the estimates (II.3).

The existence of singularities of the Borel transform connected to the large momentum behaviour seems to be independent from the existence of a Landau ghost for real g. Also, if we assume the existence of an ultraviolet stable fixed point for real values of g, under a simplifying hypothesis, a slight modification of arguments due to Khuri[29] and based on the renormalization group shows the existence of a Landau ghost for complex values of g, i.e. the function $\Gamma_4(P|g)$ has a singularity on the curve $\text{Im } g = O(\text{Re } g^2)$ when $P \sim \exp\left[1/(\beta' \text{Re}(g))\right]$.

If we denote by r(p) the position of the nearest singularity to the origin of $\Gamma_4(P|g)$ in the right complex plane, we find

$$r(p) \sim \frac{1}{\beta' \ \ln P} + O \ \left(\frac{\ln \ln P}{\ln^2 P}\right) \qquad \text{(IV.3)}$$

Eq. (IV.3) suggests that the large momentum behaviour of the Borel transform $G_4(P|b)$ of the 4 point Green function $\Gamma_4(P|b)$ is given by

$$C_4(P|b) \simeq P^{b\beta'} \qquad \text{(IV.4)}$$

Eq. (IV.4) can be derived starting from the representation (I.4) for the Borel transform, by shifting part of the integration path to the right, picking the contribution of the Landau ghost and neglecting the rest. Eq. (IV.4) becomes exact in the leading log.approximation.

The increase in the large momentum region of C_4 is the origine of the new singularities of the Borel transform. The argument is quite general. The D-S equations for the Green functions induce similar equations for the Borel transform. Indeed, if we slightly change our notation and write

$$\Gamma(g) = \int \exp(-b/g) C(b) db \qquad \text{(IV.5)}$$

the following convolution theorem holds

$$\mathcal{B}\left|f(g)h(g)\right|(b) =$$

$$\int_0^b \int_0^b db_1\, db_2\, \delta(b-b_1-b_2)\, \mathcal{B}\left[f(g)\right](b_1 - \mathcal{B}\left[h(g)\right](b_2) \qquad \text{(IV.6)}$$

where β stands for "Borel transform of".

The Borel transformed D-S equation can be obtained by simple use of the convolution theorem (IV.6). Let us see an example. A typical contribution to $\Gamma_6(0|g)$ (i.e. the six point function at zero momentum) is given by

$$\Gamma_6(0|g) = \int \frac{d^4p}{(p^2+m^2)^3} \left|\Gamma_4(P,0|g)\right|^3 \qquad \text{(IV.7)}$$

$\Gamma_4(P, 0|g)$ being a short notation for $\Gamma_4(P,-P\ 0,0|g\ )$. The Borel transform statement is

$$C_6(0|b) = \int \frac{d^4p}{(p^2+m^2)^3} I(P|b)$$

$$I(P|b) = \iiint_0^b C_4(P,0|b_1) C_4(P,0|b_2) C_4(P,0|b_3) \delta(b_1+b_2+b_3-b)\, db_1 db_2 db_3 \qquad \text{(IV.8)}$$

If the large momentum behaviour of $C_4(P\ 0|b)$ is given by eq. (IV.4), the integral in eq. (IV.8) becomes divergent for $b = 2\beta'$. This argument shows that the Borel transforms of the Green functions have a sequel of singularities at the points $b = 2\beta' n$, n being a positive integer. In this derivation we have neglected the difference between asymptotic behaviour when all the momenta are large and when only a subset of the momenta is large; it is possible that some of the found singularities are absent.

A two line derivation of this singularities of the Borel may be interesting. Let us consider the Callan Symanzik equation 30,31 for $\Gamma^6(0\ g)$ in which the anomalous dimension has been taken equal to zero and terms of order g^3 have been neglected in the $\beta(g)$ function (these approximations are harmless for our aims)

$$\left[\beta'\ g^2 \frac{d}{dg} + 2\right] \Gamma^6(0|b) = \Delta\Gamma^6\ (0|b) \tag{IV.9}$$

The Borel transformed statement is

$$\left[2-\beta' b\right]\ C^6(0|b) = \Delta C^6(0|b) \tag{IV.10}$$

This equation may be interpreted either as a pole of $C^6(0|b)$ at $b = 2/\ \beta'$, or as a sum rule $(\Delta C^6(0|\ 2/\beta') = 0)$ for ΔC^6. If we disregard the second interpretation we find the wanted result. This argument may be dangerous as far as the precise position of the singularity is concerned, but it should be considered as an indication for the existence of singularities of the Borel transform on the real positive axis.

It may be useful to study a similar problem. Let us consider the $g\phi^4$ interaction in less than 4 dimensions $(D = 4-\varepsilon)$. The renormalized coupling constant spans the range $0-g_c(\varepsilon)$ $(\beta_\varepsilon(g_c(\varepsilon)) = 0;\ \ g_c(\varepsilon) = O(\varepsilon))$. Arguments similar to the previous ones 32,33 suggest that when $g \to g_c(\varepsilon)$, $\Gamma_6(g)$ behaves as $(g_c(\varepsilon)-g)^{2(1-\varepsilon)/\beta'_c}$ $\left(\frac{d}{dg}\ \beta_\varepsilon(g)\ \Big|_{g=g_c(\varepsilon)} = \beta'_c = O(\varepsilon)\right)$.
A more refined analysis shows that a singularity is present, but it is slightly weaker (the coefficient of the term linear in ε in the exponent is wrong)[34].

A singularity when $g = O(\varepsilon)$ implies that the Kth order of the perturbative expansion increases as $(1/\varepsilon)^K$ at large K. This result may seem to be in variance with the well known fact that all the coefficients of the renormalized perturbative expansion have a finite limit when $\varepsilon \to 0$. This contradiction is removed if there is a cross over region for $\varepsilon K \sim O(1)$ such that

$$G_K \sim K! \qquad (\varepsilon K) << 1$$

$$G_K \sim (1/\varepsilon)^K \qquad (\varepsilon K) >> 1 \tag{IV.11}$$

In the limit $\varepsilon \to 0$ only the first region survives and we reproduce asymptotic estimates similar to (IV.2).

We have seen explicitly the existence of a relation between the singularities of the Borel transform for positive b and the behaviour of the Green function in the critical region ($g \sim g_c(\varepsilon)$). We note en passant that it is very likely that the Borel transform of the ε expansion for the critical exponents is free of these singularities and the results of ref. 2 should be correct. In principle, the critical exponents can be computed by using the conformally invariant self consistency conditions[35-39], in which there is no trace of a possible Landau ghost.

Concluding this review of the asymptotic estimates for renormalizable interactions, we note that there are three main problems which wait for a solution.

1) To compute the position, nature and strength of these singularities of the Borel transform in the right complex b plane as it has been done for the singularities in the left plane.
2) To find a relation between the true Borel transform and the perturbative one (the presence of singularities on the real positive axis destroys the Borel summability).
3) To understand if a two step Borel transform may work in this case (the presence of a Landau ghost in the complex g plane implies that the Borel transform can be only asymptotic).

V. GAUGE THEORIES

It is believed that hadrons are composed by a multiplet of coloured quarks interacting via coloured gluon exchange. It would be quite interesting to extend the results of the previous section to gauge theories. Let us consider for simplicity a pure Yang-Mills theory. The existence of the instanton, i.e. a solution of the classical Euclidean equations of motion, implies the presence of tunneling between different vacua which are classified by a topological winding number[40,41]. There are non perturbative effects proportional to $\exp -A_I/g^2$, (A_I/g^2 is the instanton action); they reflect in singularities of the Borel transform with respect to g^2 having the form $\theta(b-A_I)$.

As explained in section III only those solutions of the classical equations of motion which satisfy periodic boundary conditions are relevant for the estimates of the high orders of the perturbative expansion. In gauge theories, the instanton field goes to zero at infinity and it is not clear in which sense it does or does not satisfy periodic boundary conditions. We suggest that a solution A(x) of the classical equation of motion contributes to the asymtptotic estimates only if we can find a sequence of functions $A_K(x)$, having compact support, such that

$$\lim_{K\to\infty} A_K(x) = A(x)$$

$$\lim_{K\to\infty} \mathcal{A}(A_K) = \mathcal{A}(A) \qquad (V.1)$$

$\mathcal{A}[A]$ being the action functional. We think that condition (V.1) reflects the fact that at any finite order in the perturbative expansion we can always compute the diagrams in a finite volume and perform the infinite volume limit only at the end.

Using the inequality $|F\tilde{F}| \leq F^2$ it is possible to prove that the single instanton field does not satisfy condition (V.1) and consequently it does not contribute to the asymptotic estimates. The relevant fields are the instanton-anti-instanton configurations. They are not solutions of the equations of motion but in the limit of large distances they form a family of quasi solutions. The nearest singulairity of the perturbative Borel transform is at $b = 2\mathcal{A}_I$ provided that in the zero topological charge section there are no true solutions of the classical equation of motion having an action smaller than $2\mathcal{A}_I/g^2$. Borel summability is lost. We are not arrived at the end of the troubels. Gauge theories describe the renormalizable interaction of massless particles; in the leading logarithmic approximation we find a Landau ghost in the ultraviolet region when g^2 is negative (as implied by asymptotic freedom) and in the infrared region when g^2 is positive. The arguments of the previous section suggest the presence of singularities of the Borel transform having the form of cuts starting from the points $b = \pm 2\, n/\beta'$. Of course some of these cuts may be absent.

the meaning of these singularities is rather transparent in this case. In the approximation in which we retain only the first non zero contribution in the $\beta(g^2)$ function and we neglect anomalous dimensions, we find

$$G(P,g) = 1/p^2\; G(\mu, \bar{g}^2(g^2, P/\mu)) \qquad (V.2)$$

where

$$\bar{g}^{-2}(g^2, P/\mu) = g^{-2} - \beta' \,\ln P/\mu \qquad (V.3)$$

For simplicity, we are considering the case of a two point function

$$G(P,g) = \int_0^\infty \frac{1}{p^2+m^2} P(m^2,g)\,dm^2 \qquad (V.4)$$

We denote by $B(P,b)$ the Borel transform of $G(P,g)$

$$G(P,g) = \int_0^\infty db\, \exp -b/g\; B(P,b) \qquad (V.5)$$

Using eq. (V.3) we find

$$G(P,g) = \frac{1}{p^2} \int_0^\infty d\alpha \, \exp(-2\alpha/\beta' g) B(\mu, 2\alpha/\beta')(P^2/\mu^2)^{-\alpha} \quad (V.6)$$

Singularities at integer α correspond to negative integer powers of p^2 in the asymptotic region. These terms look like mass corrections. They are welcome if we think that gluons are confined and a mass gap is automatically generated as in the two dimensional non linear σ model[46].

As in the previous section the Borel transform may be only asymptotic. Only a Borel representation of infinitesimal order (eqs. (III.13-14) may be valid. The singularities in the complex g plane accumulate toward the origine, leaving no singularity free sector. As suggested by 't Hooft[9] we can write a two step Borel representation

$$F(P,s) = \int_0^\infty \frac{db \, s^{2b/\beta'}}{\Gamma(2b/\beta'+1)} B(P,b)$$

$$G(P,g) = \int_0^\infty ds \, e^{-s} F(P, s \exp(-2/\beta' g)) \quad (V.7)$$

Using the approximate relation (V.3), we find

$$G(P,g) = \frac{1}{\mu^2} \int_0^\infty ds \, F(\mu, s \exp(-2/\beta' g)) \exp(-sK^2/\mu^2)$$

$$F(\mu, s \exp(-2/\beta' g)) = \int_0^\infty dm^2 \, \rho(m^2, g) \exp -sm^2/\mu^2 \quad (V.8)$$

As it can be seen from eq. (V.7), the function has a non convergent expansion in powers of $1/\ln s$ around $s = 0$. The singularities of $F(\mu,s)$ are quite interesting. A sequence of narrow resonance equidistant in M^2 produces a singularity of the function $F(\mu,s)$ on the imaginary s axis. Oscillations in the hadronic spectrum reflect in singularities of $F(\mu,s)$. In order to obtain WKB type approximations[47] to the hadronic masses, we need to know the singularities of $F(\mu,s)$. Unfortunately, the representation(V,7)is useless as long as we do not know how to compute the Borel transform $B(\mu,b)$. We only know the perturbative Borel transform $BP(\mu,b)$ and it is not evident if the singularities of $F^P(\mu,s) \equiv \int_0^\infty \frac{db \, s^{2b/\beta'}}{\Gamma(2b/\beta'+1)} \mathrm{Re}\, B^P(\mu,b)$ coincide with those of $F(\mu,s)$.

This lengthy analysis is necessary if we want to use the perturbative expansion to obtain numerical and reliable results for strong interactions. This aim may seem hopeless. However, when the physics is quite clear and many orders of the perturbation expansion have been computed, (as in the theory of the critical exponents[47,48]) extremely good results have been obtained also for strong coupling[15,49]. The difficulties we face in gauge theories are connected to our lack of physical comprehension of the problem.

The reader may think that in spite of our efforts, the situation is still a mess. That is true. However, in these last few years a progress has been made: we do not yet know how to get correct answers for a ield theory of strong interactions, but we begin to understand which are the right questions to ask.

REFERENCES

1. L.N. Lipatov, JEPT 72 (1977) 411
2. E. Brezin, J.C. Le Guillou, J. Zinn-Justin, Phys.Rev. D15 (1977) 1544
3. G. Parisi, Phys.Lett. 66B (1977) 167
4. E. Brezin, J.C. Le Guillou, J. Zinn-Justin, Phys.Rev. D15 (1977) 1558
5. C. Itzykson, G. Parisi, J.B. Zuber, Phys.Rev.Letters 38 (1977) 306
6. G. Parisi, Phys.Lett. 66B (1977) 382
7. C. Itzykson, G. Parisi, J.B. Zuber, Saclay preprint DPh/T 77/27 (Phys.Rev. D to be published)
8. R. Balian, C. Itzykson, G. Parisi, J.B. Zuber, Saclay preprint DPh/T 77/92
9. G. 't Hooft, Lectures given at Erice, July 1977
10. B. Lautrup, Phys.Lett. 69B (1977) 109
11. S. Graffi, V. Grecchi, B. Simon, Phys.Lett. 32B (1970) 631
12. J.D. Eckmann, J. Magnen, R. Seneor, Comm. Math. Phys. 39 (1975) 251
13. P. Renuard, preprint Ecole Polytechnique Palaiseau, n°A247 1076 (1976)
14. J.J. Loeffel "Workshop on Pade Approximants", Ed. D. Bessis, J. Gilewiczond, P. Merry CEA (1976)
15. J.C. Le Guillou, J. Zinn-Justin, Phys.Rev.Lett. 39 (1977)95
16. G. Parisi, Phys.Lett. 65 B (1977) 329
17. G. Parisi, Lectures given at the Cargèse Summer Institute (1976)

18. J.S. Langer, Ann. of Phys. 41 (1967) 108
19. H. Lieb, Rev. Mod. Phys. 49 (1977) 137
20. See for example L. Landau, F. Lifschitz, "Mécanique Quantique", Ed. Mir, Moscou (1970)
21. A.A. Belavin, A.M. Poliakov, A.S. Schwartz, Yu S. Tyupkin, Phys.Lett. 59B (1977) 85
22. G. 't Hooft, Phys.Rev.Lett. 37 (1976) 8
23. A.M. Poliakov, Nucl. Phys. B120 (1977) 429
24. E. Brezin, G. Parisi, J. Zinn-Justin, Saclay Preprint DPh/T 77/18 (Phys.Rev. D to be published)
25. T. Banks, C.M. Bender, T.T. Wu, Phys.Rev. D8 (1973) 3346
26. L.D. Landau, A.A. Abrikosov, I.M. Khalatnikow, Doklady Acad. Nauk URSS 95 (1954) 773
27. C.M. Bender, T.T. Wu, Phys.Rev.Lett. 37 (1976) 117
28. G. Parisi, Phys.Lett. 68B (1977) 361
29. N.N. Khury, Phys.Rev. D12 (1975) 2298 and Rockfeller Univ. Preprint COO 2232B 131
30. C.G. Callan, Phys.Rev. D2 (1970) 1541
31. K. Symanzik, Comm.Math.Phys. 18 (1970) 227
32. G. Parisi, Lectures given at the Cargèse Summer School (1973), Columbia University Preprint
33. G. Parisi, Nuovo Cimento, 21A (1974) 179
34. K. Symanzik, DESY preprint 77/05
35. A.A. Migdal, Sov.Phys. JEPT 28 (1969) 1036
36. A.M. Poliakov, Sov.Phys. JEPT 20 (1969) 533
37. A.M. Poliakov, JEPT Letters 12 (1970) 381
38. G. Parisi, L. Peliti, Lett.Nuov. Cimento 2 (1971) 627, Phys.Lett. 41A (1972) 331
39. G. Mack, L. Todorov, Phys.Rev. D8 (1973) 1764
40. C. Callan, R. Dashen, D. Gross, Phys.Lett. 63B (1976) 334
41. R. Jackiw, C. Rebbi, Phys.Rev.Lett. 37 (1976) 172
42. K. Symanzik, Nuovo.Cim.Lett. 6(1973) 420
43. G. 't Hooft, Marseille Conference June 1972 (unpublished)
44. H.D. Politzer, Phys.Rev.Lett. 30 (1973) 1346
45. D.J. Gross, F. Wilczek, Phys.Rev.Lett. 30 (1973) 1343
46. E. Brezin, J. Zinn-Justin, Phys.Rev. 36 (1976) 691
47. R. Balian, C. Bloch, Annals of Phys. 85 (1974) 514
48. K.G. Wilson, J. Kogut, Phys.Reports 12C (1974) 75
49. C.A. Baker, B.C. Nikel, M.S. Green, D.I. Meiron, Phys.Rev. Lett. 36 (1976) 1351.

Journal of Statistical Physics, Vol. 19, No. 3, 1978

Critical Exponents and Large-Order Behavior of Perturbation Theory

E. Brézin[1] and G. Parisi[2,3]

Received May 24, 1978

The principles of the recent calculations of critical exponents from three- and two-dimensional field theory are reviewed. They rely on the Callan–Symanzik equations, diagram calculations, and on the characterization of the asymptotic behavior of perturbation series at large order. We then present new results concerning the normalization of the large-order behavior.

KEY WORDS: Critical exponents; perturbation theory; asymptotic behavior; Callan–Symanzik equations.

1. INTRODUCTION

Field theoretic techniques applied to the problem of critical phenomena have recently been able to produce very accurate values for the three-dimensional critical indices.[(1)] In addition to all the previously available information, they involve recent progress concerning the quantitative characterization of the asymptotic orders of the perturbation series, when the order goes to infinity.[(2)] In this article we report some new results concerning this large-order problem, but it seemed to us that it might be useful at this stage to summarize the various logical steps combined in these 3D calculations. This article thus contains two very different parts. In the first, we expose without any derivation the set of principles underlying these calculations. In the second, we present the computation of a Fredholm determinant which occurs when one looks for the absolute normalization of the coefficients of very large orders of perturbation theory.

[1] Service de Physique Théorique, CEA–Saclay, Gif-sur-Yvette, France.
[2] Ecole Normale Supérieure, Laboratoire de Physique Théorique, Paris, France.
[3] On leave of absence from INFN, Frascati.

0022-4715/78/0900-0269$05.00/0

2. PRINCIPLES OF THE THREE-DIMENSIONAL CALCULATIONS OF CRITICAL INDICES FROM FIELD THEORY

It is convenient to distinguish the following steps:

1. Use of Wilson's renormalization group theory.
2. Wilson's Feynman graph approach, which leads to the understanding that the Callan–Symanzik equations govern the scaling properties near T_c and may be used to generate the ϵ expansion.
3. Direct use of the Callan–Symanzik equations in three dimensions.
4. Systematic calculation of several orders of the perturbation series (up to six loops) in three dimensions.
5. The obtaining of the large-order behavior of the perturbation series by instantons. They govern the instability of the vacuum when the coupling constant becomes attractive and reveal the nature of the divergence of the series.
6. Summation techniques based on the large-order information, which lead to accurate estimates from the perturbation series though the series is divergent and the fixed-point value of the expansion parameter is of order unity.

2.1. Step 1

The general ideas of Wilson's renormalization group approach are presented in great detail in several books and articles[3] and need not be repeated here.

2.2. Step 2

Similarly, the basis of the use of Callan–Symanzik equations in the problem of critical phenomena has been exposed at length elsewhere.[4] However, it may be useful to restate here the following features. In the Landau–Ginzburg–Wilson theory, an N-component order parameter $\boldsymbol{\phi} = (\phi_1, \ldots, \phi_N)$ is introduced. Its spatial distribution in d-dimensional space $\boldsymbol{\phi}(x)$ is weighted by a probability measure

$$\exp(-S\{\phi\})$$

with

$$S\{\phi\} = \int d^d x \, \{\tfrac{1}{2}(\nabla\phi)^2 + \tfrac{1}{2} r_0 \boldsymbol{\phi}^2 + \tfrac{1}{4} u_0 (\boldsymbol{\phi}^2)^2\} \tag{1}$$

In addition, if $\phi(x)$ is decomposed into Fourier components, the original lattice spacing a of the underlying magnetic model forbids variation of $\phi(x)$ with a wavelength smaller than a and therefore cuts off the wavenumbers at a value $\Lambda \sim 1/a$. The M-point correlation functions given by the expectation value [with the respect to the normalized weight $\exp(-S)$] $\langle\phi_{\alpha_1}(x_1)\cdots\phi_{\alpha_M}(x_M)\rangle$ depend on r_0, u_0, Λ. It is convenient to eliminate r_0 at the benefit of the correlation length ξ. Near the critical point ξ is very large and we are interested in correlation functions in the regime $|x_i - x_j| \gg a$, $\xi \gg a$. In addition, u_0 is a dimensional parameter proportional to a^{d-4} and thus it becomes very large in the limit of interest in less than four dimensions. It is therefore convenient to introduce a renormalized coupling constant g instead of u_0 and to modify the scale of the correlation functions. This may be done in the following way. Let

$$\delta(p_1 + \cdots + p_M)G_M(p_1, \alpha_1;\ldots;p_M, \alpha_M) = \int \langle\phi_{\alpha_1}(x_1)\cdots\phi_{\alpha_M}(x_M)\rangle_c \exp[i(p_1x_1 + \cdots + p_Mx_M)]\, dx_1\cdots dx_M \quad (2)$$

the M-point, connected (i.e., it contains only connected diagrams) correlation function in momentum space. We define the correlation length ξ and the field strength Z by parametrizing the behavior of G_2 near zero momentum by

$$G_2(p, \alpha; -p, \beta) = Z\frac{\delta_{\alpha\beta}}{1/\xi^2 + p^2 + O(p^4)} \quad (3)$$

and the dimensionless renormalized coupling constant g by

$$G_4(0, \alpha; 0, \beta; 0, \gamma; 0, \delta) = \xi^{4+d}2Z^2g(\delta_{\alpha\beta}\delta_{\gamma\delta} + \delta_{\alpha\gamma}\delta_{\beta\delta} + \delta_{\alpha\delta}\delta_{\beta\gamma}) \quad (4)$$

The renormalized correlation functions defined as

$$G_M^{(R)} = Z^{-M/2}G_M \quad (5)$$

may be expressed in terms of ξ and g (instead of u_0 and r_0) and they have a finite limit when the lattice spacing a vanishes.

One of the main advantages of this parametrization is that the dimensionless renormalized coupling g remains finite[4] also when the bare dimensionless coupling $u_0\xi^{4-d}$ goes to infinity.

These renormalized functions $G_M^{(R)}$ satisfy a first-order partial differential equation, first obtained by Callan and Symanzik[(6)] in the following way. Let us vary the temperature scale r_0, or equivalently the correlation length ξ with fixed values for u_0 and the lattice spacing a. The coupling constant g

[4] This has been rigorously proved for $N = 1$ in Ref. 5, using the Lebowitz inequality.

defined by (4) is modified by this variation. Therefore we perform the derivation

$$Z^{-M/2}\xi \frac{d}{d\xi}\bigg|_{u_0,a} Z^{M/2}G_M^{(R)}$$

and obtain from (5)

$$\left\{\xi \frac{\partial}{\partial \xi} + \xi \frac{\partial g}{\partial \xi}\bigg|_{u_0,a} \frac{\partial}{\partial g} + \frac{M}{2}\,\xi \frac{\partial}{\partial \xi}\bigg|_{u_0,a} \ln Z\right\} G_M^{(R)} = Z^{-M/2} \frac{\partial r_0}{\partial \xi}\bigg|_{u_0,a} \frac{\partial G_M}{\partial r_0}$$

or

$$\left\{\xi \frac{\partial}{\partial \xi} - \beta(g) \frac{\partial}{\partial g} - \frac{M}{2}\,\eta(g)\right\} G_M^{(R)} = \Delta G_M^{(R)} \tag{6}$$

in which we have defined

$$\beta(g) = -\xi \frac{\partial g}{\partial \xi}\bigg|_{u_0,a} \quad \text{and} \quad \eta(g) = -\xi \frac{\partial}{\partial \xi}\bigg|_{u_0,a} \ln Z \tag{7}$$

For momenta large compared with the inverse correlation length ξ^{-1} (but of course small compared to the inverse lattice spacing Λ, which has gone to infinity) the right-hand side is negligible compared to the left-hand side (technically this is true order by order in a double expansion in powers of g and $4 - d$). The reason is that $\partial G_M / \partial r_0$ contains only diagrams with one propagator squared which fall off more rapidly when the external momenta go to infinity. Therefore in this regime $a^{-1} \gg |p_i| \gg \xi^{-1}$, $G_M^{(R)}$ goes to $G_{M,\mathrm{as}}^{(R)}$, which is a solution of

$$\left\{\xi \frac{\partial}{\partial \xi} - \beta(g) \frac{\partial}{\partial g} - \frac{M}{2}\,\eta(g)\right\} G_{M,\mathrm{as}}^{(R)} = 0 \tag{8}$$

In a field theory g can be chosen arbitrarily. In this problem it is expressed in terms of r_0, u_0, and a. In the limit $u_0^{1/(4-d)}a \to 0$ one shows that

$$g \to g^*: \qquad \beta(g^*) = 0 \tag{9}$$

Consequently $G_{M,\mathrm{as}}^{(R)}(\xi, g^*)$ is proportional to $\xi^{M\eta(g^*)/2}$. For the two-point function, noting that $G_2^{(R)}(p, \xi, g^*)$ is dimensionally of the form $(1/p^2)f(p\xi)$, this leads to

$$G_2^{(R)}(p, \xi) \underset{p\xi \gg 1}{\sim} C(p\xi)^{\eta(g^*)}/p^2 \tag{10}$$

from which one sees that $\eta(g^*)$ is the usual η exponent. All the scaling laws and properties have been derived from this procedure. It relies on the existence of a fixed point g^* such that $\beta(g^*) = 0$ and its stability requires

$\beta'(g^*) > 0$. It is a crucial and implicit assumption that the renormalized Green's functions are finite for $g = g^*$. Near four dimensions, if $d = 4 - \epsilon$,

$$\beta(g) = -\epsilon g + \frac{N+8}{6} g^2 + O(\epsilon g^2, \epsilon^3), \qquad \eta(g) = \frac{N+2}{72} g^2 + O(\epsilon g^2, \epsilon^3) \tag{11}$$

and thus there is a stable fixed point

$$g^* = \frac{6\epsilon}{N+8} + O(\epsilon^2), \qquad \eta(g^*) = \frac{N+2}{2(N+8)^2} \epsilon^2 + O(\epsilon^3) \tag{12}$$

Higher orders in this direction have led to the calculations of the critical exponents as ϵ series. It is now known that these series are divergent.[14] However, they give very reasonable values for $\epsilon = 1$, either if truncated after two or three terms, or if resummed adequately.

2.3. Callan–Symanzik Equations in Three Dimensions

The ϵ expansion has been introduced for two purposes. It allows one to define a critical theory by neglecting the right-hand side of Eq. (6). Second, it gives a small parameter for the fixed point g^* and thus for the critical exponents. The $1/N$ expansion has the same properties and allows one to define a critical theory directly in three dimensions. It has been proposed in Ref. 6 that the existence of a three-dimensional critical theory implies that the right-hand side of (6) should be globally negligible for large $p\xi$, even if it is not true order by order in g. This may be supported by the ϵ expansion, which together with the short distance expansion allows one to show that, to all orders in ϵ,

$$\Delta G_2^{(R)}/G_2^{(R)} \underset{p\to\infty}{\sim} Ap^{-1/\nu} + Bp^{-(1-\alpha)/\nu} \tag{13}$$

Therefore, extrapolating to three dimensions, as long as α remains smaller than one, the right-hand side of (6) can be globally neglected. The problem is reduced to a direct computation of $\beta(g)$ and $\eta(g)$, and then to the search of the fixed point g^* defined by $\beta(g^*) = 0$, $\beta'(g^*) > 0$, which is now a finite number, and finally to the computation of $\eta = \eta(g^*)$. Similar equations hold for ν and the other critical exponents.

2.4. Three-Dimensional Calculations

After the promising initial calculations of Ref. 7 a very extensive program of calculating the power series expansion of the functions $\beta(g)$, $\eta(g)$, and $\eta_4(g)$, from which one deduces η and ν, was initiated by Nickel[8] and extended in Ref. 9. All Feynman diagrams involving at most six loops, that is, a priori 18-dimensional integrals, have been computed, together with their symmetry factor, which depends on N, the dimension of the order parameter. In order to decrease the number of integrals for a given diagram, these authors used the fact that one can compute analytically in three dimensions any one-loop subgraph with an arbitrary number of lines leaving the loop. The technical details are presented in Ref. 10. The total number of diagrams is 1142. Similarly these authors have computed the same functions in two dimensions up to four loops. The direct use of these results to calculate critical exponents is not possible. The series are badly divergent and the fixed point is a finite number. However, the control of the nature of the divergence of the series turns out to be a powerful tool for handling this problem.

2.5. Large Order Behavior of Perturbation Series

Perturbation series in many problems of quantum mechanics or field theory are asymptotic but divergent for any value of the coupling constant. The origin in the complex coupling constant plane is an essential singularity and the first indication of this fact was given by Dyson,[(11)] who argued that in quantum electrodynamics, if one changes the sign of the fine structure constant α, the vacuum would become unstable since electron–positron pairs would be pulled out of the vacuum. This argument has recently been made quantitative and it can be applied to many problems.

Let us first discuss a very simple example from quantum mechanics. Consider the one-dimensional anharmonic oscillator

$$H = \tfrac{1}{2}(p^2 + x^2) + gx^4 \tag{14}$$

and imagine that we want to calculate the ground-state energy in perturbation

$$E(g) = \tfrac{1}{2} + \sum_1^{\infty} g^K E_K \tag{15}$$

The E_K have been computed up to $K = 150$ (Ref. 12); $E_1 = 0.75$, $E_2 = -2.625$, $E_3 = 20.8125$, $E_4 = -241.289, \ldots$, but E_{75} is already of order 10^{144}. The series (15) looks divergent. This, and its rate of divergence, can be shown in the following way. Let us express $E(g)$ as the zero-temperature limit of the free energy

$$E(g) = \frac{1}{2} + \lim_{\beta\to\infty} - \frac{1}{\beta} \log \frac{\operatorname{Tr} e^{-\beta H}}{\operatorname{Tr} e^{-\beta H_0}} \tag{16}$$

and express the partition function by the Feynman–Kac formula

$$F(g) = \frac{\operatorname{Tr} e^{-\beta H}}{\operatorname{Tr} e^{-\beta H_0}} \propto \int \mathscr{D}x(\tau) \exp\left\{-\int_0^\beta \left[\frac{1}{2}(\dot{x}^2 + x^2) + gx^4\right]\right\} dt \tag{17}$$

in which we integrate over all periodic paths such that $x(0) = x(\beta)$. The function $F(g)$ is analytic in the complex g plane cut along the negative real axis and for large g behaves like $g^{1/3}$. The coefficients of its expansion in powers of g, $F(g) = \sum F_K g^K$, may be expressed as the moments of the discontinuity of $F(g)$, i.e.,

$$F_K = \int_C \frac{dg}{2i\pi} g^{-(K+1)} F(g) \tag{18}$$

in which the contour C encloses the negative real axis. If we substitute the representation (17) of F, we obtain for F_K an integral over g and paths, which for large K can be evaluated by the saddle-point method. The saddle points are given by the conditions

$$\begin{pmatrix} \delta/\delta g \\ \delta/\delta x(t) \end{pmatrix} \left\{ K \log g + \int_0^\beta dt \left[\frac{1}{2} \dot{x}^2 + \frac{1}{2} x^2 + g x^4 \right] \right\} = 0$$

which are

$$\ddot{x}_c = x_c + 4 g_c x_c^3 \tag{19}$$

$$K/g_c = -\int_0^\beta dt\, x_c^4(t) \tag{20}$$

From the second equation we see that g_c is, as expected, negative, and through the rescaling

$$x_c = (-g_c)^{-1/2} y_c \tag{21}$$

these equations become

$$\ddot{y}_c = y_c - 4 y_c^3 \tag{22}$$

$$g_c = -(1/K) \int_0^\beta dt\, y_c^4 \tag{23}$$

which show that g_c is infinitesimal for K large. It is convenient to introduce a mechanical analog to depict Eq. (22), which represents the motion in time of a particle located at $y_c(t)$ moving in the potential $V = (-\frac{1}{2} y_c^2 + y_c^4)$ (Fig. 1). We look for periodic solutions, and an elementary calculation shows

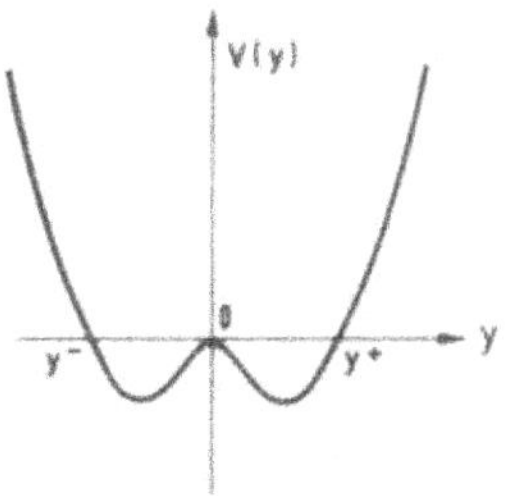

Fig. 1

that the leading contribution is given by the paths of period β of minimal action since

$$S(x_c, g_c) = K \log g_c + \int_0^\beta dt\,(\tfrac{1}{2}\dot{x}_c^2 + \tfrac{1}{2}x_c^2 + g_c x_c^4) \tag{24}$$

and

$$\exp[-S(x_c, g_c)] = [\exp(K \ln K - K)]/\left[-\int^\beta y_c^4\, dt\right]^K \tag{25}$$

A periodic path in this potential corresponds to a fixed energy, and in the large-β limit the lowest action corresponds to the periodic path that starts infinitesimally close to the origin, goes to y_+ (or y_-), and comes back:

$$y_c(t) = \pm\sqrt{\tfrac{1}{2}}\,[1/\cosh(t - t_0)] \tag{26}$$

$$\int_{-\infty}^{+\infty} y_c^4\, dt = \tfrac{1}{3} \tag{27}$$

It has been named "instanton" or "pseudoparticle" in the literature. We thus obtain

$$F_K \underset{K\to\infty}{\sim} K!\,(-3)^K \tag{28}$$

Fluctuations around the saddle point (22)–(23) may be systematically calculated. The parameter of this saddle-point expansion is $1/K$. Equation (22) is invariant under time translation, and the origin t_0 on the periodic trajectory is arbitrary. One may integrate properly over t_0; this may be done by the method of collective coordinates. The result is

$$F_K \underset{K\to\infty}{\sim} \beta\Gamma(K + \tfrac{1}{2})(-3)^K(6/\pi^{3/2})[1 + O(1/K)] \tag{29}$$

We now have to take the logarithm of $F(g)$ and divide by $-1/\beta$ to obtain $E(g)$. At large order, this is very easy. Indeed

$$\begin{aligned}&\log(1 + F_1 g + \cdots + F_K g^K + \cdots)\\ &\quad = F_1 g + \cdots + g^K[F_K - F_1 F_{K-1} + F_{K-2}(F_2 - F_1^2) + \cdots]\end{aligned}$$

and since F_K grows like $K!$, F_{K-1}/F_K is of order $1/K$, F_{K-2}/F_K of order $1/K^2$, etc. Therefore we can neglect all these terms. A diagrammatic way of visualizing this property consists in noticing that the diagrams that contribute to the free energy are only the connected ones and at large orders there is only a fraction $1/K$ of disconnected diagrams. This completes the asymptotic calculation of E_K:

$$E_K \underset{K}{\sim} -\Gamma(K + \tfrac{1}{2})(-3)^K(6/\pi^{3/2})[1 + O(1/K)] \tag{30}$$

The same result can be obtained also by using the traditional WKB method.[12] However, it is possible to show that in the one-dimensional quantum mechanical case, the approach presented here and the WKB method are deeply connected and they give identically the same answer. This makes manifest the divergence of the perturbation series of $E(g)$.

Let us apply the same procedure to the Landau–Ginzburg–Wilson system. The M-point functions are given by the functional integral

$$G_M(x_1, \alpha_1; \dots; x_M, \alpha_M) = \frac{\int \mathscr{D}\varphi \varphi_{\alpha_1}(x_1) \cdots \varphi_{\alpha_M}(x_M) \exp(-S\{\varphi\})}{\int \mathscr{D}\varphi \exp(-S\{\varphi\})} \tag{31}$$

$$S(\varphi) = \int d^d x \, \{\tfrac{1}{2}(\nabla \boldsymbol{\varphi})^2 + \tfrac{1}{2}m^2 \boldsymbol{\varphi}^2 + \tfrac{1}{4} g m^{4-d} (\boldsymbol{\varphi}^2)^2 + \tfrac{1}{2}(r_0 - m^2)\boldsymbol{\varphi}^2\} \tag{32}$$

in which m is the inverse of the magnetic susceptibility χ. The Kth order of the expansion of G_M in powers of g is obtained by a contour integral as in Eq. (17), and we apply the saddle-point method to the integrations over g and $\boldsymbol{\varphi}(x)$,

$$\begin{pmatrix} \delta/\delta g \\ \delta/\delta \varphi^\alpha(x) \end{pmatrix} \left\{ K \log g + \int d^d x \, [\tfrac{1}{2}(\nabla \boldsymbol{\varphi})^2 + \tfrac{1}{2} m^2 (\boldsymbol{\varphi}^2) + \tfrac{1}{4} g m^{4-d} (\boldsymbol{\varphi}^2)^2] \right\} = 0 \tag{33}$$

The mass counter-term of Eq. (32) is only relevant at the subleading order, which is discussed below. The variational equations are then

$$\frac{K}{g_{0c}} + \tfrac{1}{4} m^{4-d} \int d^d x \, (\boldsymbol{\varphi}_c^{\,2})^2 = 0, \qquad \Delta \varphi_c^{\,\alpha} = m^2 \varphi_c^{\,\alpha} + g_c m^{4-d} (\boldsymbol{\varphi}_c^{\,2}) \varphi_c^{\,\alpha}$$

or with the rescaling

$$\varphi_c^{\,\alpha}(x) = \frac{m^{(d-1)/2}}{(-g_c)^{1/2}} \Phi_c^{\,\alpha}(mx) \tag{34}$$

we obtain

$$g_c = -\frac{1}{4K} \int (\Phi_c^{\,2})^2 \, d^d x \tag{35}$$

$$\Delta \Phi_c^{\,\alpha} = \Phi_c^{\,\alpha} - 4 \Phi_c^{\,2} \Phi_c^{\,\alpha} \tag{36}$$

Here again g_c is negative and it goes to zero when K goes to infinity. At the saddle point, an easy calculation gives

$$\left(\frac{1}{g_c}\right)^K \exp[-S(\varphi_c)] = \left[-\frac{4}{\int (\boldsymbol{\Phi}_c^{\,2})^2 \, d^d x}\right]^K \exp(K \log K - K) \tag{37}$$

showing a behavior very similar to that of the anharmonic oscillator. The leading contribution is thus given by the instanton, the solution of (36) that minimizes the integral of $(\Phi^2)^2$. This solution is spherically symmetric,[13] nodeless, and corresponds to a fixed direction in internal space

$$\Phi_c^{\alpha} = u^{\alpha} f(x), \qquad \mathbf{u}^2 = 1 \tag{38}$$

$$\left(\frac{d^2}{dx^2} + \frac{d-1}{x}\frac{d}{dx}\right) f(x) = f(x) - f^3(x) \tag{39}$$

At large distance $f(x)$ decreases as $e^{-x} x^{-(d-1)/2}$. The solution has been determined by numerical computation in two and three dimensions.[14]

This shows that any Green's function of the theory has an expansion in powers of g which at large order behaves as $g^K \Gamma(K + b) a^K c$. The number a is the same for all Green's functions independent of the number of external legs of the values of the external momenta:

$$a = -\left[\frac{1}{4}\int (\Phi_c^2)^2\, d^d x\right]^{-1} \tag{40}$$

The number b requires a more detailed calculation; for an M-point function

$$b = \tfrac{1}{2}(M + N + d - 1) \tag{41}$$

The constant c is a function of all the variables, the external momenta, etc. It is calculated in the second part of this article for the functions that appear in the Callan–Symanzik equations.

2.6. Large-Order Behavior and Summation Techniques

We have seen in the previous section that perturbation theory is divergent for any value of the coupling constant. However, with the use of the large-order information, it is possible in various ways to obtain convergent algorithms. Let us study as an example the problem of the ground-state energy of the anharmonic oscillator of Eqs. (14)–(15) for a coupling constant equal to unity. If we use the truncated series for $g = 1$ the result is absurd. However, since we know that E_K grows as $K!$, it is indicated that we perform a Borel transformation

$$E(g) = \int_0^{\infty} dt\, e^{-t} \sum_0^{\infty} (tg)^K \frac{E_K}{K!} \tag{42}$$

(in which we recover the ordinary perturbation series if the summation and integration are interchanged). There is actually a proof[15] that $E(g)$ is indeed given by (41). The Borel transformation of $E(g)$

$$f(b) = \sum_0^{\infty} b^K \frac{E_K}{K!} \tag{43}$$

is analytic in a circle of radius 1/3 as indicated by (28) and the singularity closest to the origin in the b plane is at $b = -1/3$. The function $f(b)$ is furthermore analytic in the vicinity of the real, positive axis. If we replace $f(b)$ by

$$f_L(b) = \sum_0^L b^K \frac{E_K}{K!} \tag{44}$$

we recover the perturbation series truncated at order L. But we can replace the function $f_L(b)$ by a rational function, a Padé approximant, before performing the integration (42). In Ref. 1 the authors have taken as an example $L = 6$; the error between the approximate value of $E(1)$ and the exact one (which can be obtained by variational methods) is 10^{-3}.

However, we can use more than just the $K!$ of E_K and notice that Eq. (30) implies that $f(b)$ has a square root branch point at $b = -1/3$. Assuming analyticity in the whole b plane cut along the negative axis from $-1/3$ to $-\infty$, we can map the whole b plane in the interior of a circle by the conformal mapping

$$z = [(1 + 3b)^{1/2} - 1]/[(1 + 3b)^{1/2} + 1] \tag{45}$$

The natural representation of the corresponding function of z is a Taylor expansion and the knowledge of L coefficients E_K determines the L first coefficients a_K of the representation

$$\tilde{f}_L(b) = \sum_0^L a_K z^K(b) \tag{46}$$

which is then transformed by (42). With the same value $L = 6$ the error drops to 3×10^{-4}. It is thus manifest that the large-order behavior is a useful systematic guide to transforming the knowledge of the low-order coefficients into a modified convergent scheme.

These ideas have been developed and applied with success in Ref. 1 to the calculation of the critical exponents. The problem is exactly of the same nature. The fixed point is of order unity. The series diverges in the same fashion. However, the critical exponents have been computed with an accuracy which is higher than that achieved by all previous methods.

3. NORMALIZATION OF THE LARGE-ORDER ESTIMATES

Previously we have shown that the coefficients of all the Green's functions at large order behave as

$$g^K \Gamma(K + b) a^K c[1 + O(1/K)] \tag{47}$$

The calculation of a relies on the value of the classical action for the instanton solution. We present here the calculation of b and c for the functions that appear as coefficients in the Callan–Symanzik equations, from which the critical exponents are computed. The technique involves the calculation of the contribution to the functional integral of the Gaussian fluctuations near the instanton solution. The main body of the calculation is to evaluate the determinant of the quadratic form of the fluctuations around the instanton, namely

$$\Delta_L = \det\left\{1 + \frac{1}{-\Delta + 1}\,[-3\Phi_c^2(x)]\right\} \tag{48}$$

$$\Delta_T = \det\left\{1 + \frac{1}{-\Delta + 1}\,[-\Phi_c^2(x)]\right\} \tag{49}$$

in which we have distinguished between longitudinal fluctuations in the direction of a given classical solution and transverse fluctuations along perpendicular directions in internal space; Δ_T is needed only if $N \neq 1$.

Some caution is, however, needed in order to take into account properly the zero modes. Indeed the classical equation (36) has an infinite set of lowest action solutions which differ from one another by a translation, or an $O(N)$ rotation in internal space. It follows from these invariances that (i) the operator $[-\Delta + 1 - 3\Phi_c^2(x)]$ has a zero eigenvalue d-fold degenerate, the corresponding eigenvectors being $\partial\Phi_c(x)/\partial x_\alpha$, $\alpha = 1, 2, \ldots, d$; (ii) the operator $[-\Delta + 1 - \Phi_c^2(x)]$ also has a zero mode corresponding to the eigenfunction Φ_c itself.

These modes should be properly quantized, by using the collective coordinates method,[16] which performs the integral exactly (without Gaussian approximation) in the direction in which the action $S\{\Phi\}$ remains constant. The result is then the product of a simple Jacobian corresponding to the collective coordinate change of variable and of the determinants of the operators restricted to the subspace orthogonal to the zero eigenmodes $\Delta_L{}^{\perp}, \Delta_T{}^{\perp}$.

This gives for the $2M$-point function at zero external momenta

$$[G_{2M}(0, n;\ldots; 0, n)]_K = [\exp(-S_{\rm cl})]J\left[\frac{4KI_1{}^2}{I_4}\right]^M \frac{\Gamma(M + 1/2)\Gamma(N/2)}{\sqrt{\pi}\,\Gamma(M + N/2)}$$
$$\times (\Delta_L{}^{\perp})^{-1/2}(\Delta_T{}^{\perp})^{-(N-1)/2}\{\exp[-\tfrac{1}{2}(N + 2)I_2G_2{}^F(0)]\} \tag{50}$$

in which

$$\exp(-S_{\rm cl}) = \frac{K!}{(2\pi K)^{1/2}}\left(\frac{4}{I_4}\right)^K\left[1 + O\left(\frac{1}{K}\right)\right] \tag{51}$$

$$I_p = \int d^dx\,[\Phi_c(x)]^p \tag{52}$$

and the Jacobian

$$J = \frac{\sqrt{2}\,\pi^{(N-1)/2}}{\Gamma(N/2)} K^{[(N+d)/2]-1}(2\pi)^{-d/2-(N-1)/2} \tag{53}$$

The mass counterterm in the action is responsible for the last factor in (50) and it involves the divergent integral

$$G_2{}^F(0) = \int \frac{d^d q}{(2\pi)^d} \frac{1}{q^2+1} \tag{54}$$

which will be discussed below.

The calculation of the integrals I_p is done by numerical integration and the problem is reduced to that of computing the $\Delta^\perp$. We found it convenient to introduce the function

$$D(z) = \det\left[1 - \frac{3z}{-\Delta+1}\Phi_c{}^2(x)\right] \tag{55}$$

from which we recover $\Delta_L{}^\perp$ and $\Delta_T{}^\perp$ as

$$\bar{D}(1) = \lim_{z\to 1} \frac{D(z)}{(1-z)^d} = \Delta_L{}^\perp \lim_{z\to 1} \frac{\det\{1 - [3z/(-\Delta+1)]\Phi_c{}^2\}}{(1-z)^d} \tag{56}$$

in which the last determinant is d over d in the subspace of the longitudinal zero modes $\partial_\alpha \Phi_c$. This leads to

$$\Delta_L{}^\perp = \bar{D}(1)\left[\frac{\frac{1}{4} d I_4}{I_6 - I_4}\right]^d \tag{57}$$

Similarly we obtain

$$\Delta_T{}^\perp = \bar{D}(1/3)/4 \tag{58}$$

$$\bar{D}(1/3) = \lim_{z\to 1/3} [D(z)/(1-3z)] \tag{59}$$

This determinant $D(z)$ may be expressed in terms of the eigenvalues[17] λ_n defined by the equation

$$(-\Delta\psi_n + \psi_n) = \lambda_n 3\Phi_c{}^2(x)\psi_n \tag{60}$$

as

$$D(z) = \prod_n \left(1 - \frac{z}{\lambda_n}\right) \tag{61}$$

in which it is understood that an m-fold degenerate eigenvalue appears m times in the product. In two or more dimensions this infinite product is divergent. The mass counterterm of Eq. (50) is also infinite, but since

$$\sum_n (1/\lambda_n) = G_2{}^F(0)3I_2 \tag{62}$$

as may be checked by expanding $D(z)$ at first order in z, there is an exact cancellation in less than four dimensions when we combine the mass counterterm and the divergent part of $D(z)$ at $z = 1$. We can thus eliminate the counterterm contribution from (50), absorb it into $D(z)$, and define the renormalized convergent determinant

$$D_R(z) = \prod_n \left(1 - \frac{z}{\lambda_n}\right) e^{z/\lambda_n} \tag{63}$$

The convergence of the infinite product holds provided that

$$\sum_n 1/\lambda_n{}^2 < \infty$$

which is true below four dimensions since the sum on the left-hand side is related to the one-loop diagram with four external legs, which is finite.

In order to perform the numerical computation of $D_R(z)$ we can calculate explicitly the lowest λ_n's and perform the product (63). However, it is important to improve this procedure by evaluating the asymptotic behavior of the large λ_n's. Let us discuss in some detail the three-dimensional case.

All the eigenvalues are real and positive, since the operator $(-\Delta + 1)^{-1/2}\Phi_c{}^2(x)(-\Delta + 1)^{-1/2}$ is Hermitian and positive. Let $N(\lambda)$ be the number of eigenvalues smaller than λ. Equivalently $N(\lambda)$ may be defined as the number of eigenvalues of the Schrödinger problem

$$H = p^2 - 3\lambda\Phi_c{}^2(x) \tag{64}$$

with an energy smaller than minus one. The asymptotic behavior of $N(\lambda)$ for λ large may thus be obtained from this last remark, since it is known that the asymptotic number $B(V)$ of bound states in a large attractive potential $-V$ is correctly given by the Thomas–Fermi approximation as[18]

$$B(V) \sim (1/6\pi^2) \int d^3x\, V^{3/2}(x) + O(V^{1/2}) \tag{65}$$

In the present problem since $V(x) = 3\lambda\Phi_c{}^2(x)$ this gives

$$N(\lambda) \underset{\lambda \to \infty}{\sim} (3^{1/2}/2\pi^2) I_3 \lambda^{3/2} \equiv C\lambda^{3/2} \tag{66}$$

The method will thus consist in using this asymptotic information together with the explicitly computed lowest eigenvalues. This procedure turns out to be quite useful in order to improve the convergence, as will be shown below. Specifically, this is done in the following way. Let $\rho(\lambda) = dN/d\lambda$ be the exact density of eigenvalues and $\rho_{as}(\lambda) = \frac{3}{2}C\lambda^{1/2}$. Then from (63) we may write

$$\begin{aligned} D_R(z) &= \exp \int_0^\infty d\lambda\, \rho(\lambda)\left[\ln\left(1 - \frac{z}{\lambda}\right) + \frac{z}{\lambda}\right] \\ &\simeq \exp \int_0^{\bar\lambda} d\lambda\, \rho(\lambda)\left[\ln\left(1 - \frac{z}{\lambda}\right) + \frac{z}{\lambda}\right] \\ &\quad \times \exp\left\{\frac{3}{2} C \int_{\bar\lambda}^\infty d\lambda\, \lambda^{1/2}\left[\ln\left(1 - \frac{z}{\lambda}\right) + \frac{z}{\lambda}\right]\right\} \end{aligned} \tag{67}$$

which, for large $\bar\lambda$, reduces to

$$D_R(z) = \left\{\prod_{\lambda_n < \bar\lambda} \left(1 - \frac{z}{\lambda_n}\right) \exp \frac{z}{\lambda_n}\right\} \exp\left(-\frac{3}{2} C z^2 \bar\lambda^{-1/2}\right). \tag{68}$$

Given the numerical value

$$C = (3^{1/2}/2\pi^2) I_3 = 2.78083 \tag{69}$$

this gives a correction factor to the truncated product which is about 50% for $\bar\lambda \sim 50$ at $z = 1$. Furthermore, in order to avoid numerical oscillations when $\bar\lambda$ passes through an eigenvalue, the correction factor may be replaced by an asymptotically equivalent one, namely

$$D_R(z) = \left\{\prod_{n=1}^{N} \left(1 - \frac{z}{\lambda_n}\right) \exp \frac{z}{\lambda_n}\right\} \exp\left(-\frac{3}{2} C^{4/3} N^{-1/3} z^2\right) \tag{70}$$

in which it is understood that the λ_n are ordered increasingly with n. The lowest 900 λ_n counted with their multiplicity have been computed by solving the Schrödinger equation (64) expanded in partial waves up to $l_{\max} = 14$, which corresponds to a $\bar\lambda$ of about 48. In Table I we reproduce the results for the lowest λ_n up to $\bar\lambda = 20$.

Table I. Eigencouplings of $(-\Delta + 1 - 3\lambda\Phi_c^2)$ up to $\bar\lambda = 20$

$l = 0$	1	2	3	4	5	6	7	8
0.33333	1.00000	2.11863	3.69159	5.71487	8.18630	11.1049	14.4703	18.2822
1.38758	2.46066	4.02308	6.06067	8.55922	11.5114	14.9137	18.7645	
3.11042	4.57446	6.56661	9.05841	12.0251	15.4529	19.3353		
5.49860	7.34474	9.75643	12.6936	16.1213				
8.55053	10.7730	13.5965	16.9715					
12.2653	14.8599	18.0894						
16.6423	19.6059							

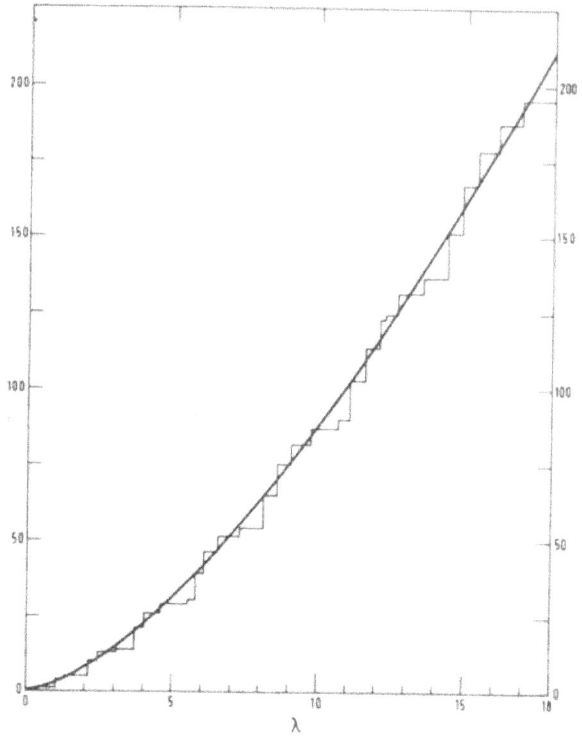

Fig. 2

In order to see how the asymptotic regime (66) is reached, we have represented the true $N(\lambda)$ in Fig. 2 and compared it with $N_{as}(\lambda)$. The results are

$$\bar{D}_R(1) = 10.544 \pm 0.004 \tag{71a}$$

$$\bar{D}_R(1/3) = 1.4571 \pm 0.004 \tag{71b}$$

in which the errors come from the extrapolation from $N = 900$ to infinity.

Putting the results into Eq. (50) for the four-point function at zero momentum, we obtain

$$g = g_0 - g_0^2 \frac{N+8}{8\pi} + \cdots + \omega_K g_0^K + \cdots \tag{72}$$

$$\omega_K \underset{K\to\infty}{\sim} (-)^{K+1} a^K \Gamma(K+b) c\left[1 + O\left(\frac{1}{K}\right)\right] \tag{73}$$

with

$$a = 4/I_4 \tag{74}$$

$$b = 3 + \tfrac{1}{2}N \tag{75}$$

$$c = \frac{2^{N/2+2}3^{-3/2}\pi^{-2}}{\Gamma(\frac{1}{2}N+2)}\left(\frac{I_1^{\,2}}{I_4}\right)^2\left(\frac{I_6}{I_4}-1\right)^{3/2} \times [\bar{D}_R(1)]^{-1/2}[\bar{D}_R(1/3)]^{-(N-1)/2} \tag{76}$$

Inverting (72) for g_0 in terms of g, we obtain

$$g_0 = g + g^2\frac{N+8}{8\pi} + \cdots + \delta_K g^K + \cdots \tag{77}$$

$$\delta_K \underset{K\to\infty}{\sim} -\omega_K\left(\exp - \frac{N+8}{8\pi a}\right)\left[1 + O\left(\frac{1}{K}\right)\right] \tag{78}$$

The field strength renormalization is asymptotically (i.e., for large K) negligible, and thus we can use the formula

$$\tilde{\beta}(g) = -g\left(1 + g\frac{d}{dg}\ln\frac{g_0}{g}\right)$$

This leads to

$$\tilde{\beta}(g) = -g + \frac{N+8}{8\pi}g^2 + \cdots + g^K\tilde{\beta}_K + \cdots \tag{79}$$

with

$$\tilde{\beta}_K = (-)^K a^K\Gamma(K+b+1)c\exp - \frac{N+8}{8\pi a} \tag{80}$$

Using, as in Ref. 9, the normalization

$$g = \frac{8\pi}{N+8}V \tag{81}$$

$$\tilde{\beta}(g) = \frac{N+8}{8\pi}\beta(V) \tag{82}$$

we find

$$\beta(V) = -V + V^2 + \cdots + \beta_K V^K + \cdots \tag{83}$$

$$\beta_K \underset{K\to\infty}{\sim} (-A)^K\Gamma(K+B)C \tag{84}$$

Table II. Fredholm Determinants and Integrals

d	$\bar{D}_R(1)$	$\bar{D}_R(1/3)$	I_1	I_4	I_6	H_3
3	10.544 ± 0.004	1.4571 ± 0.0001	31.691522	75.589005	659.868352	13.563312
2	135.3 ± 0.1	1.465 ± 0.001	15.10965	23.40179	71.08023	9.99118

with

$$A = 32\pi/(N + 8)I_4 \tag{85}$$

$$B = 4 + \tfrac{1}{2}N \tag{86}$$

$$C = \frac{(N + 8)2^{N/2-1}3^{-3/2}}{\pi^3\Gamma(2 + \frac{1}{2}N)} \left(\frac{I_1{}^2}{I_4}\right)^2 \left(\frac{I_6}{I_4} - 1\right)^{3/2} \times [\bar{D}_R(1)]^{-1/2}\left[\bar{D}_R\left(\frac{1}{3}\right)\right]^{-(N-1)/2} \exp\left(-\frac{N+8}{32} I_4\right) \tag{87}$$

A similar calculation in two dimensions gives, if

$$g = \frac{4\pi}{N + 8} V \tag{88}$$

and

$$\tfrac{1}{2}\beta(V) = -V + V^2 + \cdots + \beta_K V^K + \cdots \tag{89}$$

Table III. Parameters Characterizing the Asymptotic Behavior

	$d = 3, N = 0$	$d = 3, N = 1$	$d = 3, N = 2$
A	0.1662460	0.14777422	0.1329968
B	4	4.5	5
C	$(8.5489 \pm 16\cdot 10^{-4}) \times 10^{-2}$	$(3.9962 \pm 6\cdot 10^{-4}) \times 10^{-2}$	$(1.6302 \pm 3\cdot 10^{-4}) \times 10^{-2}$
	$d = 3, N = 3$	$d = 2, N = 1$	
A	0.12090618	0.238659	
B	5.5	4	
C	$(5.9609 \pm 10^{-3}) \times 10^{-3}$	$(4.886 \pm 5\cdot 10^{-4}) \times 10^{-2}$	

Table IV. Comparison of the Low-Order Calculations and of the Asymptotic Behavior for the Function $\beta(g)$

	$d = 3, N = 0$		$d = 3, N = 1$		$d = 3, N = 2$		$d = 3, N = 3$		$d = 2, N = 1$	
K	$(-)^K\beta_K$	β_K	$(-)^K\beta_K$	β_K	$(-)^K\beta_K$	β_K	$(-)^K\beta_K$	β_K	$(-)^K\beta_K$	β_K
2	1	30.15×10^{-2}	1	15.91×10^{-2}	1	7.85×10^{-2}	1	3.66×10^{-2}	1	14.6×10^{-2}
3	0.4398	13.29×10^{-2}	0.4225	7.00×10^{-2}	0.4029	3.40×10^{-2}	0.3832	1.54×10^{-2}	0.7162	7.32×10^{-2}
4	0.3899	10.13×10^{-2}	0.3511	5.26×10^{-2}	0.3149	2.50×10^{-2}	0.2829	1.11×10^{-2}	0.9308	5.69×10^{-2}
5	0.4473	8.74×10^{-2}	0.3765	4.48×10^{-2}	0.3179	2.11×10^{-2}	0.2703	9.23×10^{-3}	1.5824	5.07×10^{-2}
6	0.6339	8.27×10^{-2}	0.4955	4.20×10^{-2}	0.3911	1.95×10^{-2}	0.3126	8.41×10^{-3}	—	—
7	1.0349	8.13×10^{-2}	0.7497	4.09×10^{-2}	0.5524	1.88×10^{-2}	0.4149	8.03×10^{-3}	—	—
C		8.55×10^{-2}		4.00×10^{-2}		1.63×10^{-2}		5.96×10^{-3}		4.89×10^{-2}

a formula for β_K analogous to (84) with

$$A = 16\pi/(N+8)I_4 \tag{90}$$

$$B = 7/2 + N/2 \tag{91}$$

$$C = \frac{(N+8)2^{(N-3)/2}\pi^{-5/2}}{\Gamma(2+\frac{1}{2}N)} \left(\frac{I_1{}^2}{I_4}\right)^2 \left(\frac{I_6}{I_4} - 1\right) \times [\bar{D}_R(1)]^{-1/2} \left| \bar{D}_R\left(\frac{1}{3}\right) \right|^{-(N-1)/2} \exp\left(-\frac{N+8}{16\pi} I_4\right) \tag{92}$$

In order to complete this calculation we give the numerical results of Table II, where

$$I_2 = (1 - \tfrac{1}{4}d)I_4 \tag{93}$$

$$I_3 = I_1 \tag{94}$$

$$H_3 = \int d^dx\, x^2 \Phi_c{}^3(x) \tag{95}$$

This leads to the results of Table III.

These asymptotic formulas for K large are now compared with the explicit calculations of Ref. 9. Table IV gives the β_K provided by perturbation theory, and also these numbers divided by their leading asymptotic estimates

$$\bar{\beta}_K = |\beta_K|/\Gamma(K+B)A^K \tag{96}$$

For $K = 7$, $d = 3$, $\bar{\beta}_K$ is within 5% of its asymptotic limit for $N = 0$ or 1. However, the agreement gets worse for larger values of N, indicating a slower approach to the asymptotic limit. This is to be connected with the crossover to the large-N regime, in which the coefficients of the powers of $1/N$ are analytic at $g = 0$.

The critical exponents are calculated from two additional series

$$\eta_4(V) = -V\frac{(N+2)}{(N+8)} + \cdots + V^K\gamma_K + \cdots \tag{97}$$

$$\eta(V) = \frac{V^2(N+2)}{2(N+8)^2} + \cdots + V^K\delta_K + \cdots \tag{98}$$

Table V. The Normalization of the Large-Order Behavior of $\eta_4(g)$ and $\eta(g)$

	$d=3, N=0$	$d=3, N=1$	$d=3, N=2$	$d=3, N=3$	$d=2, N=1$
C'	1.0107×10^{-2}	6.2991×10^{-3}	3.0836×10^{-3}	1.2813×10^{-3}	1.049×10^{-2}
C''	2.8836×10^{-3}	1.7972×10^{-3}	0.8798×10^{-3}	0.3656×10^{-3}	3.468×10^{-3}

Table VI. Comparison of the Asymptotic Behavior and of the Low-Order Calculations for the Function $\eta_4(g)$

	$d=3, N=0$		$d=3, N=1$		$d=3, N=2$		$d=3, N=3$		$d=2, N=1$	
K	$(-)^K\gamma_K$	$\tilde{\gamma}_K$	$(-)^K\gamma_K$	$\tilde{\gamma}_K$	$(-)^K\gamma_K$	$\tilde{\gamma}_K$	$(-)^K\gamma_K$	$\tilde{\gamma}_K$	$(-)^K\gamma_K$	$\tilde{\gamma}_K$
1	25×10^{-2}	2.506×10^{-2}	33.33×10^{-2}	43.11×10^{-3}	40×10^{-2}	25.1×10^{-3}	45.45×10^{-2}	13.06×10^{-3}	0.667	11.64×10^{-2}
2	6.25×10^{-2}	1.881×10^{-2}	7.41×10^{-2}	11.79×10^{-3}	8×10^{-2}	6.28×10^{-3}	8.26×10^{-2}	3.02×10^{-3}	0.250	3.65×10^{-2}
3	3.58×10^{-2}	1.081×10^{-2}	4.43×10^{-2}	7.34×10^{-3}	4.95×10^{-2}	4.18×10^{-3}	5.26×10^{-2}	2.12×10^{-3}	0.233	2.38×10^{-2}
4	3.44×10^{-2}	0.893×10^{-2}	3.95×10^{-2}	5.91×10^{-3}	4.08×10^{-2}	3.23×10^{-3}	4.00×10^{-2}	1.57×10^{-3}	0.323	1.97×10^{-2}
5	4.09×10^{-2}	0.799×10^{-2}	4.44×10^{-2}	5.28×10^{-3}	4.38×10^{-2}	2.90×10^{-3}	4.13×10^{-2}	1.41×10^{-3}	—	—
6	5.97×10^{-2}	0.779×10^{-2}	6.03×10^{-2}	5.11×10^{-3}	5.56×10^{-2}	2.77×10^{-3}	4.91×10^{-2}	1.32×10^{-3}	—	—
C'		1.01×10^{-2}		6.26×10^{-3}		3.08×10^{-3}		1.28×10^{-3}		1.05×10^{-2}

Table VII. Comparison of the Asymptotic Behavior and the Low-Order Calculations for the Function $\eta(g)$

	$d = 3, N = 0$		$d = 3, N = 1$		$d = 3, N = 2$		$d = 3, N = 3$		$d = 2, N = 0$	
K	$(-)^K\delta_K$	$\breve{\delta}_K$	$(-)^K\delta_K$	$\breve{\delta}_K$	$(-)^K\delta_K$	$\breve{\delta}_K$	$(-)^K\delta_K$	$\breve{\delta}_K$	$(-)^K\delta_K$	$\breve{\delta}_K$
2	9.25×10^{-3}	13.95×10^{-3}	10.97×10^{-3}	9.60×10^{-3}	11.85×10^{-3}	5.58×10^{-3}	12.24×10^{-3}	29.09×10^{-4}	3.40×10^{-2}	24.8×10^{-3}
3	-0.77×10^{-3}	-1.40×10^{-3}	-0.91×10^{-3}	-0.98×10^{-3}	-0.99×10^{-3}	-0.58×10^{-3}	-1.02×10^{-3}	-3.08×10^{-4}	0.20×10^{-2}	1.24×10^{-3}
4	1.59×10^{-3}	2.89×10^{-3}	1.80×10^{-3}	2.01×10^{-3}	1.84×10^{-3}	1.17×10^{-3}	1.79×10^{-3}	5.97×10^{-4}	1.14×10^{-2}	4.88×10^{-3}
5	0.66×10^{-3}	1.03×10^{-3}	0.65×10^{-3}	0.66×10^{-3}	0.59×10^{-3}	0.35×10^{-3}	0.504×10^{-3}	1.64×10^{-4}	—	—
6	1.41×10^{-3}	1.66×10^{-3}	1.39×10^{-3}	1.12×10^{-3}	1.25×10^{-3}	0.80×10^{-3}	1.090×10^{-3}	3.44×10^{-4}	—	—
C''		2.88×10^{-3}		1.80×10^{-3}		0.88×10^{-3}		3.66×10^{-4}		3.47×10^{-3}

from the zero of $\beta(V)$:

$$\beta(V^*) = 0, \qquad \beta'(V^*) > 0 \tag{99}$$

$$\eta_4(V^*) = \eta + (1/\nu) - 2 \tag{100}$$

$$\eta(V^*) = \eta \tag{101}$$

Similar calculations lead to the asymptotic formulas

$$\gamma_K \underset{K \to \infty}{\sim} (-)^K A^K \Gamma(K + B')C' \tag{102}$$

$$\delta_K \underset{K \to \infty}{\sim} (-)^K A^K \Gamma(K + B'')C'' \tag{103}$$

with

$$B' = \tfrac{1}{2}(d + N + 5) \tag{104}$$

$$B'' = \tfrac{1}{2}(d + N + 3) \tag{105}$$

$$C' = \begin{cases} C\dfrac{N+2}{N+8} 8\pi \dfrac{I_2}{I_1^2} & \text{in three dimensions} \\[2ex] C\dfrac{N+2}{N+8} 4\pi \dfrac{I_2}{I_1^2} & \text{in two dimensions} \end{cases}$$

and

$$C'' = C' \frac{2H_3}{I_1 d(4 - d)}$$

in which the I_p and H_3 have been defined in Eqs. (52) and (95) and tabulated in Table II. The results are given in Table V and compared with the perturbation series in Tables VI and VII.

ACKNOWLEDGMENTS

It is a pleasure to thank C. Itzykson, J.-C. Le Guillou, J. Zinn-Justin, and J.-B. Zuber for numerous stimulating discussions.

REFERENCES

1. G. A. Baker, B. G. Nickel, M. S. Green, and P. I. Meiron, *Phys. Rev. Lett.* **36**:1351 (1976); J. C. Le Guillou and J. Zinn-Justin, *Phys. Rev. Lett.* **39**:95 (1977); G. A. Baker, B. S. Nickel, and P. I. Meiron, Saclay Preprint 77-39.
2. C. S. Lam, *Nuovo Cimento* **55**:258 (1968); L. U. Lipatov, *Zh. Eksp. Teor. Fiz.* **72**:411 (1977); E. Brézin, J. C. Le Guillou, and J. Zinn-Justin, *Phys. Rev. D* **15**:1544 (1977).

3. K. Wilson and J. Kogut, *Phys. Rep.* **12C**:75 (1974); K. Wilson, *Rev. Mod. Phys.* **47**:773 (1975); in *Phase Transitions and Critical Phenomena*, Vol. VI, C. Domb and M. S. Green, eds. (Academic, New York, 1976); S. K. Ma, *Modern Theory of Critical Phenomena* (Benjamin, Reading, Mass., 1976).
4. E. Brézin, J. C. Le Guillou, and J. Zinn-Justin, in *Phase Transitions and Critical Phenomena*, Vol. VI, C. Domb and M. S. Green, eds. (Academic, New York, 1976); see also C. Di Castro, *Nuovo Cimento Lett.* **5**:69 (1972); K. Symanzik (unpublished), as quoted in G. Mack, *Lecture Notes in Physics*, Vol. 17 (Springer Verlag, Berlin, 1973); P. K. Mitter, *Phys. Rev. D* **7**:2927 (1973); B. Schroer, *Phys. Rev. B* **8**:4200 (1973).
5. J. Glimm and A. Jaffe, *Notes of the 1976 Cargèse Summer School.*
6. K. Symanzik, *Comm. Math. Phys.* **18**:227 (1970); J. Callan, *Phys. Rev. D* **1**:1541 (1971).
7. G. Parisi, *Notes of the 1973 Cargèse Summer School.*
8. B. G. Nickel, Critical Exponents via the ϵ expansion: A Study of Terms beyond ϵ^3, Oxford Preprint (1974), unpublished.
9. B. G. Nickel, D. I. Meiron, and G. A. Baker, Compilation of 2-pt and 4-pt Graphs, University of Guelph Report (1977); Evaluation of Series for Critical Exponents for a ϕ^4 model in 3 and 2 Dimensions, *Phys. Rev.*, to be published.
10. B. G. Nickel, *J. Math. Phys.* **19**:542 (1978).
11. F. J. Dyson, *Phys. Rev.* **85**:631 (1952); J. S. Langer, *Ann. Phys. (N. Y.)* **41**:108 (1967).
12. T. Banks, C. M. Bender, and T. T. Wu, *Phys. Rev. D* **8**:3346 (1973).
13. G. Parisi, *Phys. Lett.* **67B**:167 (1977); C. Itzykson, G. Parisi, and J. B. Zuber, *Phys. Rev. Lett.* **38**:306 (1977); J. R. Klauder, *Acta Phys. Austriaca Suppl.* **11**:341 (1973); E. M. Stein, *Singular Integrals and Differentiability Properties of Functions* (Princeton Univ. Press, Princeton, N.J., 1970); S. Coleman, V. Glaser, and A. Martin, *Comm. Math. Phys.* **58**:211 (1978).
14. E. Brézin, J. C. Le Guillou, and J. Zinn-Justin, *Phys. Rev. D* **15**:1558 (1977).
15. S. Graffi, V. Grecchi, and B. Simon, *Phys. Lett.* **32B**:631 (1970); J. D. Eckmann, J. Magnen, and R. Sénéor, *Comm. Math. Phys.* **39**:251 (1975).
16. J. Zittartz and J. S. Langer, *Phys. Rev.* **148**:741 (1966); J. L. Gervais and B. Sakita, *Phys. Rev. D* **11**:2943 (1975); V. E. Korepin, P. P. Kulish, and L. D. Faddeev, *JETP Lett.* **21**:138 (1975).
17. C. Itzykson, G. Parisi, and J. B. Zuber, *Phys. Rev. D* **16**, 996 (1977).
18. V. Glaser, H. Grosse, A. Martin, and W. Thirring, in *Studies in Mathematical Physics, Essays in Honor of V. Bargman*, E. Lieb *et al.*, eds. (Princeton Univ. Press, Princeton, N.J., 1976).

PHYSICS REPORTS (Review Section of Physics Letters) 49, No. 2 (1979) 215–219. North-Holland Publishing Company

XI. The Borel Transform and the Renormalization Group

G. PARISI*

Laboratoire de Physique Théorique de l'Ecole Normale Supérieure', Paris, France

Abstract:

We study the consequences of the renormalization group on the large momenta behaviour of the Borel transform of the Green's functions. We discuss the implications on the structure of the ultraviolet singularities of the Borel transform.

* On leave of absence from INFN, Frascati (Italy).
' Laboratoire Propre du C.N.R.S., associé à l'Ecole Normale Supérieure et à l'Université de Paris-Sud.
Postal address: 24, rue Lhomond, 75231 Paris Cedex 05, France.

It is quite difficult to study directly the solutions of the Dyson-Schwinger (D–S) equations of motion in a renormalizable field theory. Most of our non perturbative information come from the use of the renormalization group equations.

In this note we apply the renormalization group equations to the study of the large momenta behaviour of the Borel transform of the Green's functions, and we check that the results we find are consistent with the Borel transformed D–S equations. As a consequence the Borel transform is ultraviolet singular on the positive real axis in non asymptotically free field theories. The Borel summability of the Green's functions is no more valid.

Let us consider the $g(\phi^2)^2$ theory in dimension $D = 4$, where ϕ is an N component scalar field. We define the Borel transform $B(b)$ of the function $A(g)$ by:

$$A(g) = \sum_{K=0}^{\infty} A_K g^{K+1}, \qquad B(b) = \sum_{K=0}^{\infty} (A_K/K!)b^K. \tag{1}$$

If A_K grow like $K!$ $B(b)$ is analytic in the neighbourhood of the origin. If the function $A(g)$ has "good" analyticity properties, the following relation holds [1]

$$A(g) = \int_0^{\infty} \exp(-b/g)B(b)\,db. \tag{2}$$

Vice versa, if the function $B(b)$ can be analytically continued on the positive real axis up to infinity, the function defined by eq. (2) (provided that the integral converges) is the Borel sum of the asymptotic expansion, eq. (1).

The renormalization group equations for the Green's functions are

$$\left[-p\frac{\partial}{\partial p} + \beta(g)\frac{\partial}{\partial g} + d_N + \delta_N(g)\right]\Gamma_N(p,g) = 0. \tag{3}$$

Let us work in the approximation in which

$$\beta(g) = \beta_2 g^2, \qquad \delta_N(g) = 0. \tag{4}$$

The renormalization group equations for the Borel transform $C_N(p,b)$ of $\Gamma_N(p,g)$ are:

$$\left[-p\frac{\partial}{\partial p} + \beta_2 b + d_N\right]C_N(p,b) = 0. \tag{5}$$

The solution is simply given by:

$$C_N(p,b) = r_N(b)p^{d_N}(p/\mu)^{\beta_2 b}. \tag{6}$$

μ being a normalization point.

The deviation (proportional to $(p/\mu)^{\beta_2 b}$) of the large momenta behaviour of the Green's function from the canonical value will play a crucial role in the rest of the paper. It is easy to check that, if we remove the approximation, eq. (4), we find, in the large p limit, the same result as eq. (6), apart from logarithmic corrections.

Indeed the Borel transform of eq. (3) is

$$\left[-p\frac{\partial}{\partial p} + \beta_2 b + d_N\right]C_N(p,b) + \int_0^b [\tilde{\beta}(b-b')b' + \tilde{\delta}_N(b')]\,C_N(p,b')\,db' = 0. \tag{7}$$

where

$$
\begin{aligned}
[\beta(g) - \beta_2 g^2]/g^2 &= -g\beta_3 + O(g^2) = \int_0^\infty \exp(-b/g)\tilde{\beta}(b)\,db, \\
\delta_N(g) &= -\delta_N^1 g + O(g^2) = \int_0^\infty \exp(-b/g)\tilde{\delta}_N(b)\,db.
\end{aligned}
\tag{8}
$$

In the large momenta region, eqs. (7) and (8) imply that

$$
C_N(p, b) \underset{p\to\infty}{\to} \tilde{r}_N(b)p^{d_N}(p/\mu)^{\beta_2 b}(\ln p/\mu)^{-(\beta_3 b + \delta_N^1)/\beta_2}
\tag{9}
$$

The large momenta behaviour of $C_N(p, b)$ given by eq. (9) is in agreement with the arguments of refs. [2-3].

Eq. (9) does not imply the presence of a Landau ghost for positive real values of g, e.g. the function $\beta(g)$ has a zero at $g = \beta_3/\beta_2$ if $\tilde{\beta}(b) = \beta_3$; in this case the large p limit can be studied using the standard fixed point renormalization group analysis [4].

It is crucial to realize that the behaviour of $\Gamma_N(p, g)$ for large p depends on the analytic properties of $C_N(p, b)$ as function of both variables and not on the behaviour of $C_N(p, b)$ in the limit $p \to \infty$, considered as function of p only. Also, eq. (6) is compatible with the absence of a Landau ghost: e.g., if $r_N(b) = \exp(-b^2)$, we find:

$$
\Gamma_N(p, g) = p^{-d_N} \int_0^\infty \exp[-b^2 - b/g + \beta_2 b \ln(p/\mu)]\,db.
\tag{10}
$$

In this case the perturbative expansion is asymptotic to the $\Gamma_N(p, g)$ function (as given by eq. (10)) only in the sector $\mathrm{Re}\, g \geqslant 0$. The theory is not asymptotically free for negative g: in this case we cannot apply the arguments of refs. [2, 5] on the existence of a Landau ghost in a theory which is asymptotically free only for negative g.

The large momenta behaviour of the Green's functions recalls non-renormalizable theories. In order to study the effects on the D-S equations, it may be convenient to write directly the Borel transformed D-S equations, using the convolution theorem for the Borel transform:

$$
\begin{aligned}
B[F_1 \ldots F_N] &= \int_0^b db_1 \ldots \int_0^b db_N f_1(b_1)\ldots f_N(b_N)\delta(b_1 + \ldots + b_N - b), \\
B[F_i] &= f_i(b),
\end{aligned}
\tag{11}
$$

where B stands for "Borel transform of".

Using dimensional analysis it can be checked that the δ function in eq. (11) assures the self-consistent reproduction of the factor $(p/\mu)^{\beta_2 b}$ in the D–S equations. However, for large $\beta_2 b$ ($\beta_2 b \geqslant 2$), we find ultraviolet divergences, which, according to the B.H.P. theorem [6] are proportional to the insertion of local operators. More precisely, when $\beta_2 b = 2$, there are logarithmically divergent diagrams* whose divergence may be compensated by adding the operators of dimension 6 as

* The existence of ultraviolet singularities for integer $\beta_2 b$ has been pointed out in ref. [7].

counterterms in the Lagrangian, the coefficients of the counterterms being proportional to $\exp(-1/\beta_2 g)$. From this point of view, non asymptotically free field theories differ from non renormalizable field theories in that, in the first case, the coefficients of the counterterms are exponentially small, while in the second case they are proportional to a power of the coupling constant. Using the renormalization group, a direct analysis [8] shows that the singularities at $b\beta_2 = 2$ are cuts, e.g., if for simplicity we assume that ϕ^6 is the only operator of dimension 6, (i.e. we neglect operator mixing), we find:

$$C_6(p,b) \approx \Gamma(1-\alpha)(2-\beta_2 b)^{\alpha-1}, \qquad \beta_2 b \approx 2, \qquad \alpha = 2\beta_3/\beta_2^2 + \gamma_6^1/\beta_2, \tag{12}$$

where $C_6(p,b)$ is the Borel transform of the 6-point vertex function $\Gamma_6(p,g)$ and the anomalous dimension of the operator ϕ^6 is $\gamma_6(g) = \gamma_6^1 g + O(g^2)$.

The contribution of this ultraviolet singularity to the large order behaviour of the perturbative expansion is given by

$$\Gamma_6(p,g) \approx \sum_K K!\, K^{-1+\alpha} g^K (\tfrac{1}{2}\beta_2)^K. \tag{13}$$

Let us compare these results with the semi-classical arguments [9–12]. The instantons induce a singularity of the Borel transform at $b = b_I = -16\pi^2$. The ultraviolet singularities of the Borel transform are at $b = b_u = 2/\beta_2 = -6b_I/(N+8)$, the theory being $O(N)$ invariant.

For $N > -2$, $|b_u| < |b_I|$ and the ultraviolet singularities dominate the large order behaviour of the perturbative expansion; for $N < -8$ the ultraviolet singularities are on the negative axis; for $N = -14$, they are on the top of the instantons; while for $N < -14$, the instanton is on the cut produced by the ultraviolet singularities. In this region ($N < -14$) the instanton contribution computed in ref. [11] is ultraviolet divergent: the ultraviolet singularities are the nearest to the origin and they shield the instanton singularities. In this case it would be nonsense to try to consider the effects of the instantons, while neglecting the contributions of the ultraviolet singularities.

Let us verify the correctness of eq. (13) in the large N limit. For convenience, we rescale the coupling constant g in such a way that $\beta_2 = 1$. In this case the instanton contribution is negligeable. It is convenient to indicate by a wavy line the geometrical sum of all the bubble diagrams; in fig. 1, we show the relevant diagrams contributing to $\Gamma_6(p,g)$ in the large N limit. In the massive theory at $p = 0$, we find:

$$\Gamma_6(0,g) = \int \frac{d^4 q}{(q^2+m^2)^3} \frac{(1+gF(q))^3}{(1+g\Pi(q))^3}. \tag{14}$$

A simple computation shows that:

$$\Pi(p) \to -\ln p + R_1, \qquad F(p) \to -\ln p + R_2, \qquad R_1 - R_2 \neq 0. \tag{15}$$

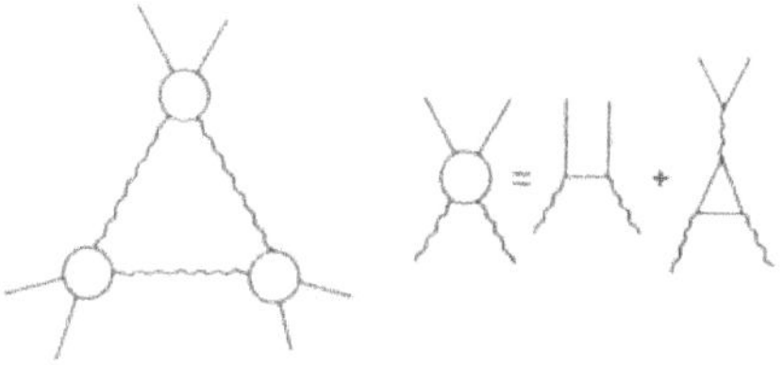

Fig. 1. Relevant diagram for the large order behaviour in the limit $N \to \infty$.

R_1 and R_2 depend separately on the way the coupling constant is defined, but $R_1 - R_2$ is independent from the renormalization scheme.

Expanding (14) in powers of g, we find:

$$\Gamma_6(0, g) \approx \sum_K K!\, K^{-4} 2^{-K} g^K (R_1 - R_2)^3. \tag{16}$$

In the limit $N \to \infty$, $\beta_3/\beta_2^2 \to 0$ and $\gamma_6^1/\beta_2 \to 3$; eq. (16) is in perfect agreement with eq. (13). In order to get the correct result it is crucial to consider all the diagrams relevant at this order, the contribution of a single diagram being proportional to eq. (13) with α set equal to zero.

The techniques here described can be extended also to the study of the infrared divergences [3]. However, at the present moment, we are far from understanding how to use the perturbative expansion to compute the Green's functions for the coupling constant not too small in the cases when the Borel summability is no more valid and the Borel transform is singular on the real axis.

The author is grateful to Prof. K. Symanzik for his collaboration in the study of the limit $N \to \infty$. He also acknowledges many illuminating discussions with E. Brézin, C. Itzykson, J. Zinn-Justin and J.B. Zuber.

References

[1] S. Graffi, V. Grecchi and B. Simon, Phys. Lett. 32B (1970) 631.
[2] G. Parisi, Lectures given at the Cargèse Summer Institute (1977).
[3] G. Parisi, On infrared singularities, Nucl. Phys. B, to be published.
[4] K. Symanzik, Comm. Math. Phys. 23 (1971) 61;
K. Wilson, Phys. Rev. D3 (1971) 1818.
[5] N.N. Khuri, Phys. Rev. D12 (1975) 2298.
[6] N.N. Bogoliubov, O.S. Parasiuk, Acta Math. 97 (1957) 227;
K. Hepp, Comm. Math. Phys. 2 (1966) 301.
[7] G. 't Hooft, Lectures given at Erice (1977);
B. Lautrup, Phys. Lett. 69B (1977) 109;
P. Olesen, Phys. Lett. 73B (1977) 327.
[8] G. Parisi, Singularities of the Borel transform in renormalizable theories, Phys. Lett. 76B (1978) 65.
[9] C.S. Lam, Nuovo Cimento 55 (1968) 258.
[10] L.N. Lipatov, JETP 72 (1977) 411.
[11] E. Brézin, J.C. Le Guillou and J. Zinn-Justin, Phys. Rev. D15 (1977) 1544, 1558.
[12] C. Itzykson, G. Parisi and J.B. Zuber, Phys. Rev. Lett. 38 (1977) 306.

SINGULARITIES OF THE BOREL TRANSFORM IN GAUGE THEORIES

G. PARISI[†]

Laboratoire de Physique Théorique de l'Ecole Normale Supérieure[‡]

ABSTRACT

The singularities of the Borel transform in gauge theories are studied using a few modes approximation. The asymptotic freedom of the theory implies the presence of infrared singularities on the positive real axis; these singularities cannot be computed using semi-classical methods.

In the past the perturbative expansion has been the main tool used to study quantum field theory. Only recently the interest shifted towards nonpertubative phenomena [1,2]. In order to study nonpertubative effects it is convenient to introduce the Borel transform with respect to the coupling constant α:

$$\begin{aligned} A(\alpha) &= \sum_{k=0}^{N} A_k \alpha^k + O(\alpha^{N+1}) \ \forall N , \\ B_p(b) &= \sum_{k=0}^{\infty} A_k/k!\, b^k . \end{aligned} \tag{1}$$

If the A_k's grow like $k!$ $B_p(b)$ is an analytic function in a neighbourhood of the origin.

Let us suppose that the function $A(\alpha)$ has the representation

$$A(\alpha) = 1/\alpha \int_0^{\infty} db B(b) \exp(-b/\alpha) . \tag{2}$$

Intuitively $B(b)$ is the number of field configurations having an action near to b, in the units in which α has been set to 1.

For small b we have:

$$B_p(b) = B(b) . \tag{3}$$

If Eq. (3) holds for all positive b, $B(b)$ being an analytic function in a neighbourhood of the positive real axis, the function $A(\alpha)$ can be computed starting from

[†]On leave of absence from INFN, Frascati (Italy)

[‡]Laboratoire Propre du C.N.R.S., associé à l'Ecole Normale Supérieure e à l'Université de Paris Sud,

Postal address: 24, rue Lhomond, 75231 PARIS CEDEX 05

its perturbative expansion, as it happens in the ϕ^4 theory in less than 4 dimensions [3].

If nonpertubative phenomena are present, Eq. (3) does not hold: in the presence of instantons with topological quantum numbers,

$$B(b) - B_p(b) = \vartheta(b - A_I)(b - A_I)^\gamma ,$$

(A_I being the action of the instanton for $\alpha = 1$), the function $B_p(b)$ is regular at $b = A_I$ and the instantons produce a singularity in $B_p(b)$ only at $b = 2A_I$ [4]. It is also possible that the function $B_p(b)$ has a cut on the real positive axis, starting from the point b_c, and Eq. (3) does not hold only for $b > b_c$ where the analytic continuation of the function $B_p(b)$ is no more real.

The best suited technique to investigate the singularities of the Borel transform consists in approximating the full functional integral with an integral over a finite dimensional manifold of classical field configurations, using a Gaussian approximation for the integration in the directions transverse to the manifold [5]. It is convenient to consider manifolds which have the same symmetry of the Lagrangian (e.g. translational invariance).

In a massive superrenormalizable theory in this approximation the Borel transform has singularities whose positions depend on the choice of the manifold; the semiclassical results of Lipatov can be obtained by integrating near those classical configurations which produce the nearest singularity to the origin.

If the Lagrangian is renormalizable, the classical theory is approximately scale invariant at least in the short distance region and we must use manifolds which are also dilatation invariant.

Let us see what happens in the case of pure gauge theories (the inclusion of massive fermions will not change our analysis). Let us normalize the coupling constant in such a way that:

$$\begin{aligned} B(\alpha) &= -\alpha^2 , \\ [\alpha(p)]^{-1} &= [\alpha(\mu)]^{-1} - \ln(p/\mu) + O(1) , \end{aligned} \tag{4}$$

where $B(\alpha)$ is the function appearing in the renormalization group equations and $\alpha(p)$ is the effective momentum dependent running coupling constant ($\alpha(\mu) \equiv \alpha$, being the subtraction point).

The contribution to vacuum energy of a manifold of field configurations having classical action S, radius ρ and centered at x, is approximatively given by:

$$\begin{aligned} &\int d^4x \int \frac{d\rho}{\rho^5} f(S) \exp[-S/\alpha(\rho^{-1}/\mu)] \\ &= \int d^4x \int \frac{d\rho}{\rho^5} f(S)(\rho\mu)^S \exp(-S/\alpha) , \end{aligned} \tag{5}$$

where the factor $f(S)(\rho\mu)^S$ [6] comes from the Gaussian integration. The ρ integral is infrared divergent for $\rho \geq 4$. A similar effect can be found by coupling the gauge fields to a massive fermionic field. Let us compute the contribution of the same gauge field configuration to the current-current correlation function; one finds:

$$\langle J(y)J(z)\rangle \underset{y\to z}{\rightarrow} \int \frac{d\rho}{\rho^{5-S}} \exp(-S/\alpha) f'(S)(y-z)^{-2} h(\rho/|y-z|) , \qquad h(\infty) = 1 . \tag{6}$$

In the derivation of Eq. (6) we have used the following expression for the current-current correlation function in the presence of an external field:

$$\langle J(x)J(0)\rangle \underset{x\to 0}{\rightarrow} \frac{1}{x^6} + \frac{1}{x^2}F^2(0) + \dots , \qquad F^2(0) = \mathrm{Tr}\,[F_{\mu\nu}F_{\mu\nu}] . \tag{7}$$

Also in this case the ρ integration is divergent for $S > 4$. It is easy to see that if we integrate over a smooth manifold of fields having actions ranging from zero to infinity, the corresponding Borel transform will be singular at $b = 4$. The presence of branch cut singularities of the Borel transform arising from the renormalization process have been found also using more conventional techniques based on the Dyson-Schwinger equations. The approach presented in this note throws new light on the previously found results [7].

The integration over a manifold of field configuration having an action greater than 4 is not defined in this approximation; the perturbative expansion has an intrisic ambiguity of order $\exp(-4/\alpha)$ and it does not make sense to compute the contribution of field configurations having an action greater than 4 if we do not understand how to compute the function $B(b)$ for $b > 4$ (it is not clear why the proposal $B(b) = \mathrm{Re}\, P_b(b)$ should work). Notice that $\exp(-4/\alpha)$ corresponds to a correction proportional to μ^4/p^4 [8].

In these units the action of the instanton in a SU(N) theory is 11/3 N (11 for $N = 3$) and its contribution is infrared divergent; indeed for $N \geq 2$, the singularity of the Borel transform produced by the instanton sits on the cut produced by the divergence of the integral over the dilatations; in order to compute the effects of the instanton we must disentangle the two contributions.

At this point the reader may suspect that the singularities of the Borel transform at $b = 4$ are artefacts of our approximation: we cannot use Eqs. (5-6) in the region where $\alpha(\rho^{-1}/\mu) > 1$: the corrections to the Gaussian approximation cannot be neglected and excluded volume effects become important (we cannot put more than one fluctuation at the same place). However a simple computation shows that cutting off the integration over sizes

$$\rho > \rho_c \equiv \exp(4/\alpha).\mu$$

does not modify the singularities of the Borel transform at $b = 4$.

The final picture is quite similar to that arising from [9] a dense gas of istantons-antiinstantons pairs: at the semiclassical the vacuum is filled by fluctuations of the fields — however, the action of each fluctuation may be much smaller than the action of the instanton. This difference is very important in the limit $N \to \infty$ where the contribution of the instantons becomes negligible.

Is this picture able to explain quark confinement?

The question is not simple, but if we try to be speculative, we can suggest a proof going though the following steps:

a) There is no order parameter and there is no long range correlation among the local excitations

b) The correlation functions of the gauge field go exponentially to zero, i.e. a mass gap is generated (A mechanism which may give a mass to the gluon is the reflection of the gluon field on the domains walls which separate two different fluctuations)

c) The fluctuations are strong enough to make

$$\left\langle \exp\left[\int_{x_1}^{x_2} A_\mu(x) dx_\mu\right]\right\rangle = 0, \quad |x_1 - x_2| > \rho \, ,$$

when we mediate over all the types of local fluctuations taken with their weight.

d) From points a, b and c, it follows that

$$\left\langle \rho\left\{\exp\left[-\int_C A_\mu(x) dx_\mu\right]\right\}\right\rangle \simeq \exp[-LT] \, , \tag{8}$$

the contour C being a rectangle of sizes L and T; indeed the integral in Eq. (8) would be equal to zero neglecting the correlations between the gauge field A, which are exponentially small.

It is possible that points a and b may be proved in the same way as in the two dimensional sigma model, while the proof of point c is much more delicate.

The reader will notice that we have gathered all the main ingredients of the droplet model for quark confinement of Glimm and Jaffe [10], without using either istantons or merons. I think that the speculative arguments on quark confinement presented in this note are quite promising as far as they can be applied also in the limit $N \to \infty$ or in dimensions less than 4.

References

1. A. A. Belavin, A. M. Poliakov, A. S. Schwartz, Yu S. Tyupkin *Phys. Lett.* **59B**, 85 (1977).
2. L. N. Lipatov, *JEPT* **72**, 441 (1977).
3. J. D. Eckmann, J. Magnen, R. Seneor, *Comm. Math. Phys.* **39**, 251 (1975).
4. E. Brezin, G. Parisi, J. Zinn-Justin, *Phys. Rev.* **D16**, 408 (1977); E. B. Bogomolny, V. A. Fateyev, *Phys. Lett.* **71B**, 93 (1977).
5. J. L. Gervais, A. Neveu, M. A. Virasoro, *LPTENS* 77/17 (1977)

6. G. 't Hooft *Phys. Rev.* **D14**, 3432 (1976); A. A. Belavin, A. M. Poliakov, *Nucl. Phys.* **B123**, 429 (1977).
7. G. 't Hooft, Lectures given at Erice (1977); B. Lautrup, *Phys. Lett.* **69B**, 109 (1977); P. Olesen, *Phys. Lett.* **73B**, 327 (1977); G. Parisi, Lectures given at the Cargese Summer Institute (1977); "Singularities of the Borel Transform in Renormalizable Theories" *Phys. Lett.* to be published; "On Infrared Singularities" *Nucl. Phys.* to be published; "The Borel Transform and the Renormalization Group" *Phys.* Reports to be published.
8. L. Baulieu, J. Ellis, M. K. Gaillard, W. J. Zakrzewski, CERN preprint TH. 2482 (1978).
9. C. Callan, R. Dashen, D. Gross, *Phys. Lett.* **66B**, 343 (1977).
10. J. Glimm, A. Jaffe, Rockefeller University Preprint HUTMP-77/B54 (1978).

Nuclear Physics B150 (1979) 163–172

ON INFRARED DIVERGENCES

G. PARISI *
Laboratoire de Physique Théorique de l'Ecole Normale Supérieure, Paris, France **

Received 10 April 1978
(Revised 25 September 1978)

We study the structure of infrared divergences in superrenormalizable interactions. We conjecture that there is an extension of the Bogoliubov-Parasiuk-Hepp theorem which copes also with infrared divergences. We discuss the consequences of this conjecture on the singularities of the Borel transform in a massless asymptotic free field theory. The application of these ideas to gauge theories is briefly discussed.

1. Introduction

The celebrated Bogoliubov-Parasiuk-Hepp [1] theorem states that in a polynomial field theory all ultraviolet divergences may be eliminated by subtracting, from the appropriate vertex functions, a polynomial in the external momenta: this procedure is equivalent to the introduction of local operators as counterterms in the Lagrangian. This theorem is also valid for non-renormalizable interactions where the number of needed counterterms is not bounded.

The aim of this paper is to discuss the possibility of constructing a similar theorem concerning infrared divergences in off-shell Green functions in the Euclidean region at no exceptional momenta. In sect. 2 of this paper, we conjecture that infrared divergences may be reabsorbed by adding new operators as counterterms in the Lagrangian: these new operators U_i may be multilocal, i.e.,

$$U_i = \int d^D x_1 \ldots d^D x_N \, O_1(x_1) \ldots O_N(x_N) P(x_K - x_j) \,, \tag{1}$$

where $O_K(x)$ are local operators and P is a polynomial in the x's variables.

After the introduction of the multilocal operators U_i as counterterms infrared divergences are absent; on the other hand the Lagrangian is no longer local and the Green functions do not satisfy fundamental physical requirements (e.g., the cluster

* On Leave of absence from INFN, Frascati, Italy.
** Laboratoire Propre du CNRS, associé à l'Ecole Normale Supérieure et à l'Université de Paris-Sud. Postall address: 24 rue Lhomond, 75231 Paris Cedex 05 (France).

decomposition). Therefore, the reader may think that the subject of this paper has only a purely academic interest.

However, in sect. 3, we show that there is a deep connection between the singularities of the Borel transform on the positive real axis in a massless asymptotically free field theory and the operators U_i: the positions of the singularities are related to the canonical dimensions of the operators U_i. In sect. 4, we apply these ideas to gauge theories and we find the structure of the singularities of the Borel transform.

In appendix A, we study some simple diagrams whose infrared divergences are proportional to the insertion of multilocal operators. In appendix B, we give an alternative definition of the multilocal operators using a large-distance expansion. Multilocal operators appear to be the "dual" of local operators.

2. The conjecture

If the interaction is superrenormalizable, naïve power counting arguments imply that the perturbative expansion in a massless field theory is infrared divergent [2]; however finite results may be obtained by using the dimensional or the analytic renormalization. At any finite order of the perturbative expansion, the Green functions are meromorphic functions of the dimension (D) of space-time, the only possible singularities being isolated poles for real rational dimensions. If D is irrational, an infrared divergence-free perturbative expansion can be constructed, using a dimensional regularization. We may wonder which is the structure of counterterms which corresponds to this procedure.

It is well-known that infrared divergences arise from the "bad" behaviour of the Green functions at infinity. For example there are diagrams which give a finite contribution in momentum space, but their Fourier transform is divergent because the singularities are too strong at zero momenta, e.g., the Fourier transform of $K^{-\alpha}$ is finite only if $\alpha < D$. We propose to call this category of diagrams "superficially infrared divergent"; indeed they can be made finite by adding to them a polynomial in x (i.e., delta functions in momentum space). This happens automatically if we define the diagrams by analytic continuation in the dimensions.

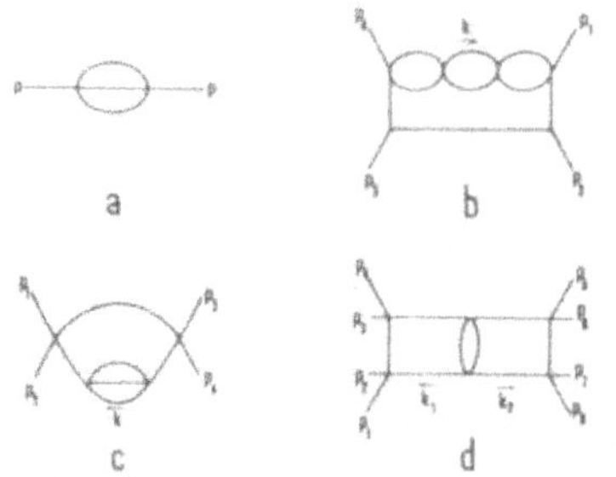

Fig. 1. Some infrared divergent diagrams.

We conjecture that, if all internal infrared subdivergences of a diagram are regularized by subtracting the divergent part, the diagram may be only superficially infrared divergent, and therefore it can be renormalized in the same way as the other diagrams; subtracting from the diagram a polynomial in the x variable is equivalent to adding to the Lagrangian a multilocal counterterm U_i defined in eq. (1).

Let us consider in more detail the massless $g\phi^4$ interaction in dimensions $D = 4 - \epsilon$. The theory is super-renormalizable, and it is infrared divergent, in perturbation theory. As an example, simple Feynman diagrams are shown in fig. 1. In table 1, we list some of the dimensions at which they are singular and the corresponding counterterms. The coefficient of the counterterm U_i is $g^{K_i(\epsilon)}$. From dimensional analysis

$$K_i(\epsilon) = -\lambda_i/\epsilon \equiv [U_i]/[g] . \qquad (2)$$

λ_i is the canonical dimensions of U_i. In the last column of table 1, we find the explicit expression for $K_i(\epsilon)$. In appendix A we show in detail that the infrared singularities of the diagrams of fig. 1 are proportional to the insertion of multilocal counterterms.

As can readily be seen, all operators have the form (1). Moreover the only local operators present are ϕ^3 and ϕ^2. ϕ^4 and ϕ are absent. The absence of ϕ^4 can be understood diagrammatically. On the other hand ϕ is diagrammatically an admissible counterterm; however its zero-momentum intersection is divergent ($1/K^2$ is infinite at $K = 0$). The infrared singularities are proportional to $\Box\phi$, which is equal to $g\phi^3$, from the equations of motion.

We stress that, for an arbitrary choice of the infrared counterterms, the Green functions do not satisfy the cluster decomposition and the resulting theory is not physically acceptable. In this case, as well as in non-renormalizable theories, we can

Table 1

	D	U_i	$K_i(\epsilon)$
a	3	$\int d^Dx\, \phi^2(x)$	$2/\epsilon$
b	3	$\int d^Dx\, d^Dy\, \phi^2(x)\, \phi^2(y)$	$4/\epsilon$
b	2.5	$\int d^Dx\, d^Dy\, \phi^2(x)\, \phi^2(y)(x-y)^2$	$6/\epsilon$
c	$\frac{10}{3}$	$\int d^Dx\, d^Dy\, \phi^3(x)\, \phi^3(y)$	$2/\epsilon + 1$
d	$\frac{8}{3}$	$\int d^Dx\, d^Dy\, d^Dz\, \phi^3(x)\, \phi^3(y)\, \phi^2(z)$	$4/\epsilon + 1$

In the first column, we indicate the diagram of fig. 1 to which we refer. In the second column there is one of the dimensions at which the diagram is infrared divergent. In the third column, we find the corresponding multilocal operator, $K_i(\epsilon)$, which is indicated in the last column, is the power of the coupling constant which multiplies the counterterm in the Lagrangian.

ask the question whether there exists an appropriate choice (maybe unique) of the counterterms such that all the pathology of the Green functions disappears. In some cases (e.g., ϕ^4 interaction with $2 < D < 4$) the answer to this question is positive, as can be explicitly seen using the $1/N$ expansion. However, the aim of this paper is to characterize the infrared divergences which appear in the off-shell Green functions in perturbation theory; the problem of constructing an infrared (ultraviolet) finite theory starting from an infrared (ultraviolet) divergent perturbative expansion is quite interesting, but it goes beyond the framework of perturbation theory and it will not be discussed here.

3. Infrared Borel singularities

In four dimensions, massless renormalizable interactions are free of infrared divergences in perturbation theory. This may not be true outside perturbation theory. More precisely, it is possible that the Borel transform of the Green functions is infrared divergent on the real axis; if the theory is asymptotically free, these infrared singularities are on the positive real axis [3]: the Borel summability of the perturbative expansion is no longer valid *.

We will show the existence of a strict connection between the infrared Borel singularities in renormalizable theories, and the singularities present in the perturbative expansion of a superrenormalizable massless field theory.

Let us consider the $g\phi^4$ theory in $D = 4$. We denote by $\bar{g}(g, p)$ the effective running coupling constant at the momentum p as a function of the coupling constant g at the subtraction point μ. In the leading logarithmic approximation, we find that

$$\bar{g}(g, p) = \sum_{K=1}^{\infty} g^{K+1} \beta_2^K \ln^K \frac{p}{\mu} . \tag{3}$$

The Borel transform of $\bar{g}$ is:

$$C(b, p) = \sum_{K=0}^{\infty} \frac{(\beta_2 b)^K}{K!} \ln^K \frac{p}{\mu} = \left(\frac{p}{\mu}\right)^{b\beta_2} . \tag{4}$$

Note that

$$\bar{g}(g, p) = \int_0^{\infty} \mathrm{d}b \exp(-b/g) \cdot C(b, p) . \tag{5}$$

The behaviour of $C(b, p)$ in the leading logarithmic approximation recalls a non-renormalizable theory for $\beta_2 b > 0$ and a superrenormalizable theory for $\beta_2 b < 0$.

* We disregard the singularities of the Borel transform due to the instantons.

Arguments can be given [4] to show that eq. (3) is also valid outside the leading logarithmic approximation. The main tool is the convolution theorem for Laplace-Borel transforms *

$$\widetilde{f \cdot g} = \tilde{f} * \tilde{g} , \tag{6}$$

where the tilde indicates Borel transform, the star indicates convolution:

$$\tilde{f} * \tilde{g}(b) = \int_0^b db_1 \, db_2 \, \delta(b - b_1 - b_2) \tilde{f}(b_1) \tilde{g}(b_2) . \tag{7}$$

Starting from eq. (7), we can transform the standard Dyson-Schwinger integral equations in equivalent equations for the Borel transform. It is easy to check that the ansatz (4) reproduces itself self-consistently as an effect of the δ function in the convolution. For small $|\beta_2 b|$ no divergences are present; however if we increase $|\beta_2 b|$ new divergences appear. Ultraviolet divergences, for positive $\beta_2 b$ [3–7] may be classified in terms of local operators, thus leading to the results of ref. [6]; similarly, if we denote by λ_i the canonical dimensions of the multilocal operators which are able to compensate infrared divergences, ($\lambda_i = -2n$, n positive integer), the Borel transform has infrared singularities at the points

$$b = \lambda_i / \beta_2 . \tag{8}$$

Explicit verifications of this phenomenon can be found by considering the $1/N$ expansion for the $O(N)$ symmetric theory [5–8].

There is also an alternative method to find the positions of the singularities (both infrared and ultraviolet) of the Borel transform (i.e., to find the dimensions of the relevant multilocal or local operators). It is based on the observation that the contribution of the instantons to the singularities of the Borel transform becomes divergent when the instanton singularity crosses with the infrared or ultraviolet singularities [8]. The method is quite efficient when we can shift the position of the singularities of the Borel transform by changing the symmetry group under which the theory is invariant.

4. Gauge theories

As we have seen in the previous sections, the conjectured structure of infrared singularities in superrenormalizable field theories has quite interesting consequences on the analyticity properties of the Borel transform in renormalizable field theories.

The main body of this paper has been dedicated to the ϕ^4 theory; however the ideas here presented may also be applied to other interactions, in particular also to gauge theories.

* The Borel transform is the Laplace transform with respect to the inverse of the coupling constant.

Let us consider a pure Yang-Mills field theory in four dimensions. We denote by g, A_μ and F^2, the coupling constant, the gauge field and the Lagrangian density, respectively. We want to find the singularities of the Borel transform with respect to $\alpha \equiv g^2$. Ultraviolet singularities are on the negative axis at $b = 2/\beta_2$ (β_2 is negative, the theory being asymptotically free). In order to find the position of the infrared singularities, as in the ϕ^4 theory, we must study the structure of the admissible multilocal counterterms which are needed to compensate infrared divergences in less than four dimensions in the gauge invariant sectors.

For the counterterms the following prototypes are possible *:

(a) $\int d^D x\, d^D y\, F^2(x) F^2(y)(x-y)^2$,

(b) $\int d^D x\, d^D y\, F(x) F(y)$,

(c) $\int d^D x\, d^D y\, F(x) P[\exp i \int_x^y dz_\mu\, A_\mu(z)]\, F(y)$,

(d) no admissible operator exists as far as there are no infrared divergences.

It is not evident which of the four possibilities is correct; it is clear that each of them has serious drawbacks:

(a) is gauge invariant, but it is hard to understand diagrammatically. Notice that in the $g\phi^4$ theory, the ϕ^4 operator is not present in the multilocal counterterms,

(b) is not gauge invariant,

(c) is gauge invariant; however the choice of the path going from x to y in the exponent seems to be arbitrary.

(d) is rather unexpected, and it would require a rather large amount of cancellations.

The puzzle may be solved if we exclude possibilities (a) and (d) and if we realize that (b) and (c) differ one from the other only for higher-order corrections in g; we conclude that (b) is an admissible operator of dimension 4 for $D = 4$ at zero order in g; how higher orders in the coupling constant conspire to maintain gauge invariance (i.e., which path is chosen in (c)), is a dynamical problem which we are unable to solve. Using the duality between local and multilocal operators which is studied in appendix B, from the absence of local gauge-invariant operators of dimension less than 4, we argue that the highest dimension of a gauge invariant admissible multilocal counterterm is -4. Therefore there are indications that the operator (b) (one of its gauge invariant extensions) is dual to F^2 and it is the relevant multilocal operator of maximal dimension. As a consequence of this statement, the infrared singularities of the Borel transform are on the positive real axis at $b = -4/\beta_2$.

Let us apply these results to the SU(N) case. We normalize the coupling constant

* We denote by P the path ordering, the sum over the space, time and colour indices has not been indicated.

in such a way that $\beta_2 = -1$, i.e.,

$$\alpha(p) = 1/\ln(p/\mu) + O(\ln \ln(p/\mu)/\ln^2(p/\mu)) .$$

With this normalization the ultraviolet, infrared and instanton singularities start from $b = -2, 4, \frac{11}{3}N$, respectively. For $N > \frac{12}{11}$, the singularities due to instantons are shielded by the infrared singularities and for many processes the effects of the instantons are uncomputable (they are formally written as infrared divergent integrals).

Also in the limit $N \to \infty$ the perturbative expansion is not Borel summable. At the present stage of ignorance the best we can do to extract information from the perturbative expansion for the function $A(\alpha)$ is to perform the split

$$A(\alpha) = A_{\mathrm{P}}(\alpha) + A_{\mathrm{NP}}(\alpha), \qquad A_{\mathrm{NP}}(\alpha) = O(\exp(-4/\alpha)),$$

$$A_{\mathrm{P}}(\alpha) = \int_0^4 \exp(-b/\alpha)\, B(b)\, \mathrm{d}b = \sum_{K=1}^{\infty} \alpha^K A_K$$

$$\times \left[1 - \exp(-4/\alpha) \sum_{j=0}^{K-1} 1/j!\,(4/\alpha)^j\right] , \tag{9}$$

where $B(b)$ is the Borel transform of $A(\alpha)$.

$A_{\mathrm{P}}(\alpha)$ is computable from its perturbative expansion, (the series in the r.h.s. of the equation would be convergent in absence of the singularities of the Borel transform at -2; it may be converted to a convergent expansion by doing a conformal mapping in the integral) and our ignorance of the large-distance behaviour of the theory is concentrated in $A_{\mathrm{NP}}(\alpha)$ which is of order μ^4/p^4, (apart from possible terms proportional to $\ln(p/\mu)$). This separation in perturbative and non-perturbative pieces is preliminary to any approximate computation of non-perturbative effects and is essential to avoid double counting.

The author is grateful to L. Baulieu, E. Brezin, H. Epstein, J. Iliopoulos, L. Maiani, J. Zinn-Justin and J.B. Zuber for illuminating discussions.

Appendix A

In this appendix we study the infrared divergences of the diagrams of fig. 1 and we show that they are proportional to the insertion of the operators in table 1.

The self-energy diagram (a) is equal to

$$\Sigma_R(p) = \Sigma(p) - \Sigma(0), \qquad \Sigma(p) \propto p^{2-2\epsilon} . \tag{A.1}$$

The subtraction $\Sigma(0)$ is needed to maintain the theory massless. If $\epsilon \geqslant 1$, $\Sigma(0)$

is infrared divergent; the corresponding operator is the mass insertion $\int \phi^2(x)\, d^D x$. This case is quite standard and it is fully discussed in ref. [2].

Diagram (b) is proportional to

$$\int \frac{d^{4-\epsilon}K}{K^{3\epsilon}} F(p_i, K)\,,$$

$$F(p_i, K) = [(p_1 + K)^2 (p_1 + p_2 + K)^2 (p_1 + p_2 + p_3 + K)^2]^{-1}\,, \tag{A.2}$$

where p_i ($i = 1, 4$) are the incoming external momenta. The integral in eq. (A.2) may be divergent at $K = 0$; indeed it behaves as

$$\frac{1}{1-\epsilon} F(p_i, K)|_{K=0}\,, \qquad \epsilon \sim 1\,,$$

$$\frac{1}{\frac{3}{2}-\epsilon} \Delta_K F(p_i, K)|_{K=0}\,, \qquad \epsilon \sim \tfrac{3}{2}\,,$$

$$\frac{1}{2-\epsilon} \Delta_K^2 F(p_i, K)|_{K=0}\,, \qquad \epsilon \sim 2\,, \tag{A.3}$$

where

$$\Delta_K = \sum_{\mu=1}^{D} \left(\frac{\partial}{\partial K_\mu}\right)^2 .$$

For $\epsilon > 1$ the integral is defined by analytic regularization. The divergent terms in eq. (A.3) correspond to the insertion of the operators

$$\int d^D x\, d^D y\, \phi^2(x)\, \phi^2(y)\,,$$

$$\int d^D x\, d^D y\, \phi^2(x)\, \phi^2(y)(x-y)^2\,,$$

$$\int d^D x\, d^D y\, \phi^2(x)\, \phi^2(y)(x-y)^4\,. \tag{A.4}$$

Similarly, we obtain for diagram (c):

$$\int \frac{d^{4-\epsilon}K}{K^{2+2\epsilon}} \frac{1}{(p_1 + p_2 + K)^2} \simeq \frac{1}{\frac{2}{3}-\epsilon} \frac{1}{(p_1 + p_2)^2}\,, \qquad \epsilon \sim \tfrac{2}{3}\,, \tag{A.5}$$

which is the insertion of the operator $[\int \phi^3(x)\, d^D x]^2$. Diagram (d) is given by

$$\int \frac{d^{4-\epsilon}K_1}{K_1^2} \frac{d^{4-\epsilon}K_2}{K_2^2} \frac{F(p_i, K_j)}{(K_1 + K_2)^\epsilon}\,, \qquad i = 1, 8, \quad j = 1, 2\,,$$

$$F(p_i, K_j) = [(p_1 + p_2 + K_1)^2 (\sum_{i=1}^{4} p_i + K_1)^2$$

$$\times (p_5 + p_6 + K_2)^2 (\sum_{i=5}^{8} p_i + K_2)^2]^{-1} . \tag{A.6}$$

The integral may be divergent when both K_1 and K_2 are small. For $D \sim \frac{8}{3}$ we obtain

$$\frac{1}{\frac{4}{3} - \epsilon} F(p_i, K_j)|_{K_j=0} + \ldots, \tag{A.7}$$

where the dots stand for the symmetric contribution coming from the region of integration where both $K_1 + \Sigma_{i=1}^{4} p_i$ and $K_2 + \Sigma_{i=5}^{8} p_i$ are small. Both terms are proportional to the insertion of the multilocal operator

$$\int d^D x\, d^D y\, d^D z\, \phi^3(x)\, \phi^3(y)\, \phi^2(z) .$$

Appendix B

The multilocal operators U_i have been introduced as counterterms in the perturbative expansion. On more general grounds they can be introduced as the "dual" of local operators. Let us consider the Wilson expansion for the product of N fields in the massless scale invariant case:

$$\langle \phi(\lambda x_1) \ldots \phi(\lambda x_N) \phi(y_1) \ldots \phi(y_M) \rangle$$

$$\xrightarrow[\lambda \to 0]{} \sum_i \lambda^{-Nd_\phi + d_{O_i}} C_i(x_1 \ldots x_N) \langle O_i(0) \phi(y_1) \ldots \phi(y_M) \rangle$$

$$= \sum_i \lambda^{-Nd_\phi + d_{O_i}} \langle \phi(x_1) \ldots \phi(x_N) U_i \rangle \langle O_i(0)\, \phi(y_1) \ldots \phi(y_M) \rangle . \tag{B.1}$$

The last equality is the definition of the operator U_i dual to O_i; their dimensions satisfy the relation

$$d_{O_i} + d_{U_i} = 0 . \tag{B.2}$$

Eq. (B.1) can also be written in the form of a large-distance expansion

$$\langle \phi(x_1) \ldots \phi(x_N) \phi(\lambda y_1) \ldots \phi(\lambda y_M) \rangle$$

$$\xrightarrow[\lambda \to \infty]{} \sum_i \lambda^{Md_\phi + d_{U_i}} \langle \phi(x_1) \ldots \phi(x_N) U_i \rangle \langle \phi(y_1) \ldots \phi(y_M) O_i(0) \rangle . \tag{B.3}$$

The validity of eq. (B.3) is a straightforward consequence of the Wilson expansion and the scale invariance of the theory.

Starting from the large-distance expansion, it is possible to show that infrared divergences are proportional to the operator U_i; however, only heuristic arguments can be given to prove that they are multilocal, i.e., that they can be written as products of local operators. We recall that multilocality of the operators U_i has not been used in the derivation of the position of the singularities of the Borel transform: the results of the paper on this particular point seem to be rather sound. Indeed the infrared singularities of the Borel transform are the reflex of the non-perturbative phenomenon discussed in ref. [9]: a local operator O_i (e.g., F^2 in gauge theories) develops a non-zero expectation value

$$\langle O_i \rangle \propto \mu^{d_{O_i}} \exp(d_{O_i}/(\beta_2 \alpha)) .$$

References

[1] N.N. Bogoliubov and O.S. Parasiuk, Acta Math. 97 (1957) 227;
S. Weinberg, Phys. Rev. 118 (1960) 838;
K. Hepp, Comm. Math. Phys. 2 (1966) 301;
M. Bergère and Y.S. Lam, Comm. Math. Phys. 39 (1974) 1.
[2] K. Symanzik, Lectures given at the 1973 Cargèse Summer Institute.
[3] G. 't Hooft, Lectures given at Erice, 1977.
[4] G. Parisi, Lecture given at the 1977 Cargèse Summer Institute.
[5] P. Olesen, Phys. Lett. 73B (1977) 327.
[6] G. Parisi, Phys. Lett. 76B (1978) 65.
[7] B. Lautrup, Phys. Lett. 69B (1977) 109.
[8] G. Parisi, LPTENS preprint 78/15, Phys. Reports, to be published.
[9] A. Vainshtein, V. Zakharov and M. Shifman, ZhETF Pisma 27 (1978) 68;
M. Shifman, A. Vainshtein, M. Voloshin and V. Zakharov, Phys. Lett. 77B (1978) 80.

QUARTIC OSCILLATOR*

R. Balian
Service de Physique Théorique, CEA-Saclay, BP n°2, 91190 Gif-sur-Yvette, France

G. Parisi
I.N.F.N., Frascati, Italy

A. Voros+
Service de Physique Théorique, CEA-Saclay, BP n°2, 91190 Gif-sur-Yvette, France

Abstract : On the example of the semi-classical expansion for the levels of the quartic oscillator $-(d^2/dq^2)+q^4$, we show how the complex WKB method provides information about the singularities of the Borel transform of the semi-classical series. In this problem there occurs a tunneling effect between complex turning points, by which those singularities generate exponentially small, yet detectable, corrections to the energy levels.

Résumé : Sur l'exemple du développement semi-classique des niveaux de l'oscillateur quartique $-(d^2/dq^2)+q^4$, nous montrons comment la méthode BKW complexe renseigne sur les singularités de la transformée de Borel de la série semi-classique. Dans ce problème, les singularités engendrent, par un effet tunnel entre points tournants complexes, des corrections aux niveaux d'énergie qui sont exponentiellement petites et cependant détectables.

One of the oldest recipes to make numerical sense out of an asymptotic (divergent) expansion like $\sum_{k=0}^{\infty} F_k x^{-k}$ for a function $F(x)$ $(x\to\infty)$ is, for any x large but fixed, to sum the series up to the point $k=K$ where the general term $F_k x^{-k}$ attains its minimum modulus as a function of k [2]. The resulting sum $F^*(x)=\sum_{k=0}^{K} F_k x^{-k}$ is taken as an estimate for the exact value $F(x)$, and the first neglected term allegedly provides an estimate for the error : $|F(x)-F^*(x)| \lesssim \varepsilon = |F_{K+1}x^{-K-1}|$.

Such a procedure relies on a faith supported by experience, but it has no rigorous derivation from the general definition of an asymptotic series. In some cases, like the Stirling series for $\log\Gamma(x)$ (x real), it can indeed be justified. But we intend to show that many interesting series exhibit a different type of behaviour ; on one hand they may lead to a systematic discrepancy $F(x)-F^*(x)$ much larger than the "apparent uncertainty" ε measuring the minimum fluctuation of the sequence of partial sums. On the other hand, the asymptotic representation of F(x) may contain additional terms that are *subdominant*, i.e. exponentially small when compared to all terms

* Dedicated to the centennial of the instanton [1]
+ Member of CNRS

of the initial *(dominant)* series. These new terms are asymptotically negligible and seem ill-defined mathematically, but their numerical contributions must be retained as it precisely compensates for the observed discrepancy from the dominant series. We have been surprised not to find this elementary observation in classic textbooks [2,3] (it was noted independently of us by J. Zinn-Justin[4]).

In general, this raises the following questions about the asymptotic representation $\Sigma F_k x^{-k}$ for a given function F(x) :

- how to define subdominant contributions if and when they exist ? (possibly by analytic continuation to regions in x whose they become dominant, but this may not always be explicitly feasible).
- when is a subdominant contribution numerically relevant ? The answer to this question results from a competition as $x \to \infty$ between the size of the subdominant series (essentially of its leading term) and the size of the *smallest term* of the dominant series, which is governed by the *large order behaviour* of its coefficients.
- how do subdominant contributions influence the currently used Borel summation procedure ?

Following the approach we introduced in ref.[5], we shall relate these questions to the study of the analyticity properties of a Fourier-Laplace transform of the function F(x) . We shall first treat the more or less explicit and illustrative example of expansions generated by the saddle-point method, but our main application will be the study of the semi-classical expansion for the levels of the quartic oscillator.

1. The case of saddle-point expansions.

As a typical example, we shall study asymptotic expansions generated by the method of steepest descent[3]. We consider an integral along a complex path without endpoints :

$$F(x) = \int e^{-x\,\varphi(u)}\,du \tag{1.1}$$

where $x > 0$, and $\varphi(u)$ is an analytic function that makes the integral (1.1) convergent and that has only isolated simple critical points (or saddle-points) where $\varphi'(u) = 0$, $\varphi''(u) \neq 0$. In the saddle-point method, the integration path is first distorted to a path of steepest descent. Then for $x \to +\infty$, each saddle-point u_j encountered by this new path formally produces an independent asymptotic contribution to (1.1) of the form $e^{-x\,\varphi(u_j)}\left(\sum_{k=0}^{\infty} F_k^{(j)}\, x^{-k-1/2}\right)$. Assume for simplicity that $u_0 = 0$ is a saddle-point with $\varphi(0) = 0$, and that all other saddle-points u_j satisfy $\mathrm{Re}\,\varphi(u_j) > 0$). Then :

$$F(x) \sim \sum_{k=0}^{\infty} F_k^{(0)}\, x^{-k-1/2} + \sum_j e^{-x\,\varphi(u_j)} \left(\sum_k F_k^{(j)}\, x^{-k-1/2}\right) \tag{1.2}$$

We are precisely in the case of a dominant series (the first one, produced by the saddle-point $u = 0$) supplemented by exponentially small contributions. But (1.2) is up to now a formal expression that requires a suitable interpretation.

One method to understand and to build the expression (1.2) is to take $\varphi(u) = s$ as the integration variable. This transforms (1.1) into :

$$F(x) = \int_C e^{-xs} \rho(s)\, ds \tag{1.3}$$

Thus, F is the Laplace transform along a suitable path C of the function $\rho(s) = \left.\frac{du}{d\varphi}\right|_{\varphi=s}$. $\rho(s)$ is multivalued, the critical values $s_j = \varphi(u_j)$ are its branch points (of the $(s-s_j)^{-1/2}$ type). When we take the path of steepest descent in (1.1), the contour C, which lies on the Riemann surface of $\rho(s)$, gets pushed as far to the right as permitted by the branch points (Fig.1).

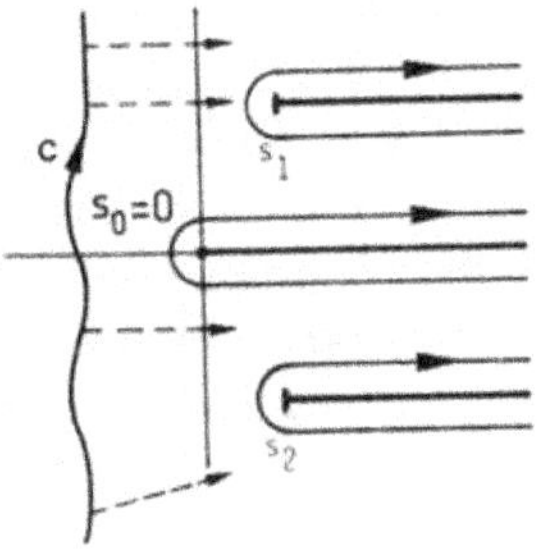

Fig.1

By expanding $\rho(s) = \sum_{n=-1}^{\infty} \frac{\rho_{n/2}^{(0)} s^{n/2}}{2\Gamma(\frac{n}{2}+1)}$ around $s_o = 0$, we then find the coefficients of the dominant expansion as :

$$F_k^{(0)} = \rho_{k-1/2}^{(0)} \tag{1.4}$$

This means that

$$\Delta_o \rho(s) = \rho(s+i0) - \rho(s-i0) = \sum_0^{\infty} \frac{\rho_{k-1/2}^{(0)} s_+^{k-1/2}}{\Gamma(k+1/2)}, \tag{1.5}$$

the *discontinuity* of ρ across the cut from $s = 0$, is a *Borel transform* of the dominant series (s_+^{α} is the function $s_+^{\alpha} \equiv s^{\alpha}$ for $s > 0$, $s_+^{\alpha} \equiv 0$ for $s < 0$ [6]). The same analysis applies to the other cuts : the series multiplying $e^{-x\,\varphi(u_j)}$ in (1.2) results from expanding the discontinuity $\Delta_j\rho$ of $\rho(s)$ across the cut from $s_j = \varphi(u_j)$. Finally the function F itself can be reconstructed from the integral representation

$$F(x) = \int_0^{\infty} e^{-xs} \Delta_o \rho(s)\, ds + \sum_j \int_{s_j}^{\infty} e^{-xs} \Delta_j\, \rho(s)\, ds \tag{1.6}$$

This exact formula shows how each contribution to F(x), whether dominant or subdominant, exists independently of the others.

We now want to compare the subdominant terms (at least the largest ones) in (1.2) with the size ε of the smallest term $F_K^{(0)} x^{-K-1/2}$ in the dominant series. Its rank K satisfies

$$|F^{(0)}_{K+1} / F^{(0)}_{K}| \sim x \tag{1.7}$$

and should be large for large x . Thus, K and ε are asymptotically governed by the *large order behaviour* of $F_k^{(0)}$, itself controlled by the radius of convergence $|s_\varepsilon|$ of the series (1.5) : here s_ε is the singularity of $\rho(s)$ closest to the origin and lying on either of the two sheets associated with the branch point $s = 0$. From the relations (1.4) and $|\rho^{(0)}_{k-1/2}| / \Gamma(k+\frac{1}{2}) \propto |s_\varepsilon|^{-k}$, we get :

$$|F_k^{(0)}| \propto (k+\frac{1}{2}) |s_\varepsilon|^{-k} \tag{1.8}$$

Returning to Eq.(1.7), we approximately find

$$K \sim x|s_\varepsilon| \quad \text{and} \quad \varepsilon \propto \exp(-x|s_\varepsilon|) \tag{1.9}$$

By comparison, the largest subdominant contribution to (1.2) is produced by the *leftmost* singularity s_δ encountered (besides $s = 0$) as the contour C is pushed to the right. The leading contribution of s_δ is $\delta \propto \exp(-xs_\delta)$. Although δ is exponentially small since $\mathrm{Re}\, s_\delta > 0$, we can distinguish two generic situations :

(i) <u>$|s_\varepsilon| < \mathrm{Re}\, s_\delta$</u> . Subdominant terms can be neglected in the numerical scheme of the introduction (they nevertheless *remain in the Borel summation formula* (1.6)). In the extreme case $\mathrm{Re}\, s_\delta = +\infty$, there are strictly no subdominant contributions, as when the function F(x) is Borel summable in the usual sense.

(ii) <u>$|s_\varepsilon| > \mathrm{Re}\, s_\delta$</u> . A systematic deviation is introduced by the dominant term, which should be corrected by including the subdominant contribution δ. This correction becomes relatively more and more important as $x \to \infty$, since $|\delta/\varepsilon| \to \infty$. A frequent occurrence of this case is when the *same* branch point plays both roles : $s_\varepsilon = s_\delta$, and $\mathrm{Im}\, s_\delta \neq 0$ (as for the Bessel function $J_0(z)$ when $\mathrm{Re}\, z > 0$ [5]). Another example is the integral (1.1) taken on the real axis with $\varphi(u) = 36u^2 - 20u^3 + 3u^4$: the subdominant saddle-point is $u = 3$ ($s_\delta = 27$), whereas the saddle-point $u = 2$ ($s_\varepsilon = 32$) controls the large order behaviour (Fig.2, showing the cuts and the contour C in this case, demonstrates that some realistic cases may be harder to describe than the ideal situation pictured on Fig.1).

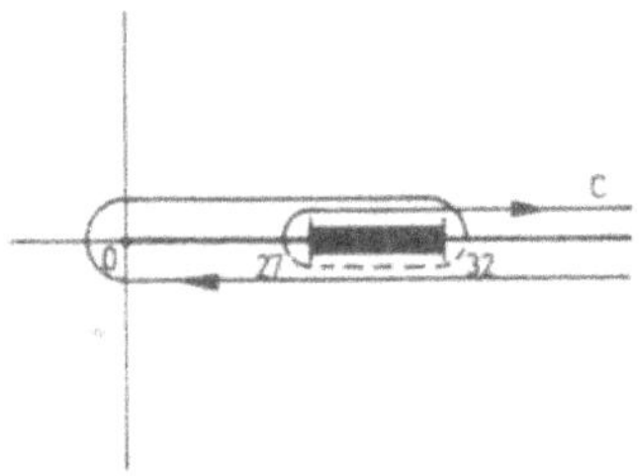

Fig.2

A refinement of Eq.(1.8) has become very popular with the recent advent of instanton calculations[7] : it consists in computing a *large order expansion* for the coefficients F_k , in the form

$$F_k \underset{k\to\infty}{\sim} \Gamma(k+\alpha)\ r^{-k} \left(d_o + \frac{d_1}{k} + \frac{d_2}{k^2} + \dots\right)$$

with all constants to be determined. This question is actually answered by a hundred year-old theorem of Darboux[1], discussed in ref.[3] : if an analytic function $f(s) = \sum_{k=0}^{\infty} f_k\ s^k/k!$ has its nearest singularity at s_ε with a local expansion $f(s) \sim \sum_j c_{\ell_j} (s-s_\varepsilon)^{\ell_j}$ ($\{\ell_j\}$ is an increasing sequence of real numbers, and we place the cut for the functions z^{ℓ} on $[0,\infty)$), then the dominant large order behaviour of the Taylor coefficients f_k is :

$$f_k \underset{k\to\infty}{\sim} -\pi^{-1}\ s_\varepsilon^{-k} \sum_j c_{\ell_j}\ \sin(\pi\ell_j)\ \Gamma(\ell_j+1)\ (-s_\varepsilon)^{\ell_j}\ \Gamma(k-\ell_j) \tag{1.10}$$

In terms of the *discontinuity* of f across its cut :

$$\Delta f(s) = f_+(s) - f_-(s) = \sum_{\ell_j} f_{\ell_j}^{(\varepsilon)} (s-s_\varepsilon)_+^{\ell_j} / \Gamma(\ell_j+1) \tag{1.11}$$

$\left(f_{\ell_j}^{(\varepsilon)} = (1-e^{2\pi i\ell_j})\ \Gamma(\ell_j+1)\ c_{\ell_j}\right)$, Eq.(1.10) becomes :

$$f_k \sim (2i\pi)^{-1}\ s_\varepsilon^{-k} \sum_j f_{\ell_j}^{(\varepsilon)}\ (-s_\varepsilon)^{\ell_j}\ e^{-i\pi\ell_j}\ \Gamma(k-\ell_j) \tag{1.12}$$

which has the advantage of allowing positive integral powers ℓ_j in the discontinuity ; for negative integers, $(s-s_\varepsilon)_+^{\ell_j} / \Gamma(\ell_j+1)$ in Eq.(1.11) must now be interpreted as the generalized function $\delta^{(-\ell_j-1)}(s-s_\varepsilon)$ [6].

From Eq.(1.4) we see that to get the large order behavior of $F_k^{(0)}$ we must apply Darboux's theorem to $f(s) = \Delta_0\ \rho(s)$. The fact that this function admits not a Taylor but a Puiseux series is only a slight complication : we must separate the odd and even parts of ρ as a function of $s^{1/2}$. The result is that if $\rho(s) \sim \sum_{n=-1}^{\infty} \rho_{n/2}^{(\varepsilon)} (s-s_\varepsilon)^{n/2} / 2\Gamma(\frac{n}{2}+1)$ around $s = s_\varepsilon$, the coefficients of the series (1.5) have the large order expansion :

$$\rho_{k-1/2}^{(0)} \sim (2i\pi)^{-1}\ s_\varepsilon^{-(k-1/2)} \sum_{\ell=0}^{\infty} \rho_{\ell-1/2}^{(\varepsilon)}\ (-s_\varepsilon)^{\ell-1/2}\ e^{-i\pi(\ell-1/2)}\ \Gamma(k-\ell) \tag{1.13}$$

where the singularity s_ε is now located on the *Riemann surface* of $\rho(s)$: Eq.(1.13) is then unambiguous as long as the two cuts (from $s = 0$ and $s = s_\varepsilon$) stay disjoint (if there are several branch points at the same distance $|s_\varepsilon|$, their contributions add up).

This example (1.1) suggests the following conclusion, probably connected with

Dingle's interpretation of asymptotic expansions[3] : the best understanding of a general function F(x) is achieved by expressing it as a Laplace transform of a ramified analytic function $\rho(s)$, the integral being taken along some path on the Riemann surface $\mathcal{S}$ of $\rho(s)$. Some branch points of $\rho(s)$, depending on the global structure of $\mathcal{S}$, then contribute to the asymptotic expressions of F(x) through the expansions of the corresponding discontinuities of $\rho(s)$. In turn, the large order behaviour of each such contribution is a reflection of the first orders in the corresponding expansion at (a) neighboring branch point(s) also specified by the global structure of $\mathcal{S}$. The discontinuity expansions can often be computed perturbatively ; moreover if we can fully resum the discontinuities from their power series, and either express the Laplace integral for F(x) in terms of the discontinuities alone as in Fig.1, or recover $\rho(s)$ itself by a dispersion relation, then we have succeeded to resum the function F by a generalization of the Borel method.

The main difficulty of this program is, for a given F(x), to know whether a suitable $\rho(s)$ exists, and to understand its analytic structure. If F(x) is given by an integral like (1.1) but now *multidimensional*, we can take $\rho(s) = \int \frac{du}{s-\varphi(u)}$.

The branch points of $\rho(s)$ are again the critical values of φ (the saddle-point values), but the analytic structure of $\rho(s)$ is very hard to describe in detail[8].

Now, our main concern lies with functions F(x) that admit semi-classical (or perturbative) expansions in quantum mechanics and in field theory. Such expansions present some analogy with the previous case, as they can be generated by Feynman path integrals which formally look like (1.1), with x replaced by $\hbar^{-1}$ (or the inverse coupling constant g^{-1}), the integration variable u by the trajectories in configuration space, and $\varphi(u)$ by the classical action function. We therefore expect the previous results to hold (qualitatively at least) : there should exist an *analytic* function $\rho(s)$ on a Riemann surface $\mathcal{S}$, such that F(x) be a Laplace transform of $\rho(s)$; the branch points of $\rho(s)$ should be the stationary values of the action and should thus arise from *each classical trajectory* (real or complex) ; in principle, semi-classical methods should provide the discontinuity expansions of $\rho(s)$ at the branch points (a semi-classical description of the *global topology* of $\mathcal{S}$ would be also highly desirable, but this stands as an open question). This description of quantum mechanics in terms of analytic functions of the variable s (conjugate of $1/\hbar$) has been given in ref.[9].

Our analysis of exponentially small contributions to asymptotic expansions means here that tunneling effects (or instanton effects) should sometimes be numerically relevant even in cases when they are subdominant. Moreover they should become relatively more and more important as $\hbar$ (or g) $\to 0$, since we have seen that $|\delta/\varepsilon| \to \infty$.

A last interesting consequence of the Riemann surface structure is that the large orders of a semi-classical expansion at one branch point are governed by the leading orders of an expansion *of the same nature* at another branch point. This reciprocity property is probably universal for semi-classical expansions : it was noted in a completely different way by Dingle in the case of the WKB expansion for wave functions[3]. It provides a new insight into the semi-classical nature of large order analysis[7].

2. Semi-classical eigenvalues of the quartic oscillator.

The quartic oscillator is the one-dimensional quantum system described by the Hamiltonian :

$$\hat{H}(\hbar,g) = -\hbar^2 \frac{d^2}{dq^2} + gq^4 \qquad (g>0) \quad . \tag{2.1}$$

This simple Hamiltonian, although not (yet) exactly solvable, has been extensively studied[10-12,29]. We shall be concerned here with its eigenvalues. Because of the obvious scaling property ($\approx$ denotes unitary equivalence) :

$$\hat{H}(\hbar,g) \approx \hbar^{4/3} g^{1/3} \hat{H}(1,1) \tag{2.2}$$

(under the change of variables $q \to g^{-1/6} \hbar^{1/3} q$), we can restrict ourselves to the case $g=1$. But in view of a semi-classical study of (2.1), instead of letting $\hbar=1$ immediately as suggested by Eq.(2.2) we prefer to keep it as an explicit expansion parameter for some time. The classical Hamiltonian corresponding to (2.1) with $g=1$ is $H=p^2+q^4$. Its classical trajectory for a given energy E is the solution of $\frac{dq}{dt} = 2\sqrt{E-q^4}$, namely a Jacobi elliptic function[13] :

$$q(t) = E^{1/4} \, \mathrm{cn}(2\sqrt{2}\, E^{1/4} t \mid \tfrac{1}{2}) \quad . \tag{2.3}$$

Let us first rewrite the WKB expansion for the eigenfunctions and eigenvalues of a general one-dimensional Schrödinger equation with an analytic potential V(q) :

$$\left(-\hbar^2 \frac{d^2}{dq^2} + V(q)\right) \psi(q) = E\,\psi(q) \quad . \tag{2.4}$$

The solutions of (2.4) in the complex q plane are known to have the exact form[14] :

$$\psi = u^{-1/2} \exp \frac{i}{\hbar} \int u \, dq \quad , \tag{2.5}$$

where $u(q,\hbar)$ is a solution of

$$u^2 - p^2 = \hbar^2 \, (u^{-1/2})'' \, u^{1/2} \quad , \tag{2.6}$$

$p(q) = \pm(E-V(q))^{1/2}$ is the classical momentum, and $' = \frac{d}{dq}$.

For each determination of the function p(q) , Eq(2.6) (which is *exact*) can be formally solved in powers of $\hbar^2$, as :

$$u(q) = u(q\,,p(q)) = p + \sum_{n=1}^{\infty} \hbar^{2n}\, u_{2n}(q,p) \quad . \tag{2.7}$$

The terms u_{2n} can be computed recursively : they are polynomial functions of $V'(q)$, $V''(q)$, ... and of p^{-1}, odd in p. For instance :

$$u_2 = \frac{V''(q)}{8p^3} + \frac{5V'(q)^2}{32p^5}$$

An alternative method uses the representation : $\psi = \exp \frac{i}{\hbar}\int v\, dq$ and $v = p + \sum_{n=1}^{\infty} (i\hbar)^n v_n$ [15,12]. This results in a somewhat simpler recursion relation

$$v_n = \frac{1}{2p}\left(v'_{n-1} - \sum_{k=1}^{n-1} v_k\, v_{n-k}\right) \quad , \quad v_o = p \tag{2.8}$$

By comparing the two expressions of ψ we see that $u_{2n} = (-1)^n\, v_n$, but the remarkable structure (2.5) of ψ is harder to understand by this method.

A third way of computing the expansion (2.7) is by means of a closed (i.e. non-recursive) formula[11,16].

In all methods, complexity increases rapidly with the order n. Practical calculations are best performed by computer in symbolic computing languages[17,11-12,16].

From the knowledge of the WKB solution (2.5) we can now derive an eigenvalue quantization condition correct to all orders in $\hbar$, by an argument of Wentzel and Dunham[15,18]. We take $V(q)$ to be a simple-well potential with two (real) turning points $q_- < q_+$ as in Fig.3, and we define the analytic function $p(q) = (E - V(q))^{1/2}$ in the complex q plane cut along $[q_-,q_+]$, with the values of $\varphi = \operatorname{Arg} p$ as indicated on Fig.4 (thus neglecting the effects of complex turning points).

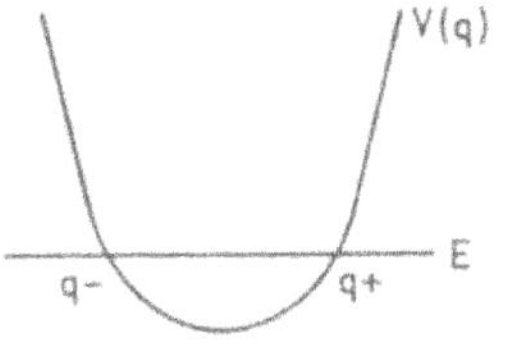

Fig.3

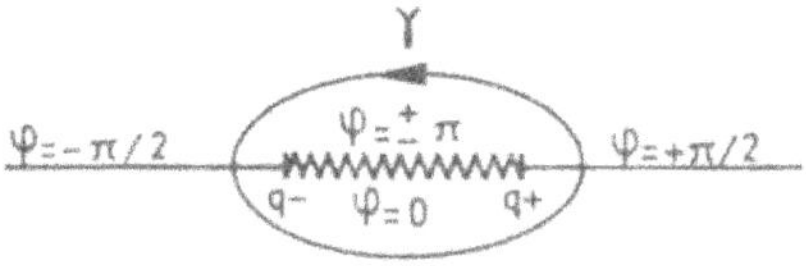

Fig.4

Then the eigenfunction ψ admits the asymptotic form :

$$\psi \sim u(q,p)^{-1/2} \;\exp \frac{i}{\hbar} \int_{q_o}^{q} u(q'\,,p(q'))\, dq' \tag{2.9}$$

in the whole complex q plane away from the cut, since (2.9) satisfies the correct boundary conditions for $q \to \pm\infty$. Let γ be a positive contour encircling the cut (Fig.4). The *exact* eigenfunction ψ is analytic and has an integral number k of zeros, all of them on the interval (q_-,q_+), hence it satisfies :

$$\oint_{\gamma} \frac{\psi'}{\psi}\, dq = 2\pi i k \qquad .$$

The substitution of (2.9) then yields (we refer to [18,12] for details) :

$$\oint_{\gamma} u\, dq = \oint_{\gamma} p\, dq + \sum_{n=1}^{\infty} \hbar^{2n} \oint_{\gamma} u_{2n}\, dq = 2\pi(k+\tfrac{1}{2})\,\hbar \tag{2.10}$$

which is an eigenvalue condition of the Bohr-Sommerfeld type, but with all corrections in powers of $\hbar$ included.

We note that the expressions (2.5) and (2.10) for the eigenfunctions and eigenvalues only differ from their usual lowest order approximations by the replacement everywhere of the classical momentum $p(q)$ by the solution u of Eq.(2.6). Besides, the construction of u to any finite order in $\hbar$ is *purely algebraic* (and u only depends on $\hbar^2$).

We have given here a rapid derivation of the result (2.10). We do not know yet whether and when (2.10) gives a truly asymptotic expansion of the eigenvalues. Existing works only concern leading order estimates[19,20] and they suggest that this is a difficult question. But in spite of its formal derivation Eq.(2.10) appears to be the correct asymptotic expansion in all cases of interest.

We now specialize our results to the case of the potential $V(q) = q^4$, following ref.[12] but with slightly different normalizations. The recursion relation (2.8) now yields :

$$v_1 = -\frac{q^3}{p^2}$$

$$v_2 \;(= -u_2) = -\left(\frac{3q^2}{2p^3} + \frac{5q^6}{2p^5}\right)$$

$$v_3 = -\left(\frac{3q}{2p^4} + \frac{27q^5}{2p^6} + \frac{15q^9}{p^8}\right)$$

$$v_4 \;(= u_4) = -\left(\frac{3}{4p^5} + \frac{339q^4}{8p^7} + \frac{663q^8}{4p^9} + \frac{1105q^{12}}{8p^{11}}\right) \qquad .$$

More generally :

$$v_n = \sum_{\ell=0}^{[3n/4]} v_{n,\ell}\; q^{3n-4\ell}\; p^{-3n+2\ell+1} \tag{2.11}$$

(the precise powers involved, and the summation upper bound $[3n/4]$ (integer part of $3n/4$) can be understood by dimensional analysis).

In the quantization condition (2.10) now written as :

$$2\pi(k+\tfrac{1}{2})\,\hbar = \sum_{0}^{\infty} \hbar^{2n} (-1)^n \oint_{\gamma} v_{2n}\, dq = \sum_{0}^{\infty} \hbar^{2n}\, \sigma_n \tag{2.12}$$

we can evaluate the loop integrals in terms of the Euler Beta function :

$$\sigma_n = (-1)^n E^{\frac{3}{4}(1-2n)} \sum_{\ell=0}^{[3n/2]} v_{2n,\ell}\, B\left(-3n+\ell+\frac{3}{2}, \frac{3n}{2}-\ell+\frac{1}{4}\right) \qquad (2.13)$$

In particular, letting $c = \Gamma(1/4) = 3.62561\ldots$

$$\sigma_o = \oint_\gamma p\,dq = E^{3/4} B\left(\frac{3}{2}, \frac{1}{4}\right) = \frac{c^2}{3}\sqrt{\frac{2}{\pi}}\, E^{3/4} = \sigma \qquad (2.14)$$

is the classical action around the closed orbit (2.3) of energy E. In terms of σ, the quantization condition (2.12) takes the final form :

$$2\pi(k+\frac{1}{2})\hbar = \sum_{n=0}^{\infty} b_n\, \sigma^{1-2n}\, \hbar^{2n} \qquad (2.15)$$

$$b_n = \frac{(-1)^n}{2}\left(\frac{3}{c^2}\sqrt{\frac{\pi}{2}}\right)^{1-2n} \sum_{\ell=0}^{[3n/2]} v_{2n,\ell}\, B\left(-3n+\ell+\frac{3}{2}, \frac{3n}{2}-\ell+\frac{1}{4}\right) \qquad (2.16)$$

We give a few details about the practical computation of the b_n. The use of a formal computer algebraic system to solve the WKB recursion equations yields results in closed arithmetic form, but becomes extremely time-consuming as the order increases. With the REDUCE language[21] we found the iteration of Eq.(2.6) slightly faster than other methods, but we nevertheless stopped at b_{16}. For a more efficient evaluation of b_n for large n, we shifted to the ordinary numerical computation (in FORTRAN) of the coefficients $v_{n,\ell}$ in Eq.(2.11) recursively on n, followed by the evaluation of formula (2.16). To compensate for the errors (increasing with n) caused by huge cancellations between terms of Eq.(2.16), we worked with the multiple-precision arithmetic package MULTILONG [22]. A 30 minute IBM 360-91 computer run produced the b_n up to $n = 53$ in ordinary double-precision.

We empirically found that the sequence of signs of the b_n was + − + + − − − + + − − ... and that for large n :

$$|b_n| \sim (2n-2)!\; 2^n \times C \quad , \qquad C \simeq 0.63 \qquad (2.17)$$

(these facts will be explained in section 4). We give the list (Table 1) of the computed values of $b'_n = 2^{-n} b_n/(2n-2)!$. The first exact b_n are (with $c = \Gamma(1/4)$) :

$$b_o = 1 \qquad\qquad b_1 = -\frac{\pi}{3}$$

$$b_2 = \frac{11\, c^8}{10368\, \pi^2} \qquad\qquad b_3 = \frac{4697\, c^8}{466560\, \pi}$$

$$b_4 = -\frac{390065\, c^{16}}{501645312\, \pi^4} \qquad\qquad b_5 = -\frac{53352893\, c^{16}}{1934917632\, \pi^3} .$$

It follows from Eq.(2.16) that :

$$b_{2n} = R_{2n}\, c^{8n}\, \pi^{-2n} \qquad\qquad b_{2n+1} = R_{2n+1}\, c^{8n}\, \pi^{-2n+1} .$$

where $\{R_n\}$ is a sequence of *rational* numbers, which we do not know how to generate by any simpler law.

3. An investigation of the semi-classical series.

The object of our study will be the semi-classical series (2.15). We first note that this quantization condition only involves powers of $\hbar/\sigma$, in relation with the scaling property (2.2). By formal inversion of the series (2.15), we get an eigenvalue formula in the form of an expansion for *large* k, which now depends trivially on $\hbar$:

$$\sigma_k = \frac{c^2}{3}\sqrt{\frac{2}{\pi}}\,E_k^{3/4} = 2\pi\hbar\,(k+\tfrac{1}{2})\left[1+\frac{c_1}{(k+1/2)^2}+\frac{c_2}{(k+1/2)^4}+\ldots\right]$$

Thus, in the original series (2.15) (which we found simpler to study), we can let $\hbar = 1$ and keep $1/\sigma$ as the expansion parameter. The power series

$$\sigma + \frac{b_1}{\sigma} + \frac{b_2}{\sigma^3} + \ldots = 2\pi\,(k+\tfrac{1}{2}) \tag{3.1}$$

now formally defines the quantum number as a (continuous) function of the classical action. However (3.1) certainly diverges everywhere because of the law (2.17). We are thus faced with the problem of interpreting the series (3.1) and/or resumming it.

We shall first compare the exact eigenvalues with the successive approximations defined by (3.1) : let $E_k^{(j)}$ be the solution of :

$$\sigma(E) + \frac{b_1}{\sigma(E)} + \ldots + \frac{b_j}{\sigma(E)^{2j-1}} = 2\pi(k+\tfrac{1}{2})$$

such that $\sigma(E_k^{(j)}) \sim 2\pi(k+\frac{1}{2})$ for $k\to+\infty$: $E_k^{(j)}$ is a reasonable j-th order estimate of the exact k-th eigenvalue E_k. Table 2a shows $E_k^{(j)}$ for a sample of values of k and j. For given k, $E_k^{(j)}$ exhibits the typical behaviour of the j-th partial sum of an asymptotic series : it first seems to "converge" rapidly while the successive terms of the series decrease, then it "blows up" without limit. As stated in the introduction, we take as "best" estimate E_k^* the approximation $E_k^{(j)}$ for which $|E_k^{(j+1)} - E_k^{(j)}|$ attains its minimum value ε, and we compare ε with the actual error $|E_k-E_k^*|$. From Table 2a, we find important discrepancies between the best estimates and the exact values : $|E_k-E_k^*| / \varepsilon(k)$ is much larger than 1 (and increases with k).

As in section 1, we want to interpret this phenomenon by the occurrence of subdominant corrections to the quantization condition (it is also a sign that the series (3.1) is not Borel summable in the simplest sense).

To understand the origin of subdominant corrections, we adopt the viewpoint that the Bohr-Sommerfeld rule can be formally derived from a Feynman path integral formula for the trace of the Green's function[23,24]

$$G(E) = \mathrm{Tr}(\hat{H}-E)^{-1} = \frac{i}{\hbar}\int_0^{\infty} dt\, e^{iEt/\hbar}\, \mathrm{Tr}\, e^{-i\hat{H}t/\hbar}$$
$$= \frac{i}{\hbar}\int_0^{\infty} dt\, e^{iEt/\hbar} \int_{q(t)=q(0)} \{\mathcal{D}q\}\, e^{iS\{q\}/\hbar} \qquad (3.2)$$

where $S\{q\} = \int_0^t \left[\frac{1}{4}\left(\frac{dq}{d\tau}\right)^2 - V(q(\tau))\right] d\tau$ is the classical action, and the path integral only involves paths with $q(t) = q(0)$ .When (3.2) is evaluated semi-classically by the stationary phase method, only the (real) closed classical trajectories of energy E contribute, as they are the stationary points of $S\{q\} + Et$. The contributions of one such closed trajectory C_E as it is traversed 1,2,3,... times in both directions form a geometric series of ratio $\exp \frac{i}{\hbar}\oint_{C_E} p\, dq$ which sums up to :

$$G(E) \sim \frac{T(E)}{\hbar} \tan\left(\frac{1}{2\hbar}\oint_{C_E} p\, dq\right)$$

where $T(E) = \frac{d}{dE}\oint_{C_E} p\, dq$ is the primitive period of C_E . The poles of G(E) (the eigenvalues) are then given by the Bohr-Sommerfeld rule $\oint p\, dq = 2\pi(k+\frac{1}{2})\hbar$. Corrections in powers of $\hbar$ are in principle contributions to the integral (3.2) from the paths fluctuating around the (same) stationary trajectories (see ref.[24] for a review of these methods). The quantization rule (3.1) for the quartic oscillator precisely involves the action σ of the real periodic classical trajectory (2.3).

But we know, from the experience of finite-dimensional integrals of the form $\int dq\, e^{iS(q)/\hbar}$, that if S is analytic we can push the integration contour into the region $\mathrm{Im}\, S(q) > 0$ to uncover contributions from *complex* stationary points (saddle-points), each of them of the form $e^{iS_0/\hbar} \times$ [power series] . We claim that such terms give *numerically relevant* contributions, and that in particular, they account for the observed discrepancy in the series (3.1). Precisely, the complex saddle-points in the path integral (3.2) are the *closed orbits* of energy E of the classical Hamiltonian in *complex* coordinates[9]. For $H = p^2+q^4$, the complex trajectory $q_E(t) = E^{1/4}\, \mathrm{cn}(2\sqrt{2}E^{1/4}t \mid \frac{1}{2})$ has *a lattice* Λ *of periods* in the t plane[13], generated by $T_1 = (T+iT)/2$ and $T_2 = (-T+iT)/2$, where

$$T = \frac{E^{-1/4}}{2\sqrt{2}}\, 4K(1/2) = \frac{E^{-1/4}}{2\sqrt{2\pi}}\, \Gamma(1/4)^2 \qquad (3.3)$$

is the primitive real period (Fig.5).

A complex closed orbit is associated with every time path $\lambda\Big|_{t_1}^{t_2}$ with $t_2 - t_1 \in \Lambda$; it can also be represented as a loop on the two-sheeted Riemann surface of the complex momentum $p = (E-q^4)^{1/2}$, whose branch points are the four classical turning points $E^{1/4}$, $iE^{1/4}$, $-E^{1/4}$, $-iE^{1/4}$ (Fig.6). In the saddle-point method, we keep the contribution of each complex period along which $\mathrm{Im}\oint p\, dq \geq 0$, instead of just the multiples of the real period as before. The actions $\oint p\, dq$ along the periods also form a lattice in the complex s plane, generated by

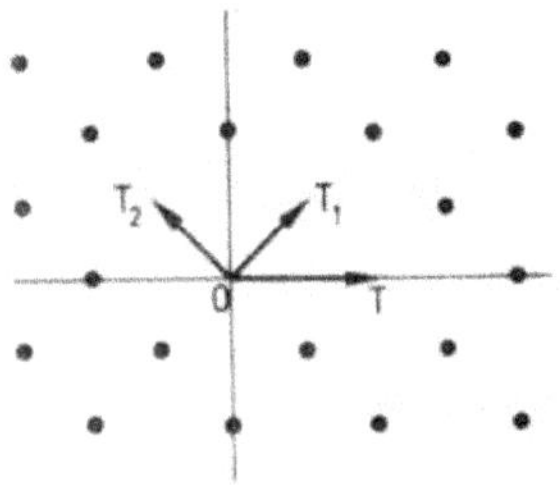

Fig.5

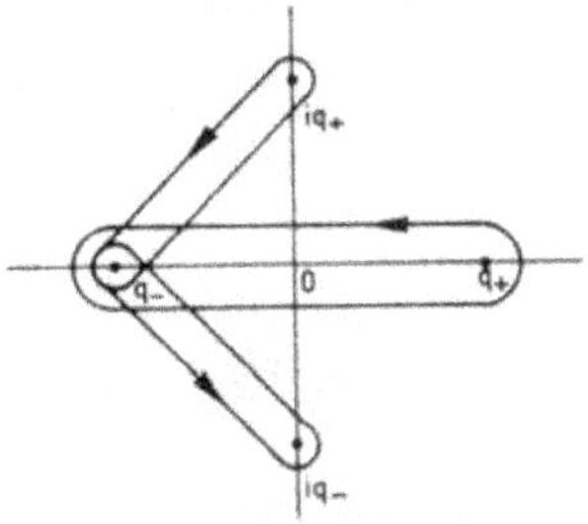

Fig.6

$$S_1 = \int_0^{T_1} p(t)\,dq(t) = \frac{\sigma+i\sigma}{2} \quad \text{and} \quad S_2 = \int_0^{T_2} p(t)\,dq(t) = \frac{-\sigma+i\sigma}{2} \tag{3.4}$$

We expect on general grounds that trajectories of periods T_1+mT ($m \in \mathbb{Z}$) will contribute terms of the order $\delta = e^{-\sigma/2\hbar}$ to the quantization rule (higher lying periods would contribute still smaller terms $e^{-\sigma/\hbar},\ldots$). But the same trajectories (and their opposites) also produce the singularities of ϕ : $s_\varepsilon = \pm i(S_1+m\sigma)$ that lie closest to the dominant ones $m\sigma$ (at a distance $\sigma/\sqrt{2}$). Accordingly, the apparent accuracy of the semi-classical series is $\varepsilon \propto e^{-\sigma/\sqrt{2}\hbar} \ll \delta$. We are thus in the case where subdominant contributions *should be meaningful*. But instead of making a saddle-point expansion of (3.2), we shall rather compute them by WKB connexion formulas in the complex q plane [19,25], closely following[26].

We first place cuts in the complex q plane to define a single-valued function $p(q) = (E-q^4)^{1/2}$. For instance, we may choose the cuts and the values of $\operatorname{Arg} p = \varphi$ as in Fig.7. We then plot the *Stokes lines*, which are the lines starting from each turning point q_0, along which $\int_{q_0}^{q} p\,dq$ is real.

We initially neglect correction terms in powers of $\hbar$. In each region $\alpha = 1,2,\ldots,8$ limited by the Stokes lines, the (real) eigenfunction ψ has the asymptotic form $\psi = p^{-1/2} e^{-i\pi/4}\left(A_\alpha e^{i\int p\,dq} + B_\alpha e^{-i\int p\,dq}\right)$. The WKB matching conditions relate the values of the constants A_α, B_α from one region to the adjacent one. We shall follow the changes of the constants on a distant path lying far from the cuts (dashed line on Fig.7). In region 3, because of the boundary condition for $q \to -\infty$:

$$\psi(3) = p^{-1/2} e^{-i\pi/4} \exp i\int_{q_-}^{q} p\,dq \quad .$$

In crossing to region 1, the turning point q_- is involved alone. The usual one

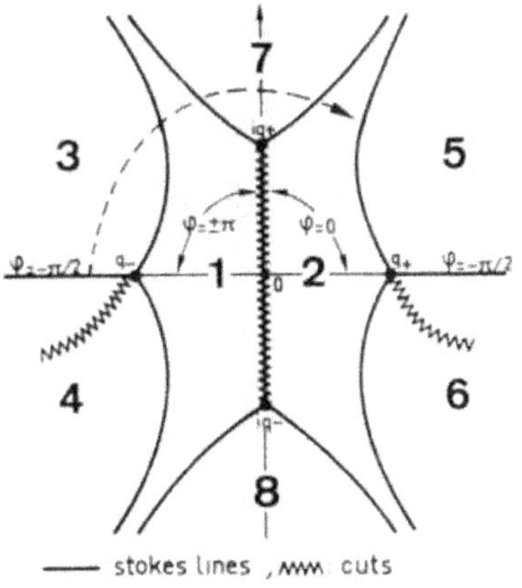

Fig.7

turning-point connexion formula implies :

$$\psi(1) = p^{-1/2}\left(e^{-i\pi/4} \exp\int_{q_-}^{q} p\,dq - e^{i\pi/4} \exp -i\int_{q_-}^{q} p\,dq\right)$$

which we rewrite as

$$\left.\begin{aligned} \psi(1) &= |p|^{-1/2}\left(A_1 \exp i\int_{o}^{q} p\,dq + A_1^* \exp -i\int_{o}^{q} p\,dq\right) \\ \text{where}\quad A_1 &= \exp i\left(\frac{\pi}{4}-\frac{\sigma}{4}\right) \quad , \quad \frac{\sigma}{4} = -\int_{q_-}^{o} p\,dq \end{aligned}\right\} \qquad (3.5)$$

Similarly, starting from $\psi(5) = \pm p^{-1/2}\, e^{-i\pi/4} \exp i\int_{q_+}^{q} p\,dq$:

$$\left.\begin{aligned} \psi(2) &= |p|^{-1/2}\left(A_2 \exp i\int_{o}^{q} p\,dq + A_2^* \exp -i\int_{o}^{q} p\,dq\right) \\ \text{with}\quad A_2 &= \pm\exp i\left(\frac{\pi}{4}-\frac{\sigma}{4}\right) \ . \end{aligned}\right\} \qquad (3.6)$$

If there were no other turning points, we could simply write $A_2 = A_1^*$, taking into account the fact that p(q) becomes -p(q) across the cut. But actually, to match the expressions (3.5) and (3.6), we still have to cross the system of Stokes lines associated with the turning points iq_+ and iq_- (Fig.8). The correct matching conditions must be the same as when the wave encounters an "underdense potential barrier" (Fig.9), which precisely has the turning point structure of Fig.8. Namely[26] :

$$A_2 = \sqrt{1+e^{-\sigma}}\; e^{-i\delta(\sigma)}\, A_1^* - ie^{-\sigma/2} A_1 \qquad (3.7)$$

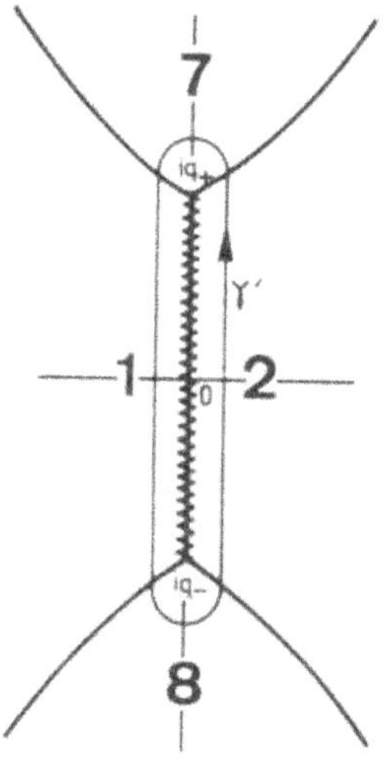

Fig.8

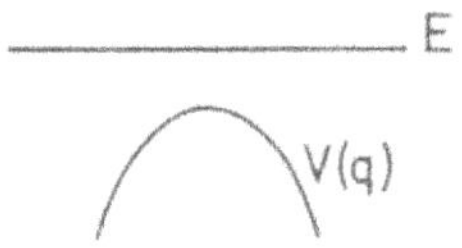

Fig.9

where the quantity $e^{-\sigma}$ arises from the contour integral $\exp i\int_{\gamma'} p\,dq$ (Fig.8), and $\delta(\sigma)$ is a phase angle which the matching procedure cannot determine in general[26]. Consistency between Eqs.(3.5), (3.6) and (3.7) leads to the quantization condition $\sqrt{1+e^{-\sigma}}\cos(\frac{\sigma}{2}-\delta(\sigma))+e^{-\sigma/2}=0$, or equivalently :

$$2\pi(k+\frac{1}{2}) \;=\; \sigma-\delta(\sigma)-2(-1)^k \,\mathrm{Arctg}\, e^{-\sigma/2} \tag{3.8}$$

The presence of an unknown $\delta(\sigma)$ in (3.8) seems most annoying ; fortunately we can argue that in our approach of expanding for $\sigma\to\infty$, it is consistent to let $\delta(\sigma)\equiv 0$ while incorporating power corrections in σ^{-1} into (3.8). First, in all known cases[26], $\delta(\sigma)\to 0$ as $\sigma\to\infty$, hence it can contribute only power corrections in σ^{-1} , or subdominant terms. Concerning powers of σ^{-1} , we have neglected some of them in our matching procedure, but we have already included *all of them* consistently in formula (3.1), hence $\delta(\sigma)$ is redundant in this respect. As for corrections to (3.8) in powers of $e^{-\sigma/2}$, it is reasonable to expect that in the approximation where the two turning points iq_- and iq_+ are treated as separate, the leading term (of order $e^{-\sigma/2}$) should come out correctly ; this implies that no contribution of order $e^{-\sigma/2}$ arises from $\delta(\sigma)$ in Eq.(3.8). A different argument is that for the comparison potential having no other complex turning points that the two on Fig.8 (i.e. the inverted parabolic barrier[26]), $\delta(\sigma)\equiv \mathrm{Arg}\,\Gamma(\frac{1}{2}+\frac{i\sigma}{2\pi})-\frac{\sigma}{2\pi}(\log\frac{\sigma}{2\pi}-1)$ has an asymptotic expansion $\delta\sim\frac{\pi}{12\sigma}+\frac{7\pi^3}{360\sigma^3}+\cdots$ which is known to be accurate to order $e^{-\sigma}$ when truncated at its smallest term.

Thus, in the following, we shall consistently leave out all those terms of order $e^{-\sigma}$ that might arise from $\delta(\sigma)$, or equivalently from the imperfections of our matching procedure (cooperative effects of all four turning points were overlooked). We are entitled however to include power terms in σ^{-1} . The quantization rule will then involve powers of σ^{-1} and of $e^{-\sigma/2}$ considered as independent variables.

The simplest way to write such a quantization rule is to draw the power contributions in σ^{-1} from (3.1) and the power contributions in $e^{-\sigma/2}$ from (3.8), neglecting cross terms :

$$2\pi(k+\frac{1}{2}) \;=\; \sigma+\frac{b_1}{\sigma}+\frac{b_2}{\sigma^3}+\ldots-2(-1)^k\,\mathrm{Arctg}\,e^{-\sigma/2} \quad . \tag{3.9}$$

As a numerical test, we consider the sequences of approximations obtained by solving Eq.(3.9) to increasing orders in $1/\sigma$ for various values of k (Table 2b). Comparison with Table 2a shows that the subdominant term in (3.9) explains indeed most of the discrepancy between the previous best estimates E_k^* and the exact values E_k .

A more consistent synthesis of Eqs.(3.1) and (3.8) should also involve terms of higher orders in $e^{-\sigma/2}$ and σ^{-1} . From the above discussion, we expect that the matching procedure, extended to higher orders in σ^{-1} , will yield safely all contributions of order $e^{-\sigma/2}\sigma^{-n}$. This is easily done if we note, as in section 2, that the analytic structure of the WKB wave functions, and hence the matching coefficients, are the same at higher orders in $\hbar$ as for the leading orders ; the only difference is that the WKB expressions are "dressed up" with power corrections in $\hbar$ (or $1/\sigma$), through the replacement everywhere of the function $p(q)$ by the expansion $u(q)$ of Eq.(2.7). If we now remember that in Eq.(3.8), the dominant term σ stands for the contour integral $\oint_\gamma p\,dq$ along the contour γ of Fig.4 (cf. Eq.(2.10)), whereas $e^{-\sigma/2}$ stands for $\exp i \oint_{\gamma'} p\,dq$ along the contour γ' of Fig.8, we see that the full-bodied quantization rule must be :

$$2\pi(k+\tfrac{1}{2}) = \oint_\gamma u\,dq - 2(-1)^k \operatorname{Arctg}\left(\exp \frac{i}{2}\oint_{\gamma'} u\,dq\right) \tag{3.10}$$

(this formula holds for any problem in which the "important" turning points have the same structure as in Fig.7). In the case of $V(q) = q^4$, we take the expressions (2.11) for $u_{2n} = (-1)^n v_{2n}$ and (2.13) for $\sigma_n = \oint_\gamma u_{2n}\,dq$, and readily find that $\oint_{\gamma'} u_{2n}\,dq = (-1)^{n+1} i\sigma_n$, hence the explicit $1/\sigma$-expansion of (3.10) reads :

$$2\pi(k+\tfrac{1}{2}) = \sigma + \frac{b_1}{\sigma} + \frac{b_2}{\sigma^3} + \frac{b_3}{\sigma^5} + \ldots - 2(-1)^k \operatorname{Arctg} e^{-\frac{1}{2}\left(\sigma - \frac{b_1}{\sigma} + \frac{b_2}{\sigma^3} - \frac{b_3}{\sigma^5}\ldots\right)} \tag{3.11}$$

The sequences of approximations obtained by solving Eq.(3.11) with both expansions truncated to the same increasing orders, are listed in Table 2c. *All discrepancies have now disappeared* : the errors on the best estimates have become of the same order as the series fluctuations. This test is sensitive enough to give a (heuristic) check for the term of order $e^{-\sigma/2}$ with its first power correction, namely $\exp\left(-\frac{1}{2}(\sigma - b_1/\sigma)\right)$ (an indirect check of the following powers will follow from the large order behaviour of the series (3.1) : see next section). However, our present computations are not accurate enough to check the validity of Eq.(3.11) to order $e^{-\sigma}$.

4. The Borel transform of the semi-classical expansion.

The following question now arises about the divergent series (3.1) : is there a natural way to sum it to a smooth function $\bar{F}(\sigma)$ such that the equation :

$$\bar{F}(\sigma_k) = 2\pi(k+\tfrac{1}{2}) \quad , \quad \sigma_k = \sigma(E_k) \tag{4.1}$$

yields the *exact* eigenvalue E_k for each k (Fig.10) ? Such an $\bar{F}(\sigma)$ would appear as an exact version of the Thomas-Fermi distribution[9]. Actually this is a very tricky question : on one hand, the existence of such a $\bar{F}(\sigma)$ is strongly suggested by the numerical accuracy of the semi-classical series, but the asymptotic series alone defines $\bar{F}(\sigma)$ only up to $\mathcal{O}(\sigma^{-\infty})$ (terms decreasing faster than any power of σ^{-1}). On the other hand, the exact relation (4.1) only defines $\bar{F}$ *at* the eigenvalues ; it does suggest an analytic relationship between energy and quantum number, but to our knowledge we cannot significantly define such a unique relationship for non-integral quantum numbers. It will thus be useful to switch to another function $F(\sigma)$, which is discontinuous but has an intrinsic meaning in the quantum theory. We define it as a "staircase" function (Fig.10) :

$$F(\sigma) = 2\pi\ \mathrm{Tr}(\sigma-\hat{I}) = 2\pi \sum_{k=0}^{\infty} \theta(\sigma-\sigma_k) \tag{4.2}$$

where θ is the Heaviside step function, $\sigma(E)$ is the action function as in Eq.(2.14) and $\hat{I} = \sigma(\hat{H})$ is the same function of the *operator* $\hat{H}$. Assuming that we have a smooth function $\bar{F}(\sigma)$ satisfying (4.1), we can recontruct $F(\sigma)$ from the following "bootstrap" relation :

$$F(\sigma) = 2\pi \sum_{k=-\infty}^{+\infty} \theta\left(\bar{F}(\sigma) - 2(k+\tfrac{1}{2})\pi\right) \tag{4.3}$$

which, by virtue of the Fourier expansion

$$2\pi\left(\sum_{k=-\infty}^{+\infty} \theta(x-2(k+\tfrac{1}{2})\pi)\right) - x = i \sum_{\substack{m=-\infty \\ m\neq 0}}^{+\infty} \frac{(-1)^m}{m}\, e^{-imx} ,$$

is equivalent to :

$$F(\sigma) = \bar{F}(\sigma) + i \sum_{m\neq 0} \frac{(-1)^m}{m}\, e^{-im\bar{F}(\sigma)} \tag{4.4}$$

We have thus realized an explicit decomposition of $F(\sigma)$ as a sum of a smooth term $\bar{F}(\sigma)$ and of oscillating terms[9]. Moreover, the oscillating terms are completely determined by the smooth term $\bar{F}(\sigma)$.

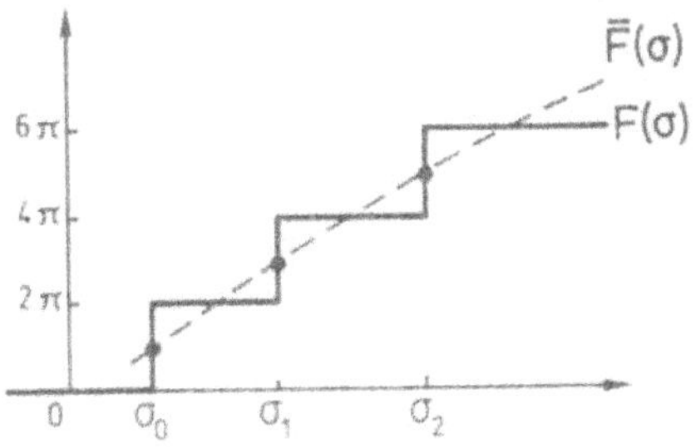

Fig.10

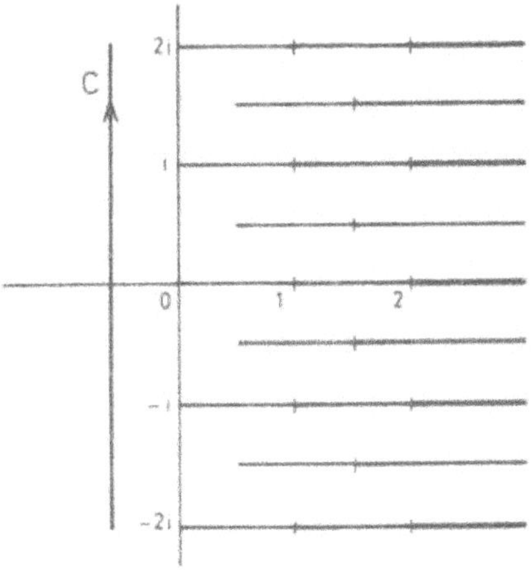

Fig.11

We now interpret this result in terms of the Laplace transform $\rho(s)$ of $F(\sigma)$ which, by Eq.(4.2), is simply :

$$\rho(s) = \frac{i}{s}\ \mathrm{Tr}\ \exp(s\hat{I}) = \frac{i}{s}\sum_{k=0}^{\infty} e^{s\sigma_k} \qquad (4.5)$$

Eq.(4.5) defines ρ as a holomorphic function in the half-plane $\mathrm{Re}\, s < 0$, such that :

$$F(\sigma) = \int_C e^{-\sigma s}\,\rho(s)\, ds \qquad (4.6)$$

along the contour C shown in Fig.11.

As in section 1, by pushing the contour C to the right we can express $F(\sigma)$ as a sum of contributions from the singularities s_j of $\rho(s)$, each discontinuity $\sum \rho_n^{(j)} (s-s_j)_+^n / \Gamma(n+1)$ producing a term $e^{-s_j\sigma}(\sum \rho_n^{(j)}\, \sigma^{-n-1})$.

On the other hand, by substituting into (4.4) the expansion $\bar{F}(\sigma) \sim \sum b_j \sigma^{1-2j}$ (with subdominant terms neglected), we explicitly get :

$$F(\sigma) \sim \sigma + \frac{b_1}{\sigma} + \frac{b_2}{\sigma^3} + \ldots + \sum_{m\neq 0} e^{-im\sigma}\left[\frac{(-1)^m}{m}\exp\left(-im\left(\frac{b_1}{\sigma} + \frac{b_2}{\sigma^3} + \ldots\right)\right)\right] \qquad (4.7)$$

This means that $\rho(s)$ has singularities at *all integral points* on the imaginary axis. The discontinuity $\Delta_0\rho(s)$ from $s=0$ is directly responsible for the Thomas-Fermi term $\bar{F}(\sigma)$, of which it is the Borel transform :

$$\Delta_0\rho(s) = \sum_{j=0}^{\infty} b_j\ s_+^{2j-2} / \Gamma(2j-1) \qquad (4.8)$$

The discontinuities $\Delta_{im}\rho$ from the other points $s = im$ $(m \in \mathbf{Z})$ are found by identifying :

$$(-1)^m \frac{i}{m}\ \exp\left(-im\left(\frac{b_1}{\sigma} + \frac{b_2}{\sigma^3} + \ldots\right)\right) \equiv \sum \rho_\ell^{(im)}\ \sigma^{-\ell-1} \quad .$$

An equivalent expression uses the integration operator $\left(\frac{d}{ds}\right)^{-1}$ [6] :

$$\begin{aligned}\Delta_{im}\rho(im+s) &= \frac{i(-1)^m}{m}\exp\left(-im\sum_1^{\infty} b_j \left(\frac{d}{ds}\right)^{1-2j}\right)\delta(s) \\ &= (-1)^m\left(\frac{i}{m}\,\delta(s) + b_1\theta(s) - \frac{im}{2}\,b_1^2\, s_+ + \ldots\right)\end{aligned} \qquad (4.9)$$

In conclusion : $\rho(s)$ has singularities distributed periodically, in order to build up the jumps of the function $F(\sigma)$; moreover, the discontinuity at $s=0$ *generates* the other discontinuities at $s = im$ in the explicit fashion that we have just described.

We shall now see how the same thing happens with the *complex* singularities of $\rho(s)$. Actually, because of the alternative formula :

$$F(\sigma) = \frac{1}{\pi}\int_0^{\sigma} \mathrm{Im}\ \mathrm{Tr}(\hat{I} - \sigma' - i0)^{-1}\, d\sigma' \quad ,$$

we can also express $F(\sigma)$ as a path integral similar to (3.2), and deduce that the singularities of $\rho(s)$ are the actions of classical periodic orbits, multiplied by i : they form a lattice generated by $\frac{-1+i}{2}$ (not by $(-1\pm i)\frac{\sigma}{2}$ as in Eq.(3.4), because we now Laplace transform with respect to σ instead of $1/\hbar$). Only the singularities with positive real parts should appear on the first sheet, whose tentative structure is shown on Fig.11. The singularities at the points im $(m \in \mathbb{Z})$ arise from the real orbit traversed m times, and they were described by Eq.(4.9). We now consider the *next* row of singularities, those with $\mathrm{Re}\, s = +\frac{1}{2}$. By the same argument of contour deformation they correspond to the corrections of order $e^{-\sigma/2}$ to the function $\bar{F}(\sigma)$ in the quantization condition. Eq.(3.11) does not express directly $\bar{F}(\sigma)$ since k appears on both sides, but if we substitute once $(-1)^k = \sin \frac{\bar{F}(\sigma)}{2} + \mathcal{O}(e^{-\sigma/2})$ in Eq.(3.11), insert the resulting modified $\bar{F}(\sigma)$ into Eq.(4.4) and collect all terms of order $e^{-\sigma/2}$, we find :

$$F(\sigma) \sim \sum_0^\infty b_j \sigma^{1-2j} + \sum_{m\neq 0} e^{-im\sigma} \frac{(-1)^m}{m} \exp\left\{-im \sum_1^\infty b_j \sigma^{1-2j}\right\}$$
$$+ \sum_{m=-\infty}^{+\infty} e^{-i(m+\frac{1}{2})\sigma - \sigma/2} \left(-2i(-1)^m \cdot \exp\left\{-i(m+\tfrac{1}{2}) \sum_1^\infty b_j \sigma^{1-2j}\right\} \cdot \exp\left\{-\frac{1}{2}\sum_1^\infty (-1)^j b_j \sigma^{1-2j}\right\}\right)$$
$$+ \mathcal{O}(e^{-\sigma}) \qquad (4.10)$$

The difference between Eqs.(4.10) and (4.7) is precisely the contribution from the singularities with $\mathrm{Re}\, s = \frac{1}{2}$:

$$\sum_{m=-\infty}^{+\infty} e^{-i(m+\frac{1}{2})\sigma - \sigma/2} \int_0^\infty \Delta_{\left(\frac{1}{2}+i(m+\frac{1}{2})\right)} \rho\left(\frac{1}{2}+i(m+\tfrac{1}{2})+s\right) e^{-\sigma s}\, ds \qquad (4.11)$$

By identification with (4.10), we find :

$$\Delta_{\left(\frac{1}{2}+i(m+\frac{1}{2})\right)} \rho\left(\frac{1}{2}+i(m+\tfrac{1}{2})+s\right) = -2i(-1)^m \exp\left\{-\sum_{j=1}^\infty \left[\frac{(-1)^j}{2} + (m+\tfrac{1}{2})i\right] b_j \left(\frac{d}{ds}\right)^{1-2j}\right\} \delta(s) \qquad (4.12)$$

We thus see that those discontinuities are *also generated by the* $s = 0$ *discontinuity*, as they involve the same coefficients b_j. Clearly, by further using Eq.(3.11) to all orders in $e^{-\sigma/2}$ we could similarly express all the other discontinuities of $\rho(s)$ on its first sheet. However, we cannot ascertain that Eq.(3.11) is correct to higher orders in $e^{-\sigma/2}$ as it stands, so we have not pursued that computation. We rather believe that a better understanding of all the discontinuities should involve some group action of the lattice of classical periods (real and complex) on the quantum function $\rho(s)$. An interesting consequence of this lattice structure is that it no longer makes sense to isolate $\bar{F}(\sigma)$ as "the purely non-oscillating" term in $F(\sigma)$ (the subdominant terms *are* oscillatory) and moreover, any tentative definition of $\bar{F}(\sigma)$ from $\rho(s)$ will raise problems of Borel summability, as each cut carries an infinite sequence of new branch points (Fig.11). Our analysis with Eq.(4.4) remains valid anyhow : it only involves an asymptotic representation for *some* $\bar{F}(\sigma)$ satisfying (4.1).

We now give an application (and a check) of Eq.(4.12) by considering the large order behavior of the b_j . The latter arise as coefficients in the expansion of the discontinuity $\Delta_o\rho(s)$ (Eq.(4.8)), therefore their large order behavior is controlled by the nearest singularities of $\rho(s)$, namely the four points $s_\varepsilon = \frac{(\pm 1 \pm i)}{2}$. Two of those points : $\frac{1\pm i}{2}$, lie on the first sheet and the corresponding discontinuities are given by Eq.(4.12) with $m=0$ and -1 . We now want to apply the large order formula (1.13). At first sight it is inadequate since $\rho(s)$ has logarithmic discontinuities whereas (1.13) involved square-root discontinuities. Fortunately, relations like (1.13) between various discontinuities of $\rho(s)$ are preserved if we multiply $F(\sigma)$ by any factor σ^α , i.e. if we apply a fractional derivative $(d/ds)^\alpha$ [6] to all discontinuities of $\rho(s)$. By selecting an (arbitrary) *half-integer* for α, we can transform $\rho(s)$ to a function $\rho'(s)$ with the desired structure (at least formally), apply (1.13) to ρ' , and express the result in terms of the original function ρ . The discontinuity $\Delta_\varepsilon\rho(s_\varepsilon+s) = \sum_{k=0}^{\infty} \rho_k^{(\varepsilon)} s_+^k / \Gamma(k+1)$ at $s=s_\varepsilon$ gives the following contribution to the large order terms in $\Delta_o\rho(s) = \sum_{k=0}^{\infty} \rho_k^{(0)} s_+^k / \Gamma(k+1)$:

$$\rho_k^{(0)} \sim (2i\pi)^{-1} s_\varepsilon^{-k} \sum_{\ell=0}^{\infty} \rho_\ell^{(\varepsilon)} s_\varepsilon^\ell \Gamma(k-\ell) \tag{4.13}$$

In our case, the contribution of the two points $s_\varepsilon = \frac{1\pm i}{2}$ to the large order behavior of $b_j = \rho_{2j-1}^{(0)}$ has the form :

$$b_j \sim \Gamma(2j-1) \frac{2^{j+1/2}}{\pi} \cos\left(j\frac{\pi}{2} + \frac{3\pi}{4}\right) \left(1 + \frac{\alpha_1}{2j-2} + \frac{\alpha_2}{(2j-2)(2j-3)} + \cdots\right)$$

We cannot directly evaluate the contribution of the two points $s_\varepsilon = \frac{-1\pm i}{2}$ on the other sheet but, because the original Hamiltonian (2.1) is even in M , we expect the discontinuities of $\rho(s)$ to be symmetrical under $s\to -s$; this is also necessary in order that $\rho_{2j}^{(0)} \sim 0$ at the end . In particular, the contribution of the two points $\frac{-1\pm i}{2}$ to the large orders of b_j should be equal to that of the points $\frac{1\pm i}{2}$ and should add to it (indeed, the empirical law (2.17) shows that the previous formula for b_j was too small by a factor 2). The final correct formula is thus

$$b_j \sim \Gamma(2j-1) \frac{2^{j+3/2}}{\pi} \cos\left(j\frac{\pi}{2} + \frac{3\pi}{4}\right) \left(1 + \frac{\alpha_1}{2j-2} + \frac{\alpha_2}{(2j-2)(2j-3)} + \cdots\right) \tag{4.14}$$

where, due to Eqs.(4.12-13), the α_ℓ admit the generating function :

$$\sum_0^\infty \alpha_j t^j = \exp\left(\frac{b_1 t}{2} + \frac{b_2 t^3}{2^2} - \left(\frac{b_3 t^5}{2^3} + \frac{b_4 t^7}{2^4}\right) + \frac{b_5 t^9}{2^5} + \frac{b_6 t^{11}}{2^6} - \cdots\right) \tag{4.15}$$

$$\left(\alpha_1 = \frac{b_1}{2} = -\frac{\pi}{6}\ , \quad \alpha_2 = \frac{b_1^2}{8} = \frac{\pi^2}{72}\ , \quad \alpha_3 = \frac{b_1^3}{48} + \frac{b_2}{4} = -\frac{\pi^3}{1296} + \frac{11\ \Gamma(1/4)^8}{41472\ \pi^2} \cdots\right)$$

Table 3 shows a good agreement between the values of α_ℓ ($\ell \le 7$) derived from Eq.(4.15) and those found by a numerical fit on the b_j' of Table 1. Moreover (4.14) predicts the correct *sign* of b_j (starting from $j = 2$). All this confirms the validity of formula (4.14), and indirectly of the correction of order $e^{-\sigma/2}$ (with power terms included) in the quantization condition (3.11).

We further note that a curious reciprocity law seems to apply to semi-classical expansions. On one hand (and on general grounds), the large order behavior of $\rho_j^{(0)}$ is governed by the first terms $\rho_j^{(\varepsilon)}$ of the nearest discontinuity (with the notations of section 1). On the other hand, and this is a peculiarity of the WKB expansion, all the discontinuities with $s \neq 0$ are generated by the one at $s = 0$. So that on the whole, the behavior of b_j $(j \to \infty)$ is governed by the first b_j themselves, as shown by Eqs.(4.14-15). It is interesting to note that a similar property was derived in a different way by Dingle, again in the case of a semi-classical expansion, this time for the wave function near a turning point (cf.[3], chap.14.3, Eq.(27)).

To conclude, let us discuss possible generalizations of the foregoing results. Apart from the particularly simple dependence of $\bar{F}(\sigma)$ on σ, the reconstruction of $F(\sigma)$ from $\bar{F}(\sigma)$ and the subsequent analysis of the singularities of $\rho(s)$ on $\text{Re}\, s = 0$ are probably valid for arbitrary *one-dimensional, single-well* potentials, but the analytic continuation to $\text{Re}\, s > 0$ will be very difficult except for a general *quartic* potential, in which case the classical trajectories are still elliptic functions. For multi-dimensional but *integrable* systems, our analysis can probably be transposed using a full set of action variables.

As for non-integrable systems, it follows from the general analysis of Balian and Bloch[9] that the singularities of $\rho(s)$ lie at the points $i\times$(actions along periodic orbits) as before, but the actual structure of periodic orbits is unknown and probably quite complicated. A related family of investigations concerns the Laplace operator Δ either in a bounded cavity[27], or on a compact Riemannian manifold[28]. In both cases, the analytic function under study is $\rho(s) = \text{Tr} \exp s\sqrt{-\Delta}$, $t = is$ is now the actual time variable and the singularities of ρ on $\text{Re}\, s = 0$ lie at the *periods* of classical closed orbits ; the corresponding discontinuities obey explicit formulas much more complicated than Eq.(4.9)[28]. An additional difficulty in non-integrable systems is that the singularities may be dense on the imaginary axis ; even without that, the problem of analytically continuing $\rho(s)$ across $\text{Re}\, s = 0$ has not been raised so far.

Acknowledgements : One of us (A.V.) has benefited from helpful discussions with C.M. Bender (who communicated his results of Ref.[12] before publication), M.V. Berry and J. Zinn-Justin. We are also grateful to J. Lascoux and T.T. Wu for stimulating discussions, to R. Schaeffer who helped us derive Eq.(3.9), and finally to Mrs. N. Tichit, Mrs. C. Verneyre and R. Conte for their assistance in the computer calculations.

References

1. G. Darboux, J. Math. 4 (1878) 5, 377.
2. H. Poincaré, Acta Math. 8 (1886) 295 ; W. Wasow, "Asymptotic Expansions for Ordinary Differential Equations" (Wiley, New York, 1965) ; F.W.J. Olver, "Asymptotics and Special Functions" (Academic Press, New York, 1974).
3. R.B. Dingle, "Asymptotic Expansions : their Derivation and Interpretation" (Academic Press, London, 1973).
4. J. Zinn-Justin, Princeton 1978 Lecture Notes (unpublished).
5. R. Balian, G. Parisi and A. Voros, Phys. Rev. Lett. 41 (1978) 1141.
6. I.M. Gelfand, G.E. Shilov, "Generalized Functions", vol.I (Academic Press, 1968).
7. C.M. Bender, T.T. Wu, Phys. Rev. D7 (1973) 1620 ; L.H. Lipatov, JETP 72 (1977) 411 ; E. Brézin, J-C. Le Guillou and J. Zinn-Justin, Phys. Rev. D15 (1977) 1554, 1558 ; G. Parisi, Phys. Lett. 66B (1977) 167.
8. J. Leray, Bull. Soc. Math. Fr. 87 (1959) 81 ; D. Fotiadi, M. Froissart, J. Lascoux and F. Pham, Topology 4 (1965) 159.
9. R. Balian, C. Bloch, Ann. Phys. 63 (1971) 592 ; 85 (1974) 514.
10. For instance : L.I. Schiff, Phys. Rev. 92 (1953) 766 ; C. Schwartz, Ann. Phys. 32 (1965) 277 ; C.E. Reid, J. Molec. Spectrosc. 36 (1970) 183 ; P.M. Mathews, K. Eswaran, Lett. Nuovo Cimento 5 (1972) 15 ; F.T. Hioe, E.W. Montroll, J. Math. Phys. 16 (1975) 1945.
11. A. Voros, Thèse, Université Paris-Sud (Orsay, 1977).
12. C.M. Bender, K. Olaussen and P.S. Wang, Phys. Rev. D16 (1977) 1740.
13. A. Erdélyi et al., "Higher Transcendental Functions" vol.2 (Bateman Manuscript Project, McGraw Hill, New York, 1953) ; W. Magnus, F. Oberhettinger, R.P. Soni, "Formulas and Theorems for the Special Functions of Mathematical Physics" (Springer Verlag, 1966) ; M. Abramowitz, I.A. Stegun, "Handbook of Mathematical Functions"(Dover, New York).
14. For instance : A. Messiah, "Mécanique Quantique" vol.I, ch.6 (Dunod, Paris, 1959 ; English Translation : North-Holland, 1961) ; N. Fröman, Ark. för Fysik 32 (1966) 541.
15. G. Wentzel, Z. Phys. 38 (1926) 518 ; J.L. Dunham, Phys. Rev. 41 (1932) 713.
16. A. Voros, Ann. Inst. H.Poincaré 26A (1977) 343.
17. J.A. Campbell, J. Comput. Phys. 10 (1972) 308 ; 15 (1974) 413 and refs. therein.
18. N. Fröman, P.O. Fröman, J. Math. Phys. 18 (1977) 96.
19. N. Fröman, P.O. Fröman, "JWKB-Approximation, Contributions to the Theory" (North-Holland, Amsterdam, 1965).
20. E.C. Titchmarsh, "Eigenfunctions Expansions" vol.I (Oxford Univ. Press, 1961) ; V.P. Maslov, "Théorie des Perturbations et Méthodes Asymptotiques" (Dunod, Paris, 1972) ; J.P. Eckmann, R. Sénéor, Arch. Rational Mechanics 61 (1976) 153.
21. A.C. Hearn, "REDUCE User's Manual" (University of Utah, 1973).
22. P. Bonche, M. Froissart, J-F. Renardy, "Une Chaîne de Programmes d'Arithmétique à Longueur Variable sur IBM-360" (Note CEA-N-1247, Saclay, 1970).
23. M.C. Gutzwiller, J. Math. Phys. 12 (1971) 343 ; R. Dashen, B. Hasslacher, A. Neveu, Phys. Rev. D10 (1974) 4114.
24. A. Neveu, Rep. Progr. Phys. 40 (1977) 709.
25. J. Heading, "An Introduction to Phase-Integral Methods" (Methuen, London, 1962).
26. M.V. Berry and K.E. Mount, Rep. Progr. Phys. 35 (1972) 315.
27. R. Balian, C. Bloch, Ann. Phys. 69 (1972) 76.
28. Y. Colin de Verdière, Comptes Rendus Acad. Sci. 275 (1972) 805 and 276 (1973) 1517 ; J. Chazarain, Inventiones Math. 24 (1974) 65 ; J.J. Duistermaat and V.W. Guillemin, Inventiones Math. 29 (1975) 39.
29. G. Parisi, "Trace Identities for the Schrödinger Operator and the WKB Method", Preprint LPTENS 78-9 (Ecole Normale Supérieure, Paris, March 1978).

n	b'_n	n	b'_n	n	b'_n
0	-	18	0.62690462618350	36	-0.63187742695115
1	-0.52359877559830	19	0.62743981578357	37	-0.63200858341836
2	0.40119597249644	20	-0.62792002257294	38	0.63213268306740
3	0.49832256904969	21	-0.62835082264362	39	0.63225028032241
4	-0.61771117216663	22	0.62874081905774	40	-0.63236187301556
5	-0.61443022626847	23	0.62909616407778	41	-0.63246790944398
6	0.59570714354132	24	-0.62942091956050	42	0.63256879437991
7	0.60522456343627	25	-0.62971871161799	43	0.63266489422160
8	-0.61578165645471	26	0.62999285275772	44	-0.63275654143407
9	-0.61744255110996	27	0.63024609031867	45	-0.63284403838861
10	0.61789138060000	28	-0.63048070231369	46	0.63292766069161
11	0.61975452512628	29	-0.63069865940814	47	0.63300766008082
12	-0.62185534397455	30	0.63090167915462	48	-0.63308426695388
13	-0.62300181061251	31	0.63109124785922	49	-0.63315769258223
14	0.62392636719298	32	-0.63126865722127	50	0.63322813105504
15	0.62483581881424	33	-0.63143503956354	51	0.63329576099071
16	-0.62563788089519	34	0.63159139189047	52	-0.63336074704796
17	-0.62631163052424	35	0.63173659438058	53	-0.63342324126315

Table 1 - Computed values of $b'_n = b_n \times 2^{-n} / (2n-2)!$

k = 0 E_0 = 1.060

j	(a)	(b)	(c)
0	0.87*	1.00*	1.00*
1	0.98*	1.09*	1.08*
3	0.79	1.01	1.00
5	1.40	1.43	1.42

k = 3 E_3 = 11.644 745 51

j	(a)	(b)	(c)
0	11.611 525 3	11.611 501 7	11.611 501 7
1	11.644 989 5	11.644 966 4	11.644 967 0
3	11.644 765 8	11.644 742 7	11.644 743 3
5	... 768 2	... 745 2	... 745 7
7	... 767 9 *	... 744 9 *	... 745 4 *
9	... 768 1 *	... 745 0 *	... 745 6 *
11	... 767 9	... 744 8	... 745 4

k = 6 E_6 = 26.528 471 183 682

j	(a)	(b)	(c)
0	26.506 335 510 963	26.506 335 513 306	26.506 335 513 306
1	26.528 512 551 758	26.528 512 554 070	26.528 512 554 040
3	26.528 471 147 158	26.528 471 149 470	26.528 471 149 441
5	... 181 652	... 183 964	... 183 935
7	... 181 390	... 183 703	... 183 673
9	... 181 401 *	... 183 713 *	... 183 684 *
11	... 181 399 *	... 183 712 *	... 183 682 *

Table 2 - Semi-classical estimates for a sample of energy levels of the quartic oscillator, using : a) the standard WKB series (3.1) ; b) the same with an exponential term added (3.9) ; c) the same with the complete subdominant expansion added (3.11). The best estimates (up to the computed order) are shown by * . Exact levels are borrowed from Reid[10] and Ref.[12] (the latter contains a misprint for E_6).

ℓ	α_ℓ (theory)	α_ℓ (fit)
0	1	0.999 999 99
1	- 0.523 598 776	- 0.523 598 7
2	0.137 077 839	0.137 08
3	0.778 467 349	0.778 49
4	- 0.416 999 718	- 0.416 3
5	-11.850 079	-11.9
6	6.564 854	6.57 (±0.02)
7	442.947	437 (±3)

Table 3 - Coefficients α_ℓ of the large order expansion (4.14) for b_j as given by theory (Eq.(4.15)) and by a numerical fit based on the values of Table I . Discrepancies are compatible with computer noise.

TRACE IDENTITIES FOR THE SCHROEDINGER OPERATOR AND THE WKB METHOD

G. Parisi

Abstract

Trace identities are derived for the operator $p^2 + x^q$, q positive even. For q=4 these identities can be used to estimate the ground state energy within an accuracy of about 10^{-4}.

1. Introduction

Trace identities for differential operators play an important role in many domains of mathematical physics[1-5]. In this paper, we consider the differential operator

$$H = p^2/2 + x^q = -\frac{\hbar^2}{2}\left(\frac{d}{dx}\right)^2 + x^q \qquad (1)$$

i.e. the Schroedinger operator with potential x^q, q positive even.

This case differs from the one studied by Faddeev[5] in that the potential does not go to zero at infinity. In Section 2 of this paper we present a formal proof of the trace identities for H and we point out the relation between trace identities and the WKB expansion for the eigenvalues of H. In Section 3, we show how trace identities can be used to improve the WKB method for computing the ground state energy.

2. The trace identities

Let us consider the operator (1) where p and x satisfy the canonical commutation relations

$$[p, x] = i\hbar \qquad (2)$$

(In the following $\hbar$ will be used as an expansion parameter which will finally be set to 1). H is a positive self-adjoint operator with discrete spectrum $\{\lambda_n\}$.

We define the following functions: the trace of the resolvent,

$$R(E) = Tr(1/(H+E)) = \sum_{0}^{\infty}{}_n 1/(E+\lambda_n) = \int_0^{\infty} dt\, G(t) \exp(-Et)$$
$$G(t) = Tr\left[\exp(-tH)\right] \tag{3}$$

the Mellin transform of R(E),

$$M(s) = \int_0^{\infty} E^{-s} R(E)\, dE \tag{4}$$

and the Zeta function of the operator H,

$$Z(s) = Tr(H^{-s}) = \sum_{0}^{\infty}{}_n \lambda_n^{-s} . \tag{5}$$

After simple manipulations we obtain:

$$Z(s) = \frac{1}{\pi} \sin(\pi s) M(s) . \tag{6}$$

The analytic structure of the functions M(s) and Z(s) is quite interesting. Notice that if M(s) is not singular for integer s, (s = m), Z(m) must be zero.

Using the WKB method we find the asymptotic expansion[6, 7]:

$$\lambda_n \underset{n\to\infty}{\longrightarrow} N^{2q/(q+2)} (C_o + C_1/N^2 + C_2/N^4 \dots) \qquad N = n + \frac{1}{2} \tag{7}$$

Eq. (7) implies that Z(s) is a meromorphic function with simple poles at the points:

$$s_k = (\frac{1}{2} + \frac{1}{q})(1 - 2k) \qquad k = 0, 1, 2, \dots . \tag{8}$$

The residuum of Z(s) (Z_k) at $s = s_k$ is an algebraic function of $C_o, C_1, C_2 \dots C_k$.

We want to derive a similar result for the function M(s). We need to know the large E behaviour of the function R(E), which can be obtained by studying the small t behaviour of G(t) using semiclassical methods. Indeed we know that

$$\mathrm{Tr}\, f(p,x) = \frac{\hbar}{2\pi} \int dp\, dx\, f(p,x) \tag{9}$$

if all p operators are on the left of the x operator. Using the Baker-Hausdorff formula for $\exp - tA - tB$, $A = p^2/2$, $B = x^q$, all x operators may be carried to the right of the p operators, thus producing an expansion in powers of $\hbar$ (small t):

$$G(t) \underset{t \to 0}{\simeq} \sum_0^{\infty}{}_k B_k \hbar^{2k-1} t^{-s_k}/\Gamma(-s_k + 1). \tag{10}$$

Equivalently:

$$R(E) \underset{E \to \infty}{\simeq} \sum_0^{\infty}{}_k B_k \hbar^{2k-1} E^{-1+s_k}. \tag{11}$$

Therefore, the only poles of $M(s)$ for $s < 1$ are at the points $s = s_k$; their residua being B_k. Of course the following relation holds:

$$Z_k = \sin(\pi s_k)/\pi B_k . \tag{12}$$

We now have at our disposal all the results needed to find the trace identities; they are

$$\begin{aligned} Z(m) &= 0 && \text{if } m \notin \{s_k\}, \\ Z(m) &= (-1)^m B_k && \text{if } m \in \{s_k\}, \end{aligned} \tag{13}$$

where m is a negative integer. If the second case is realized, the pole of $Z(s)$ at $s = s_k$ is absent ($C_k = 0$).

As an example, let us consider the harmonic oscillator $q = 2$. In this case, we find:

$$\begin{aligned} \lambda_n &= n + \frac{1}{2}, \qquad s_k = 1 - 2k, \\ Z(s) &= \sum_0^{\infty}{}_n 1/(n+\tfrac{1}{2})^s = (2^s - 1)\, \zeta(s), \end{aligned} \tag{14}$$

where $\zeta(s)$ is the Riemann ζ function.

Our results imply that $Z(n) = 0$, n being a negative even integer, i.e. we have recovered the trivial zeros of the $\zeta(s)$ function.

3. A numerical application

As stressed by Dikii[3], trace identities are useful for numerical purposes also if they involve the value of Z(s) in a region where the representation (5) does not converge.

Let us define the WKB approximant of order k,

$$W_n^{(k)} = N^{2q/(q+2)} \left[\sum_0^k {}_i C_i N^{-2i} \right] , \qquad N = n + \frac{1}{2} . \tag{15}$$

We can write

$$Z(s) = \sum_0^\infty {}_n (\lambda_n^{-s} - (W_n^{(j)})^{-s}) + \sum_0^\infty {}_n (W_n^{(j)})^{-s} . \tag{16}$$

The first sum is convergent for $s > s_{j+1}$ while the second sum may be calculated by explicit analytic continuation in m. When m is a negative integer, the trace identities (if $m \notin \{s_k\}$) can be written as:

$$\sum_0^\infty {}_n \left[\lambda_n^{-m} - (W_n^{(j)})^{-m} \right] = - \sum_0^\infty {}_n (W_n^{(j)})^{-m} \tag{17}$$

provided that $m > s_{j+1}$.

The identities (17) can be used to evaluate the energies of the lower states by approximating the convergent sum in the l.h.s. of (17) with a finite number of terms.

Let us consider the case q - 4. The simplest approximation is

$$\lambda_o - W_o^{(1)} = - \sum_0^\infty {}_n W_n^{(1)} . \tag{18}$$

A more refined approximation is

$$\lambda_o + \lambda_1 - W_o^{(2)} - W_o^{(2)} = - \sum_0^\infty {}_n W_n^{(2)} ,$$

$$\lambda_o^2 + \lambda_1^2 - \left[W_o^{(2)} \right]^2 - \left[W_o^{(2)} \right]^2 = - \sum_0^\infty {}_n \left[W_n^{(2)} \right]^2 . \tag{19}$$

We use the values[6, 7]

$$C_o = 3^{4/3}\pi^2 \left[\Gamma(1/4)\right]^{-8/3}, \qquad C_1 = C_o/9\pi ,$$
$$C_2 = -C_o\left[5/(2^3 3^4 \pi^2) + 11/(2^9 3^5 \pi^6)(\Gamma(1/4))^8\right]. \qquad (20)$$

Solving for the λ's, we find in the first and second approximations: $\lambda_o = 0.6603$ and $\lambda_o = 0.6677$, $\lambda_1 = 2.3941$ respectively; the exact values are $\lambda_o = 0.6680$ and $\lambda_1 = 2.3936$ (Ref. 8).

For comparison, we report on Table I, the values of $W_n^{(o)}$ $W_n^{(1)}$ $W_n^{(2)}$ for $n = 0, 3$. The values of λ_o and λ_1 - using two trace identities - are more accurate than the second order WKB approximants $W^{(2)}$ of a factor 300 and 15 respectively.

Table I - $W_n^{(0)}$, $W_n^{(1)}$ and $W_n^{(2)}$ are the zeroth, the first and the second order WKB approximants respectively. $\overline{W}_0^{(1)}$ has been computed using the first trace identity as indicated in the text, $W_n^{(1)} = W_n^{(1)}$ for $n \neq 0$. Similarly, $\overline{W}_0^{(2)}$ and $\overline{W}_1^{(2)}$ have been computed using the first two trace identities, $W_n^{(2)} = W_n^{(2)}$ for $n > 1$. λ_n are the exact values.

n	$W_n^{(0)}$	$W_n^{(1)}$	$\overline{W}_n^{(1)}$	$W_n^{(2)}$	$\overline{W}_n^{(2)}$	λ_n
0	0.5463	0.6235	0.6603	0.5960	0.6677	0.6680
1	2.3636	2.4007	2.4007	2.3993	2.3941	2.3936
2	4.6705	4.6970	4.6970	4.6966	4.6966	4.6968
3	7.3148	7.3359	7.3359	7.3357	7.3357	7.3357

The reader interested in high energy physics has quite likely remarked the strong similarity of the method here described with the use of superconvergence sum rules in the Regge theory[9].

Acknowledgements

The author is grateful to C. Itzykson, J. Lascoux and A. Voros for stimulating discussions. He also acknowledges S. Graffi for useful suggestions.

References

1) I. M. Gelfand, Uspekhi Mat. Nauk 11, 191 (1956), MR 18; 129.
2) L. A. Dikii, Izv. Akad. Nauk SSSR Ser. Mat. 19, 187 (1955), MR 17; 619.
3) L. A. Dikii, Uspekhi Mat. Nauk 13, 111 (1958), MR 20; 6655.
4) I. M. Gelfand and B. M. Levitan, Dokl. Akad. Nauk SSSR 88, 593 (1953), MR 13; 588.
5) L. D. Faddeev, Dokl. Akad. Nauk SSSR 115, 878 (1957), MR 20; 1029.
6) A. Voros, Doctorat Thesis, Saclay preprint.
7) C. M. Bender, K. Olaussen and P. S. Wang, Phys. Rev. D16, 1740 (1977).
8) F. Hioe and E. Montroll, J. Math. Phys. 16, 1945 (1975).
9) V. De Alfaro, S. Fubini, G. Furlan and C. Rossetti, Phys. Rev. Letters 8, 576 (1966).

INFN, Laboratori Nazionali di Frascati, 00044 Frascati, Italy

Part II. Disordered Systems

COURSE 6

AN INTRODUCTION TO THE STATISTICAL MECHANICS OF AMORPHOUS SYSTEMS

Giorgio PARISI

Università di Roma II (Tor Vergata) and Laboratori Nazionali INFN (Frascati),
C.P. 13 I.00044 Frascati,
Italy

J.-B. Zuber and R. Stora, eds.
Les Houches, Session XXXIX, 1982—Développements Récents en Théorie des Champs et Mécanique Statistique/Recent Advances in Field Theory and Statistical Mechanics

Contents

1. Introduction

In recent years we have seen serious progress in the theoretical understanding of the properties of amorphous material (for a review see ref. 1). Many points of principle have been completely (or at least partially) understood. A great part of this progress is due to the combined use of the renormalization group and the "replica" method [2, 3], which allows us to use the field theory language for a class of problems which are not strictly field theory. Here I will try to present some of the recent results for an audience of people working mainly on field theory and on statistical mechanics of non-amorphous systems.

In these lectures we will study the application of the techniques of standard statistical mechanics to random systems, i.e. to systems whose Hamiltonian is not translational invariant and depends locally on random parameters (e.g. electrons in an amorphous material). For reasons of time I will not discuss the phenomenological side (i.e. applications of these ideas to real systems), but I will concentrate on a few theoretically simple examples; magnetic systems in a random magnetic field, spin glasses and the Schrödinger equation with a random potential. In all these cases I will try to stress the limits of our present understanding of the problems.

A few words will be spent on related arguments: the relations between super-symmetric field theories and stochastic differential equations.

2. Random magnetic field (mean field approximation)

It is well known that magnetic systems are well described near the critical point by the following effective hamiltonian [4]

$$\beta H[\phi, h] = \int_V \mathrm{d}^D x[\tfrac{1}{2}(\partial_\mu \phi(x))^2 + \tfrac{1}{2}m^2\phi^2(x) + \tfrac{1}{4}g\phi^4(x) - h(x)\phi(x)], \tag{2.1}$$

where the dependence from the temperature is concentrated in m^2 and $h(x)$ is the magnetic field at the point x. The h-dependent free energy is given by:

$$F[h] = -\frac{1}{V}\ln\int d[\phi]\exp(-H[\phi,h]) \tag{2.2}$$

(we have set for simplicity β to 1).

Standard statistical mechanics deals with the problem of computing F for a given h. Here we will suppose that h is not known, but it is a random variable with probability distribution $d\mu[h]$. We are interested in computing the average value of the free energy:

$$\bar{F} = \int d\mu[h]F[h]. \tag{2.3}$$

Standard thermodynamical argument implies that (if the probability distribution $d\mu[h]$ does not contain too long range correlations) in the infinite volume limit the free energy of any magnetic field configuration with non-zero probability coincides with $\bar{F}$. More precisely:

$$\overline{F^2} - (\bar{F})^2 \equiv \int d\mu[h]F^2[h] - \left(\int d\mu[h]F[h]\right)^2 = O(V^{-1}). \tag{2.4}$$

We can imagine realizing this model by doping a normal ferromagnet with magnetic impurities whose spin is fixed in a random direction by stereo-chemical constraints. If the feedback of the magnetic interaction on the position of the impurities can be neglected, the probability distribution $d\mu[h]$ is not modified by the ϕ field (as in eq. (2.3)): h is quenched. The other possibility (h annealed) corresponds to

$$\bar{F}_{an} = -V^{-1}\ln\int d\mu[h]\,d[\phi]\exp(-H[\phi,h]). \tag{2.5}$$

The replica trick (which we will discuss later on) allows us to obtain the properties of the quenched system as the analytic continuation of those of a class of annealed systems [2, 3].

Let us introduce some notation: the bar denotes the average over h and the brackets $(\langle\ \rangle)$ denote the thermodynamical average over ϕ. There are two kinds of correlation functions: the normal ones:

$$\chi(x_1 - x_2) = \overline{\langle\phi(x_1)\phi(x_2)\rangle - \langle\phi(x_1)\rangle\langle\phi(x_2)\rangle} \tag{2.6}$$

and the randomness induced ones:

$$G(x_1 - x_2) = \overline{\langle\phi(x_1)\rangle\langle\phi(x_2)\rangle}. \tag{2.7}$$

G will be different from zero also when the spontaneous magnetization $M = \overline{\langle\phi\rangle}$ is equal to zero.

Now we will discuss the mean field approximation:

$$VF[h] = H[\phi_h, h], \tag{2.8}$$

where ϕ_h is the field configuration which minimizes $H[\phi, h]$; ϕ_h satisfies the following differential equation [5]:

$$-\Delta\phi_h(x) + m^2\phi_h(x) + g\phi_h^3(x) = h(x), \tag{2.9}$$

whose solution for m^2 positive, is unique as can be seen from convexity arguments.

In this approximation we find:

$$\begin{aligned} M &= \bar{\phi}_h, \\ G(x_1 - x_2) &= \overline{\phi_h(x_1)\phi_h(x_2)}, \\ \chi(x_1 - x_2) &= \langle x_1|(-\Delta + m^2 + 3g\phi^2)^{-1}|x_2\rangle. \end{aligned} \tag{2.10}$$

We will now consider the white noise limit.

$$d\mu[h] = d[h]\exp\left(-\int d^D x \frac{h^2(x)}{2\lambda}\right), \tag{2.11}$$

in other words:

$$\begin{aligned} \overline{h(x)h(y)} &= \lambda\delta^D(x-y), \\ \overline{h(x)h(y)h(z)h(t)} &= \overline{h(x)h(y)}\;\overline{h(z)h(t)} + 2 \text{ permutations}. \end{aligned} \tag{2.12}$$

If m^2 is positive, eq. (2.9) can be written as:

$$\phi(x) = \int d^D y\, D(x-y)[h(y) - g\phi^3(y)], \tag{2.13}$$

where

$$D(x-y) = \langle x|(-\Delta + m^2)^{-1}|y\rangle.$$

At $g = 0$ one immediately finds in momentum space

$$\begin{aligned} G(p) &= \lambda/(p^2+m^2)^2, \\ \chi(p) &= 1/(p^2+m^2). \end{aligned} \tag{2.14}$$

The expansion of ϕ in powers of g can be done using eq. (2.13) recursively: in this way, as proved by Symanzik [6], one generates all possible tree diagrams as can be seen in fig. 1.

Similarly the diagrams contributing to the connected correlation function (χ) are shown in fig. 2.

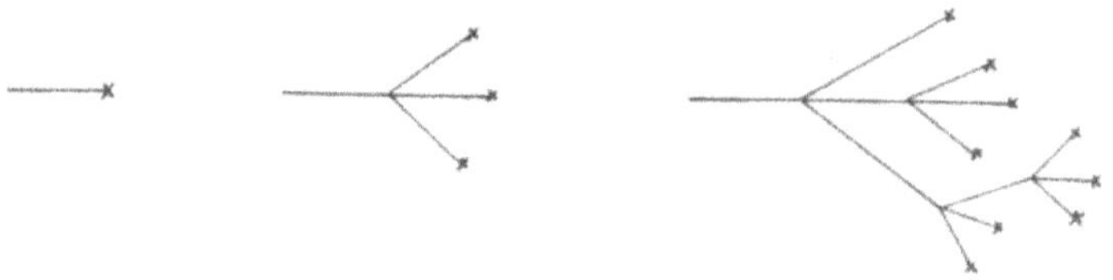

Fig. 1. Diagrams entering in the expansion of ϕ in powers of g. The line and the cross stand for the free propagator and the magnetic field respectively.

Fig. 2. Diagrams for the expansion of $-\Delta+m^2+3g\phi^2$ in powers of g.

Fig. 3. Diagrams for the expansion of G in powers of g.

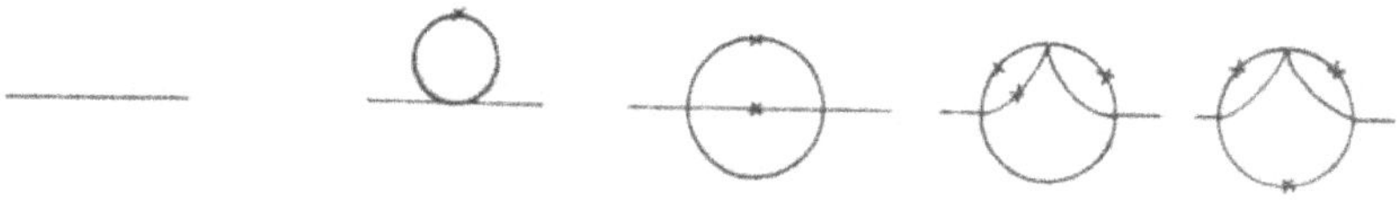

Fig. 4. Diagrams for the expansion of χ in powers of g.

When we average over h field and we take care of the gaussianity of the h distribution we find that all the crosses must coincide at pairs of two. The diagrams contributing to G and χ are shown in figs. 3 and 4 respectively, where the full line is $(p^2+m^2)^{-1}$ and the crossed line is $\lambda(p^2+m^2)^{-2}$. Let us discuss the general rules to obtain all possible diagrams. It is clear that if we neglect the crosses we obtain all possible diagrams of the ϕ^4 theory. We have only to understand where to put crosses. In the case of G the crosses must be such that if we cut the diagram along the corsses, we remain with two disconnected diagrams without loops.

Similarly the crosses for χ must be such that if we cut the diagram along the crosses, we remain with one connected diagram, without loops.

Now let us state the main theorem [5, 7]. Let us consider a conventional field theory with coupling constant g (let us set $\lambda=1$ for the moment) in dimensions $d = D-2$ with hamiltonian

$$H = \int d^d x \left[\tfrac{1}{2}(\partial_\mu \phi)^2 + \tfrac{1}{2}m^2\phi^2 + \tfrac{1}{2}g\phi^4\right]. \tag{2.15}$$

The correlation function will be defined as usual:

$$C(x-y) = \int d[\phi]\, \phi(x)\phi(y) \exp(-H) \Big/ \int d[\phi] \exp(-H). \tag{2.16}$$

At all orders in g we get:

$$\begin{cases} \tilde{\chi}(p) = \tilde{C}(p), \\ G(x) = C(x), \end{cases} \tag{2.17}$$

where $\tilde{\chi}(p)$ and $\tilde{C}(p)$ are the Fourier transforms of $\chi(x)$ and $C(x)$ done in D and d dimensions respectively.

In other words we have

$$\tilde{C}(p) = \int \frac{d\rho(\mu^2)}{\mu^2 + p^2} = \tilde{\chi}(p),$$

$$\tilde{G}(p) = \int \frac{d\rho(\mu^2)}{(\mu^2 + p^2)^2}. \tag{2.18}$$

The proof is very simple after some non-trivial preliminaries which date back to more than twenty years ago [8].

Let us use the Schwinger representation for the propagator:

$$(p^2 + m^2)^{-1} = \int_0^\infty d\alpha \exp[-\alpha(p^2 + m^2)],$$

$$(p^2 + m^2)^{-2} = \int_0^\infty d\alpha\, \alpha \exp[-\alpha(p^2 + m^2)]. \tag{2.19}$$

If we associate an α_i to each line of the diagram, the contribution of a conventional Feynman diagram can be written as

$$\int dp_i \prod_V \delta(p) \int \prod_L d\alpha_i \exp[-\alpha_i(p^2 + m^2)], \tag{2.20}$$

where $\prod_V \delta(p)$ stands for all the δ functions which are needed to enforce momentum conservation (one for each vertex).

The diagrams with crosses can be obtained by multiplying the integrand in eq. (2.20) with

$$\prod_{i\in\{S\}} \alpha_i, \tag{2.21}$$

where S is the set of lines which have crosses. The integration over the p's can be done independently of the set $\{S\}$. We finally find for the contribution of all topological equivalent diagrams with different crosses:

$$\int \prod_L \mathrm{d}\alpha_i \left(\sum_{\{S\}} \prod_{i\in\{S\}} \alpha_i \right) \mathscr{D}(\alpha)^{-D/2} \exp[-N(\alpha)/\mathscr{D}(\alpha)p^2]. \tag{2.22}$$

The contribution of the same diagram in a d-dimensional field theory would be

$$\int \prod_L \mathrm{d}\alpha_i \, \mathscr{D}(\alpha)^{-d/2} \exp[-N(\alpha)/\mathscr{D}(\alpha)p^2], \tag{2.23}$$

$\mathscr{D}(\alpha)$ and $N(\alpha)$ are polynomials in α. Let us now consider the self energy diagrams for $\tilde{\chi}(p)$. If we use the magic formula [8]

$$\mathscr{D}(\alpha) = \sum_{\{S\}} \prod_{i\in\{S\}} \alpha_i, \tag{2.24}$$

which has been proved a long time ago, we see that eqs. (2.22) and (2.23) coincide when $D = d+2$. In the same way one gets the result for G.

We have proved eq. (2.17) at all orders in g for m^2 positive. If the equivalence holds beyond perturbation theory and the mean field approximation is justified, we would have proved the equality of the critical exponents of the random magnetic field model in dimension D with those of a pure system in dimension $d = D-2$. In particular as far as the Ising model in dimension 1 has no transition, we would argue that in the random field model no transition is present in $D = 3$ [9]. However this conclusion is not justisfied.

Indeed let us see what are qualitatively the predictions in the (λ, m^2) plane for $D > 3$: there is a transition which corresponds to the breaking of the $\phi \to -\phi$ symmetry which for $\lambda > 0$ is shifted to negative m^2. However all the previous arguments were done at $m^2 > 0$ and some care must be taken to extrapolate in the region where $m^2 < 0$. In order to see better the difficulties let us start again from eq. (2.10) for $m^2 > 0$, and use the identity

$$\delta(\phi - \phi_h) = \det(-\Delta + m^2 + 3g\phi^2)\delta((-\Delta + m^2 + g\phi^2)\phi - h). \tag{2.25}$$

We get:

$$G(x,y) = \int d[\phi]\delta(\phi-\phi_h)\phi(x)\phi(y)d\mu[h],$$

$$= \int d[\phi]\phi(x)\phi(y)\det(-\Delta+m^2+3g\phi^2)$$

$$\times \exp\left\{-\int d^D x \frac{(-\Delta\phi+m^2\phi+g\phi^3)^2}{2\lambda}\right\}. \tag{2.26}$$

That is correct if $m^2 > 0$ and the stochastic differential equation has only one solution. Let us consider the case in which there are many solutions:

$$\phi_h^i(x), \quad i = 1,\ldots,n[h]$$

ordered with increasing action. The physical quantity is

$$G^P(x-y) = \int d\mu[h]\phi_h^1(x)\phi_h^1(y), \tag{2.27}$$

while if we call G^A the output of eq. (2.26) we get (see Parisi and Sourlas [10], similar conclusions were reached later by Ceccoti and Girardello [10]):

$$G^A(x-y) = \int d\mu[h] \sum_{i=1}^{n(h)} \phi_h^i(x)\phi_h^i(y)$$

$$\times \operatorname{sign}[\det(-\Delta+m^2+3g(\phi_h^i)^2)], \tag{2.28}$$

as can be seen going backward and using the relation:

$$\delta(ax-1) = |a|^{-1}\delta(x-a^{-1}).$$

Now G^A is obviously analytic in m^2 at $m^2 = 0$ and coincides with the result of the $(D-2)$-dimensional field theory. There are no reasons which imply that $G^A = G^P$ for $m^2 < 0$.

An analysis of the zero-dimensional case is useful. We want to find the number ϕ which is the minimum of the function

$$F[\phi,h] = \tfrac{1}{2}m^2\phi^2 + \tfrac{1}{4}g\phi^4 - h\phi, \tag{2.29}$$

where h has the distribution probability: $\exp -\frac{1}{2}h^2 dh$. The correct result for $\overline{\langle\phi\rangle^2}$ is

$$G^P = \int d\phi \exp[-(m^2\phi+g\phi^3)^2/2](m^2+3g\phi^2)\theta(m^2+g\phi^2)\phi^2, \tag{2.30}$$

while in G^A the θ function would be missing. The singular term in G^P around $m^2 = 0$ is given by $(-m^2)^{5/2}\theta(-m^2)$.

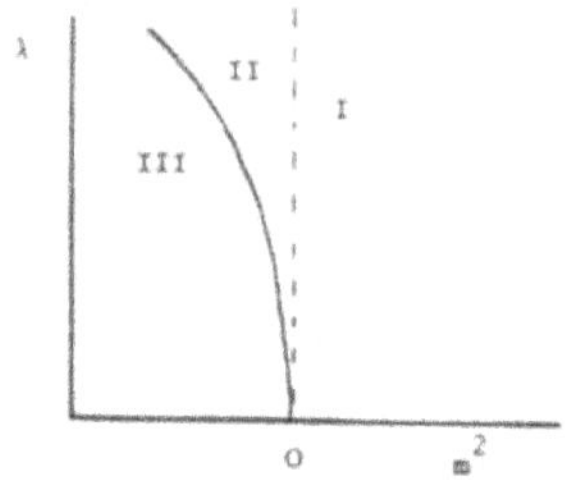

Fig. 5. Schematic phase diagram in the λ, m^2 plane. In regions I and II the spontaneous magnetization is zero, while it is different from zero in region III. The dotted line is at $m^2 = 0$.

The disappointing conclusion is that the dimensional reduction is not valid in the region II of fig. 5, so it does not hold at the transition. The physical solution has a singularity at $m^2 = 0$ which is not present in the dimensionally reduced formulae. The singularity at $m^2 = 0$ follows from general principles: indeed there is a rigorous theorem by Griffith [11] whose generalized heuristic version says: "The free energy of a system with quenched random impurities is singular (C^∞ but not analytic) at the point where the pure system undergoes a phase transition". The physical explanation of this surprising result is that we can have a region of space as large as we want (with very small probability) which is impurity free: the zeros of the partition function can be as near as we want to the real axis and this fact produces the singularity.

We can hope however that the difference between G^P and G^A will be of order $\exp(-1/g)$ in high dimensions, so that the ε-expansion in $6-\varepsilon$ survives as an asymptotic series. Let us sketch the road to estimate the difference. It is clear that we need to know the minimum value of $\int d^D x h^2(x)$ for which there are more than one solution to the eq. (2.9). Let us call $\bar{h}(x)$ the field which minimizes $\int h^2(x) d^D x$ with the constraint that there is more than one solution, and let us call $\bar{\phi}(x)$ the solution which disappears when we decrease $\int h^2(x) d^D x$. Let us see what happens for $h' = \bar{h} + \delta h$. We have:

$$\begin{cases} \displaystyle\int d^D x (h'(x))^2 = \int d^D x \bar{h}^2(x) + 2\int d^D x \bar{h}(x)\delta h(x), \\ (-\Delta + m^2 + 3g\bar{\phi}^2)\delta\phi = \delta h, \end{cases} \tag{2.31}$$

where $\phi' = \bar{\phi} + \delta\phi$ is a solution of eq. (2.9) in the field h'.

$\delta\phi$ should not exist for a δh such that

$$\int d^D x \bar{h}(x) \delta h(x) \neq 0;$$

this is impossible only if:

$$(-\Delta + m^2 + 3g\phi^2)\bar{h} = 0. \tag{2.32}$$

We are therefore interested in finding non-trivial solutions of eqs. (2.32) and (2.9) together, i.e. to

$$(-\Delta + m^2 + 3g\phi^2)(-\Delta\phi + m^2\phi + g\phi^3) = 0. \tag{2.33}$$

In other words we must find the stationary points of

$$S[\phi] = \int d^D x (-\Delta\phi + m^2\phi + g\phi^3)^2. \tag{2.34}$$

We have reduced the problem to the estimation of some instanton like solution, however there are still tricky points (e.g. the choice of the boundary condition in the region II) so we shall not push the evaluation of the effect of the Griffith's singularities up to the end. In §4 we shall see how to obtain eq. (2.33) in the framework of the replica approach.

3. An intermezzo on supersymmetry

We will mention the relation between stochastic differential equations and supersymmetry.

Let us consider eq. (2.26). By introducing fermions ψ and $\bar{\psi}$ it can be written as [5]:

$$G(x-y) = \int d[\phi] d[\psi] d[\bar{\psi}] \exp\left[-\int d^D x S(\phi, \bar{\psi}, \psi)\right] \phi(x)\phi(y),$$

$$S(\phi, \bar{\psi}, \psi) = (-\Delta\phi + m^2\phi + g\phi^3)^2/2\lambda + \bar{\psi}(-\Delta + m^2 + 3g\phi^2)\psi. \tag{3.1}$$

The theory is supersymmetric, i.e. there is a transformation which mixes fermions and bosons and leaves the action invariant. This has been shown explicitly in refs. 5 and 10 where some comments are made on the meaning of supersymmetry breaking. The existence of this supersymmetry allows us to prove the dimensional reduction in a compact way [5, 12]. Unfortunately the theory is not appealing from the point of view of a particle physicist: its fermions have spin 0. However in 2 dimensions it is possible to construct from stochastic differential equations fermions of spin $\frac{1}{2}$.

Let us consider the following system,

$$\partial_1\phi_1+\partial_2\phi_2+g(\phi_1^2-\phi_2^2) = \eta_1,$$
$$-\partial_1\phi_2+\partial_2\phi_1+2g\phi_1\phi_2 = \eta_2, \tag{3.2}$$

$\eta_1(x)$ and $\eta_2(x)$ are uncorrelated white noises. After some manipulations (similar to the previous ones) we find that

$$\overline{\phi_i(x)\phi_j(y)} = \int d\phi_1 d\phi_2 d\psi d\bar{\psi} \exp\left[-\int d^2x S(\phi_1,\phi_2,\psi,\bar{\psi})\right]\phi_i(x)\phi_j(y), \tag{3.3}$$

where

$$S(\phi_1,\phi_2,\psi,\bar{\psi}) = \bar{\psi}[\partial\!\!\!/+g(\phi_1+\gamma_5\phi_2)]\psi$$
$$+\tfrac{1}{2}g^2(\phi_1^2+\phi_2^2)^2+\tfrac{1}{2}(\partial_\mu\phi_1)^2+\tfrac{1}{2}(\partial_\mu\phi_2)^2. \tag{3.4}$$

The model coincides with the two-dimensional supersymmetric Wess Zumino model (for a review see ref. 13). It is natural to pose the following question: "which are the supersymmetric theories that can be written in terms of local stochastic differential equations?". Moreover it was suggested in ref. 10 that the breaking of supersymmetry is related to the existence of an even number of solutions of the associated stochastic differential equation. This could be an alternative way to use topological theorems [14] to study supersymmetry breaking. The field seems to be rather promising but the solution to these problems does not seem to be easy.

4. Random magnetic field (the replica method)

In order to study the random magnetic field model beyond the mean field approximation it is convenient to use the so-called replica trick [3]. The main idea is the following: we define

$$F_n = -\frac{1}{\beta n V}\ln Z_n,$$

$$Z_n = \int d\mu[h]\{Z[h]\}^n = \overline{Z^n},$$

$$Z[h] = \int d[\phi]\exp(-\beta H[\phi,h]). \tag{4.1}$$

It is evident that

$$\bar{F} = \lim_{n\to 0} F_n, \tag{4.2}$$

and that for integer n

$$Z[h]^n = \prod_{a=1}^{n} Z[h] = \int \prod_{a=1}^{n} \mathrm{d}[\phi_a] \exp\left\{-\sum_{a=1}^{n} H[\phi_a, h]\right\}. \tag{4.3}$$

The trick consists in using eq. (4.3) for integer n to define an analytic function of n (F_n) and finally to compute $F_0 = \bar{F}$. The method is very safe in perturbation theory where the dependence of F_n on n is a polynomial, while some difficulties are present (depending on the problem) in the non-perturbative region.

Let us start by studying what happens in perturbation theory and why the mean field approximation of the previous section is justified near the transition.

The main advantage of the replica approach is the possibility of integrating immediately over the h field. We are left with the following effective hamiltonian (see refs. 4, 7 and for a review ref. 15)

$$H_{\mathrm{ef}}[\phi] = \int \mathrm{d}^D x \sum_{a=1}^{n} \Big[\tfrac{1}{2}(\partial_\mu \phi_a)^2 + \tfrac{1}{2} m^2 \phi_a^2 - \tfrac{1}{2}\lambda \sum_{b=1}^{n} \phi_a \phi_b + \tfrac{1}{4!} g \phi_a^4 \Big]. \tag{4.4}$$

Let us first consider the case $g = 0$. The propagator $G_{ab} = \langle \phi_a \phi_b \rangle$ satisfies in momentum space the following equation:

$$(p^2 + m^2) G_{ab}(p) - \lambda \sum_{a} G_{ab}(p) = \delta_{ab}, \tag{4.5}$$

whose solution is given by

$$G_{ab}(p) = \delta_{ab}/(p^2 + m^2) + \lambda/(p^2 + m^2)(p^2 + m^2 - n\lambda). \tag{4.6}$$

Generally speaking the theory is invariant under the group P_n of permutation of n objects; from symmetry arguments it follows that:

$$G_{ab}(p) = \delta_{ab} \chi + G. \tag{4.7}$$

It is easy to verify that the χ and the G introduced in eq. (4.7) coincide with those introduced in eqs. (2.6) and (2.7), in the limit $n \to 0$. It is also clear that the nature of the transition is very different for $n \neq 0$ and $n = 0$. In the first case only G develops a pole at the transition, while in the second case χ and G become critical together. We can now start to do the expansion in powers of g: for generic n the diagrams have a multiplicity factor which is a polynomial in n. The continuation at $n = 0$ does not present problems.

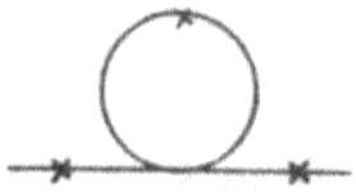

Fig. 6. A diagram having a multiplicity proportional to n.

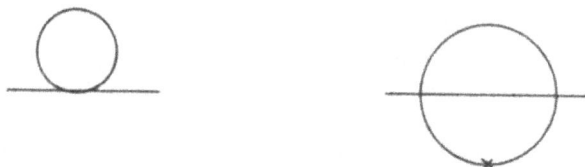

Fig. 7. Diagrams for χ not included in the mean field approximation.

In accordance with the notation of the previous sections we indicate with a full line the free χ and by a crossed line the free G. Neglecting diagrams which have zero weight when $n \to 0$ (see fig. 6), one finds the same diagrams of the previous section plus other diagrams with less crosses (see fig. 7). However it is evident by dimensional counting that the leading infrared divergent diagrams will have the maximum number of crosses; that justifies the mean field approximation where only these diagrams have been considered. The argument (which produces an upper critical dimension equal to 6) is good near 6 dimensions. Decreasing the dimensions, in particular near 4 dimensions, the other diagrams start to be infrared divergent. A careful analysis of the anomalous dimensions of ϕ^4-like operators is needed (e.g. ref. 16) to decide if these extra diagrams have the effect of changing the critical exponents and destroying dimensional reduction. This problem may be solved in the ε expansion (for $\varepsilon \sim 2!$) or in the loop expansion, using the standard techniques for computing anomalous dimensions of composite operators [17].

Let us now try to see what kind of non-perturbative effects are expected. As is clear from Zinn-Justin's lectures in this Volume, we have to look for instantons, i.e. for localized stationary points of the effective hamiltonian (4.4) [18].

The stationary point equations are:

$$-\Delta\phi_a + m\,\phi_a + g\phi_a^3 = \lambda \sum_{b=1}^{n} \phi_b; \tag{4.8}$$

now we must make an ansatz of the form of the solution: if we consider

$$\phi_a(x) = f(x), \quad a = 1, \dots, j, \tag{4.9}$$

we find the standard equation:

$$-\Delta f + (m^2 - \lambda j) f + g f^3 = 0, \tag{4.10}$$

which for $D > 1$ does not have localized solutions for $g > 0$.

Let us now consider a more complicated form for the solution:

$$\phi_a(x) = f_1(x) \quad a = 1, \dots, j; \quad \phi_a(x) = f_2(x) \quad a = j+1, \dots, n. \tag{4.11}$$

In the limit $n \to 0$ the number of ϕ's equal to f_1 and f_2 will be respectively j and $-j$. (The appearance of a negative number of objects is one of the disturbing features of the replica approach.)

Eq. (4.8) now becomes

$$-\Delta f_1 + m^2 f_1 + g f_1^3 = -\Delta f_2 + m^2 f_2 + g f_2^3 = \lambda j (f_1 - f_2). \tag{4.12}$$

This equation does not have a completely clear meaning: it says that there is a magnetic field such that eq. (2.9) has two solutions, but why must the magnetic field be proportional to the solutions?

However in the limit $j \to \infty$ we recover something familiar; indeed if we call $j(f_1 - f_2) = h$, in the limit $j \to \infty$ (if h remains finite) we find:

$$-\Delta f_1 + m^2 f_1 + g f_1^3 = h, \quad (-\Delta + m^2 + 3g f_1^2) h = 0, \tag{4.13}$$

which coincides with eqs. (2.9) and (2.32).

Of course we can consider also other solutions: the number of φ equal to f_0, f_1, f_2 being n, j, $-j$, respectively.

It is clear that there is no bound to the number and to the form of the solutions we can consider: when the replica symmetry group P_n is broken, (instantons have a definite orientation in the replica space), Pandora's box is opened, and we gain nothing by closing it again.

We shall not try to classify the solutions and to interpret them from the physical point of view: we limit ourselves to a few comments.

Instantons in the replica space seem to be connected to Griffith singularities [18]; a careful analysis of the dependence of their contribution as a function of the various parameters g, λ, m^2 and D would be welcome.

Formally in the limit $g \to 0$, $\lambda \to \infty$ at $g\lambda$ fixed the mean field approximation of § 2 is correct. Instantons here have the rôle of breaking the dimensional reduction and of allowing the possibility of a transition at $D = 3$. Contrary to what happens in quantum mechanics

[17] instantons seem to restore (not to destroy) the breaking of the symmetry $\phi \leftrightarrow -\phi$.

There is also a peculiar feature of the instantons for $n = 0$ that I would like to stress. The contribution to the hamiltonian of a P_n symmetric configuration goes to zero when $n \to 0$, while the contribution of an instanton (let us consider the case $j = 1$ for definiteness) remains finite. However there is no contradiction with the fact that the free energy F_n defined in eq. (4.1) remains finite when $n \to 0$. Indeed for small n we have:

$$Z_n \simeq \exp(-n\tilde{F}_0) + w\exp(-H_1), \tag{4.14}$$

where βV has been set to 1 for simplicity, $\tilde{F}_0$ is the free energy without instantons, H_1 is the value of the hamiltonian for the instanton configurations and w is a weight factor which contains the number of possible orientations of the instantons in internal space. In the case $j = 1$ we have to pick one direction from n and we obtain a term proportional to:

$$\binom{n}{1} = \frac{n!}{1!(n-1)!} = \frac{\Gamma(n+1)}{\Gamma(2)\Gamma(n)} \sim n, \quad n \to 0. \tag{4.15}$$

We finally obtain a finite result when $n \to 0$

$$n^{-1} \ln Z_n \to -F_0 + \exp(-H_1) \tag{4.16}$$

as it should be.

Although in the dilute gas approximation instantons are well separated and independent of one another, it is not clear if (when the instanton density increases) the instanton–instanton interaction will not be such as to orient all the instantons in the same direction in replica space, producing a spontaneous breaking of the replica symmetry at the global level.

Summarizing, we understand quite well what happens in the region I of fig. 1, but when we start to move left by decreasing the temperature, we pretty soon loose control of the situation.

5. Spin glasses (the naive approach)

Spin glasses have been the object of an intensive study in the last decade, from both the theoretical and experimental points of view. There are good reviews of the subject [19]. Here I will consider only a very simple theoretical model (the Edward–Anderson (EA) model [3]) and I will spend most of the time discussing the infinite range version of this model (the Sherrington–Kirkpatrick (SK) model [20]) where the mean field approximation is supposed to be exact.

The hamiltonian of the EA model is as follows:

$$H[\sigma] = - \sum_{\{i,k\}} \sigma_i \sigma_k J_{ik} - h \sum_i \sigma_i, \tag{5.1}$$

where h is the external magnetic field, σ_i are Ising spins (± 1) defined on a d-dimensional lattice. The sum on i and k runs on all the possible nearest neighbour pairs of the lattice and $J_{i,k}$ are independent random variables which have a gaussian probability distribution:

$$\overline{J_{i,k}} = 0, \quad \overline{J^2_{i,k}} = d^{-1/2}. \tag{5.2}$$

As before, we are interested in computing the quenched average of the correlation functions and of the free energy.

The infinite range SK model has the same hamiltonian, eq. (5.1), where the index i now goes from 1 to N, the sum over i and j runs on all the possible $N(N-1)/2$ pairs of spins, the $J_{i,k}$ are still gaussian random variable with zero mean and covariance:

$$\overline{J^2_{i,k}} = N^{-1}. \tag{5.3}$$

The thermodynamic limit is obtained when $N \to \infty$. The SK model is essentially the EA model restricted on a single cell of a hypertriangular lattice in the limit $d \to \infty$.

The quenched free energy for $h = 0$ is invariant under the local gauge transformation:

$$J_{i,k} \to -J_{i,k}, \quad \forall k; \qquad \sigma_i \to -\sigma_i. \tag{5.4}$$

Quantities which are not invariant under the transformation (5.4) are obviously zero, at $h = 0$.

Apart from quantities like internal energy, susceptibility and so on, it is interesting to consider the following two quantities

$$q_{\mathrm{EA}} = \overline{\langle\sigma_i\rangle^2}, \quad \chi_R = \sum_k \overline{(\langle\sigma_i\sigma_k\rangle - \langle\sigma_i\rangle\langle\sigma_k\rangle)^2}, \tag{5.5}$$

where the bar denotes the average over the $J_{i,k}$ and the brackets the thermodynamic average over the Ising spins at fixed J.

Although the definition of q_{EA} seems clear, we shall see later on that it is not as unambigous as it looks.

One of typical features of spin glasses is that, for a given choice of $J_{i,k}$, is very difficult to find the ground state, i.e. the configuration of spins σ_i which minimize the hamiltonian. Indeed as a consequence of the randomness of the $J_{i,k}$, it is not possible to fix the σ_i's in such a way that

$$J_{i,k}\sigma_i\sigma_k > 0, \quad \forall i, k. \tag{5.6}$$

If eq.(5.6) could be satisfied the product of all J_{ik} along a closed loop (the Toulouse–Wilson loop [21]) should be positive, but that is impossible, the J's having zero mean. (If the product of the J's along a loop is negative, the loop is said to be frustrated; for the application of the idea of frustration in a different context see ref. 22.)

We must decide which bond must be frustrated, (i.e. is $\sigma_i\sigma_k J_{i,k} < 0$). Different arrangement of the frustrated bonds may differ very little in energy but correspond to very different spin configurations. Briefly, in more than two dimensions all known algorithms for finding the ground state of an N-site spin glass take a time proportional to $\exp aN$ [23]. This multiplicity of groundstates, which are nearly equivalent from the energetic point of view, is responsible for many of the strange properties of spin glasses.

For fixed J's we expect that in the high temperature phase $\langle\sigma_i\rangle = 0$ while, for $T < T_c$, $\langle\sigma_i\rangle$ is different from zero and is oriented in the direction of one of the possible ground states. As far as $\overline{\langle\sigma_i\rangle}$ is obviously zero, the natural order parameter for characterizing the transition is q_{EA}.

At first sight no transition is expected for $h \neq 0$ and χ_R should always be finite (excluding the point $T = T_c$, $h = 0$). However we shall see that this is not the case; the phase diagram is the one shown in fig. 8: χ_R becomes infinite on the transition line which separates regions I and II [24] (and it will remain infinite in the whole of region II). The order

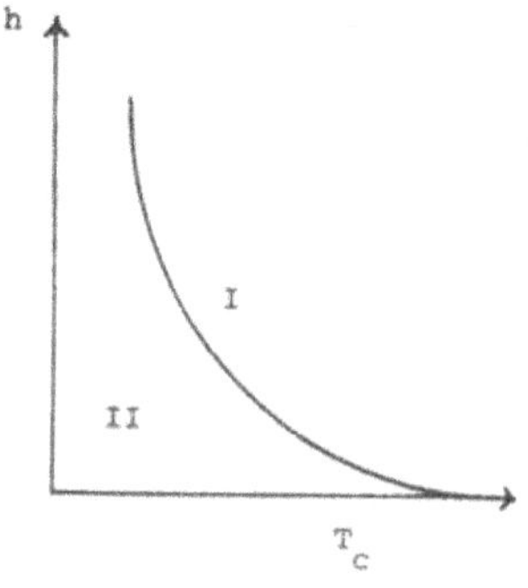

Fig. 8. Region I is the high temperature phase. Region II is the glassy phase where irreversible phenomena are present. The line separating the two regions is the AT line.

parameter characterizing the properties of region II will not be a number but a function $q(x)$ defined on the interval 0–1 [25]. A very slow approach to equilibrium is expected to be present in region II.

In order to see how all this happens, let us start by looking at the solution of the SK model, using the replica trick. With the same notation as the previous section we find after the integration over the $J_{i,k}$ [20]:

$$Z_n = \sum_{\{\sigma_i^a\}} \exp\left\{\frac{\beta^2}{4N} \sum_{i=1}^{N} \sum_{k=1}^{N} \sum_{a=1}^{n} \sum_{b=1}^{n} \sigma_i^a \sigma_i^b \sigma_k^a \sigma_k^b + h\beta \sum_{i=1}^{N} \sum_{a=1}^{n} \sigma_i^a\right\}$$

$$= \int_{-\infty}^{+\infty} \mathrm{d}Q_{ab} \exp[-NA(Q)],$$

$$A(Q) = -\frac{n\beta^2}{4} + \frac{1}{4} \sum_{a=1}^{n} \sum_{b=1}^{n} Q_{ab}^2$$

$$-\ln\left\{\mathrm{tr}\left[\exp\left(\frac{\beta}{2} \sum_{a=1}^{n} \sum_{b=1}^{n} Q_{ab} S_a S_b + \beta h \sum_{a=1}^{n} S_a\right)\right]\right\} \tag{5.7}$$

where the indices a and b label the different replicas of the spin system; Q_{ab} is an $n \times n$ matrix, zero on the diagonal, and tr stands for the sum over the 2^n possible values of the n Ising spin variables S_a.

Up to now, we have done legal operations. When N goes to infinity we would like to use the saddle point method and write:

$$\bar{F} = -\frac{1}{\beta} \lim_{n \to 0} \frac{1}{n} \left[\min_{Q} A(Q)\right]. \tag{5.8}$$

The meaning of eq. (5.8) is no clearer than that of a Delphic oracle: should we find the minimum at $n \neq 0$ and analytically continue to $n = 0$? No. The analytic continuation of a minimum may be a maximum. The minimum should be found directly at $n = 0$. However the number of variables corresponding to the Q_{ab} is $n(n-1)$ which is negative for $0 < n < 1$: the concept of a minimum of a function depending on a negative number of variables is rather subtle! Everybody would say that the minimum of $n^{-1} \mathrm{Tr}\, Q^2$ is at $Q = 0$, however if we set

$$Q_{ab} = q \quad \forall a, b \quad (Q_{a,a} = 0),$$

we find

$$n^{-1} \mathrm{Tr}\, Q^2 = (n-1) q^2. \tag{5.9}$$

The point $q = 0$ is (for $n < 1$) a maximum as a function of q. The solution to this apparent paradox is quite simple: the condition that the hessian matrix

$$H_{a,b;c,d} = \frac{\partial^2 A}{\partial Q_{ab}\,\partial Q_{cd}}, \tag{5.10}$$

has positive eigenvalues does not imply that $\langle x|H|x\rangle$ is positive, if $|x\rangle$ belongs to a negative dimensional space (e.g. the trace of the identity is equal to the dimension of the space). A moment of reflection is needed to understand that the necessary condition for the use of the saddle point method is that all the eigenvalues of the hessian must be non-negative. This also guarantees that all the susceptibilities, which are positive definite, are positive indeed.

The final interpretation of eq. (5.8) is the following: we must consider all possible analytic families of matrices $Q^{(n)}_{ab}$, which may depend on real parameters q_i or integer parameters m_i [25]. An analytic family is an infinite set of matrices (one for each n multiple of n_0), such that the P_n invariant quantities, e.g.:

$$\operatorname{Tr} Q^k \quad \text{or} \quad \sum_{a=1}^{n}\sum_{b=1}^{n} (Q_{ab})^k,$$

are analytic functions of n. For each analytic family we should compute the analytic continuation in n up to $n = 0$ of the function $A(Q)$ and of the eigenvalues of the hessian. The final result will be given by that analytic family (hopefully unique) whose elements are stationary points of $A(Q)$, for all n multiples of n_0, and the eigenvalues of the hessian analytically continued at $n = 0$ are non-negative. As far as one can construct analytic families of matrices which depend analytically on the integer parameters m_i, one is allowed also to consider non-integer values of the m_i's.

While it is not clear if this interpretation gives the correct result, it does make the problem well defined from the mathematical point of view but very hard to control from the practical point of view: the space of all analytical families is very large. Up to now the only approach has been to construct ansatzs. Let us see how this works.

The first case we study is:

$$Q_{ab} = q, \quad a \neq b. \tag{5.11}$$

After some simple algebraic manipulations we find that:

$$A(q) = -\frac{\beta^2}{4}(1+q^2) - \frac{1}{(2\pi)^{1/2}}\int_{-\infty}^{+\infty} \mathrm{d}z\{\exp[-z^2/2]\ln[2\mathrm{ch}(\beta q^{1/2} z + \beta h)]\}. \tag{5.12}$$

The hessian will have one eigenvector in the one-dimensional space (5.11). One can immediately check that for $n < 1$ the corresponding eigenvalue is positive if $A(q)$ is a maximum (not a minimum!) as function of q.

One finally finds at $h = 0$:

$$q = 0, \quad T > T_c = 1,$$

$$q_{EA} = q = \int_{-\infty}^{+\infty} \frac{dz}{(2\pi)^{1/2}} \{\exp[-z^2/2] \mathrm{th}^2(\beta q^{1/2} z)\} \neq 0, \quad T < 1,$$

$$U(T) \to -\tfrac{1}{2}\sqrt{\pi} = -0.798, \quad q(T) \sim 1 - \sqrt{\pi}\, T \quad (T \sim 0),$$

$$C(T) \sim T \quad (T \sim 0),$$

$$\chi(0) \sim \sqrt{\pi}/2, \quad S(0) = -1/2\pi, \tag{5.13}$$

where U, C, χ and S are the internal energy, the specific heat, the susceptibility and the entropy. Now the Monte Carlo results tell us that $U(0) \simeq -(0.76\text{–}0.77)$ in small but definite disagreement with eq. (5.13) and the specific heat is quadratic in T; on the other hand the dependence of $q(T)$ is qualitatively correct (apart from the fact that Monte Carlo data suggest $q(T) \sim 1 - aT^2$). Unfortunately the entropy becomes negative at low T and a negative Ising system entropy cannot be tolerated. The situation is more or less elucidated by fig. 9.

We notice that the curve at $q \neq 0$ for $T < T_c$ is definitely better than the $q = 0$ curve and coincides with the experimental line from Monte Carlo up to $T \simeq 0.4T_c$; it is too small near $T \simeq 0$. The shape of the

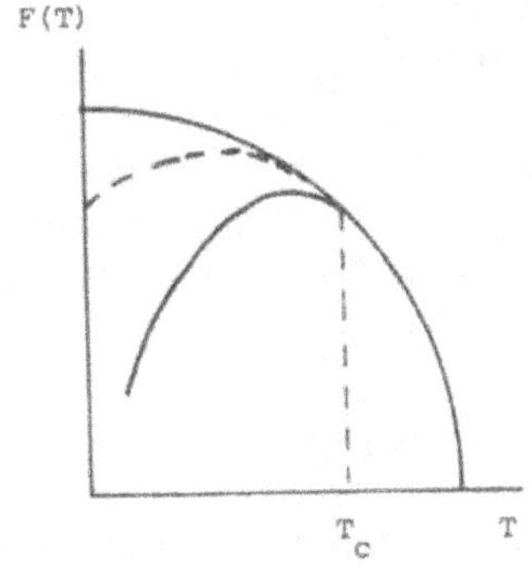

Fig. 9. The lower curve is the free energy for $q = 0$ (i.e. the analytic continuation of the high temperature free energy), the higher curve is the free energy of the computer simulations, the dashed curve in the middle is the free energy given by eq. (5.13).

curves suggests that we have not found the correct saddle point and that if we add extra parameters, they must be such that we have to maximize A as a function of them.

This point has been elucidated by De Almeida and Thouless [24, 26], who computed the other eigenvalues of the hessian: they found an eigenvalue crossing zero on the AT line of fig. 8. They also noticed that χ_R is proportional to the inverse of that eigenvalue, so it becomes infinite on the transition line. The previous computation is therefore wrong in region II of fig. 8, and eq. (5.11) is not the correct choice. Unfortunately eq. (5.11) is the only possible ansatz which is P_n symmetric, we need therefore to break spontaneously the replica symmetry. This approach will be the subject of next section. Let us close this section with a few remarks: The internal energy and the susceptibility at $h = 0$ are given by:

$$U = -\tfrac{1}{4}\beta\,(1-\tilde{\mathrm{Tr}}\,Q^2), \quad \chi = \beta(1-\tilde{\mathrm{Tr}}\,Q),$$

$$\tilde{\mathrm{Tr}} = \lim_{n\to 0}\frac{1}{n}\,\mathrm{Tr}. \tag{5.14}$$

Using eq. (5.11) we get:

$$U = -\tfrac{1}{4}\beta(1-q^2), \quad \chi = \beta(1-q). \tag{5.15}$$

The expression for χ has been derived in full generality by Fisher [27]. Indeed from the linear response theory we know that:

$$\frac{\chi}{\beta} = \sum_k \langle\sigma_i\sigma_k\rangle^c = \sum_k [\langle\sigma_i\sigma_k\rangle - \langle\sigma_i\rangle\langle\sigma_k\rangle]. \tag{5.16}$$

After the average over the J's, the term with $i \neq k$ gives a zero contribution because of the Z_2 gauge invariance. Only the term with $i = k$ survives. If we recall that $\sigma_i^2 = 1$, we get eq. (5.15) for χ.

It is clear however that the Monte Carlo data [20, 28] do not satisfy eq. (5.15); $(1-q \sim T^2, \chi \simeq 1)$: the linear response theory does not hold and this is the signal for the breaking of the replica symmetry.

6. Spin glasses (breaking the replica symmetry)

It is clear that we must now find an ansatz different from eq. (5.11). In doing that we will be guided by the following principles [25]:

(a) We should try to remain in the space spanned by the eigenvalues which become negative below the AT line. An explicit computation shows that this condition implies:

$$\sum_b (Q_{ab} - Q_{a'b}) = 0, \quad a \neq a'. \tag{6.1}$$

(b) If we also want to maximize A with respect to the new parameters we shall introduce, we must remain in the space where

$$\tilde{\mathrm{T}}\mathrm{r}\, Q^2 \leqslant 0. \tag{6.2}$$

(c) We should also avoid the possibility

$$\tilde{\mathrm{T}}\mathrm{r}\, Q^2 = \infty. \tag{6.3}$$

A possible way to implement conditions (a) and (c) consists of dividing the n replicas in n/m groups of m replicas. (Of course n must be multiple of m.) We set $Q_{ab} = q_0$ if a and b belong to the same group $Q_{ab} = q_1$ if a and b belong to different groups. (Q_{aa} is always zero.) In other words

$$\begin{aligned} Q_{ab} &= q_0 \quad \text{if } I(a/m) = I(b/m), \\ Q_{ab} &= q_1 \quad \text{if } I(a/m) \neq I(b/m), \end{aligned} \tag{6.4}$$

where $I(x)$ is an integer valued function: its value is the smallest integer greater than or equal to x. Eq. (6.4) provides us with an example of an analytic family of matrices, depending on the parameters q_0, q_1 and m.

It is evident that:

$$\tilde{\mathrm{T}}\mathrm{r}\, Q^2 = (1-m)q_0^2 - mq_1^2 \tag{6.5}$$

is not negative definite if $m > 1$; we must maximize it with respect to q_0 and minimize it with respect to q_1: this automatically leads to a free energy worse than the one obtained in the previous section [29, 30]. However if $0 < m < 1$, condition (6.2) is enforced (obviously m is no more an integer, but we are allowed to do this). After some tedious algebra (it has been remarked that the breaking of the replica symmetry has originated a subculture of very complicated arithmetics [31]) we get:

$$\begin{aligned} A(q_0, q_1, m) = &-\frac{\beta^2}{4}[1 + mq_1^2 + (1-m)q_0^2 - 2q_0^2] \\ &- \int \mathrm{d}p(z)\frac{1}{m}\ln\left\{\int \mathrm{d}p(y)\,\mathrm{ch}^m[\beta(q_1^{1/2}z + (q_0-q_1)^{1/2}y)]\right\}, \\ \mathrm{d}p(z) \equiv &\exp(-z^2/2)\mathrm{d}z/(2\pi)^{1/2}. \end{aligned} \tag{6.6}$$

Maximizing A with respect to q_0 and q_1 and m (restricted to the interval 0–1) we obtain the following surprising results [32]: the curves for V, C, χ and q_{EA} (assuming $q_{EA} = \max Q_{ab} = q_0$) are in very good agreement with the Monte Carlo data (e.g. $V(0) = -0.7652$); the free energy is obviously higher than that obtained in the previous section. The entropy at zero temperature has collapsed from $S(0) \simeq -0.16$ to $S(0) \simeq -0.01$.

We are clearly on the right track! In order to generalise eq. (6.4), let us do some unusual group theory. Eq. (6.4) correspond to breaking the P_n group in the following way

$$P_n \to (P_m)^{\otimes n/m} \otimes P_{n/m}. \tag{6.7}$$

Indeed we can permute both the replicas inside each group (and this leads to the product of n/m times P_m) and the groups among themselves (this leads to $P_{n/m}$). In the limit $n \to 0$ we have the following pattern of symmetry breaking [32, 33]:

$$P_0 \to (P_m)^{\otimes 0} \otimes P_0. \tag{6.8}$$

In other words, P_0 contains itself as a subgroup! It is clear now that we can go on and repeat the same operation many times: we introduce a set of integer numbers m_i ($i = 0, \ldots, k+1$), such that $m_0 = 0$ and $m_{k+1} = n$ and m_i/m_{i-1} is an integer (for $i = 1, \ldots, k+1$). We can divide the n replicas in n/m_k groups of m_k replicas, each group of m_k replicas is divided in m_k/m_{k-1} groups of m_{k-1} replicas and so on. The matrix Q will be given by:

$$Q_{ab} = q_i \quad \text{if } I\left(\frac{a}{m_i}\right) \neq I\left(\frac{b}{m_i}\right)$$

and

$$I\left(\frac{a}{m_{i+1}}\right) = I\left(\frac{b}{m_{i+1}}\right), \quad i = 0, \ldots, k, \tag{6.9}$$

and the q_i's are a set of $k+1$ real parameters. For $k = 1$ we recover the previous example and for $k = 0$ we recover the unbroken symmetry theory.

An easy computation shows that:

$$-\tilde{\mathrm{T}}\mathrm{r}\, Q^2 = \sum_{i=1}^{k} (m_i - m_{i+1}) q_i^2. \tag{6.10}$$

Condition (6.2) is satisfied only if

$$0 \leqslant m_{i+1} \leqslant m_i \leqslant 1. \tag{6.11}$$

From now on let us assume that condition (6.11) is valid.

For each value of k, one can compute the free energy by maximizing it with respect to the q's and the m's. An explicit computation shows that, near T_c, $F^{(k)}$ contains a term proportional to

$$(T - T_c)^5/(2k+1)^4;$$

we are naturally led to consider the case $k \to \infty$. In order to keep track of the parameters q_i and m_i it is convenient to consider the function:

$$q(x) = q_i, \quad m_{i+1} < x < m_i. \tag{6.12}$$

There is a one-to-one correspondence between the piecewise constant functions with k discontinuities and the parameters q_i and m_i. In the limit $k \to \infty$, $q(x)$ becomes a generic L^2 function on the interval 0–1.

The hierarchical construction (6.9) has produced in a strange way an order parameter which is a function. The utility of the parameterization (6.12) can be easily seen from the formulae valid at $h = 0$:

$$-\tilde{\mathrm{T}}\mathrm{r}\, Q^2 = \int_0^1 \mathrm{d}x\, q^2(x),$$

$$U = -\frac{\beta}{2}\int_0^1 (1 - q^2(x))\mathrm{d}x, \quad \chi = \beta \int_0^1 \mathrm{d}x(1 - q(x)). \tag{6.13}$$

We will argue in the next section that

$$q_{\mathrm{EA}} = \max_x q(x) \equiv q_{\mathrm{M}}. \tag{6.14}$$

If $q(x)$ is not a constant the Fischer relation is not satisfied.

Let us now compute the free energy: after some calculations one arrives to the surprising result

$$-\beta F = \max_{q(x)} A[q],$$

$$A[q] = -\tfrac{1}{4}\beta^2\left\{1 + \int_0^1 q^2(x)\mathrm{d}x + 2q(1)\right\} - a[q],$$

$$a[q] = f(0, h), \tag{6.15}$$

where the function $f(x, y)$ satisfies the following differential equation:

$$\frac{\partial f}{\partial x} = -\frac{1}{2}\frac{\mathrm{d}q}{\mathrm{d}x}\left[-\frac{\partial^2 f}{\partial y^2} + x\left(\frac{\partial f}{\partial y}\right)^2\right] \tag{6.16}$$

with the boundary condition:

$$f(1, y) = \ln[2\,\mathrm{ch}(\beta y)]. \tag{6.17}$$

Eq. (6.15) is correct only if $q(0) = 0$, otherwise

$$a[q] = \int_{-\infty}^{+\infty} \mathrm{d}p(z) f'(0, h + \sqrt{q(0)}\, z). \tag{6.18}$$

The generic shape of the function $q(x)$ which maximises eq. (6.15) is shown in fig. 10. There are arguments which suggest that $q(0) \equiv q_m = 0$ for $h = 0$; more precisely:

$$q(0) \sim |h|^{2/3}. \tag{6.19}$$

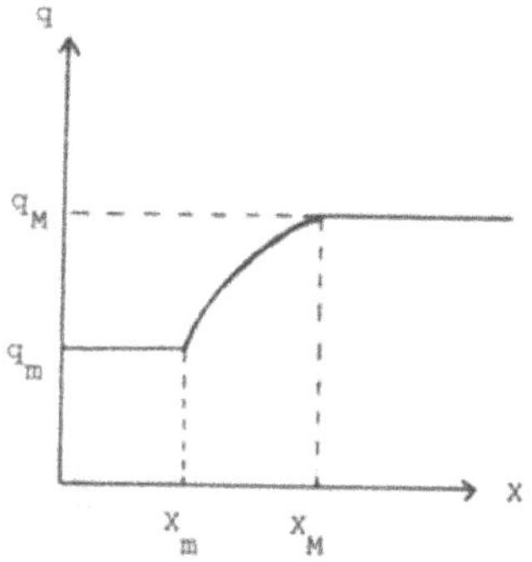

Fig. 10. The generic shape of the function $q(x)$ in region II of fig. 9; $q(x)$ is a constant in the two regions $x < x_m$ and $x > x_M$.

A long argument shows that eq. (6.19) implies that

$$\chi = 1 - \mathrm{O}(h^{4/3}).$$

When we cross the AT line $x_m \to x_M \neq 0$. The following semiempirical rules are exact, or well satisfied, in region II of fig. 8, (in region I $q(x)$ is obviously a constant) [34]:

$$q_m(\beta, h) = q_m(h),$$
$$q_M(\beta, h) = q_M(\beta),$$
$$q(x, \beta, h) = q(x\beta), \quad x_m < x < x_M. \tag{6.20}$$

Numerical investigations support the hypothesis that $S(0) = 0$ in this scheme; the ground state energy estimated is $U(0) \simeq -0.7633 \pm 10^{-4}$. Apart from the region of very small temperature, the results are very similar to those obtained for $k = 1$.

Before discussing in the next section the physical implications of the replica symmetry breaking we notice that (let us consider $h = 0$ for simplicity) eqs. (6.15) and (6.17) may be simplified by introducing the inverse function $x(q)$. We obtain:

$$
\begin{aligned}
A[x] &= -\frac{\beta^2}{4}\left\{(1+q_{\mathrm{M}})^2 - \frac{2}{\beta}\int_0^{q_{\mathrm{M}}} \chi_A(q) q \,\mathrm{d}q\right\} - a[x], \\
a[x] &= g(q,0), \quad g(q_{\mathrm{M}}, y) = \ln[2\mathrm{ch}(\beta y)], \\
\frac{\partial g}{\partial q} &= -\frac{1}{2}\left[-\frac{\partial^2 g}{\partial y^2} + \chi_A(q)\left(\frac{\partial g}{\partial y}\right)^2\right], \\
\chi_A &= \beta x(q), \quad \chi = \beta(1-q_{\mathrm{M}}) + \int_0^{q_{\mathrm{M}}} \chi_A(q)\,\mathrm{d}q, \\
U &= -\frac{1}{2}\left\{\beta(1-q_{\mathrm{M}}) + \int_0^{q_{\mathrm{M}}} \chi_A(q)\,\mathrm{d}q\right\}.
\end{aligned} \tag{6.21}
$$

Eq. (6.21) for the function g can be associated to a stochastic differential equation [33, 35]:

$$
\begin{aligned}
\frac{\mathrm{d}w}{\mathrm{d}q} &= \eta(q) - \tfrac{1}{2}\chi_A(q)\mathrm{th}(\beta w), \\
\overline{\eta(q_1)\eta(q_2)} &= \delta(q_1 - q_2), \\
g(0,0) &= \overline{g(q_{\mathrm{EA}}, w(q_{\mathrm{EA}}))} = \overline{\ln[\mathrm{ch}(2\beta w(q_{\mathrm{EA}}))]}, \\
w(0) &= 0,
\end{aligned} \tag{6.22}
$$

where η is a white noise and $w(q)$ is an η dependent random function and the bar denotes the average over η.

It is satisfactory that we end up with a stochastic differential equation to control a random system; we must admit however that the derivation is not very direct. Let us make a short speculative comment on eq. (6.22). The relation

$$
\chi = \chi_{\mathrm{Fisher}} + \int_0^1 \chi_A(q)\mathrm{d}q
$$

suggests that $\chi_A(q)$ parametrizes the anomalous part of the susceptibility. If $\chi_A(q) = 0$, the replica symmetry and the linear response theory would both be exact. We now try to interpret pictorially eq. (6.22). Suppose we add a new spin to a system of N spins. Let us now turn on the interaction of the new spin with the old spins by coupling it to only

$N(q) \equiv Nq/q_{\rm EA}$ spins. The force $h_{\rm ef}(q)$ acting on this spin is

$$h_{\rm ef}(q) = \sum_{k=1}^{N(q)} J_{N+1,k}\langle\sigma_k\rangle. \tag{6.23}$$

We would like to identify $w(q)$ with $h_{\rm ef}(q)$. If the expectation values $\langle\sigma_k\rangle$ are not changed by the addition of the new spin, $h_{\rm ef}(q)$ is a random gaussian variable with covariance q ($\mathrm{d}h_{\rm ef}/\mathrm{d}(q)$ is white noise) and we recover eq. (6.22) for $\chi_A = 0$. If the addition of a new spin at $N+1$ changes discontinously the values of $\langle\sigma_k\rangle$ we should write [36]

$$\mathrm{d}h_{\rm ef}/\mathrm{d}q = \eta(q) + \Delta h_{\rm ef}(q), \tag{6.24}$$

where $\Delta h_{\rm ef}(q)$ is the variation of $h_{\rm ef}$ due to the process of jumping between different thermodynamical ground states. It is reasonable to suppose that

$$\Delta h_{\rm ef} = \langle\sigma_{N+1}\rangle\chi_A(q), \tag{6.25}$$

where $\langle\sigma_{N+1}\rangle$ is given by $\mathrm{th}(\beta h_{\rm ef})$ and $\chi_A(q)$ parametrizes the violations of the linear response theory. This derivation of the eq. (6.22) is rather suggestive but might not be very sound. It is however tempting to identify the probability distribution of $w(q_{\rm EA})$ with that of the equilibrium

$$h_{\rm ef}^{\rm e} \equiv \sum_{1}^{N} J_{N+1,k}\langle\sigma_k\rangle.$$

It is amusing to note that eq. (6.20) implies at $T = 0$

$$\chi(q) \sim 1/(1-q)^{1/2}. \tag{6.26}$$

An easy computation shows that, if $w(q_{\rm EA})$ and $h_{\rm ef}(q_{\rm EA})$ are identified:

$$P(h_{\rm ef} \simeq |h_{\rm ef}| \quad \text{for } h_{\rm ef} \sim 0,\ T = 0, \tag{6.27}$$

as is expected from theoretical and numerical arguments [20, 37, 38].

We close this section recalling that very recently all the eigenvalues of the hessian matrix have been computed [39, 40]. Their number is infinite: zero is an accumulation point. Some eigenvalues are zero (χ_R is likely to be infinite) but none are negative, confirming the correctness of the ansatz (6.9) and opening the road to the computation of the corrections to the saddle point approximation, also for the original EA model in d dimensions.

7. Spin glasses (the physical interpretation)

The key result of the previous section was the failure of the linear response theory (i.e. $\chi \neq \beta(1-q_{EA})$).

The only possible explanation is the following: due to the large number of ground states, the direction in which the system is magnetized changes completely when the magnetic field is changed, in a finite volume system we expect this abrupt change on steps of $N^{-1/2}$ in the magnetic field [25, 30, 41, 43]. We expect that the dependence of the magnetization on the field is the one shown in fig. 11. It is clear why, when $N \to \infty$, the linear response theory gives the wrong result. A very nice verification of this picture has been given in ref. 38: they have computed at $T = 0$ the magnetization (for H positive) both in the true ground state and in a metastable state near the $H = 0$ ground state; they find $m \sim H$ and $m \sim H^2$ respectively, see fig. 11. The second result agrees with the linear response theory which predicts

$$\chi_{LR} = (1-q_{EA})/T \sim T.$$

In order to clarify the situation it is better to come back to the precise

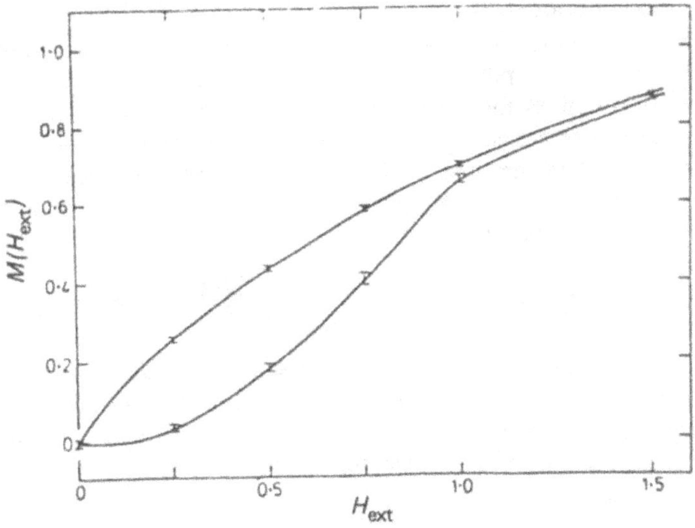

Fig. 11. Magnetization versus external field for $N = 200$. The upper curve is for equilibrium states; the lower curve shows the effect of applying a field to the zero-field ground state and then descending to a nearly local energy minimum.

definition of $q_{EA} = \overline{\langle \sigma_i \rangle^2}$. Indeed we must define the magnetization $m_i \equiv \langle \sigma_i \rangle$ for a given choice of the J's.

If we consider a real system (or a computer simulation) the magnetizations m_i are defined [41] by

$$m_i = \frac{1}{t} \int_0^t \mathrm{d}\tau \sigma_i(\tau), \tag{7.1}$$

where $\sigma_i(\tau)$ is the value that σ_i takes at the time τ and t is a large (macroscopic) but not too large observation time. For example in a d-dimensional ferromagnetic system of size L, t must satisfy the conditions

$$t_m \ll t \ll t_M \approx t_m \exp(L^{d-1}), \tag{7.2}$$

where t_m is the microscopic relaxation time, e.g. one Monte Carlo step. When we change the initial conditions (e.g. $\sigma_i(\tau)$ at $\tau = 0$) we may obtain different results: in the Ising ferromagnetic case below T_c there are essentially two possibilities ($m_i > 0$ or $m_i < 0$); it is important to note that, if the initial state is disordered, the approach to equilibrium for quantities invariant under the global Z_2 is slow (there are corrections proportional to powers of t) while one needs a time t at least proportional to the volume ($t > L^d$) in order to establish a translationally invariant state.

What do we expect for a spin glass? There will be many minima of the thermodynamic potential which are separated by very high walls. At relative short times the system will remain near one minimum, later on at a very large (macroscopic?) time the system will start jumping from one minimum to the other one, by thermodynamic tunnel effects [42, 43].

Let us denote by $m_i^{[\alpha]}$ the expectation value of σ_i in the state labelled by α. We have approximately

$$q(t) \equiv \frac{1}{V} \sum_{i \in V} \left[\frac{1}{t} \int_0^t \sigma_i(t')\mathrm{d}t' \right]^2 = \frac{1}{V} \sum_{i \in V} \left[\sum_{\alpha=1}^{M[t/t_M]} (m_i^{[\alpha]})^2 / M[t/t_M] \right], \tag{7.3}$$

where t_M is a macroscopic time $M[1] = 1$ and when $t \to \infty$ the sum runs over all possible states of the system. In this language $q(t_M) = q_{EA}$ while $q(\infty)$ is obviously much smaller ($q(\infty)$ is zero at $h = 0$). In the same spirit we can define a time dependent susceptibility:

$$\chi(t) = [m_{h+\Delta h}(t) - m_h(t)]/\Delta h,$$

$$m(t) = \frac{1}{tV} \sum_{i \in V} \int_0^t \mathrm{d}\tau \sigma_i(\tau), \tag{7.4}$$

where the two systems with different magnetic fields coincide at $t = 0$.

$\chi(t)$ will be an increasing function of t, because the jumping from one state to another tends to increase the difference between $m_{h+\Delta h}$ and m_h ($\chi(t_M) = \beta(1-q_{EA})$ at $h = 0$).

Now it has been found that it is possible to recover the results of the replica approach in this framework [35, 44]: as far as $q(t)$ is a monotonic decreasing function of t one can invert it and obtain t as function of q. One is led to consider the function $\chi(q)$. The final result is at $h = 0$:

$$\chi(q) = \beta(1-q_{EA}) + \int_q^{q_{EA}} \chi_A(q)\mathrm{d}q, \tag{7.5}$$

where the function $\chi_A(q)$ coincides with the one defined in the previous section.

Therefore

$$q_m = \lim_{t\to\infty} q(t)$$

and this explains why $q(0)$ is zero at $h = 0$. In this approach the interval of x ($x_m \leqslant x \leqslant x_M$) of the replica space corresponds to the region of very large times. The relation between the internal space of replicas and the real time is most mysterious. We try now to define $q(x)$, the framework of equilibrium, time independent, statistical mechanics. The only reasonable way to define $m_i^{[\alpha]}$ is to use the bootstrap equation [41]:

$$m_i^{[\alpha]} = \lim_{\varepsilon\to 0^+} \langle\sigma_i\rangle_{\varepsilon h^{[\alpha]}},$$

$$h_i^{[\alpha]} \propto m_i^{[\alpha]}, \tag{7.6}$$

where $\langle\ \rangle_{\varepsilon h^{[\alpha]}}$ stands for the expectation value in presence of an extra term in the hamiltonian

$$H = H_0 - \sum_{i\in V} \varepsilon h_i^{[\alpha]}\sigma_i. \tag{7.7}$$

Unfortunately in spin glasses we do not know the direction in which the magnetization points: eq. (7.7) seems to be useless. The way out can be found by introducing two real identical weakly coupled replicas of the same system. The global hamiltonian is

$$H^{(2)} = \sum_{i,k} J_{i,k}(\sigma_i^1\sigma_k^1 + \sigma_i^2\sigma_k^2) - 2\varepsilon\sum_i \sigma_i^1\sigma_i^2.$$

For $\varepsilon = 0$ the two replicas are decoupled, for $\varepsilon > 0$ each of the two replicas acts as an external magnetic field on the other one: both must be locked in the same state. In the limit $\varepsilon \to 0$ we find that $\langle \sigma_i^1 \sigma_i^2 \rangle$ is the average over α of $(m_i^{[\alpha]})^2$. Therefore:

$$q_{EA} = \langle \sigma_i^1 \sigma_i^2 \rangle = -\frac{d}{d\varepsilon} F^{(2)}(\varepsilon)|_{\varepsilon=0},$$

$$F^{(2)}(\varepsilon) = \lim_{V \to \infty} \frac{-1}{2\beta V} \ln \left\{ \sum_{\{\sigma\}} \exp[-\beta H^{(2)}(\varepsilon)] \right\}. \tag{7.8}$$

Now we are very happy because replicas have been introduced in a natural way. It is easy to introduce r replicas; $(a = 1, \ldots, r)$

$$H^{(r)}(\varepsilon) = \sum_{i,k} J_{i,k} \left(\sum_{a=1}^{r} \sigma_i^a \sigma_k^a \right) - \varepsilon \sum_i \left(\sum_{a,b} \sigma_i^a \sigma_i^b - r \right),$$

$$F^{(r)}(\varepsilon) = -\frac{1}{r\beta V} \ln \left[\sum_{\{\sigma\}} \exp(-\beta H^{(r)}(\varepsilon)) \right]. \tag{7.9}$$

The same arguments tell us:

$$Q(r) \equiv -\frac{d}{d\varepsilon} F^{(r)}(\varepsilon)|_{\varepsilon=0} = (r-1) q_{EA}. \tag{7.10}$$

In this way we define the function $Q(r)$ for integer r: the definition can be extended to non integer values of r in a constructive way:

$$F^{(r)}(\varepsilon) = -\frac{1}{\beta r V} \ln \int dh_i \exp \left\{ -\frac{\beta}{4} \sum_i h_i^2 - r\beta V F[\varepsilon^{1/2} h] \right\},$$

$$rF[\varepsilon^{1/2} h] = -\frac{1}{\beta V} \ln \left\{ \sum_{\{\sigma\}} \exp[-\beta H(\varepsilon^{1/2} h)] \right\},$$

$$H(\varepsilon^{1/2} h\} = \sum_{i,k} J_{i,k} \sigma_i \sigma_k + \sum_i (\varepsilon - \varepsilon^{1/2} h_i \sigma_i),$$

$$Q(r) = -\frac{d}{d\varepsilon} F^{(r)}(\varepsilon)|_{\varepsilon=0}. \tag{7.11}$$

It is easy to check that for integer r eqs. (7.9) and (7.11) coincide. The function $Q(r)$ tells us how the system behaves in the presence of an external random magnetic field. If the linear response theory is assumed to be valid, we get

$$Q(r) = -1 + \chi_{LR}/\beta + rq_{EA}, \tag{7.12}$$

by expanding eq. (7.11) in powers of ε.

The relation $Q(1)$ implies the Fischer result

$$\chi_{LR} = \beta(1 - q_{EA}). \tag{7.13}$$

On the other hand under reasonable hypotheses we can show that

$$\chi = \beta(1 - Q(0)). \tag{7.14}$$

If we stay in the glassy phase the linear response theory is no longer valid and the function cannot be linear. We want to argue that

$$Q(r) = \int_1^r q(x)\mathrm{d}x, \tag{7.15}$$

where $q(x)$ coincides with the order parameter defined in the previous section for $x < 1$, and it is equal to $q(1)$ foι $x > 1$. If the identification (7.15) is correct, it obviously implies that

$$q_{EA} = q(1) = \max_x q(x)$$

(the function $q(x)$ being monotonic).

Indeed for integer r, we should compute the quenched average of $F^{(r)}$ defined by eq. (7.10). Using the standard formalism we can write:

$$F^{(r)} = \lim_{n \to 0} F_n^{(r)},$$

$$F_n^{(r)} = \frac{1}{rn\beta V} \ln \left\{ \sum_{\{\sigma\}} \exp[-\beta H_n^{(r)}(\sigma)] \right\},$$

$$H_n^{(r)}(\sigma) = \sum_{\alpha=1}^{r} \sum_{a=1}^{n} \sum_{i,k} \sigma_i^{\alpha,a} \sigma_k^{\alpha,a} J_{i,k}$$

$$- \sum_i \left\{ \sum_{a=1}^{n} \left[\sum_{\alpha=1}^{r} \sum_{\beta=1}^{r} \varepsilon^{1/2} \sigma_i^{\alpha,a} \sigma_i^{\alpha,\beta} - r \right] \right\}. \tag{7.16}$$

In other words we couple the nr replicas in groups of r; we have introduced an explicit symmetry breaking in one of the directions in which the symmetry was spontaneously broken. This leads to the result for $Q(x)$. It would be rather interesting to find the consequences of this result on the shape of $Q(x)$ and to measure $Q(x)$ in numerical simulations. So far not much work in this direction has been done.

Although there are still some obscure points we now understand reasonably well the static properties of the SK model. There are many things that must be understood in the dynamics, especially as far as the hysteresis curves are concerned.

After the key work of De Dominicis and Kondor [40] the study of the EA model in finite dimensions is open. Unfortunately as in the previous case, the free energy would cease being analytic at a temperature $T_G > T_c$. In the whole region $T_G > T > T_c$ (T_c being the real critical temperature) the free energy is likely to be C^∞ but not analytic.

If we stick to the infinite range model, there are still many applications of this formalism which have not been mentioned (e.g. the extension of these results to the Heisenberg spins [45] and to the random anisotropy case [46]; it would also be rather interesting to compute quantities like

$$q(h_1, h_2) = \overline{\langle\sigma_i\rangle_{h_1}\langle\sigma_i\rangle_{h_2}}, \tag{7.17}$$

they can be easily measured in Monte Carlo simulations and quite likely have a peculiar behavior for $h_1 \sim h_2$ in the glassy phase; e.g.

$$q(h_1, h_2) \simeq q(h_1, h_1) - |h_1 - h_2|^\alpha.$$

The most important field we have neglected are the TAP equations [47] for the local magnetizations m_i; in a first approximation the TAP equations say that:

$$\langle\sigma_i\rangle \equiv m_i = \text{th}\left(\beta \sum_k J_{ik} m_k\right). \tag{7.18}$$

At $T = 0$, they imply for the ground state:

$$\sigma_i = \text{sign}\left(\sum_k J_{ik}\sigma_k\right). \tag{7.19}$$

A configuration satisfying eq. (7.19) is stable against the flipping of a single spin, but it is not necessarily stable with respect to the flipping of $N^{1/2}$ spins. Eq. (7.19) tells us that the configuration is a local minimum of $-H$, (a metastable state!) not a true minimum. It is interesting to note that an explicit computation tells us that the number of solutions of the TAP equation grows like [47, 48]:

$$\exp[\mu(h, T)N], \tag{7.20}$$

where $\mu(h, T)$ is different from zero below the AT line.

$$\mu(0, T) \sim O(T - T_c)^6, \quad T < T_c,$$

$$\mu(0, 0) \sim 0.19.$$

In other words the direct study of the TAP equations confirms the existence of a very large number of metastable states.

Let us now try to compare the theory (in the present rudimentary state) with real experiments; let me select arbitrarily two of the most favourable cases. The main prediction of the theory is the existence of a line in the T, h plane below which the linear response susceptibility χ_{LR} is different from the equilibrium susceptibility χ. Both susceptibilities can be measured: χ_{LR} by applying a magnetic field for a short time (10^{-3}–10^{2} seconds), χ_e by cooling the system in presence of the magnetic field. A typical experiment result is shown in fig. 12 [49].

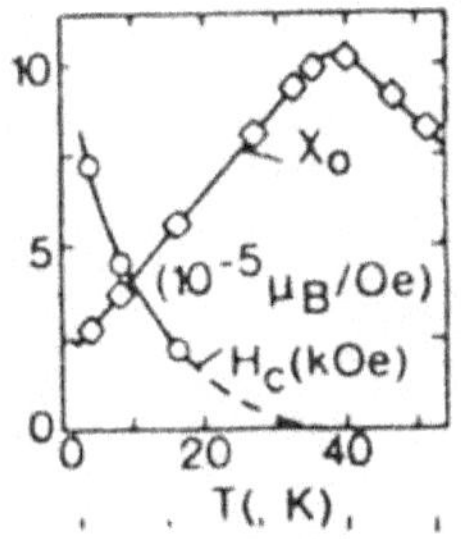

Fig. 12. H_c is the higher field (for a given T) below which irreversible phenomena are present; χ_0 is the short time ($t \simeq O(1)$ s) susceptibility. The sample is CuMn 9%.

At the quantitative level the mean field approximation predicts that below the AT line $d\chi_e/dT = 0$. Experimental results are shown in fig. 13 [50]; we are somewhat at a loss to explain why the corrections to the mean field approximation are so small. It may be that this is due to the relatively long range nature of the interaction between spins:

$$J(x, y) \simeq \frac{\sin|x-y|}{(x-y)^3}.$$

However there is another fact that impresses me more than the experimental success of the broken replica theory of spin glasses: it is the intrinsic strength of the formalism which compells us first to do manipulations whose meaning is obscure from both the mathematical and the physical point of view, and finally, just as the last formulae, gives us a theory (eqs. (6.21) and (6.22)) which explains very complicated phenomena in a simple and direct way. It is the strangest thing I have yet seen.

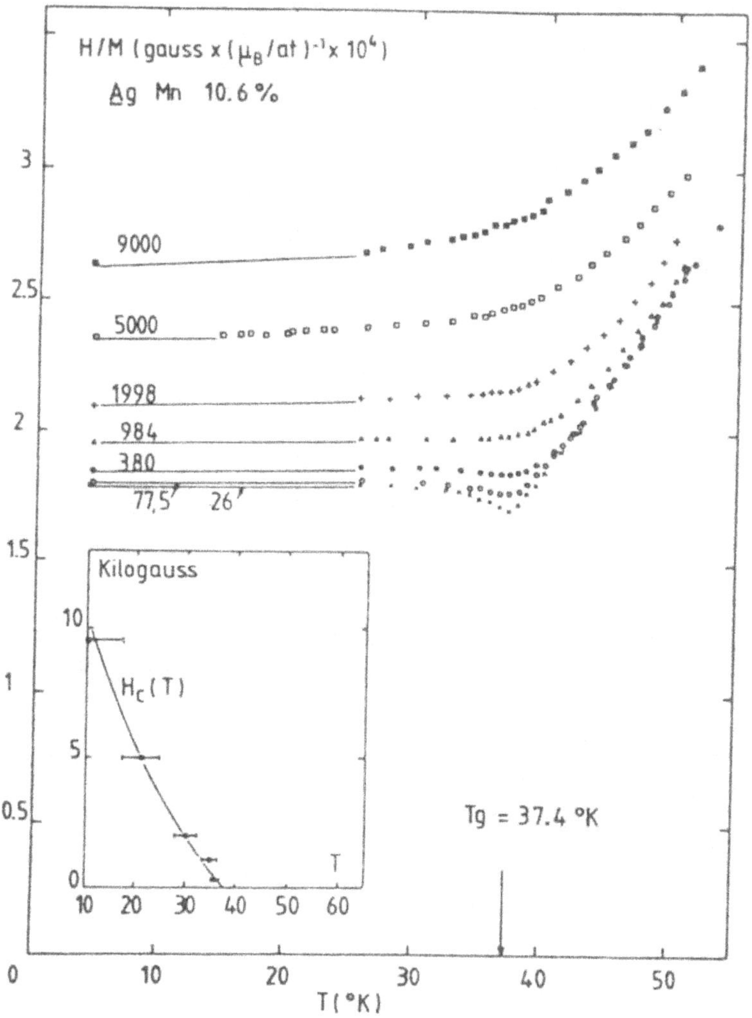

Fig. 13. The ratio H/M for AgMn 10.6% as a function of the magnetic field in Gauss, obtained by cooling the sample in the field. For $H < H_c$, $T(d/dT)(H/M)$ is zero with a good approximation.

8. Random potential model (the density of states)

Since the first paper by Anderson [51] on localization of electrons a lot of work has been done on the electronic structure of amorphous materials (for a review see refs. 1, 52–54). In recent years sophisticated field theoretical techniques have been used [55–60]. Here we want to present an introduction to current problems. If one neglects the electron–electron interaction, the problem is reduced to the computation of the levels and the eigenvectors of the Schrödinger operator:

$$H = -\Delta + V(x), \tag{8.1}$$

Here $V(x)$ is a random potential dependent on the atomic structure of the material and Δ is the d-dimensional Laplacian. In the simplest case $V(x)$ is white noise, i.e. it has a Gaussian distribution with covariance:

$$\overline{V(x)V(y)} = \lambda\delta^d(x-y); \quad \overline{V(x)} = 0,$$

where λ plays the rôle of the coupling. In these lectures we will consider only this case.

The goal consists in computing the average over V of the Green function of the operator H. From this knowledge, by filling the levels of H with electrons up to the Fermi energy, one can extract physically interesting quantities such as the density of states, the conductivity of the system and the nature of the electronic states (extended or localized).

A possible approach to this problem consists of mapping it onto the problem of computing the Green function of an appropriate field theory. Let us start by computing the density of states. We define

$$G_E(x,y|V) = \langle x|(H-E)^{-1}|y\rangle,$$
$$G_E(x-y) = \overline{G(x,y|V)}. \tag{8.2}$$

We do not indicate in an explicit way the dependence of G on λ. The density of states is given by

$$\rho(E) = \frac{1}{V}\mathrm{Tr}[\delta(H-E)] = \lim_{\varepsilon\to 0^+}\frac{1}{\pi}\mathrm{Im}[G_{E+i\varepsilon}(0)]. \tag{8.3}$$

In order to compute ρ we must set up a computational scheme for $G_E(x)$ at all x. The simplest way to set up a diagrammatical expansion would be to use the identity [61]

$$\frac{1}{H-E} = \sum_{n=0}^{\infty}\frac{1}{-\Delta-E}\left(V(x)\frac{1}{-\Delta-E}\right)^n. \tag{8.4}$$

We prefer however to reduce the problem to a field theory. Let us consider the O(n) invariant action [55, 60]:

$$A[\phi] = \int d^d x \left\{ \sum_{a=1}^{n} [\tfrac{1}{2}(\partial_\mu \phi_a)^2 + \tfrac{1}{2}M^2 \phi_a^2] + \tfrac{1}{4!}g \left[\sum_{a=1}^{n} \phi_a^2 \right]^2 \right\}, \tag{8.5}$$

where ϕ_a are n component fields. The two point correlation function of the field ϕ is given by

$$\langle \phi_a^{(x)} \phi_b^{(0)} \rangle = C(x)\delta_{ab} = \frac{\int d[\phi]\phi_a(x)\phi_b(0)\exp[-A[\phi]]}{\int d[\phi]\exp[-A[\phi]]},$$

$$\rho(E) = \lim_{n\to 0} \frac{1}{n} \sum_{a=1}^{n} \mathrm{Im}\langle(\phi_a(0))^2\rangle. \tag{8.6}$$

We shall now prove that $C(x) = G_E(x)$ if $M^2 = -E$, $g = -\lambda$ in the limit $n \to 0$. The proof is straightforward: for $g < 0$ eq. (8.6) can be written as

$$C(x)\delta_{ab} = \frac{\int d[\phi]d[V]\exp\{-A[\phi, V]\}\phi_a(x)\phi_b(0)}{\int d[\phi]d[V]\exp\{-A[\phi, V]\}},$$

$$A[\phi, V] = \int d^d x \left\{ \frac{1}{2} \sum_{1}^{n} a[(\partial_\mu \phi_a)^2 + [M^2 - V(x)]\phi_a^2(x)] - \frac{V^2(x)}{2g} \right\}. \tag{8.7}$$

This can be easily proved by integrating over the V field. Now in eq. (8.7) the ϕ integration is gaussian and the final result is:

$$C(x) = \int d\mu_n[V] G_E(x, 0|V),$$

$$d\mu_n[V] \propto d[V]\exp\left[-\int d^d x V^2(x)/2\lambda\right] \det{}^{-n/2}[-\Delta + V - E],$$

$$\int d\mu_n[V] = 1. \tag{8.8}$$

It is clear that eq. (8.8), for $n = 0$, coincides with eq. (8.2). Before going on let us make two remarks: if λ is positive, g is negative and eq. (8.6) is not defined for any n because the integral over ϕ is divergent at large ϕ because of the ϕ^4 term. Moreover at $g = 0$, the integral is not defined in the interesting region $E > 0$ (i.e. $M^2 < 0$). Of course this difficulty may be removed by declaring that the functional integral is computed in the region of M^2 and g positive and is afterwards analytically continued into the region M^2 and g negative. This problem may be bypassed by doing an explicit rotation of $\exp(-i\pi/4)$ on the fields if $\mathrm{Im}\,E < 0$ and of $\exp(i\pi/4)$ if $\mathrm{Im}\,E > 0$ [61].

The final formula is

$$C(x)\delta_{ab} = \pm i \frac{\int d[\phi]\phi_a(x)\phi_b(0)\exp[-A_\pm[\phi]]}{\int d[\phi]\exp[-A_\pm[\phi]]},$$
$$A_\pm[\phi] = \int d^Dx \left\{ \frac{\pm i}{2} \sum_{a=1}^{n} [(\partial_\mu\phi_a)^2 + M^2\phi_a^2] + \frac{g}{2}\left(\sum_{a=1}^{n}\phi_a^2\right)^2 \right\}, \tag{8.9}$$

where the sign $\pm$ is chosen depending on the sign of $\mathrm{Im}\,E = -\mathrm{Im}M^2$. The existence of two functional representations for $\mathrm{Im}\,E$, positive or negative, explains the presence of a discontinuity at $\mathrm{Im}\,E = 0$:

$$\rho(E) = \frac{1}{2\pi}\,\mathrm{Disc}[G(E)].$$

This observation will be very useful in the next section.

Those who do not like the $n \to 0$ limit, but insist on using a field theory representation, will be happy to see that in this case eq. (8.6) can be replaced by [62]:

$$G(x) = \frac{\int d[\phi_1]d[\phi_2]d[\bar\psi]d[\psi]\phi_1(x)\phi_1(0)\exp[-A]}{\int d[\phi_1]d[\phi_2]d[\bar\psi]d[\psi]\exp[-A]},$$
$$A = \frac{1}{2}\int d^Dx \left\{ \sum_{a=1}^{2} [(\partial_\mu\phi_a)^2 + M^2\phi_a^2] + \partial_\mu\bar\psi\partial_\mu\psi + M^2\bar\psi\psi \right.$$
$$\left. + g(\phi_1^2 + \phi_2^2 + \bar\psi\psi)^2 \right\}, \tag{8.10}$$

where the ψ are spin zero fermions.

Indeed if we use the usual gaussian integration to reduce the quartic term to a quadratic one, we see that the integration over the fermions produces a $\det[-\Delta + M^2 + V]$ which compensates the $\det^{-1}$ coming from the integration over the ϕ fields. (We are in the presence of a trivial supersymmetry.)

Let us go on and see what happens in the free case ($\lambda = 0$). We easily obtain

$$\rho(E) \sim \int d^DK\,\delta(K^2 - E) \sim E^{(D-2)/2}. \tag{8.11}$$

The density of levels is therefore zero for negative E, as it should be. Let us now make the so-called coherent potential approximation (CPA). In this approximation one retains only the self energy diagram shown in fig. 14 while the full propagator is used in the blob. We finally get:

$$M^2 = m^2 + \lambda \int_0^\Lambda \frac{d^D K}{m^2 + K^2} \simeq m^2 - \lambda(m^{D-2} - \Lambda^{D-2}), \tag{8.12}$$

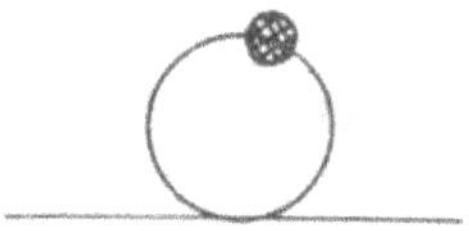

Fig. 14. The self-energy diagrams of the CPA approximation. The blob denotes the full propagator.

Λ is as ultraviolet cut-off. The CPA approximation is exact for $n \to \infty$, but let us use it at $n = 0$ for pedagogical reasons. In a normal field theory λ is negative; eq. (8.12) would imply therefore

$$\begin{aligned} &(M^2 - M_c^2) \simeq m^2, && D > 4, \\ &m^2 \sim (M^2 - M_c^2)^{2/(D-2)}, && D < 4, \\ &M_c^2 < 0. \end{aligned} \tag{8.13}$$

However in this case λ is positive, the right hand side of eq. (8.12) has the behavior shown in fig. 15 for $D < 4$. This means that one obtains:

$$\begin{aligned} &m^2 = m_c^2 + (M^2 - M_c^2)^{1/2}, \\ &m_c^2 > 0, \quad M_c^2 > 0, \\ &\rho(E) \propto (E + M_c^2)^{1/2} \theta(E + M_c^2), \quad E \sim -M_c^2. \end{aligned} \tag{8.14}$$

In fig. 16, we show the $\rho(E)$ of the free theory and of the CPA approximation.

It is rather interesting that m^2 does not become zero at the transition. One might think that the result of the CPA approximation is stable against higher order corrections; infrared divergences should be absent since the renormalized mass m^2 is non-zero at the edge of the band ($E = E_c \equiv -M_c^2$). Unfortunately the ϕ^2 propagator is a source of infrared divergences. Indeed let us define:

$$\begin{aligned} &\langle \phi_a(x)\phi_b(x)\phi_c(0)\phi_d(0)\rangle - \langle \phi_a(x)\phi_b(x)\rangle\langle \phi_c(0)\phi_d(0)\rangle \\ &= \delta_{ab}\delta_{cd}R(x) + (\delta_{ac}\delta_{bd} + \delta_{ad}\delta_{bc})\chi(x). \end{aligned} \tag{8.15}$$

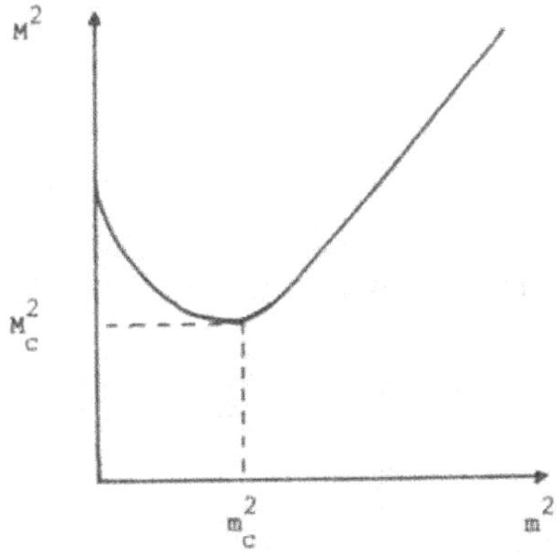

Fig. 15. M^2 versus m^2 in the CPA approximation.

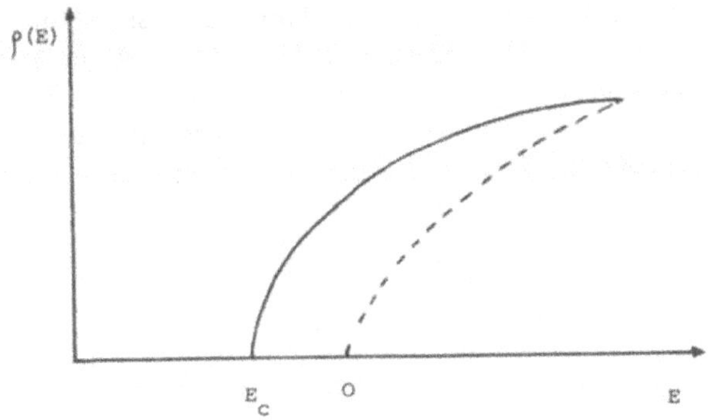

Fig. 16. The density of states in the free case (dashed line) and in the CPA approximation (full line).

It is easy to prove that for $n = 0$

$$R(x) = \overline{G_E(x, x|V)G_E(0, 0|V)} - \overline{G_E(x, x|V)}\;\overline{G_E(0, 0|V)},$$

$$\chi(x) = \overline{G_E^2(x, 0|V)}. \tag{8.16}$$

If one computes R and G in momentum space, in the CPA approximation, by summing all the diagrams shown in fig. 17 one finds

$$R(p) \sim [(M^2 - M_c^2)^{1/2} + p^2 + O(p^4)]^{-2};$$

$$\chi(p) \sim [(M^2 - M_c^2)^{1/2} + p^2 + O(p^4)]^{-1}. \tag{8.17}$$

Fig. 17. Diagrams for G (a) and χ (b) in the CPA.

This result may be checked by noticing that

$$\chi(p)|_{p=0} = -\frac{\mathrm{d}}{\mathrm{d}M^2}\lim_{n\to 0}\langle\phi^2\rangle = -\frac{\mathrm{d}m^2}{\mathrm{d}M^2}\frac{\mathrm{d}}{\mathrm{d}m^2}\int\frac{\mathrm{d}^D K}{(K^2+m^2)}$$
$$= \frac{\mathrm{d}m^2}{\mathrm{d}M^2}\int\frac{\mathrm{d}^D K}{(K^2+m^2)^2}, \qquad (8.18)$$

and that $\mathrm{d}m^2/\mathrm{d}M^2$ goes like $(M^2-M_c^2)^{-1/2}$ according to eq. (8.14).

In field theory language the ϕ^4 interaction is attractive and not repulsive as usual: a two-particle bound state is produced [59] and by decreasing the mass of the ϕ particle (m) the mass of the bound state becomes zero at $m \neq 0$. The self interaction of this bound state produces infrared divergences e.g. the diagram of fig. 18. These infrared divergences can be easily studied using the standard machinery [63, 64]: one introduces a field $q_{ab} = \phi_a\phi_b$, and derives an effective lagrangian for the field q.

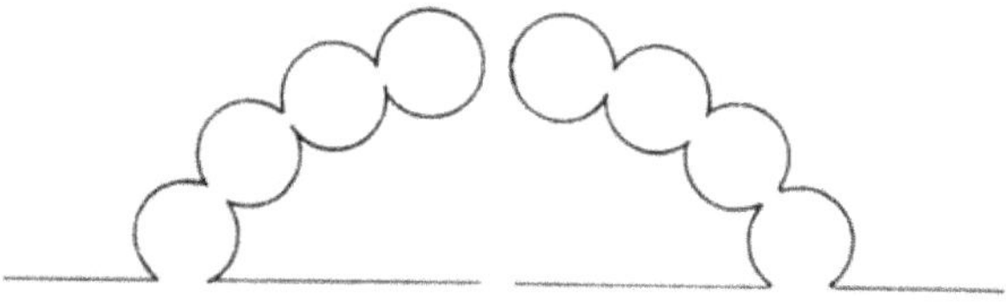

Fig. 18. Leading infrared divergent diagram for the corrections of the CPA.

An interaction proportional to q^3 is present suggesting that the corrections to the critical behavior vanish for $D > D_c = 6$ and corrections to the exponent $\frac{1}{2}$ in eq. (8.14) can be computed in powers of $\varepsilon = D_c - D$ for $D < D_c$. An explicit computation shows that as in the polymer case D_c is shifted from 6 to 8 (indeed the two theories seem to belong to the same universality class [64]).

We will not enter into details because the problem is purely academic [59]; let us summarize the situation: we are in the presence of a theory with a negative coupling constant and we suppose that the imaginary part of the Green functions starts when the mass becomes sufficiently small that a massless two-particle bound state is produced. However naive arguments tell us (as remarked by Stone) that the mass of an N particle state is:

$$M_N = Nm - \lambda(N^3 - N)/m^{3-D}. \tag{8.19}$$

In other words we expect that a three-particle bound state would become massless before the two-particle bound state. And so on. Moreover since the two-particle bound state may decay into two three-particle bound states, it becomes a resonance, its mass acquires an imaginary part and it does not become zero for real E. If we take care of only the three and two particle bound states the situation is the one shown in fig. 19. The two particle cut is shifted onto the second sheet.

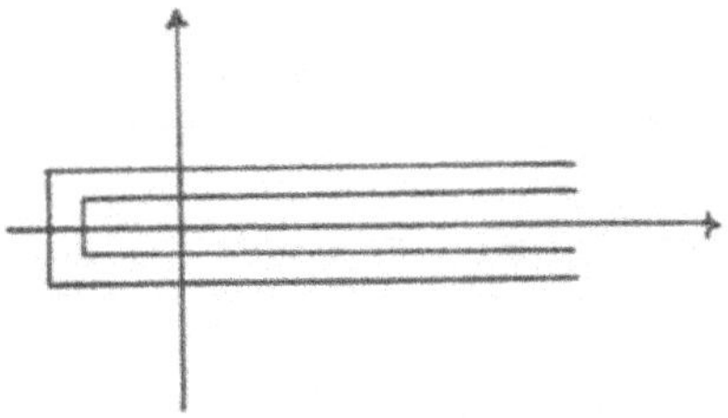

Fig. 19. The cut structure in the complex energy plane, considering only two- and three-particle bound states.

The final effect is to shift all the singularities onto the second sheet leaving a regular $\rho(E)$. The behaviour of $\rho(E)$ for $E \to -\infty$ is dominated by the condensation into the vacuum of $N(N \to \infty)$ particles, and this phenomenon is just the instanton in a different guise [65].

Indeed as shown in the lectures of Zinn-Justin in this Volume one can find the imaginary part of the Green function by looking for localized solutions of the classical equation. In our case we assume that

$$\phi_1(x) \equiv \phi(x) \neq 0;$$

we must solve the equation:

$$-\Delta\phi + M^2\phi = \lambda\phi^3. \tag{8.20}$$

The imaginary part of the Green function behaves for $\lambda \to 0$ in less than four dimensions as

$$\rho(E) \propto \frac{1}{\pi} \operatorname{Im} G_E(0) \simeq \exp[-cM^{2(4-D)/2}/\lambda] \sim \exp[-c(-E)^{(4-D)/2}/\lambda]. \tag{8.21}$$

Here we have neglected the prefactor, which is computed in ref. 66.

Eq. (8.20) can be understood directly: we want to know the probability of finding a solution of the Schrödinger equation:

$$[-\Delta + V(x)]\psi = E\psi, \tag{8.22}$$

for a given E. We must therefore minimize $\int d^D x\, V^2(x)$ with the constraint (8.22). An easy computation shows that the constrained minimum problem is equivalent to the pair of equations [67]:

$$(-\Delta + V(x) - E)\psi = 0,$$
$$V(x) \propto \psi^2(x), \tag{8.23}$$

hence eq. (8.20).

The final conclusion is that $\rho(E)$ is an analytic function which can be computed in the standard perturbative expansion when $E \to +\infty$ and in the perturbative expansion around the instanton when $E \to -\infty$. I am convinced that using appropriate resummation techniques one can

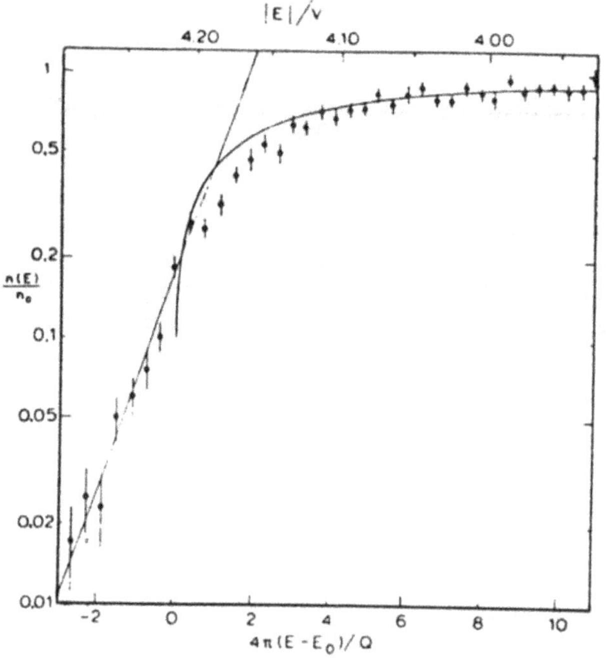

Fig. 20. Numerical results for the density of states in $D = 2$. The curve is the CPA approximation and the straight line is an exponential fit. The one instanton computation agrees with the data for $E - E_0 \sim -(2\text{-}1)$. E_0 is the edge of the band in the CPA.

obtain reasonable approximations in the whole E range. Of course $\rho(E)$ may be evaluated in two dimensions by solving the Schrödinger equation numerically on a large lattice; the results are shown in fig. 20 [68].

9. Random potential model (localisation)

Up to now we have discussed only the spectral density. It is also very important to know if the spectrum is discrete (localisation) or continuous (extended states). The "eigenstates" of the continuum spectrum may be functions that do not go to zero at infinity (Ex) (like plane waves) or functions which decay like a power, but are not L^2, in the last case the states are said to be quasi-extended (QE) or power localized.

It is very important to consider the following quantity:

$$\chi_E(x) = |G_E(x, 0|V)|^2. \tag{9.1}$$

The simple identity [69]

$$1/(H-\bar{E}) - 1/(H-E) = (\bar{E}-E)/[(H-\bar{E})(H-E)] \tag{9.2}$$

tells us that:

$$\chi_E(p)|_{p=0} = \frac{1}{\varepsilon}\,\rho(E_R), \quad E = E_R + \mathrm{i}\varepsilon. \tag{9.3}$$

We expect that in the localized region and in the extended region, respectively,

$$\chi_E(p) \sim \frac{1}{\varepsilon}\, f(p) \qquad \text{(L)},$$

$$\chi_E(p) \sim \frac{\rho(E_R)}{\varepsilon + \sigma p^2 + \mathrm{O}(p^4)} \qquad \text{(Ex)}. \tag{9.4}$$

The quantity σ can be identified with the conductivity which is given by

$$\lim_{\varepsilon\to 0}\left[-\varepsilon^2 \frac{\mathrm{d}}{\mathrm{d}p^2}\chi_{E_R+\mathrm{i}\varepsilon}(p^2)|_{p^2=0}\right]\Big/\rho(E_R).$$

If we transcribe the problem in a field theory language, using the representation (8.9) we find in the limit $n \to 0$:

$$\chi_{E_R+i\varepsilon}(x) = \frac{\int d[\phi_+]d[\phi_-]\phi^1_+(x)\phi^1_-(x)\phi^1_+(0)\phi^1_-(0)\exp(-A[\phi_+,\phi_-])}{\int d[\phi_+]d[\phi_-]\exp(-A[\phi_+,\phi_-])},$$

$$A[\phi_+,\phi_-] = \int d^D x \frac{1}{2}\sum_{a=1}^{n}\{i(\partial_\mu\phi^a_+)^2 - i(\partial_\mu\phi^a_-)^2 + iE_R[(\phi^a_+)^2-(\phi^a_-)^2] + \varepsilon[(\phi^a_+)^2+(\phi^a_-)^2]\} + \lambda\left\{\sum_{a=1}^{n}[(\phi^a_+)^2-(\phi^a_-)^2]\right\}^2. \tag{9.5}$$

Now the action in eq. (9.5) is invariant under the non-compact $O(n, n)$ group in the limit $n \to 0$, if $\varepsilon = 0$ [59]. If $\varepsilon \neq 0$, the group is broken to $O(n) \otimes O(n)$ in an explicit way. Now since

$$\langle(\phi^1_+)^2\rangle = i\overline{G_{E_R+i\varepsilon}(x, x|V)},$$
$$\langle(\phi^1_-)^2\rangle = i\overline{G_{E_R-i\varepsilon}(x, x|V)}, \tag{9.6}$$

we get:

$$\langle(\phi^1_+)^2\rangle - \langle(\phi^1_-)^2\rangle \simeq 2\pi\rho(E_R), \quad \varepsilon \to 0. \tag{9.7}$$

Therefore if $\rho(E) \neq 0$, the $O(n, n)$ symmetry is kinematically broken: that also happens in the free theory! In some sense in the localized phase the symmetry is restored as far as the symmetry breaking part of the Green functions becomes of order ε with respect to the symmetry invariant part which goes to infinity. This peculiar way of restoring the symmetry is strictly connected with the non-compactness of the model [59]. An easy exercise shows that eq. (9.3) is a Ward identity of the $O(n, n)$ symmetry and the singularity at $p^2 = 0$ in $\chi(p)$ is a Goldstone singularity [58–60]. Now as shown in Fradkin's lectures in this Volume the infrared singularities of the Goldstone bosons are well represented by a non-compact σ-model which is asymptotic and free in two dimensions. Now the instanton computation tells us that for $E \to -\infty$ the states are localized while the perturbative expansion in $D > 2$ tells us that for $E > 2$ the states are extended. A transition must happen somewhere: the simplest scenario is shown in fig. 21 (left) where we consider the $\lambda - D$ plane for a fixed positive energy E: indeed the asymptotic freedom of the model in two dimensions tells us that the extended region is unstable. However the states, if they are not extended, can be Quasi-Extended and the other scenario of fig. 21 may be possible [70]. Also if fig. 21 (left) is correct it is

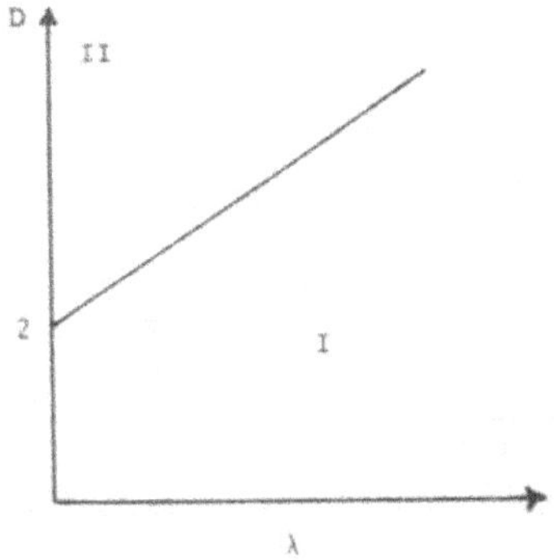

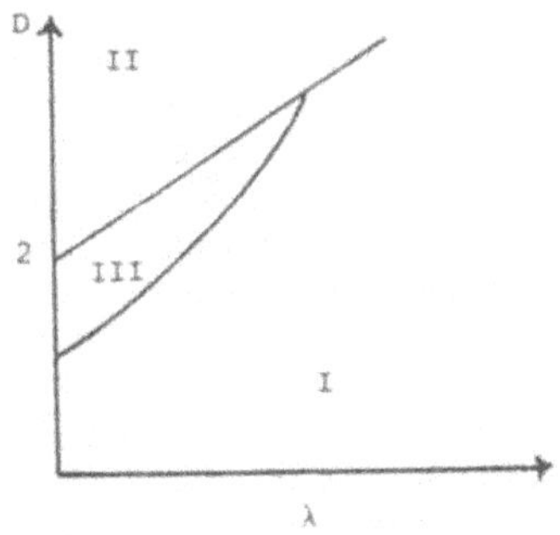

Fig. 21. On the left a possible phase diagram for a fixed positive E in the D–λ plane. In region I all states are localized, in region II all states are extended. On the right an alternative phase diagram for a fixed positive E in the D-λ plane. In region I all states are localized, in region II al states are extended and in region III all states are quasi-extended. The shape of the region III is a pure guess.

not clear if the transition in $D > 2$ from localized to extended states is second order, σ vanishing at the transition ($\sigma = 0$ in the localized region), or if it jumps at the transition, (Mott minimum metallic conductivity [71]). In the same way what happens to gauge theories in $4+\varepsilon$ dimensions (first or second order transition) is completely unknown.

This situation is somewhat frustrating: the localization transition cannot be studied in $6-\varepsilon$ or $8-\varepsilon$ dimensions because the existence of localized states is a non-perturbative phenomenon; the σ-model representation for the infrared divergences tells us that extended states do not exist in $D = 2$ but it does not tell us what exists in their place. In my opinion the situation may be clarified only by working directly in the localized phase using either the original hopping parameter expansion [51] or a model of a dense gas of interacting instantons. It is quite possible that the correct phase diagram has been already guessed. Unfortunately up to now, numerical simulations have not reached an agreement among different groups.

Acknowledgements

I am very grateful to N. Sourlas for many discussions on the subject of this lecture, to Don Petcher and Michael Stone for a careful reading of the manuscript, to Bertrand Duplantier for help in the editing and to Valerie Lecuyer for the excellent typing of a manuscript in really bad shape.

References

[1] R. Balian, R. Maynard and G. Toulouse, eds., Ill Condensed Matter (North-Holland, Amsterdam, 1979); C. Castellani, C. Di Castro and L. Peliti, eds., Disordered Systems and Localization (Springer Verlag, New York, 1981).
[2] P.G. De Gennes, Phys. Lett. 38A (1972) 336.
[3] S.F. Edwards and P.W. Anderson, J. Phys. F: 5 (1975) 965; F: 6 (1976) 1927.
[4] P. Lacour-Gayet and G. Toulouse, J. de Phys. 35 (1974) 425; Y. Imry and S.K. Ma, Phys. Rev. Lett. 35 (1975) 1399; G. Grinstein, Phys. Rev. Lett. 37 (1976) 944; A. Aharony, Y. Imry and S.K. Ma, Phys. Rev. Lett. 37 (1976) 1364.
[5] G. Parisi and N. Sourlas, Phys. Rev. Lett. 43 (1976) 744.
[6] K. Symanzik, in: Lectures in theoretical Physics, vol. III, eds., E. Brittin, B.W. Downs and J. Downs (Interscience, New York, 1961).
[7] A.P. Young, J. Phys. C: 10 (1977) L257.
[8] K. Symanzik, Prog. Theor. Phys. 20 (1958) 690 and refs. therein.
[9] D. Wallace, this volume and refs. therein.
[10] G. Parisi and N. Sourlas, Nucl. Phys. 206B (1982) 321; A. Ceccoti and L. Girardello, Harvard preprint (1982).
[11] R.B. Griffiths, Phys. Rev. Lett. 23 (1969) 17.
[12] H.S. Kogon and D.J. Wallace, J. Phys. A: 14 (1981) L522.
[13] P. Fayet and S. Ferrara, Phys. Rep. 32C (1977) 249.
[14] E. Witten, Nucl. Phys. 202B (1982) 253.
[15] T.C. Lubenski, in: Ill Condensed Matter, eds., R. Balian, R. Maynard and G. Toulouse (North-Holland, Amsterdam, 1979).
[16] F. Fucito and G. Parisi, J. Phys. A: 14 (1981) L517.
[17] G. Parisi, J. Stat. Phys. 13 (1979) 578.
[18] T.C. Lubensky and A.J. McKane, in: Disordered Systems and Localization, eds., C. Castellani, C. Di Castro and L. Peliti (Springer Verlag, New York, 1981).
[19] M. Gabay, Thèse de doctorat, unpublished, Orsay (1981); R. Rammal and J. Souletie, Spin Glasses, preprint, Grenoble (1981).
[20] D. Sherrington and S. Kirkpatrick, Phys. Rev. Lett. 35 (1975) 1792; Phys. Rev. B17 (1978) 4384.
[21] G. Toulouse, Comm. Phys. 2 (1977) 115.
[22] S. Marcus, contribution to the Sèminaire de Sémiotique Théatrale, unpublished, Urbino (1974) and P. Doreian, Mathematics and the Study of Social Relations, ch. V (Academic, New York, 1971).
[23] F. Barahona, R. Maynard, R. Rammal and J.P. Uhry, J. Phys. A: 14 (1981) 72.
[24] J.R.J. De Almeida and D.J. Thouless, J. Phys. A: 11 (1978) 983.
[25] G. Parisi, Phys. Rev. Lett. 43 (1979) 1754; J. Phys. A: 13 (1980) 1101; 1887; L115; B. Duplantier, J. Phys. A: 14 (1981) 283.
[26] A.J. Bray and M.A. Moore, Phys. Rev. Lett. 41 (1978) 1068; E. Pytte and J. Rudnick, Phys. Rev. B19 (1979) 3603
[27] K.H. Fischer, Phys. Rev. Lett. 34 (1976) 1438.
[28] G. Parisi, Phil. Mag. B41 (1980) 677.
[29] A. Blandin, M. Gabay and T. Garel, J. Phys. C: 13 (1980) 403.
[30] C. De Dominicis, T. Garel, J. de Phys. 40 (1979) L575.
[31] P.W. Anderson, in: Ill Condensed Matter, eds., R. Balian, R. Maynard and G. Toulouse (North-Holland, Amsterdam, 1979).

[32] G. Parisi, Phys. Lett. 73A (1979) 531.
[33] G. Parisi, Phys. Rep. 67C (1980) 25.
[34] J. Vannimenus, G. Toulouse and G. Parisi, J. de Phys. 42 (1981) 565.
[35] H. Sompolinsky, Phys. Rev. Lett. 47 (1981) 935; C. De Dominicis, M. Gabay and H. Orland, J. de Phys. 42 (1981) L523; J.A. Hertz, J. Phys. A, to be published.
[36] H.J. Sommers, Z. Phys. B31 (1979) 301.
[37] H.J. Sommers, Z. Phys. B33 (1979) 173 and refs. therein.
[38] F.T. Bantilan and R.G. Palmer, J. Phys. F: 11 (1981) 261 and refs. therein.
[39] J.R.L. de Almeida, R.C. Jones, J.M. Kosterlitz and D.J. Thouless, J. Phys. C: 11 (1978) L871.
[40] C. De Dominicis and I. Kondor, Phys. Rev. B23 (1983) 606; I. Kondor and C. de Dominicis, J. Phys. A: 16 (1983) L73.
[41] G. Parisi, in: Disordered Systems and Localization, eds., C. Castellani, C. Di Castro and L. Peliti (Springer Verlag, New York, 1981).
[42] K. Binder and W. Kinzel, in: Disordered Systems and Localization, eds., C. Castellani, C. Di Castro and L. Peliti (Springer Verlag, New York, 1981).
[43] A.P. Young, Phys. Rev. Lett. 49 (1982) 685.
[44] C. De Dominicis, M. Gabay and B. Duplantier, J. Phys. A: 15 (1982) L42.
[45] M. Gabay, T. Garel and C. De Dominicis, submitted to J. Phys. C; D.J. Elderfield and D. Sherrington, submitted to J. Phys. C.
[46] D. Sherrington, D.M. Cragg and D.J. Elderfield, Imperial College preprint, SST/8182/21 (1982).
[47] D.J. Thouless, P.W. Anderson and R.G. Palmer, Phil. Mag. 35 (1977) 593.
[48] C. De Dominicis, M. Gabay, H. Orland and T. Garel, J. de Phys. 41 (1980) 923; A.J. Bray and M.A. Moore, J. Phys. C: 13 (1980) L469; L907.
[49] R.W. Knitter and J.S. Kouvel, J. Magn. Magn. Mater. 21 (1980) L316.
[50] P. Monod and H. Bouchat, in: Disordered systems and Localization, eds., C. Castellani, C. Di Castro and L. Peliti (Springer Verlag, New York, 1981).
[51] P.W. Anderson, Phys. Rev. 109 (1958) 1492.
[52] D.J. Thouless, Phys. Rep. C13 (1974) 94.
[53] N.F. Mott and A.E. Davis, Electronic Processes in Non-Crystalline Materials (Pergamon Press, New York, 1971).
[54] Various authors' contributions to the proceedings of Les Houches Institute (1980) Common Trends in Particle and Condensed Matter Physics, Phys. Rep. 67 (1980).
[55] F.J. Wegner, Z. Phys. B35 (1976) 208.
[56] R. Opperman and F.J. Wegner, Z. Phys. B34 (1979) 327.
[57] S. Hikami, Prog. Theor. Phys. 64 (1980) 1466.
[58] A.M. Pruisken and L. Schäfer, Phys. Rev. Lett. 40 (1981) 490.
[59] G. Parisi, J. Phys. A: 14 (1981) 735.
[60] A.J.McKane and M. Stone, Ann. of Phys. (NY) 131 (1981) 36.
[61] S.F. Edwards, J. Phys. C: 8 (1975) 1660, D.J. Thouless, J. Phys. C: 8 (1975) 1803; A. Nitzan, K.F. Freed and D.H. Cohen, Phys. Rev. B15 (1977) 4476.
[62] G. Parisi and N. Sourlas, J. de Phys. Lett. 41 (1980) L403.
[63] A. Aharony and Y. Imry, J. Phys. C: 10 (1977) L487.
[64] T.C. Lubensky and A.J. McKane, J. de Phys. Lett. 42, (1981) L523; T.C. Lubensky, in: Disordered Systems and Localization, eds. C. Castellani, C. Di Castro and L. Peliti (Springer Verlag, New York, 1981).

[65] G. Parisi, Lectures Nores at the 1977 Cargèse Summer School (Plenum, New York, 1978).
[66] J.L. Cardy, J. Phys. C:11 1321 (1978); M.V. Sadovskii, Sov. Phys. Solid State 21 (1979) 435; A. Houghton and L. Schäfer, J. Phys. A:12 (1979)1309; E. Brézin and G. Parisi, J. Phys. C:13 (1980) L307.
[67] B.I. Halperin and M. Lax, Phys. Rev. 148 (1966) 772; J. Zittartz and J.S. Langer, Phys. Rev. 148 (1966) 741; C. Itzykson, G. Parisi and J.B. Zuber, Phys. Rev. D16 (1977) 996.
[68] D.J. Thouless and M.E. Elzain, J. Phys. C:11 (1978) 3425.
[69] B. Velicky, Phys. Rev. 184 (1969) 614.
[70] J.L. Pichard and G. Sarma, in: Disordered Systems and Localization, and refs. therein.
[71] N.F. Mott, Adv. Phys. 16 (1967) 49.

Nuclear Physics B206 (1982) 321–332

SUPERSYMMETRIC FIELD THEORIES AND STOCHASTIC DIFFERENTIAL EQUATIONS

G. PARISI

Laboratori Nazionali INFN, Frascati
and
II Università di Roma, Tor Vergata, Roma, Italy

N. SOURLAS

Laboratoire de Physique Théorique de l'Ecole Normale Supérieure, Paris, France*

Received 1 April 1982

We show that some supersymmetric models can be written as classical stochastic equations. We give examples in one and two dimensions, but in four dimensions the only rotationally invariant supersymmetric model we succeed in writing as a classical stochastic equation is the free field theory. We also discuss spontaneous breaking of supersymmetry; we show it is related to the number of solutions of the stochastic equations.

1. Introduction

It is well-known that in euclidean space bosonic field theories can be formulated in terms of probabilistic concepts; the correlation functions of the fields (Green functions) are given by

$$\langle\varphi(x_1)\ldots\varphi(x_n)\rangle = \int d\mu[\varphi]\varphi(x_1)\ldots\varphi(x_n), \tag{1.1}$$

where $d\mu[\varphi]$ is a probability measure on a functional space, which is given by

$$d\mu[\varphi] \sim d[\varphi]\exp\left\{-\frac{1}{h}S(\varphi)\right\}, \tag{1.2}$$

where $S(\varphi)$ is the classical action and the normalization is fixed by the condition

$$\int d\mu[\varphi] = 1. \tag{1.3}$$

* Laboratoire Propre du Centre National de la Recherche Scientifique, associé à l'Ecole Normale Supérieure et à l'Université de Paris Sud. Postal address: 24, rue Lhomond, 75231 Paris Cedex 05-France.

The theory may also be formulated as a stochastic differential equation by introducing an additional time t [1] (five-dimensional time) and writing the following equation:

$$\frac{\partial}{\partial t}\varphi(x,t) = -\frac{\delta S}{\delta\varphi(x,t)} + \eta(x,t), \tag{1.4}$$

where $\eta(x,t)$ is a white noise whose normalization is fixed by the Einstein relation

$$\overline{\eta(x,t)\eta(x',t')} = 2\hbar\delta(x-x')\delta(t-t'), \tag{1.5}$$

where $\overline{F[\eta]}$ denotes the average of $F[\eta]$ over the η distribution.

It is a standard result that under quite general conditions

$$\lim_{t\to\infty} \overline{\varphi(x_1,t)\ldots\varphi(x_n,t)} = \langle\varphi(x_1)\ldots\varphi(x_n)\rangle, \tag{1.6}$$

where $\varphi(x,t)$ is the $\eta(x,t)$ dependent solution of eq. (1.4) and $\langle F[\varphi]\rangle$ is given by eq. (1.1).

It may be tempting to consider stochastic equations in real space (without introducing an extra time); it is then easy to see that a supersymmetry naturally appears [2, 3].

We first illustrate this with the example of a one-dimensional system. Let $U[\varphi(x)]$ be a potential (which may depend on the derivatives of φ) and consider the associated stochastic equation

$$-U_x[\varphi] \equiv -\frac{\delta U}{\delta\varphi(x)} = \eta(x), \tag{1.7}$$

where, as usual, $\eta(x)$ is a white noise normalized to

$$\overline{\eta(x)\eta(y)} = \delta(x-y). \tag{1.8}$$

(From now on we set $\hbar = 1$.)

Let us suppose for the moment that $U[\varphi]$ is convex and that eq. (1.7) has always one (for every η) and only one solution which we denote by $\varphi_\eta(x)$. (The other cases will be discussed in sect. 4.) The expectation value of any function $F[\varphi]$ of $\varphi(x)$ is given by

$$\overline{F[\varphi]} = \int \mathcal{D}\eta \exp\left\{-\tfrac{1}{2}\int \eta^2(x)\,\mathrm{d}x\right\} F[\varphi_\eta(x)]. \tag{1.9}$$

Standard manipulations [2] allow us to write

$$\overline{F[\varphi]} = \int \mathcal{D}\eta\,\mathcal{D}\varphi \exp\left\{-\tfrac{1}{2}\int\eta^2(x)\,\mathrm{d}x\right\} F[\varphi]\,\delta\left(\frac{\delta U}{\delta\varphi(x)} + \eta(x)\right)\det[U_{x,y}]$$

$$= \int \mathcal{D}\varphi\,\mathcal{D}\psi F[\varphi] \exp - \left\{\tfrac{1}{2}\int \mathrm{d}x\,U_x^2 + \int \mathrm{d}x\,\mathrm{d}y\,\bar\psi(x)U_{x,y}\psi(y)\right\}, \tag{1.10}$$

where $U_{x,y} = \delta^2 U/\delta\varphi(x)\delta\varphi(y)$.

In eq. (1.10) we have used the standard trick of writing the determinant of $U_{x,y}$ as an integral over anticommuting variables. So we have shown that

$$\overline{F[\varphi_\eta(x)]} = \langle F[\varphi(x)]\rangle \equiv \int \mathrm{D}\varphi \mathcal{D}\psi\, \mathrm{e}^{-S(\varphi,\psi)} F[\varphi]. \tag{1.11a}$$

where

$$S(\varphi,\psi) = \tfrac{1}{2}\int \mathrm{d}x\, U_x^2 + \int \mathrm{d}x\,\mathrm{d}y\, \bar{\psi}(x) U_{x,y} \psi(y). \tag{1.11b}$$

It is very surprising to find out that the "action" $S(\varphi,\psi)$ is invariant under the supersymmetry transformations

$$\delta\varphi(x) = \bar{\varepsilon}\psi(x) + \bar{\psi}(x)\varepsilon,$$

$$\delta\bar{\psi}(x) = -\bar{\varepsilon}U_x,$$

$$\delta\psi(x) = -\varepsilon U_x, \tag{1.12}$$

where ε, $\bar{\varepsilon}$ are anticommuting parameters.

We may introduce an auxiliary field $A(x)$ in order to get supersymmetry transformations linear in the fields:

$$S'(A,\varphi,\psi) = \int \mathrm{d}x \{\tfrac{1}{2}A^2(x) + A(x)U_x(\varphi)\} + \int \mathrm{d}x\,\mathrm{d}y\, \bar{\psi}(x) U_{x,y} \psi(y). \tag{1.13}$$

As a consequence of the supersymmetry we find the following Ward identity:

$$\langle U_x \varphi(y)\rangle = \langle \psi(x)\bar{\psi}(y)\rangle. \tag{1.14}$$

It is easy to translate this Ward identity into the language of the stochastic equations

$$\langle \psi(x)\bar{\psi}(y)\rangle = \langle U_{x,y}^{-1}(\varphi)\rangle = \overline{U_{x,y}^{-1}(\varphi_\eta)}$$
$$= \overline{U_x(\varphi_\eta)\varphi_\eta(y)} = -\overline{\eta(x)\varphi_\eta(y)}. \tag{1.15}$$

This equation can easily be derived directly from the stochastic equation

$$\overline{\eta(x)\varphi(y)} = \overline{\frac{\delta\varphi(y)}{\delta\eta(x)}} = \overline{\left\{\frac{\delta\eta(x)}{\delta\varphi(y)}\right\}^{-1}} = -\overline{U_{x,y}^{-1}}.$$

In other words we can say that both the fermions and the bosons remember that they come from the same stochastic equation, and that supersymmetry precisely says that.

There have been several attempts [4, 5] to stress the similarity between euclidean field theories and the classical theory of brownian motion. In this context it is natural to ask the following questions. Up to now fermions have appeared as pure quantum mechanical effects. Under what conditions is it possible to give them a classical interpretation? Eq. (1.15) seems promising in this respect. In this approach (of stochastic equations) the fermions we get seem naturally associated with a supersymmetry, but in nature supersymmetry seems to be broken. Is it possible, in this context, to understand supersymmetry breaking? What kind of supersymmetric theories do we obtain if we start from a potential $U[\varphi]$ which is polynomial in the fields φ or, if one prefers, which has a closed algebraic expression (Nicolai's arguments [6] tell us that without this restriction the class of theories is very large)? In particular, is it possible to get spin one-half fermions and integer spin bosons in this way?

The present paper is a modest attempt in this direction. In sect. 2 we present our original model of stochastic differential equations [2] which leads to spin zero fermions. In sect. 3 we construct the two-dimensional supersymmetric $N = 2$ Wess-Zumino model. Rotation symmetry is apparently broken at the level of the stochastic equations but is recovered once we average over the random variables. We have been unable yet to generalize this to the higher dimensional Wess-Zumino models (except for the trivial case of free field theories). In sect. 4 we discuss the spontaneous breaking of supersymmetry in one-dimensional models and we make some general conjectures.

2. Stochastic equations with spin zero fermions

In this section we specialize the discussion to the case where [2]

$$U[\varphi] = \int d^D x \left\{ \tfrac{1}{2}(\partial_\mu \varphi)^2 + V(\varphi) \right\}. \tag{2.1}$$

We have studied this case because $V(\varphi) = \frac{1}{2}m^2\varphi^2 + \frac{1}{4}g\varphi^4$ is relevant for the magnetic properties of a system in a quenched random external field. Also the case $V(\varphi) = \frac{1}{2}m^2\varphi^2 + \frac{1}{3}\lambda\varphi^3$ is relevant for the properties of dilute branched polymers [7].

The associated differential equation (for the case $V = \frac{1}{2}m^2\varphi^2 + \frac{1}{4}g\varphi^4$) is

$$-\Delta\varphi + m^2\varphi + g\varphi^3 = \eta. \tag{2.2}$$

After the standard manipulations outlined in the introduction [eq. (1.10)] we get the following effective action:

$$S(A,\varphi,\psi)=\int \mathrm{d}^D x\{\tfrac{1}{2}A^2(x)+A(x)(-\Delta\varphi+m^2\varphi(x)+g\varphi^3(x))$$

$$+\bar{\psi}(x)(-\Delta+m^2+3g\varphi^2(x))\psi(x)\}. \tag{2.3}$$

The fermions $\bar{\psi}(x)$, $\psi(x)$ appearing in this equation are obviously spin zero particles. What is remarkable with this action is that it is not only invariant under the natural supersymmetry transformation (1.12) but it is also invariant under the more general one

$$\delta\varphi=-\bar{\varepsilon}_\mu x_\mu\psi, \qquad \delta A=2\bar{\varepsilon}_\mu\partial_\mu\psi,$$

$$\delta\psi=0, \qquad \delta\bar{\psi}=\bar{\varepsilon}_\mu x_\mu A+2\bar{\varepsilon}_\mu\partial_\mu\varphi, \tag{2.4}$$

where $\bar{\varepsilon}_\mu$ is an anticommuting vector.

The invariance under this larger group of transformations is responsible for the dimensional reduction in this model (as well as in the branched polymer model); i.e. the correlation functions of this model in D dimensions are the same as those generated by the action

$$S_{\mathrm{R}}(\varphi)=\int \mathrm{d}^{D-2}x\left\{\tfrac{1}{2}(\partial_\mu\varphi)^2+V(\varphi)\right\}$$

in $D-2$ dimension. This aspect of dimensional reduction is of no interest for our present purpose. The model of this section is nevertheless important because it shows how to generate spin zero fermions and will serve as a guide for the next section. It shows also that the supersymmetry of the theory may be larger than the one predicted by the general arguments of the introduction.

3. Spin one-half fermions

In order to get a hint for the interesting problem of spin one-half fermions, we consider the model of sect. 2 and integrate over the auxiliary field $A(x)$. We then get the action

$$S_{\mathrm{eff}}(\varphi,\psi)=\int \mathrm{d}^D x\{(-\Delta\varphi+V'(\varphi))^2+\bar{\psi}(-\Delta+V'')\psi\}. \tag{3.1}$$

If we write the corresponding equations of motion, we see that the kinetic part in the equation for the fermion field $\psi(x)$ is the same as in the stochastic equation we started from, while the square of this kinetic operator appears in the equation for

$\varphi(x)$. We will then try to exploit the fact that the Klein-Gordon operator is the square of the Dirac operator.

Let us consider the stochastic equation

$$\partial_\mu \gamma_\mu \varphi = \eta , \tag{3.2}$$

where γ_μ are the Dirac matrices in D dimensions, φ a classical spinor and η a random spinor whose components obey independent gaussian distributions. To this equation corresponds the effective action

$$S_0(\varphi,\psi) = \int \mathrm{d}^D x \left\{ \tfrac{1}{2} \sum_i (\partial_\mu \varphi_i)^2 + \bar\psi \partial_\mu \gamma_\mu \psi \right\} . \tag{3.3}$$

This is the $N = 2$ free Wess-Zumino model. The different components of in (3.2) have become independent scalar fields in (3.3).

Let us restrict to two dimensions now, and consider the following stochastic differential equations (we always use euclidean metric):

$$\begin{aligned} &\partial_1 \varphi_2(x) + \partial_2 \varphi_1(x) + g(\varphi_1^2 - \varphi_2^2) = \eta_1(x) , \\ &\partial_1 \varphi_1(x) - \partial_2 \varphi_2(x) + 2g\varphi_1(x)\varphi_2(x) = \eta_2(x) . \end{aligned} \tag{3.4}$$

For $g = 0$ these equations reduce to eqs. (3.2) with $\gamma_1 = \sigma_x$, $\gamma_2 = \sigma_z$ $\sigma_x, \sigma_y, \sigma_z$ being the usual Pauli matrices. By repeating the standard manipulations as outlined in the introduction we get, up to surface terms,

$$S(\varphi,\psi) = \int \mathrm{d}^2 x \left\{ \tfrac{1}{2} \sum_{i=1}^{2} (\partial_\mu \varphi_i)^2 + \tfrac{1}{2} g^2 (\varphi_1^2 + \varphi_2^2)^2 + \bar\psi \left(\partial\!\!\!/ + g(\varphi_1 + i\gamma_5 \varphi_2) \right) \psi . \right. \tag{3.5}$$

This is the $N = 2$ Wess-Zumino model (this time including the interaction) in two dimensions.

A few remarks are in order:

(A) It is possible to describe spin one-half fermions classically, through the stochastic equations (3.4). The observables, bilinear in the fermion fields, have a classical interpretation, through eq. (1.15). So one can avoid introducing fermions, provided one works in the framework of the stochastic equations.

(B) The action (3.5) is Lorentz (rotation) invariant while Lorentz invariance is not obvious in the form of the stochastic equations. In particular there are first-order derivatives of the bosonic fields in eq. (3.4) which disappear with the averaging over the noise $\eta(x)$. This is automatic for the kinetic part (the Klein-Gordon operator being the square of the Dirac operator) but there are also crossed terms between the free part and the interaction part and they disappear (for this particular form of the interaction) because they give rise to a total derivative.

As this point will become important later, let us introduce a complex notation:

$$z = x_1 + ix_2, \qquad \varphi = \varphi_1 + i\varphi_2, \qquad \eta = \eta_1 + i\eta_2.$$

Eqs. (3.4) can then be written as

$$\overline{\frac{\partial}{\partial z}\varphi} + g\varphi^2 = \eta, \tag{3.6}$$

where $\overline{A}$ is the complex conjugate of A. In this form it is obvious that in $\bar{\eta}\eta$, the cross term $(\partial\varphi/\partial z)\varphi^2 + \overline{(\partial\varphi/\partial z)\varphi^2}$ is a total derivative, and gives rise to a surface term only (which, as usual, we drop) when averaging over η. In this form it is possible to see that one can replace the interaction $g\varphi^2$ by any analytic function $f(\varphi)$ of φ and still get a Lorentz-invariant theory.

Is it possible to extend the previous results to higher dimensions? A natural way seems to generalize eq. (3.6) by introducing quaternions. Let us define

$$q = x + iy + jz + kt, \qquad \varphi = \varphi_1 + i\varphi_2 + j\varphi_3 + k\varphi_4, \tag{3.7}$$

where $i, j, k, \mathbf{1}$ form a quaternion basis with $i^2 = j^2 = k^2 = 1$, and φ_i ($i = 1, \ldots, 4$) are four scalar fields. The natural generalization of (3.6) would be

$$\overline{\frac{\partial\varphi}{\partial q}} + f(\varphi) = \eta, \tag{3.8}$$

where, in order to get Lorentz invariance with the previous mechanism, $f(\varphi)$ is an analytic function of φ (left or right differentiable, in quaternionic sense). Unfortunately, the only analytic quaternionic functions are linear. This means that eq. (3.8) defines only a free field theory. To overcome this difficulty we have also tried to introduce an interaction to eq. (3.2) in a "blind" way:

$$(\partial_\mu\gamma_\mu)_{ij}\varphi_j + A_{ijk}\varphi_j\varphi_k = \eta_i. \tag{3.9}$$

(We limited ourselves to equations giving rise to renormalizable theories.) We have imposed the conditions to the A_{ijk} that the cross terms in $\Sigma_i\eta_i^2$, where a first-order derivative of φ appears, sum up to a total derivative. The only solution to these conditions is $A_{ijk} = 0$, i.e. no interaction. At this stage we feel like wizards who succeeded in their first sorcery but are unable to do it again.

What are the reasons for this failure? A tentative explanation would be that we need a natural analytic structure of space time, which exists in two but not in four dimensions (this can perhaps be avoided by using twistors). It must be said, however, that in the four-dimensional Wess-Zumino model, the superfield has been constrained to be "left chiral". If one removes this constraint the model is much more complex. Also what seems to be natural in this framework is $N=2$ supersymmetry. To our knowledge no one has succeeded in writing the $N=2$ supersymmetric Wess-Zumino (of the Yukawa type) model with a renormalizable interaction, even at the price of considering a larger number of fields. We have also tried stochastic differential equations with constraints (e.g. more equations than unknowns) without success. We do not yet understand if we face a difficulty of principle or if this difficulty can be overcome.

4. Spontaneous supersymmetry breaking

In this section we will study spontaneous breaking of supersymmetry [8] in the framework of the stochastic differential equations. One of the most striking consequences of supersymmetry is the vanishing of the ground-state energy [9]

$$Z = \int \mathcal{D}\varphi \mathcal{D}\psi \, e^{S(\varphi,\psi)} = 1 . \tag{4.1}$$

On the contrary, when supersymmetry is spontaneously broken, the ground-state energy is strictly positive.

It is very easy to obtain eq. (4.1) in the context of stochastic differential equations (1.7).

$$\begin{aligned} 1 &= \frac{1}{c} \int \mathcal{D}\eta \exp\left\{ -\tfrac{1}{2} \int \eta^2 \mathrm{d}x \right\} = \frac{1}{c} \int \mathcal{D}\eta \mathcal{D}\varphi \delta(U_x[\varphi] + \eta) \\ &\quad \times \det\left[U_{x,y} \right] \exp\left\{ -\tfrac{1}{2} \int \eta^2 \mathrm{d}x \right\} \\ &= \frac{1}{c} \int \mathcal{D}\varphi \exp\left\{ -\tfrac{1}{2} \int U_x^2 \mathrm{d}x \right\} \det\left[U_{x,y} \right] , \end{aligned} \tag{4.2}$$

where $c \equiv \int \mathcal{D}\eta \exp\{-\frac{1}{2}\int \eta^2 \mathrm{d}x\}$. As we have stated in the introduction, there is an assumption, in order to derive (4.2), that the stochastic differential equation

$$-U_x[\varphi] = \eta \tag{4.3}$$

has one and only one solution for any η. In this section we will relate the non-fulfilment of this condition to spontaneous supersymmetry breaking.

Let us suppose that $U_x(\varphi)$ is monotonous in φ and that $-U_x(\infty) = a$, $-U_x(-\infty) = b$, $(a > b)$. So for $\eta > a$ or $\eta < b$, (4.3) does not have any solution. Then Z as defined by eq. (4.2) is given by

$$Z = \frac{1}{c}\int_b^a \mathcal{D}\eta \exp\left\{-\int \tfrac{1}{2}\eta^2 \mathrm{d}x\right\}, \tag{4.4}$$

and equals one only if $a \to \infty$, $b \to -\infty$, i.e. when (4.3) always has one solution. It is also easy to see what happens in the present case to the Ward identity, eq. (1.14) (for illustrative purposes we restrict ourselves to the zero dimensional case):

$$\langle \psi\bar{\psi}\rangle = \frac{1}{Z}\int \mathrm{d}\eta\, \mathrm{e}^{-\eta^2}(U'')^{-1}, \tag{4.5}$$

$$-\langle U_x(\varphi)\varphi\rangle = \frac{1}{Z}\int \mathrm{d}\eta\, \mathrm{e}^{-\eta^2}\eta\varphi = \frac{1}{Z}\int \mathrm{d}\eta\, \mathrm{e}^{-\eta^2}\frac{\partial\varphi}{\partial\eta} + \frac{1}{Z}\left(\mathrm{e}^{-b^2/2}\varphi(b) - \mathrm{e}^{-a^2/2}\varphi(a)\right). \tag{4.6}$$

It is the last term in (4.6) which violates the Ward identity. Let us consider the case of a system in a finite volume V and a cut off W in the field space ($-W \leqslant \varphi \leqslant W$). Then, the supersymmetry breaking term in (4.6) behaves like $(\mathrm{e}^{-b^2/2} + \mathrm{e}^{-a^2/2})W/\mathrm{e}^{cV}$. We see that the two limits $W \to \infty$ and $V \to \infty$ do not commute and that, contrary to what happens for an ordinary symmetry, spontaneous supersymmetry breaking can occur in a finite volume.

Consider now the case where (4.3) has $N(\eta)$ solutions with $N(\eta) \geqslant 1$ for every η and where there is a range in η where $N(\eta) > 1$. From the stochastic equation point of view, the natural way of defining the average over the noise distribution is the following:

$$\overline{F(\varphi_\eta)} = \frac{1}{c}\int \mathcal{D}\eta \sum_{i=1}^{N(\eta)} F\left(\varphi_\eta^{(i)}\right)\varepsilon_i \exp\left\{-\int \tfrac{1}{2}\eta^2 \mathrm{d}x\right\}, \tag{4.7}$$

where the sum Σ_i runs over all the solutions of eq. (4.3) and $\varepsilon_i = +1$ for every i. In the present case Z is given by

$$\begin{aligned} Z &\equiv \frac{1}{c}\int \mathcal{D}\eta\mathcal{D}\varphi \exp\left\{-\int \tfrac{1}{2}\eta^2 \mathrm{d}x\right\}\delta(U_x + \eta)\det\left[U_{x,y}\right] \\ &= \frac{1}{c}\int \mathcal{D}\eta \exp\left\{-\int \tfrac{1}{2}\eta^2 \mathrm{d}x\right\}\sum_{i=1}^{N(\eta)} \det\left[U_{x,y}^{(i)}\right]\Big/\left|\det\left[U_{x,y}^{(i)}\right]\right| \\ &= \frac{1}{c}\int \mathcal{D}\eta \exp\left\{-\int \tfrac{1}{2}\eta^2 \mathrm{d}x\right\}\sum_{i=1}^{N(\eta)} \mathrm{sign}\left(\det\left[U_{x,y}(\varphi^{(i)})\right]\right). \end{aligned} \tag{4.8}$$

With the definition (4.7), eq. (1.11) is not valid any more because $\det[U_{x,y}(\varphi^{(i)})]$ changes sign with φ and is not equal to its absolute value as in the case where $N(\eta)$ is always one.

It is possible to take a different average over the noise by choosing in eq. (4.7)

$$\varepsilon_i = \operatorname{sign}\left(\det\left[U_{x,y}(\varphi^i)\right]\right), \tag{4.9}$$

in which case eq. (1.11) is valid, but this choice does not seem very natural. Let us finally remark that if $\sum_{i=1}^{N(\eta)} \varepsilon_i = 1$ for every η, Z is still one and the supersymmetry Ward identity (1.14) is still valid, at least formally. So we may conclude that one has the option of the stochastic equation point of view ($\varepsilon_i = +1$) in which case supersymmetry is broken, and of a different point of view [ε_i given by eq. (4.9)] in which case supersymmetry is not broken. If $N(\eta)$ always equals one there are no different options and supersymmetry cannot be spontaneously broken.

It is also instructive to consider the following one-dimensional model, in which it has been shown by Witten [10] that dynamical supersymmetry breaking appears. Consider the Langevin equation

$$\partial_t x + \varepsilon U'(x) = \eta, \tag{4.10}$$

where $U'_x(x) \equiv (\partial/\partial x)U(x)$ and $\varepsilon = +1$ or -1. This equation leads to the following effective action:

$$S_{\text{eff}} = \int \mathrm{d}t \left\{ \tfrac{1}{2}(\partial_t x)^2 + \tfrac{1}{2}U'^2 + \bar{\psi}(\partial_t + \varepsilon U'')\psi \right\}, \tag{4.11}$$

which is the model considered by Witten [10] written in a slightly different way [11]. In passing from (4.10) to (4.11) we have dropped the "surface term"

$$I = \int_{-\infty}^{\infty} \mathrm{d}t\, \partial_t x U'(x) = \int_{-\infty}^{\infty} \mathrm{d}t \frac{\partial U(x)}{\partial t}. \tag{4.12}$$

It is convenient to consider the probability $P(x, t; x_0)$ of the solution $x_\eta(t)$ of (4.10) to take the value x at time t with the initial condition $x_\eta(t=0) = x_0$. The Fokker-Planck equation which is obeyed by $P(x, t; x_0)$ can be written in the form (see, e.g., [1]) of the Schrödinger equation

$$\frac{\mathrm{d}}{\mathrm{d}t}\mathcal{P}(x,t) = -\hat{H}\mathcal{P}(x,t),$$

$$\mathcal{P}(x,t) \equiv P(x,t)\mathrm{e}^{U(x)/2}, \tag{4.13}$$

$$\hat{H} = -\frac{1}{2}\frac{\partial^2}{\partial x^2} + \tfrac{1}{8}[U'(x)]^2 - \tfrac{1}{4}\varepsilon U''(x). \tag{4.14}$$

The $t \to +\infty$ behaviour of $P(x, t)$ is governed by the lowest eigenvalue E_0 of $\hat{H}$. $\psi_0(x) = e^{-\epsilon U(x)/2}$ is an eigenvector of $\hat{H}$ with eigenvalue $E = 0$ and because $\psi_0(x)$ is free of zeros, $E = 0$ is the lowest eigenvalue of $\hat{H}$ and therefore for $t \to +\infty$

$$P(x, t) \underset{t \to +\infty}{\to} e^{-\epsilon U(x)} / \int dx\, e^{-\epsilon U(x)}. \tag{4.15}$$

As usual, we suppose that $U(x)$ has a polynomial form $U(x) = \sum_{n=0}^{N} a_n x^n$. [We choose ϵ in eq. (4.10) such that $\epsilon a_N > 0$.]

If N is odd we see that for large enough t, all the probability $P(x, t)$ is concentrated at $x \sim -\infty$ and is zero everywhere else [as the potential is unbounded below, provided one waits long enough, the particle escapes to $-\infty$ with probability ~ 1, independently of the noise in eq. (4.10)]. So the stochastic equation won't have any solution for most of the η's and supersymmetry is spontaneously broken. If N is even $P(x, t)$ is more evenly distributed and supersymmetry is not broken. These conclusions are in agreement with Witten's analysis*.

Finally we would like to discuss the connection of our approach to Nicolai's transformation [6]. Formally one could invert the stochastic equation (4.3) and write it in the form

$$[U_x]^{-1}(\eta) = \varphi.$$

We could next introduce the free field $\varphi_F = [1/(\Box + m^2)^{1/2}]\eta$. The transformation from φ to φ_F is Nicolai's transformation. It is a non-local and non-linear (non-polynomial) transformation which can be constructed order by order in perturbation theory. The stochastic equations we have considered are on the contrary local and polynomial.

While we were writing this paper, similar suggestions on the spontaneous breaking of supersymmetry have been proposed by Cecotti and Girardello [13].

We are grateful to E. Cremmer, P. Fayet, J. Iliopoulos, C. Itzykson, L. Maiani and R. Petronzio for useful discussions.

References

[1] G. Parisi and Wu Yong-shi, Sci. Sin. 24 (1981) 483
[2] G. Parisi and N. Sourlas, Phys. Rev. Lett. 43 (1979) 744
[3] T. Banks and P. Windey, Tel Aviv University preprint
[4] E. Nelson, Phys. Rev. 150 (1966) 1079
[5] F. Guerra and P. Ruggiero, Phys. Rev. Lett. 31 (1973) 1022

* The same model has been considered by Calogero [12] as an example where non-perturbative phenomena appear.

[6] H. Nicolai, Phys. Lett. 89B (1980) 341; Nucl. Phys. B176 (1980) 419
[7] G. Parisi and N. Sourlas, Phys. Rev. Lett. 46 (1981) 871
[8] P. Fayet and J. Iliopoulos, Phys. Lett. 51B (1974) 461
[9] B. Zumino, Nucl. Phys. B89 (1975) 535
[10] E. Witten, Princeton preprint
[11] P. Salomonson and J.W. van Holten, CERN preprint
[12] F. Calogero, Nuovo Cim. Lett. 25 (1979) 533
[13] S. Cecotti and L. Girardello, Phys. Lett. 110B (1982) 39

COURSE 7

SPIN GLASSES AND OPTIMIZATION PROBLEMS WITHOUT REPLICAS

Giorgio PARISI

Università di Roma II - Tor Vergata, Rome 00173, Italy and INFN Sezione di Roma

J. Souletie, J. Vannimenus and R. Stora, eds
Les Houches, Session XLVI, 1986—Le hasard et la matière/Chance and matter

Contents

1. Introduction

In recent years much progress has been made in the field of spin glasses and optimization problems. Most of the work has been done by using the replica approach; however, the replica method is very powerful and hides the physical and mathematical assumptions one is making. In these lectures I will use the cavity approach, which can be considered as the transcription of the replica approach into more explicit terms. (For a review of the replica approach and for the original references the reader may consult refs. [1-3].)

After this short introduction the reader will find a section dedicated to the naive approach to spin glasses, which is unfortunately wrong at low temperature; in section 3 we describe our assumptions on the structure of the equilibrium states at low temperature, while section 4 contains the computation of the thermodynamical quantities using the results of section 3. Before studying optimization problems we review in section 5 the usual technique of describing linear polymers as the $n \to 0$ limit of an n-component spin model. In section 6 we use our approach to study the matching problem, while in section 7 we present our conclusions.

2. The naive approach to spin glasses

By definition an Ising spin glass with N spins has the following Hamiltonian [4]:

$$H_J[\sigma] = -\frac{1}{2} \sum_{i,k=1}^{N} J_{i,k}\sigma_i\sigma_k - B \sum_{i=1}^{N} \sigma_i, \tag{2.1}$$

where B is the external magnetic field, the σ's are the spins, which may take the values ± 1, and the J's are the couplings among the spins (we consider only two-spin interactions); the J's are not known a priori; they are determined according to a probabilistic law. Different choices of the J distribution lead to different models.

The J-dependent free energy is computed in the usual way:

$$f_J = -\frac{1}{\beta} \ln \sum_{\{\sigma\}} \exp(-\beta H_J[\sigma]). \tag{2.2}$$

For reasonable choices of the J distribution, it is expected that the free energy density $F_J = f_J/N$ does not fluctuate in the limit $N \to \infty$:

$$\overline{F_J^2} - (\bar{F}_J)^2 \to 0 \quad \text{for } N \to \infty \tag{2.3}$$

(we denote by a bar the average over the J's). Our aim is to compute the free energy $F = \bar{F}_J$ and other thermodynamic quantities in the limit $N \to \infty$.

In these notes we study the mean field approach to spin glasses [4] and we will consider only a well-known model, the Sherrington-Kirkpatrick model [5]; in this model the mean field approach is supposed to give the correct results (the mean field approach is also supposed to give the correct results for a short-range model in the limit in which the dimensions of the space go to infinity).

In the S-K model all the J's may be different from zero and they are evenly distributed random variables of order $1/N^{1/2}$, with zero mean and variance

$$\overline{J^2} = 1/N. \tag{2.4}$$

As we shall see later, the values of the higher moments of the J's do not matter in the limit $N \to \infty$, provided that $\overline{J^4} = O(1/N^2)$, as is the case for any decent probability distribution.

The very naive mean field equations we would write are the following (let us work for simplicity at $B = 0$):

$$m_i = \text{th}(\beta h_i), \quad h_i = \sum_{k=1}^{N} J_{i,k} m_k, \tag{2.5}$$

h_i being the effective field on the ith spin.

Unfortunately these equations are not correct. In order to derive the correct equations it is convenient to consider two systems with N and $N+1$ spins, respectively [6]. The Hamiltonian will be:

$$H = H_1 + H_2,$$

$$H_1 = -\frac{1}{2} \sum_{i,k=1}^{N} J_{i,k} \sigma_i \sigma_k,$$

$$H_2 = -\varphi \sigma_{N+1}, \quad \varphi \equiv \sum_{k=1}^{N} J_{N+1,k} \sigma_k. \tag{2.6}$$

In other words, we consider a system with N spins and we add an extra spin. Our aim is to compute the magnetization of the $(N+1)$th spin ($m_{N+1} = \langle \sigma_{N+1} \rangle$) as a function of the magnetization of the other N spins in the absence of this last spin. We make the crucial assumption that the connected correlation functions of the system with N spins are sufficiently small that they can be neglected at different points. This implies that the two-spin connected correlation function is given by

$$\langle \sigma_i \sigma_k \rangle_c = \delta_{i,k}(1 - m_i^2). \tag{2.7}$$

Similar formulae hold for high-order correlation functions. If we write

$$\varphi = h + \delta h, \quad h \equiv \langle \varphi \rangle = \sum_{k=1}^{N} J_{N+1,k} m_k, \tag{2.8}$$

a simple computation shows that

$$\langle (\delta h)^2 \rangle = \beta(1-q), \tag{2.9}$$

where q (the Edwards–Anderson order parameter [4]) is defined as

$$q = \frac{1}{N} \sum_{j=1}^{N} m_j^2. \tag{2.10}$$

Moreover, it can be easily shown that when $N \to \infty$

$$\langle (\delta h)^{2k} \rangle = (1-q)^k (2k-1)!!. \tag{2.11}$$

In this way we conclude that δh has a Gaussian distribution with zero mean and variance $1-q$.

The expectation value of σ_{N+1} can thus be computed as

$$\begin{aligned} m_{N+1} &= \sum_{\sigma=\pm 1} \sigma \int d(\delta h) \frac{\exp\{\beta(h+\delta h)\sigma - (\delta h)^2/[2(1-q)]\}}{[2\pi(1-q)]^{1/2}} \Bigg/ Z \\ &= \text{th}(\beta h), \\ Z &\equiv \sum_{\sigma=\pm 1} \int d(\delta h) \frac{\exp\{\beta(h+\delta h)\sigma - (\delta h)^2/[2(1-q)]\}}{[2\pi(1-q)]^{1/2}}, \end{aligned} \tag{2.12}$$

which is the result we wanted to know.

If we recompute the magnetizations of the spins after the introduction of the $(N+1)$th spin, we obtain the celebrated TAP equations [7].

Our aim is now to find a self-consistent equation for the quantity q [6]. It is evident that q must be equal to $\overline{m_{N+1}^2}$, where the average is taken over the couplings J which are present in H_2: the $(N+1)$th spin must have the same properties as the other spins. The J's appearing in H_2 and the m's are clearly independent quantities; the central limit theorem tells us that h has a Gaussian probability distribution with zero mean and variance q, as the reader can easily verify by computing the average value of the moments of h.

Using the short-hand notation

$$dP_q(h) = dh \exp[-h^2/(2q)]/(2\pi q)^{1/2}, \tag{2.13}$$

we obtain (the computation can easily be done also if the magnetic field B is different from zero)

$$q = \overline{m_{N+1}^2} = \int dP_q(h)\, \mathrm{th}^2[\beta(h+B)]. \tag{2.14}$$

The reader should note that we implicitly assume that all integrals for which no bounds are indicated run from $-\infty$ to $+\infty$.

Equation (2.14) is the same result as that of Sherrington and Kirpatrick [5], which was originally derived by using the replica approach. The solution of this equation can be easily found partially analytically and partially numerically. If $B=0$, q is equal to zero for $\beta<1$, it is equal to $\beta-1$ for β slightly greater than 1 and it goes to 1 (independently of B) when the temperature goes to zero: a transition from zero to nonzero magnetization occurs at $\beta=1$, $B=0$. On the other hand, q is always different from 0 at $B\neq 0$.

We have now in our hands all the instruments to compute the thermodynamics. The expectation value of the Hamiltonian density is in the average given by the expectation value of H_2:

$$\begin{aligned} U &= \frac{1}{2}\sum_k \langle \sigma_{N+1} J_{N+1,k} \sigma_k \rangle \\ &= -\frac{N}{2}\left(\int dP_q(h)\, h\, \mathrm{th}(\beta h) + \langle \delta h\, \sigma_{N+1}\rangle \right). \end{aligned} \tag{2.15}$$

By integration by parts we find that the first term is $-\beta q(1-q)$, while the second term is equal to $\langle (\delta h)^2\rangle = \beta(1-q)$. Collecting all the terms together we find

$$U = -\tfrac{1}{2}\beta(1-q^2). \tag{2.16}$$

Equation (2.16) is sufficient to derive the thermodynamics of the model. However, for later use it is convenient to compute directly the free energy. If we start from eq. (2.12) we get

$$Z_{N+1} = Z_N\, 2\,\mathrm{ch}(\beta h) \exp[\beta^2(1-q)/2]. \tag{2.17}$$

Therefore

$$\overline{\ln Z_{N+1}} = \overline{\ln Z_N} + \beta^2(1-q)/2 + \int \mathrm{d}P_q(h)\ln[2\,\mathrm{ch}(\beta h)]. \tag{2.18}$$

For large values of N we have

$$\overline{\ln Z_N} = N\Phi(\tilde{\beta}), \quad \tilde{\beta} = \beta[N(\overline{J^2})]^{1/2}. \tag{2.19}$$

Using the usual normalization [see eq. (2.5)], we have that the effective temperature $\tilde{\beta}$ coincides with β; however, when we add a spin without changing the couplings J, the normalization is slightly changed and we have that $\tilde{\beta} = \beta[1+1/(2N)]$. If we put everything together we find that

$$\overline{\ln Z_{N+1} - \ln Z_N} = \Phi(\beta) + \tfrac{1}{2}\beta\,\mathrm{d}\Phi/\mathrm{d}\beta. \tag{2.20}$$

The final expression for the final free energy density is thus

$$\beta F(\beta) = \beta^2(1-q)^2/4 + \int \mathrm{d}P_q(h)\ln[2\,\mathrm{ch}(\beta h)]. \tag{2.21}$$

It is a simple exercise to verify that eq. (2.21) is in agreement with eq. (2.16) and that the self-consistency condition for q [eq. (2.12)] can be obtained as a self-consistency equation for the free energy, i.e., $\partial F/\partial q = 0$.

If we now compute the internal energy as a function of the temperature, we find good agreement with the numerical simulations of the thermodynamical quantities for $T > 0.4$; however, some deviations are observed at small temperatures: this theory predicts $U(0) \approx -0.80$, while numerical simulations give [5] $U(0) \approx 0.765 \pm 0.01$.

However, worse results are obtained for the entropy, which (for any value of the magnetic field B) becomes negative for small values of the temperature; in particular at zero magnetic field we have that $S(0) \approx -0.17$. These results indicate that we have made a mistake. The most crucial hypothesis we have made is that correlations can be neglected; it is worthwhile to compute them and to check if this hypothesis is consistent.

In the high-temperature phase we find that $\langle\sigma_i\sigma_k\rangle_c$ is of order $1/N^{1/2}$, so that it is convenient to introduce the quantity

$$\chi^{(2)} = \frac{1}{N}\sum_{i\neq k}^{N} \langle\sigma_i\sigma_k\rangle_c^2. \tag{2.22}$$

The computation of $\chi^{(2)}$ can be done in a self-consistent way at high temperatures by considering two systems with N and $N+2$ spins [8]. After some work we find

$$\chi^{(2)}_{N+2} = \chi^{(2)}_N \beta^2\overline{(1-m^2)^2} + f(\beta, h), \tag{2.23}$$

where $f(\beta, h)$ is a nonvanishing given function and $\overline{(1-m^2)^2}$ is a shorthand notation for

$$\int \mathrm{d}P_q(h)\,[1-\mathrm{th}^2(\beta h)]^2 = \frac{1}{N}\sum_{i=1}^{N}(1-m_i^2)^2. \tag{2.24}$$

If $\chi^{(2)}$ has a finite limit when $N \to \infty$, we must have [8]

$$\beta^2\overline{(1-m^2)^2} \leq 1. \tag{2.25}$$

If the l.h.s. of eq. (2.25) is greater than 1, the correlations diverge exponentially when N goes to infinity and this cannot be tolerated. In the marginal case,

$$\beta^2\overline{(1-m^2)^2} = 1, \tag{2.26}$$

$\chi^{(2)}$ increases most likely as a positive power of N; in this case the picture seems to be consistent, although a more careful study would be welcome.

A simple computation [9] shows that eq. (2.25) is violated in the region shown in fig. 1 (i.e. below the so-called de Almeida–Thouless line). It is well known that for a ferromagnetic system a divergence of the susceptibility signals the presence of a transition from one equilibrium state to more than one equilibrium state. We therefore expect that in the low-temperature region more than one equilibrium state is present; this possibility has been clearly overlooked in our previous treatment. The next two sections are devoted to finding the correct results in the low-temperature region.

3. The structure of pure equilibrium states

It is well known that in a ferromagnet at low temperatures there is a spontaneous magnetization m; in other words, the system can stay in

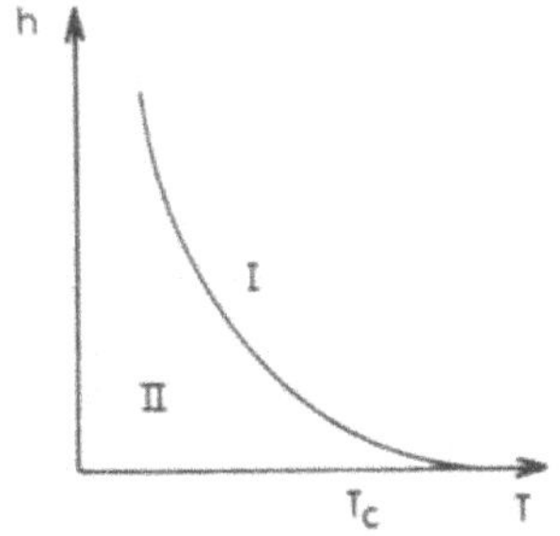

Fig. 1. Region I is the high-temperature phase, where the approach of section 2 is correct. Region II is the low-temperature phase where the approach of section 4 must be used.

two states (labeled by + or −) such that

$$\langle\sigma\rangle_+ = m, \qquad \langle\sigma\rangle_- = -m. \tag{3.1}$$

On the other hand, if we define

$$\langle\sigma_i\rangle = \sum_{\{\sigma\}} \sigma_i \exp(-\beta H) \Big/ \sum_{\{\sigma\}} \exp(-\beta H), \tag{3.2}$$

without introducing any symmetry breaking field, we have that $\langle\sigma\rangle = 0$. However, it is evident that

$$\langle\ \rangle = \tfrac{1}{2}\langle\ \rangle_+ + \tfrac{1}{2}\langle\ \rangle_-. \tag{3.3}$$

In other words, the state $\langle\ \rangle$ is not a pure state and it can be decomposed as the sum of the two pure states $\langle\ \rangle_\pm$. In these last two states the connected correlation functions go to zero at large distances, while this does not happen in the symmetric state $\langle\ \rangle$:

$$\langle\sigma_i\sigma_k\rangle \equiv \langle\sigma_i\sigma_k\rangle_c \to m^2 \tag{3.4}$$

when $i - k \to \infty$. Only pure states are important from the physical point of view, while a blind application of the formalism of statistical mechanics prefers the symmetric state.

Now what happens in a spin glass? Without loss of generality in the low-temperature region we could write

$$\langle\ \rangle = \sum_\alpha w_\alpha \langle\ \rangle_\alpha, \tag{3.5}$$

where α labels the pure equilibrium states. We must now make some hypothesis on the form of the w's and on how different states differ from one another. To this end it is convenient to introduce the overlap (q) and the distance (d) of two states [10]:

$$q_{\alpha\gamma} = \frac{1}{N}\sum_{i=1}^{N} m_i^\alpha m_i^\gamma,$$

$$d_{\alpha\gamma} = \frac{1}{N}\sum_{i=1}^{N} (m_i^\alpha - m_i^\gamma)^2 = q_{\alpha\alpha} + q_{\gamma\gamma} - 2q_{\alpha\gamma},$$

$$m_i^\alpha = \langle\sigma_i\rangle_\alpha. \tag{3.6}$$

At zero temperature, where all the m's are equal to ± 1, d becomes proportional to the usual Hamming distance (the proportionality factor being 4). We suppose that different pure states do not have macroscopically different properties and

$$q_{\alpha\alpha} = q_{\gamma\gamma} = q_{EA}, \qquad d_{\alpha\gamma} + 2q_{\alpha\gamma} = 2q_{EA}. \tag{3.7}$$

The arguments of the previous section suggest that more than one equilibrium state is present in spin glasses below the de Almeida–Thouless line; here we boldly assume that in the low-temperature phase spin glasses have infinitely many pure states [e.g., $\exp(N^{1/2})$] [6, 11] and that the w's are random (not independent) variables whose probability law is defined as follows [12].

We introduce some variables f_α, which can be interpreted as the total free energy of the state α. The w's can be computed from the f's as follows:

$$w_\alpha = \exp(-\beta f_\alpha)/\sum_\gamma \exp(-\beta f_\gamma). \tag{3.8}$$

It is evident that, if we shift all the f's by the same amount, the w's are left invariant.

We now assume that the f's are independent random variables and that the probability of finding one state in the interval $f, f+\mathrm{d}f$ is given by

$$\exp[y(f-\tilde{f})]\,\mathrm{d}f, \tag{3.9}$$

where $\tilde{f}$ is a reference free energy (roughly speaking $\tilde{f}$ is the free energy of the system) and y is a positive parameter (we also assume that $y<\beta$).

We only assume that eq. (3.9) is correct when $f-\tilde{f}=\mathrm{O}(1)$; it does not make sense when $f-\tilde{f}=\mathrm{O}(N)$, because all the pure states must have the same free energy density; moreover, if we interpret f as the free energy in a mean field approximation, eq. (3.9) is no longer correct when $f-\tilde{f}=\mathrm{O}(N)$; it should, however, become correct when $(f-\tilde{f})/N\to 0$.

Which is the probability distribution of the d's and which are the correlations between the d's and the w's? We make the very strong hypothesis that we always have (with probability 1 when $N\to\infty$)

$$d_{\alpha\gamma} \leqslant \max(d_{\alpha\beta}, d_{\gamma\beta}) \quad \forall\beta. \tag{3.10}$$

Equation (3.10) tells us that the space of states is ultrametric [12]; it amounts to saying that, for any choice of the number d, it can be decomposed into nonoverlapping clusters (of diameter d) such that α and γ belong to the same cluster if and only if $d_{\alpha\gamma}\leqslant d$. Clusters may be decomposed into subclusters of smaller diameter and so on. We could also say that a space is ultrametric if two spheres of the same diameter coincide if they have at least one point in common. The whole distribution of states is characterized if we specify the clusters and their overlaps.

Let us suppose that the values of the distance are quantized, i.e., the d's may take only the values d_j, $j=0,1,\ldots,k$ (for simplicity we suppose that the d's are a monotonous sequence). For sake of space we will write down the formulae for $k=3$, the generalization to arbitrary k being evident. At the end we will perform the limit $k\to\infty$; in this case the distance will take a continuous range of values.

If $k=3$, the distance between an arbitrary pair of states is at most d_3; states at smaller distance belong to the same supercluster; if the distance is d_1 (not d_2), they belong to the same cluster; their distance may be $d_0=0$ only if they coincide.

We will describe the probability distribution of superclusters, of clusters inside the superclusters and of the states inside the clusters; the rules are so simple that they can be easily generalized to an infinite hierarchy.

The number of superclusters goes to infinity with N; we label them with an index α_3 and we associate to each of them a "free energy" f_{α_3}; the f's are random independent variables, whose probability distribution is supposed to be

$$\exp[y_2(f_{\alpha_3}-\tilde{f})]\,\mathrm{d}f_{\alpha_3}. \qquad (3.11)$$

In each of the superclusters there are infinitely many clusters of diameter d_2; we label them with a pair of indices (α_3, α_2) and we associate to each of them a "free energy" f_{α_3,α_2}; the f's are random independent variables, whose probability distribution is supposed to be

$$\exp[y_1(f_{\alpha_3,\alpha_2}-f_{\alpha_3})]\,\mathrm{d}f_{\alpha_3,\alpha_2}; \qquad (3.12)$$

the parameter y_1 satisfies the inequality $y_2<y_1$.

The reader should note that the previous equations imply that the probability distribution of the f's associated to a cluster is

$$\exp[y_1(f-\tilde{f})], \qquad (3.13)$$

and the f's of the clusters are uncorrelated random variables, if we do not specify whether the clusters belong (or do not belong) to the same supercluster; however, there are subtle correlations among the f's which belong to the same supercluster (the same remark is also valid for the f's associated to the states); in other words, the f's and the distances are not independent variables but they are correlated; eqs. (3.11), (3.12) may be used to compute their correlations [4, 12].

Finally the states are labeled with three indices $(\alpha_3\alpha_2\alpha_1)$ and the probability distribution of the associated free energy (which are as usual random independent variables) is assumed to be

$$\exp[y_0(f_{\alpha_3\alpha_2\alpha_1}-f_{\alpha_3\alpha_2})]\,\mathrm{d}f_{\alpha_3\alpha_2\alpha_1}, \qquad (3.14)$$

where $y_1<y_0<\beta$. Integrating over the free energies of the clusters and of the superclusters one finds that the probability distribution of the free energies of the states is given by eq. (3.9) with $y \equiv y_0$.

It is evident that this construction may be generalized to any value of k. In this way we obtain a distribution of the probability of the states and of their overlaps, the only parameters being the d's and the y's. We finally let k go to infinity in such a way that the d's become dense on the interval $(0, d_M)$ $(d_M \leq q_{EA})$. In this way we obtain a monotonous decreasing function $y[d]$, which is defined on the interval $(0, d_M)$; the distribution of states depends only on this function $y[d]$. The task of the next section will be to find $y[d]$ by using the methods described in the previous section.

The reader has certainly noticed that this section contains only statements; indeed, we cannot prove that this probability distribution of the states is the correct one; however, the probability distribution described in this section has the advantage of being relatively simple and (as we will see in the next section) it is the only candidate which does not lead to contradictions.

Historically the equations of this section were derived in refs. [6, 11, 12] by using the approach of broken replica symmetry of refs. [10, 11]; however, it was realized in ref. [6] that they have the same information content as the usual assumptions that are made on the matrix $Q_{a,b}$ in the replica approach. It is therefore logically consistent to take the equations of this section as a starting hypothesis.

In this framework a very natural question is the following: If we take two states randomly (with their weights w), which is the probability $P_J(q)$ that they have overlap q (or equivalently distance $d = q_{EA} - q$)? More generally, which is the probability distribution of the function $P_J(q)$ when J changes and which are the main characteristics of the function $P_J(q)$ which do not depend on J? The precise definition of $P_J(q)$ is

$$P_J(q) = \sum_\alpha \sum_\gamma w_\alpha w_\gamma \delta(q - q_{\alpha\gamma}). \tag{3.15}$$

In order to obtain the results we want, it is convenient to introduce the function $x(q)$, which is defined as

$$x(q) \equiv y(2q_{EA} - 2q)/\beta. \tag{3.16}$$

A nontrivial computation shows that [12]

$$\overline{P_J(q)} \equiv P(q) = \mathrm{d}x/\mathrm{d}q + \delta(q - q_M)[1 - x(q_M)] + \delta(q_m - q)x(q_m),$$
$$q_M = q_{EA}, \qquad q_m \equiv q_{EA} - \tfrac{1}{2}d_M. \tag{3.17}$$

Very often one finds in the literature the function $q(x)$, which is just the inverse of the function $x(q)$ and is defined on the interval (0, 1); it is such that

$$\begin{aligned} q(x) &= q_m, && 0 < x < x(q_m), \\ x[q(x)] &= x, && x(q_m) < x < x(q_M), \\ q(x) &= q_M, && x(q_M) < x < 1. \end{aligned} \tag{3.18}$$

An explicit computation shows that the energy density is given (for the infinite-range model) by

$$U = -\tfrac{1}{2}\beta \int_0^1 dx\,[1 - q^2(x)], \tag{3.19}$$

so that the whole thermodynamics may be reconstructed from the knowledge of the function $q(x)$ [or equivalently $y(d)$] as a function of the temperature.

More detailed results may also be obtained [10], e.g.,

$$\overline{P_J(q_1)P_J(q_2)} = \tfrac{1}{2}P(q_1)P(q_2) + \tfrac{1}{2}\delta(q_1 - q_2)P(q_1). \tag{3.20}$$

Unfortunately, for reasons of space we cannot discuss further the many interesting properties of the function $P_J(q)$; the interested reader may consult the original literature [11].

4. The solution of the model

We now use the hypothesis of the previous section to compute the thermodynamics of the model. In this section we will follow ref. [6] very closely.

For simplicity let us consider the case where the distance may take only two values, 0 and d_1; consequently there are only two possible overlaps: q_0 and q_1 ($q_1 > q_0$) and only one value of y. We also set the external magnetic field to zero; a more general case will be considered later.

The strategy consists in computing the change in the free energy of the system ($\tilde{f}$) when we go from N to $N+1$ spins. However, we must not compute only the variation in the free energy of a particular state (as it was done in section 2), but we must compute how the whole distribution of states evolves [6].

We assume that each state of N spins is associated to a state of $N+1$ spins and therefore we can compute the variation in free energy of a given state when we go from N to $N+1$ spins.

We recall that for a given state the variation in the free energy contains three terms:

$$-(1/\beta)\ln[2\,\mathrm{ch}(\beta h_\alpha)], \qquad h_\alpha^{\mathrm{ef}}=\sum_k J_{N+1,k}m_k^\alpha, \tag{4.1a}$$

$$\tfrac{1}{2}\beta(1-q_{\mathrm{I.A}}), \qquad q_{\mathrm{I.A}}\equiv q_1, \tag{4.1b}$$

$$\tfrac{1}{2}U. \tag{4.1c}$$

The second term (which arises from the average of $\sigma\,\delta h$) and the third term (which is due to the variation in the value of β_{ef}) are independent of the state α and their effect can be easily taken into account. We concentrate our attention on the first term, which changes from state to state, and whose effects must be carefully computed.

We first notice that we can write

$$h_\alpha^{\mathrm{ef}}=\tilde{h}+h_\alpha, \tag{4.2}$$

where $\tilde{h}$ depends only on the J's which connect the $(N+1)$th spin with the rest of the system and the h's depend also on the particular state we consider.

Generally speaking it is immediate to see that the expectation value of $h_\alpha h_\gamma$ is equal to $q_{\alpha\gamma}$; arguments based on the central limit theorem imply that the probability distribution of the h's is a Gaussian with zero average and this Gaussian probability distribution can be reconstructed from the knowledge of the variance of the h's. In the simplified case we consider here, a short computation shows that the h's and $\tilde{h}$ are independent random variables, which have a Gaussian distribution, with zero average and variance

$$\overline{\tilde{h}^2}=q_0, \qquad \overline{(h^\alpha)^2}=q_1-q_0. \tag{4.3}$$

The probability of having a state of the system with N spins with free energy f^N is given by [see eq. (3.9)]

$$\exp[y(f^N-\tilde{f}^N)]\equiv P^N(f^N), \tag{4.4}$$

where $\tilde{f}^N$ is the reference free energy of the system with N spins.

Forgetting the second and the third terms, eq. (4.1a) implies that

$$f_\alpha^{N+1}=f_\alpha^N-(1/\beta)\ln\{2\,\mathrm{ch}[\beta(\tilde{h}+h_\alpha)]\}. \tag{4.5}$$

For only a moment we assume that $q_0 = 0$ and therefore $\tilde{h} = 0$. The probability distribution of the f_α^{N+1} is thus given by

$$P^{N+1}(f_\alpha^{N+1}) = \int dP_{q_1}(h_\alpha)\, df\, P^N(f_\alpha)$$
$$\times \delta(f_\alpha^{N+1} - f_\alpha^N + (1/\beta)\ln[2\,\mathrm{ch}(\beta h_\alpha)])$$
$$= \int dP_{q_1}(h_\alpha)\exp\{yf_\alpha^{N+1} + (y/\beta)\ln[2\,\mathrm{ch}(\beta h_\alpha)] - y\tilde{f}^N\}$$
$$= \exp[y(f_\alpha^{N+1} - \tilde{f}^{N+1}),$$
$$\tilde{f}^{N+1} \equiv \tilde{f}^N - (1/y)\ln\left(\int dP_{q_1}(h)\,[2\,\mathrm{ch}(\beta h)]^{y/\beta}\right). \tag{4.6}$$

(In the last integral the integration variable h is dummy and we do not need to use explicitly h_α). Equation (4.6) implies that f_α^{N+1} has the same exponential distribution as f_α^N, with an offset $\tilde{f}^{N+1}$ instead of $\tilde{f}^N$. The distribution of free energies is thus shifted by an amount $\Delta f \equiv \tilde{f}^{N+1} - \tilde{f}^N$; this quantity can be interpreted as the variation of the free energy of the system when we go from N to $N+1$ spins. It can be shown that this particular contribution to the variation of the free energy does not fluctuate when we change the system because the number of states is infinite (very large when N is large) and the offsets $\tilde{f}^{N+1}$ and $\tilde{f}^N$ are determined without ambiguity if we know one instance of the distribution of the $\tilde{f}^{N+1}$ and of the $\tilde{f}^N$.

Let us now see what happens when $q_0 \neq 0$. The whole argument can be repeated and one finds that for a given choice of the J's connecting the $(N+1)$th spin with the others (i.e., for a given value of $\tilde{h}$) we have

$$\Delta f(\tilde{h}) = -(1/y)\ln\left(\int dP_{q_1\ q_0}(h)\,\{2\,\mathrm{ch}[\beta(\tilde{h}+h)]\}^{y/\beta}\right). \tag{4.7}$$

We can now perform the average over the J's and we finally get

$$\overline{\Delta f} = -(1/y)\int dP_{q_0}(\tilde{h})\ln\left(\int dP_{q_1\ q_0}(h)\,\{2\,\mathrm{ch}[\beta(\tilde{h}+h)]\}^{y/\beta}\right). \tag{4.8}$$

Adding all the pieces together one finally finds

$$\beta F(q_0, q_1, y) = -(\beta^2/4)[1 + mq_1^2 + (1-m)q_0^2 - 2q_0^2]$$
$$-(1/m)\int dP_{q_0}(\tilde{h})\ln\left(\int dP_{q_1\ q_0}(h)\{2\,\mathrm{ch}[\beta(\tilde{h}+h)]\}^m\right),$$
$$m \equiv y/\beta < 1. \tag{4.9}$$

The equations for q_0 and q_1 can be obtained explicitly, as was done in section 2, or we can find them directly by imposing the conditions

$$\partial F/\partial q_0 = \partial F/\partial q_1 = 0. \tag{4.10}$$

Both approaches give the same results.

At this stage the parameter y is arbitrary; it is, however, convenient to determine it by imposing the equation

$$\partial F/\partial y = 0. \tag{4.11}$$

In this way we have obtained the same results as ref. [13], where the replica method was used and the replica symmetry was broken in a relatively simple way.

By solving explicitly eqs. (4.10), (4.11) [13], we find that $q_0 = q_1$ in the high-temperature phase and $q_0 \neq q_1$ in the low-temperature phase; in this region we still find

$$\overline{\beta^2(1-m^2)^2 - 1} < 0. \tag{4.12}$$

However, the absolute value of the l.h.s. has strongly decreased from the result obtained in the naive approach of section 2 (almost one order of magnitude). The entropy still becomes negative at low temperatures; however, its absolute value is much smaller [$S(0) \approx -0.01$ at $B = 0$].

The same computations can be done if q may take three values, i.e., $q_0 < q_1 < q_2$. In this case there is only one type of cluster (of diameter $q_2 - q_1$). The distributions of the clusters and of the states are controlled by the parameters y_1 and y_0, respectively. We will label the clusters with α_2 and the states by the pair $(\alpha_2\alpha_1)$. As in the previous case we will consider only the first term in eq. (4.1).

In this case a simple computation shows that

$$h^{\text{ef}}_{\alpha_1\alpha_2} = \tilde{h} + h_2(\alpha_2) + h_1(\alpha_2, \alpha_1), \tag{4.13}$$

where the variables h are independent Gaussian variables, $\tilde{h}$ is the same for all the states, $h_2(\alpha_2)$ is the same for all the states in the same cluster and $h_1(\alpha_2, \alpha_1)$ is the term in h^{ef} which changes when we change the state remaining in the same cluster. The variance of the h's turns out to be given by

$$\overline{(\tilde{h})^2} = q_0, \qquad \overline{(h_1)^2} = q_1 - q_0, \qquad \overline{(h_2)^2} = q_2 - q_1. \tag{4.14}$$

The distribution of the free energies of the clusters will be

$$\exp[y_1(f^N_\alpha - \tilde{f}^N)], \qquad \exp[y_1(f^{N+1}_\alpha - \tilde{f}^{N+1})], \tag{4.15}$$

for the system with N and $N+1$ spins, respectively.

Our aim is to compute the average value of $\Delta f \equiv \tilde{f}^{N+1} - \tilde{f}^{N}$. Using the same arguments as before it is easy to obtain that

$$f_{\alpha_2}^{N+1} - f_{\alpha_2}^{N} = -(1/y_0) \times \ln\left(\int \mathrm{d}P_{q_2-q_1}(h_1)\{2\,\mathrm{ch}[\beta(\tilde{h} + h_2(\alpha_2) + h_1)]\}^{y_0/\beta} \right), \tag{4.16}$$

where $f_{\alpha_2}^{N+1} - f_{\alpha_2}^{N}$ depends on α_2 through $h_2(\alpha_2)$. The shift in the normalization energy of a cluster depends only on $h_2(\alpha_2)$ and $\tilde{h}$; for given values of $h_2(\alpha_2)$ and $\tilde{h}$, this shift is not a fluctuating variable because the number of states inside a cluster is extremely large.

The same argument as before gives the final result

$$\Delta f = -(1/y_1) \int \mathrm{d}P_{q_0}(\tilde{h}) \ln\left[\int \mathrm{d}P_{q_1-q_0}(h_2) \times \left(\int \mathrm{d}P_{q_2-q_1}(h_1)\,\{2\,\mathrm{ch}[\beta(\tilde{h} + h_1 + h_2)]\}^{y_0/\beta} \right)^{y_1/y_0} \right]. \tag{4.17}$$

The reader should note that if $y_1 = y_0$ eq. (4.17) reduces to eq. (4.7).

We can now add all the terms together. The values of the parameters q and y can be found by solving the equations

$$\partial F/\partial q_0 = \partial F/\partial q_1 = \partial F/\partial q_2 = \partial F/\partial y_1 = \partial F/\partial y_2 = 0. \tag{4.18}$$

Equation (4.18) has been previously obtained using a more complex pattern of replica symmetry breaking [14]. If one solves numerically eqs. (4.18) one finds that a nontrivial solution exists where all the q's are different from each other. The zero-temperature entropy at zero magnetic field is still negative ($S \approx -0.003$), but it is rather small in absolute value.

It is clear that good results should be obtained in the limit in which the number k (of different values of the overlap) goes to infinity. After some work we obtain that in the limit $k \to \infty$ the free energy can be written as a functional of the function $x(q)$ [14]:

$$F = -\tfrac{1}{4}\beta\left(1 + q_{\mathrm{EA}} - 2\int_{q_m}^{q_{\mathrm{EA}}} \mathrm{d}q\, q x(q)\right) - \beta \int \mathrm{d}P_{q_m}(z)\, f(q_m, z), \tag{4.19}$$

where the function $f(q, h)$ satisfies the following parabolic integral equation in the interval $-\infty < h < \infty$ and $q_m < q < q_{\mathrm{EA}}$:

$$\partial f/\partial q = -\tfrac{1}{2}[\partial^2 f/\partial h^2 + x(q)(\partial f/\partial h)^2], \tag{4.20}$$

with the boundary condition

$$f(q_{EA}, h) = \ln[2\,\mathrm{ch}(\beta h)]. \tag{4.21}$$

Equations (4.19)-(4.21) coincide with those obtained using the full hierarchical replica symmetry breaking of refs. [10, 14].

The properties of the function $x(q)$ [or $q(x)$] which make F stationary [i.e., the solution of the equation $\delta F/\delta x(q)$] are well known and they will not be discussed here (the interested reader may consult refs. [1, 13-15]). We only recall that it can be proved analytically that in this case the zero-temperature entropy is exactly zero [16] (the entropy being proportional to T^2) and that in the whole low-temperature region [where $q(x)$ is not a constant function] we have [17]:

$$\beta^2\overline{(1-m^2)^2} = 1. \tag{4.22}$$

Summarizing this section we have seen that the hypothesis we have made in the previous section leads to a reasonable solution of the S-K model and the main features of this solution are in good agreement with the numerical data [18]. The form of the probability distribution we have used arises naturally in the replica approach; however, we do not have very convincing arguments which lead to such a distribution; in particular there is no simple compulsory argument which explains the most striking property of the probability distribution, i.e. its ultrametric structure.

The function $q(x)$ and the probability distribution of the effective fields h (which is not a Gaussian!) can be compared with the numerical data [18] and the agreement (see figs. 2, 3) is quite good; similar results have

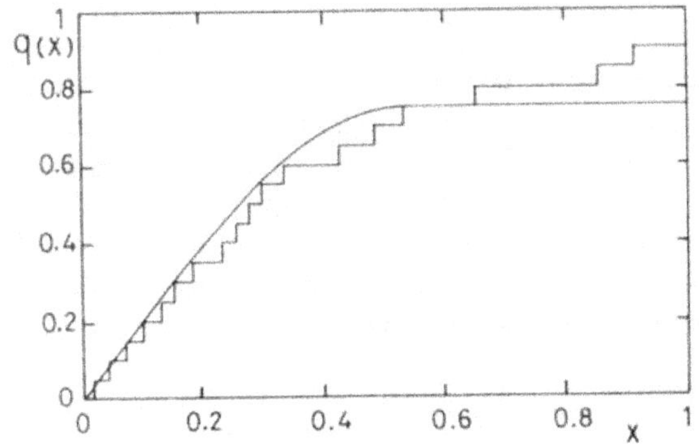

Fig. 2. The function $q(x)$, theory (full line) and numerical experiments (histogram) at $T=0.4$, $B=0$ and $N=64$ [18].

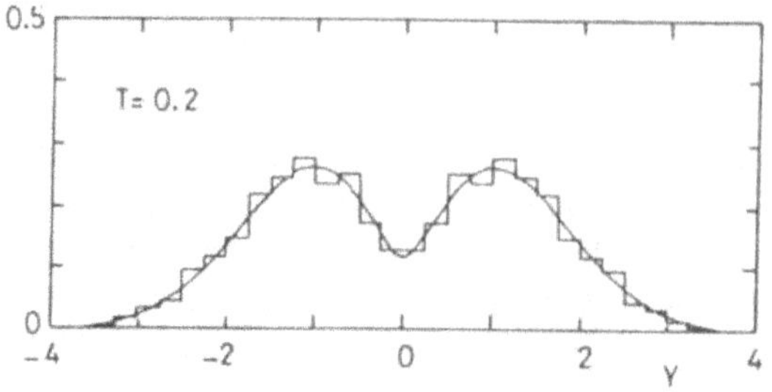

Fig. 3. Numerical results for the probability distribution for the effective field h, at $T = 0.2$, $B = 0$ and $N = 64$ [18].

also been obtained in ref. [19]. In general numerical simulations seem to indicate that the ultrametricity property is satisfied to a good approximation [20] and up to now no contradictions have been found.

A deeper theoretical understanding of the problem could be reached by considering probability distributions which are more general than those studied in these lectures; in this case we should prove that the condition of stationarity (or stability) of the free energy implies that these more general distributions reduce to the *simple* distributions used here.

5. An intermezzo on polymers

In the next two sections the ideas presented up to now will be applied to optimization problems like the matching problem or the traveling salesman.

We start by recalling some properties of the self-avoiding walk. Generally speaking, we consider N sites and we assign to each link (i, k) (from the point i to the point k) a nonnegative number $R_{i,k}$. A walk $\omega_{i,k}$ of length $L \equiv |\omega_{i,k}|$ from the point i to the point k is a path of L links going from i to k. A walk is self-avoiding if it does not intersect with itself; more precisely the number of links which start from a given point must be 1 for the endpoints and 2 for the other points of the walk.

The two-point correlation function is defined as

$$G(i, k) = \sum_{\{\omega_{i,k}\}} \mu^{|\omega_{i,k}|} \prod_{\{l,j\} \in \omega_{i,k}} R_{l,j}, \tag{5.1}$$

where the product is over all the links belonging to $\omega_{i,k}$ and the sum is over all the self-avoiding walks $\omega_{i,k}$.

In a similar way we can consider the four- (and more) point correlation function defined as

$$G(i_1, k_1, i_2, k_2) = \sum_{\{\omega^1_{i_1,k_1}\}} \sum_{\{\omega^2_{i_2,k_2}\}} \mu^{|\omega^1_{i_1,k_1}|+|\omega^2_{i_2,k_2}|} \times \prod_{\{l_1,j_1\}\in\omega^1_{i_1,k_1}} R_{l_1,j_1} \prod_{\{l_2,j_2\}\in\omega^2_{i_2,k_2}} R_{l_2,j_2}, \tag{5.2}$$

where ω^1 and ω^2 are two mutually self-avoiding walks.

Depending on the value of μ, we can distinguish two regions: $\mu < \mu_c$ where the sum is dominated by paths whose length remains finite when $N \to \infty$, and $\mu > \mu_c$ where the sum is dominated by paths whose length goes to infinity when $N \to \infty$. In this last region it is usually assumed that the correlation functions factorize as [21]

$$G(i, k) \approx m_i m_k \tilde{z}(N),$$

$$G(i_1, k_1, i_2, k_2) \approx m_{i_1} m_{k_1} m_{i_2} m_{k_2} \tilde{z}(N), \tag{5.3}$$

for points i and k which are widely separated.

The quantities m_i play the same role as the spontaneous magnetization in normal spin systems and $\tilde{z}$ is proportional to the partition function of the system, defined as

$$\tilde{z} = \sum_{\{\omega\}} \mu^{|\omega|} \prod_{\{l,j\}\in\omega} R_{l,j}, \tag{5.4}$$

where the sum is over all the self-avoiding closed paths. In the region $\mu > \mu_c$ one can define a nonzero free energy density as

$$F(\mu) = -\lim_{N\to\infty} \frac{1}{N} \ln[\tilde{z}(N)]. \tag{5.5}$$

The analogy with spin systems may be pushed much further [22]. We can consider an n-component spin system invariant under the transformations of the O(n) group (at temperature equal to $1/\mu$) with Hamiltonian [23]

$$H = -\sum_{i<k} R_{i,k} \sigma_i \cdot \sigma_k, \tag{5.6}$$

where the σ's are n-component spins normalized in such a way that

$$|\sigma_i|^2 = n. \tag{5.7}$$

We can now consider this model in the unconventional limit $n \to 0$. The reader should note that this "n" is not the "n" used in the replica approach for spin glasses: the limit $n \to 0$ must be considered as a small trick in order to avoid more cumbersome and direct considerations on the polymers. He should also note that the same results of the $n \to 0$ limit could be obtained by considering the Hamiltonian [21]

$$H = \sum_{i<k} R_{i,k}\{\bar{\sigma}_i\sigma_k + \bar{\psi}_i\psi_k\}, \tag{5.8}$$

where the σ's are usual complex numbers, while the ψ's are anticommuting variables; both variables satisfy the constraint

$$\bar{\sigma}_i\sigma_i + \bar{\psi}_i\psi_i = 0. \tag{5.9}$$

For $\mu < \mu_c$ it is easy to check that

$$\langle\sigma_i^a\sigma_k^b\rangle = \delta^{a,b}G(i, k). \tag{5.10}$$

(Similar formulae hold for higher-order correlation functions.)

For $\mu > \mu_c$ it is usual to assume that the O(n) symmetry is broken and only the σ^1's have a nonzero expectation value, which is given by

$$\langle\sigma_i^1\rangle = m_i, \tag{5.11}$$

where the m's are defined in eq. (5.3). Moreover, we have that

$$\lim_{n\to 0} [(1/n)(Z_{\text{spin}} - 1)] = \tilde{z}, \tag{5.12}$$

where $\tilde{z}$ is defined in eq. (5.4) and Z_{spin} is the partition function of the Hamiltonian (5.6).

This analogy may be pushed further if we introduce a "magnetic field" h as follows: The probability distribution of the spins is proportional to

$$\prod_i (1 + h\sigma_i^1) \exp(-\mu H[\sigma]). \tag{5.13}$$

The partition function $Z_{\text{spin}}(h)$ can be written as the partition function of a gas of open mutually self-avoiding paths, each path ω having a weight

$$h^2\mu^{|\omega|} \prod_{\{j,l\}\in\omega} R_{j,l}. \tag{5.14}$$

In the next section we will use this construction to study the matching problem.

6. The matching

In the matching problem we consider a set of N points ($i = 1, \ldots, N$) and of distances $d_{i,k}$, N being an even number. The configuration space is the set of all possible choices of the numbers $n_{i,k}$ ($n_{i,k} = n_{k,i}$, $n_{i,i} = 0$), which take only the values 0 or 1 and satisfy the constraint

$$\sum_k n_{i,k} = 1, \quad \forall i. \tag{6.1}$$

(N could be odd if we allow the r.h.s. of eq. (6.1) to be zero for only one value of i.)

It is clear that specifying the n's is equivalent to matching the points pairwise, a point being matched with one and only one other point. We associate a cost function $H[n]$ to each set of n's (or equivalently to each matching), $H[n]$ being defined in the following way:

$$H[n] = \sum_{i<k} d_{i,k} n_{i,k}. \tag{6.2}$$

A well-known and widely studied problem in optimization theory consists in finding the configuration which minimizes $H[n]$ for a given instance of the d's; there are algorithms which solve this task in a time proportional to N^3.

Our aim is different: For a given probability distribution of the d's we want to compute analytically the value that $H[n]$ takes at the minimum in the limit $N \to \infty$. A possible approach to this problem consists in using the method of statistical mechanics. To this end it is convenient to introduce a d-dependent partition function and free energy given by

$$Z = \sum_{\{n\}} \exp(-\gamma H[n]), \qquad F(\gamma) = -(1/\gamma)\ln(Z). \tag{6.3}$$

It is clear (at least for finite N) that

$$\min_{\{n\}} H[n] = \lim_{\gamma \to \infty} F[\gamma]. \tag{6.4}$$

Our aim is to compute the average (over the possible realizations of the d's) of the value of $F(\gamma)$ in the limit $N \to \infty$; in doing this we will follow quite closely ref. [24]. The results will obviously depend on the probability distribution of the d's. In the usual Euclidean matching, the points belong to a D-dimensional Euclidean space (for example, they are randomly chosen inside the unit D-dimensional cube) and their distance $d_{i,k}$ is given by

$$d_{i,k} = (l_{i,k})^{\nu}, \tag{6.5}$$

where $l_{i,k}$ is the usual Euclidean distance ($l_{i,k} = |x_i - x_k|$) and ν is a number which characterizes the model (quite often $\nu = 1$).

This model is very difficult to solve exactly (as any finite-dimensional problem); therefore we consider a simple model in which the d's are independent random variables with a probability distribution

$$P(d) \approx d^r \quad (d \to 0). \tag{6.6}$$

It is possible to prove that in this simplified model the only thing that matters is the behaviour of $P(d)$ for small d. Different models are obtained by changing the value of r.

If we go back to the Euclidean matching problem we find that the distance [defined in eq. (6.5)] has the same probability distribution as in eq. (6.6), with

$$r = D/\nu - 1. \tag{6.7}$$

However, the correlation of three distances is not zero in the Euclidean case. (Remember the triangular inequality!) The independent-distance model may be considered as an approximation to the true Euclidean matching, where we have neglected the correlations between the distances. This statement could be qualified by noticing that in the limit $D \to \infty$ at fixed r, the Euclidean matching collapses into the independent-distance matching.

For simplicity we will study here the case $r = 0$. It is convenient to introduce a variable $\beta = \gamma/N$ and to study the free energy as a function of β [25].

If we compare eq. (6.3) with the results of the previous section, it turns out that

$$Z = \lim_{h \to \infty} Z_{\text{spin}}(h)/(h^N \mu^{N/2}), \tag{6.8}$$

where we have

$$R_{i,k} = \exp(-N\beta d_{i,k}), \qquad R_{i,i} = 0. \tag{6.9}$$

In the limit $h \to \infty$ the dependence on μ is trivial and we can freely set $\mu = 1$.

Our task now is to solve the model given by eqs. (6.8), (6.9). To this end we will use the cavity approximation. We will do the same steps as in the Ising model, *mutatis mutandis.*

Let us consider a system with N spins and let us add an extra spin (as long as h is not exactly equal to infinity, the model makes sense also for odd N). As usual we assume that the N spins are not correlated and therefore their probability distribution can be factorized as

$$P[\sigma] = \prod_i p_i(\sigma_i), \quad p_i(\sigma_i) \propto 1 + h_i \sigma_i^1. \tag{6.10}$$

A detailed analysis shows that the inclusion of higher powers of σ_i^1 is irrelevant and will not change the computation; while it is nearly obvious that the $(\sigma_i^1)^k$ do not matter for $k>2$ [eq. (5.9) implies that the σ's are practically anticommuting variables], the case $k=2$ must be studied more carefully.

According to the usual rules of mean field theory, the h's must be chosen in such a way that

$$\langle \sigma_i^1 \rangle = m_i, \tag{6.11}$$

m_i being the magnetization of the ith spin [26]. In the limit of large h we have that

$$h_i = 1/m_i. \tag{6.12}$$

We have now in our hands all the tools needed to compute m_{N+1}; after some computations we find for large h

$$m_{N+1} = h \Bigg/ \left(\sum_k R_{N+1,k} m_k \right). \tag{6.13}$$

If we absorb a factor $h^{1/2}$ in the m's, we finally obtain

$$1/m_{N+1} = \sum_k R_{N+1,k} m_k. \tag{6.14}$$

Let us now derive the naive theory (as in section 2) neglecting the possibility of having more than one equilibrium state. To this end we must solve the following problem in probability theory: The variables m are independent random variables with probability distribution $P(m)$; the R's are also independent random variables such that

$$\overline{(R_{N+1,k})^p} = g(p)/N \quad (p>0), \quad g(p) = 1/(\beta p). \tag{6.15}$$

The problem consists in finding the probability distribution of the m_{N+1} [i.e. $\tilde{P}(m)$] induced by eq. (6.14) and to impose the self-consistency condition

$$\tilde{P}(m) = P(m). \tag{6.16}$$

We sketch here the solution of this nontrivial problem. We introduce the quantities:

$$M(p) = \int \mathrm{d}m\, P(m) m^p, \qquad \tilde{M}(p) = \int \mathrm{d}m\, \tilde{P}(m) m^p. \tag{6.17}$$

Obviously we must have that $M(p)=\tilde{M}(p)$. If we take eq. (6.14) to an arbitrary integer power (p), we get $\tilde{M}(p)$ for negative integer p as a function of the M's evaluated at positive integers, e.g.,

$$\begin{aligned} &\tilde{M}(-1)=g(1)M(1),\\ &\tilde{M}(-2)=[g(1)M(1)]^2+g(2)M(2),\\ &\tilde{M}(-3)=[g(1)M(1)]^3+3g(2)M(2)[g(1)M(1)]^2+g(3)M(3),\\ &\ldots \end{aligned} \tag{6.18}$$

It is convenient to introduce the generating function

$$G(x)=\sum_{p=1}^{\infty}(-1)^{p-1}g(p)M(p)\exp(px)/p!. \tag{6.19}$$

A lengthy computation shows that eqs. (6.18) are satisfied if $G(x)$ satisfies the integral equation [27]

$$G(x)=(1/\beta)\int \mathrm{d}y\, B(x+y)\exp[-G(y)], \tag{6.20}$$

where the integration runs from $-\infty$ to $+\infty$ and the function B is defined as

$$B(y)=\sum_{p=1}^{\infty}(-1)^{p-1}\exp(yp)/(p!)^2. \tag{6.21}$$

One can also prove that the internal energy (i.e. the expectation value of the Hamiltonian) is given by

$$(1/\beta)\int \mathrm{d}x\, G(x)\exp[-G(x)]. \tag{6.22}$$

In the low-temperature limit $(\beta\to\infty)$ one finds that

$$G(x)\approx W(x\beta), \tag{6.23}$$

where W is exactly given by

$$W(x)=\ln[1+\exp(x)]. \tag{6.24}$$

The internal energy at zero temperature turns out to be

$$E=\pi^2/12\approx 0.822, \tag{6.25}$$

the normalization being such that E should be equal to $\frac{1}{2}$ if each point would be matched to the nearest one.

This value for E is in agreement with numerical simulations ($E = 0.825 \pm 0.01$) [27], although a 1% discrepancy between the value in eq. (6.25) and the correct value is quite possible. Indeed, in our derivation we have assumed that there is only one equilibrium state for a system of N spins; if this hypothesis fails, we should study the problem by using the same techniques as in section 4.

The generalized matching problem ($n_{i,k} = 0, 1, \sum_k n_{i,k} = s$) [24] and the traveling salesman problem [28] can be studied in a similar way.

7. Conclusions

We have seen in sections 2–4 how to solve the S–K model without using replicas and obtain the same results as in the replica approach. The approach described in these notes has the advantage of being more explicit than the replica approach, which is much more compact. The formalism we use is rather complex in the low-temperature region where the replica symmetry is broken and many equilibrium states are present; we have assumed that their distance satisfies the ultrametric inequality.

In sections 5, 6 we have extended this approach to some optimization problems, in particular to the matching. Here, also at the naive level, the problem is much more difficult than for the spin glasses because all the moments of the magnetization are coupled with one another and an infinite number of order parameters are needed even if we consider only one equilibrium state.

The generalization of this approach to the case where many equilibrium states may exist is very interesting but unfortunately it quite difficult: we must introduce infinitely many functions of infinitely many variables. It would be very interesting to obtain numerically some information on the structure of the equilibrium state and on the presence (or absence) of an ultrametric structure. Although the analytic treatment of the matching is quite difficult, it has the advantage (over spin glasses) that its numerical study can be done much better.

Acknowledgment

It is a pleasure for me to thank Romeo Brunetti for a very careful reading of the manuscript and for interesting suggestions.

References

[1] G. Parisi, in: Recent Advances in Field Theory and Statistical Mechanics, Les Houches session XXXIX (1982), eds J.B. Zuber and R. Stora (North-Holland, Amsterdam, 1984).

[2] I. Van Hemmen and I. Morgenstern, eds., Heidelberg Colloquium on Spin Glasses (Springer, Berlin, 1984).

[3] M. Mezard, G. Parisi and M. Virasoro, Spin Glass Theory and Beyond (World Scientific, Singapore, 1987).

[4] S.F. Edwards and P.W. Anderson, J. Phys. F 5 (1975) 965.

[5] D. Sherrington and S. Kirkpatrick, Phys. Rev. Lett. 35 (1975) 1792; Phys. Rev. B 7 (1978) 983.

[6] M. Mezard, G. Parisi and M. Virasoro, Europhys. Lett. 1 (1986) 77.

[7] D.J. Thouless, P.W. Anderson and R.G. Palmer, Philos. Mag. 35 (1977) 593.

[8] M. Mezard, G. Parisi and M. Virasoro, unpublished;
for an older derivation of this equation, see: A.J. Bray and M.A. Moore, J. Phys. C 12 (1979) L441.

[9] J.R.J. De Almeida and D.J. Thouless, J. Phys. A 11 (1978) 983.

[10] G. Parisi, Phys. Rev. Lett. 50 (1983) 1946.

[11] M. Mezard, G. Parisi, N. Sourlas, G. Toulouse and M. Virasoro, J. Physique 45 (1984) 843.

[12] M. Mezard, G. Parisi and M. Virasoro, J. Physique Lett. 46 (1985) L217;
B. Derrida and G. Toulouse, J. Physique Lett. 46 (1985) L221.

[13] G. Parisi, J. Phys. A 13 (1980) 1101, 1887.

[14] G. Parisi, J. Phys. A 13 (1980) L115.

[15] J. Vannimenus, G. Toulouse and G. Parisi, J. Physique 42 (1981) 565.

[16] H.J. Sommers and W. Dupont, J. Phys. C 17 (1984) 5785.

[17] A.V. Goltev, J. Phys. C 17 (1984) L241.

[18] K. Namoto, Sapporo preprint (1986).

[19] A.P. Young, Phys. Rev. Lett. 51 (1983) 1206.

[20] R.N. Bhatt and A.P. Young, Phys. Rev. Lett. 54 (1984) 924.

[21] G. Parisi and N. Sourlas, J. Physique Lett. 41 (1980) L403.

[22] C. Aragao de Carvalho, S. Caracciolo and J. Fröhlich, Nucl. Phys. B 215 [FS7] (1983) 209.

[23] P.G. de Gennes, Phys. Lett. A 38 (1972) 339.

[24] M. Mezard and G. Parisi, Europhys. Lett., to be published.

[25] J. Vannimenus and M. Mezard, J. Physique 45 (1984) L1145.

[26] P.W. Anderson, G. Baskaran and Y. Fu, J. Stat. Phys., to be published.

[27] M. Mezard and G. Parisi, J. Physique Lett. 46 (1985) 1771.

[28] M. Mezard and G. Parisi, J. Physique 47 (1986) 1285.

SPIN GLASS THEORY

Giorgio Parisi

Dipartimento di Fisica, II Universita' di Roma
"Tor Vergata" and INFN, sezione di Roma, Italy

A spin glass is a disordered magnet: in a typical example 50% of the bonds (randomly chosen) between two spins are ferromagnetic while the others are antiferromagnetic. From the experimental point of view[1] spin glasses are characterized by the presence of a low temperature phase in which the relaxation time toward equilibrium is very large (as in real glasses). The magnetic susceptibility in the presence of a time dependent magnetic field depends on the frequency also in the region of a few Hertz.

From the theoretical point of view these properties may be connected to the practical impossibility of finding numerically the ground state of a spin glass (if the number of spins is not very small). Indeed, according to the conventional wisdom, the number of operations any algorithm needs to find the ground state increases exponentially with the volume (the problem is NP complete in more than 2 dimensions). There is an exponentially large number of spin configurations which are local ground states, in the sense that their energy increases if only one spin is flipped. Although it is very easy to find efficient algorithms for finding local minima, it is very difficult for the algorithm to find the global minimum (i.e. the true ground state).

Numerical simulations[2] suggest that different minima (of the energy at zero temperature or of the free energy at finite temperature) are macroscopically different from each other. This situation is rather unprecedented in statistical mechanics: macroscopically different configurations with the same total free energy are normally present at the point where a first order transition happens (which is an isolated point). For spin glasses this phenomenon is present for a large range of temperatures and magnetic fields, where there is no first order phase transition from the thermodynamic point of view. A strong effort has been needed to find the appropriate theoretical framework to study these systems.

The origin of this peculiar behaviour is the presence of frustration: it is not possible for all pairs of spins connected by a ferromagnetic (antiferromagnetic) bond have the same (the opposite) sign. Different terms of the Hamiltonian push in different directions, and the number of possible compromises is very high. For example in the Ising case the Hamiltonian is

$$H = -1/2 \sum_{i,k} J_{i,k} \sigma_i \sigma_k - h \sum_i \sigma_i \qquad (1)$$

where h is the magnetic field and the J's are the coupling between the spins. It is clear that it is not always possible to find configurations of the σ's such that all terms are positive.

There are many systems in which different terms of the Hamiltonian (or different constraints) are in competition with each other. This happens very frequently when the system is complex. Typical examples are real glasses, other NP complete problems like the traveling salesman or the matching problem, protein folding, biological organization, prebiotic evolution, neural networks and so on. At the present moment many of the concepts that have been developed in the study of spin glasses start to be useful also for these systems and it is quite possible that in the long run the applications of these ideas beyond solid state physics will be the most interesting ones.

In recent years much progress has been made and after much effort a self-consistent mean field approximation has been obtained[3] (sometimes this construction goes under the name of broken replica theory). This theory should be exact for weak long range forces or when the dimensions of the space become very large (fluctuations are neglected). The ideas involved in this construction are rather different from those of the mean field approximation for other models and take care of the peculiar properties of spin glasses. In this note we will not show how these ideas have been developed. We will also skip most of the technical details and concentrate our attention on a few important physical results.

We suppose a system whose Hamiltonian $H_J[\sigma]$ depends on the configuration $[\sigma]$ of the system and on some control variables J, which are distributed according to a probability distribution P[J]. For each choice of the J's we can compute the partition function

$$Z_J = \sum_{\{\sigma\}} \exp\{-\beta H_J[\sigma]\} \qquad (2)$$

and the free energy density is

$$F_J = -1/(\beta N) \ln Z_J = -1/(\beta N) \ln \{\sum_{\{\sigma\}} \exp\{-\beta H_J[\sigma]\}\}, \qquad (3)$$

where N is the total number of variables σ's. Standard statistical mechanics deals with the problem of computing F at given J. Here we suppose that the J's are not known, but that they are random variables and that their probability distribution is known. The J's are called quenched variables. Physically the necessity of computing the free energy F at fixed J's results from the fact that the changes of the J's happen on a time scale which is infinitely larger than the time scale characterizing the changes in the σ's.

We are interested in computing the average value of the free energy density:

$$F= \Sigma_J P[J]\ F_J = \overline{F_J}\ . \tag{4}$$

The replica method was originally proposed as a trick to simplify the computation of eq. (4). Later on it was found that the replica method (when the replica symmetry is broken) is very powerful in coding in a simple and compact way quite complex properties of the system.

In a nutshell the basic idea of the replica method is very simple. We start with some preliminary definitions:

$$Z_n= \Sigma_J P[J]\ (Z_J)^n = \overline{(Z_J)^n}, \tag{5}$$

$$F_n= -1/(\beta n N)\ \ln Z_n .$$

Using relations $\Sigma_J P[J] = 1$ and $A^n \approx 1+n\ \ln(A)$ for $n \approx 0$, it is evident that

$$\lim_{n\to 0} F_n \equiv F_0 = \overline{F}. \tag{6}$$

For integer n we can write:

$$(Z_J)^n = \prod_{a=1}^{n} \Sigma_{\{\sigma^a\}} \exp\{-\Sigma_{a=1,n} H[\sigma^a]\}, \tag{7}$$

where we have introduced n replicas of the same system. Indeed the partition function of n replicas of the same system is the partition function of the original system to the power n. The variables σ^a_i carry two indices: the upper one denotes the replica and goes from 1 to n and the lower one denotes the site and goes from 1 to N. At the end we must perform the two limits $n \to 0$ and $N \to \infty$ (the two limits are supposed to commute).

The replica trick consists of the following steps: we use equation (7) for integer n to define the function F_n for integer n; we extend this function to an analytic function of n and finally we compute $F_0 = \overline{F}$. If the partition function is expanded in power of β, F_n is a polynomial in n and the method is quite safe. Problems will arise in the low temperature region.

We can apply this method to the case of infinite range spin glasses (the Sherrington-Kirkpatrick model[4]) where the J's are independent Gaussian random variables with zero mean and variance 1/N:

$$\overline{J_{i,k}} = 0, \quad \overline{J_{i,k}^2} = 1/N, \quad J_{i,k} = J_{k,i}. \tag{8}$$

The thermodynamic limit is obtained when N goes to infinity. The factor 1/N in eq. (8) has been chosen in such a way that at fixed β the total energy is proportional to N and therefore the energy density is N-independent.

For fixed J's we expect that in the high temperature phase the local magnetizations $m_i \equiv \langle\sigma_i\rangle$ are different from zero at non-zero magnetic field and they become zero when the magnetic field goes to zero. On the other hand, we naively expect that in the low temperature region there should be some freezing of the spins in the position which is mostly favoured energetically, hence the m_i should be different from zero also at h=0. It is evident that m_i depends on the J's and it will be sometimes positive and sometimes negative (a detailed computation shows that $\overline{m_i} = 0$ at h=0) so it is convenient to characterize the system in terms of the quantity

$$q_{EA} \equiv 1/N \sum_{i=1}^{N} m_i^2, \tag{9}$$

i.e., the Edward Anderson order parameter[5]. It is possible to show that using a good definition of the m's q_{EA} does not depend on the J's in the limit $N \to \infty$. We can thus conclude that q_{EA} is also equal to $\overline{m_i^2}$. Summarizing, q_{EA} should be equal to zero at h=0 for a temperature greater than the critical temperature (T_c), while it should be different from zero at low temperature.

If we apply eq.(5) to the infinite range case, after some simple computations we find that

$$Z_n = \int d[Q] \exp\{-N\, A[Q]\} \tag{10}$$

$$A[Q] = -n\beta^2/4 + 1/4 \sum_{a=1,n} \sum_{b=1,n} Q^2_{a,b} - \ln Z[Q]$$

$$Z[Q] = \sum_{\{S\}} \exp\{-\beta H[Q,S]\}$$

$$H[Q,S] = \sum_{a=1,n} \sum_{b=1,n} Q^2_{a,b} S_a S_b - h \sum_{a=1,n} S_a$$

$$F_n = -1/\beta \min A[Q] \quad \Rightarrow \quad \partial A/\partial Q_{a,b} = 0,$$

where the matrix Q is an n x n symmetric matrix, with zeroes on the diagonal, and the integral is done on all non-diagonal elements of the matrix Q from $-\infty$ to ∞. The sum over {S} goes over the 2^n configurations of the variables S_a, a=1,n ($S_a = \pm 1$).

The function A[Q] is left invariant when we exchange some of the lines (or the rows) of the matrix Q and therefore the group of permutations of n elements (P_n or S_n) is a symmetry of the problem (all replicas are equivalent!). This group is often called the replica group.

For positive n the minimum of A can be found and the matrix Q has the following form

$$Q_{a,b}=q \ \forall \ a,b \ a\neq b \quad (Q_{a,a}=0 \ \forall \ a). \qquad (11)$$

This form of the matrix Q is the only one which is left invariant by the action of the replica group and it is therefore the natural solution, usually named the replica symmetric solution.

If we analytically continue the solution of the equation $dA/dq=0$, up to $n=0$, we find the following equation for q

$$q= \int dz/(2\pi)^{1/2} \exp(-z^2/2) \ \mathrm{th}^2(\beta q^{1/2}z + \beta h), \qquad (12)$$

where the integral over z goes from $-\infty$ to $+\infty$.

At zero magnetic field eq.(11) has only the solution $q=0$ for $1/\beta \equiv T > T_c=1$; for $T < T_c$ there is another solution (the physical one) which is different from zero. In conclusion there is a phase transition at $T=1$, $h=0$ while there is no transition at $h\neq 0$.

Everything seems perfect. Unfortunately, detailed computations show that the entropy (which in a discrete system is non-negative by definition) becomes negative at low temperature and at zero temperature we get $S(0)=-1/(2\pi)\approx -.17$. In this case, the replica method leads to a disaster!

It is evident that the saddle point method can be applied correctly only if the eigenvalues of the Hessian matrix $(\partial^2 A/\partial Q\partial Q)$ are non-negative, as it can be checked by computing the correction to the saddle point result(6). At low temperatures we must look for a new solution of the equation

$$\partial A/\partial Q=0, \qquad (13)$$

whose Hessian has no negative eigenvalues.

If the matrix Q does not have the form shown in eq.(11), it is not left invariant by the action of the replica group and we have to specify in detail the value of the different elements of the matrix.

We will always write formulae which are correct for integer n and make the analytic continuation at $n=0$ only at the end. In other words we will consider some matrices $Q^{[n]}{}_{a,b}$ which depend on n in a simple way. We will compute all the quantities we need for integer n and only at the end we will perform the continuation of the results at $n=0$.

The 0 x 0 matrix $Q^{(0)}{}_{a,b}$ is defined in terms of the matrices $Q^{(n)}{}_{a,b}$ for all values of n and therefore the space of all 0 x 0 matrices is an infinite dimensional space!

At the present moment only one reasonable form of the matrix is known. There is a widespread agreement that this choice is the correct one because the results are very satisfactory. It is quite possible that for systems which are different from spin glasses (e.g. for real glasses), different choices of the matrix $q_{a,b}$ are the correct ones.

A natural suggestion for the matrix Q consists in dividing the n replicas into n/m groups of m replicas. (Of course n must be a multiple of m, i.e. n/m must be an integer.) We set $Q_{a,b}=q_0$, if a and b belong to the same group, and $Q_{a,b}=q_1$, if a and b belong to different groups (we do not consider the $Q_{a,a}$'s which are identically equal to 1). In other words

$$Q_{a,b} = q_0 \quad \text{if} \quad I(a/m) = I(b/m) \tag{14}$$

$$Q_{a,b} = q_1 \quad \text{if} \quad I(a/m) \neq I(b/m),$$

where I(x) is an integer valued function whose value is the smallest integer which is greater than or equal to x.

Each line of the matrix has m-1 off-diagonal elements which are equal to q_0 and n-m which are equal to q_1 (the total number of off diagonal elements is n-1, i.e. -1 in the limit $n \to 0$). According to eq. (14) we have in the limit $n \to 0$:

$$P(q) = m\,\delta(q-q_0) + (1-m)\,\delta(q-q_1) \tag{15}$$

where the function P(q) is defined as

$$P(q) = 1/[n(n-1)/2]\Sigma_{a,b}\,\delta(q_{a,b}-q). \tag{16}$$

We will see later that the function P(q) must be non-negative and this is possible only if:

$$0 \leq m \leq 1 \tag{17}$$

It is obvious that (if we exclude the two uninteresting cases m=0 and m=1) m cannot be an integer and satisfy the inequality (17). On the other hand the limit $n \to 0$ is obtained by doing an analytic continuation in m and nothing seems to forbid that in such a process m may take non-integer values.

After some computations we find the explicit form of the function $A(q_0,q_1,m)$. It is easy to check that when m is set to 0 or to 1 we recover (as we should) the free energy of the replica symmetric approach with $q= q_0$ or q_1 respectively.

A careful analysis shows that the free energy $A(q_0,q_1,m)$ must be maximized with respect to all the variables, i.e. q_0, q_1 and m. If a maximum is found in the region 0<m<1, the results of this approach must be better than those obtained assuming unbroken replica symmetry. The true free energy takes values greater than those obtained from the unbroken replica appproach.

The function $A(q_0, q_1, m)$ can be easily computed numerically with high precision. Maximizing A with respect to the q_0, q_1, and m one finds that, in the whole region where the replica symmetry should be broken (i.e. at sufficiently low temperatures), m is neither zero or one ($m \to 0$ both for $T \to 0$ and $T \to 1$ at zero magnetic field).

The properties of the solution are very satisfactory:

a) the zero temperature internal energy at zero magnetic field is -.7652 in good agreement with the numerical data (-.765±.01).
b) the zero temperature entropy at zero magnetic field has collapsed from $S(0) \approx -.16$ to $S(0) \approx -.01$.

We are clearly on the right track although we have not found the final answer.

The final form of the matrix Q is thus the following: we introduce a set of integer numbers m_i ($i=0,\ldots,k+1$) such that $m_0=0$ and $m_{k+1}=n$ and m_i/m_{i+1} is an integer (for $i=1,\ldots,k+1$). We can divide the n replicas in n/m_k groups of m_k replicas; each group of m_k replicas is divided in m_k/m_{k-1} groups of m_{k-1} replicas and so on....

The off-diagonal elements of the matrix Q are thus given by:

$$Q_{a,b} = q_i \text{ if } I(a/m_i) \neq I(b/m_i) \text{ and } \qquad (18)$$

$$I(a/m_{i1}) = I(b/m_{i+1}),\ i = 0,\ldots,k,$$

where the q_i's are a set of k+1 real parameters. For k=1 we recover the previous example and for k=0 we recover the unbroken replica symmetry theory.

An easy computation shows that

$$P(q) = \Sigma_{i=0,k}\ (m_i - m_{i+1})\ \delta(q - q_i). \qquad (19)$$

Eq. (19) for P(q) is positive only if the m's satisfy the conditions

$$0 \le m_{i+1} \le m_i \le 1. \qquad (20)$$

Detailed arguments suggest that we should look for the maximum of the free energy as a function of the q_i's and the m_i's (from now on we will assume that conditions (20) are satisfied).

In order to keep track of the parameters q and m it is convenient to introduce the function q(x) defined as

$$q(x) = q_i \text{ if } m_{i+1} < x < m_i. \qquad (21)$$

There is a one-to-one correspondence between the piecewise constant function with k discontinuities and the parameters q and m (if eq. (21) holds).

If the function q(x) is a monotonic function (as we shall see it must be for physical reasons) the relation between the functions P(q) and q(x) is the following:

$$dx/dq= P(q) \tag{22}$$

where $q(x(q))=q$.

In the limit $k\to\infty$ the function q(x) becomes arbitrary (any reasonable function can be approximated by a piecewise constant function) and

$$\lim_{n\to 0} \{n^{-1} \Sigma_{a,b=1,n} Q^k{}_{a,b}\} = -\int_0^1 dx\, q^k(x) \tag{23}$$

In this formulation the free energy becomes a functional of q(x), (A[q]) which must be maximized with respect to q(x) and in this way we find the correct solution of the infinite range model.

The replica formalism is far from physical intuition and therefore it is convenient to recast the previous result using a more familiar language. In principle all the previous results can be obtained without using replicas. To this end we analyze the concept of a pure equilibrium state. If we consider a simple three dimensional ferromagnetic Ising model (with zero magnetic field) at a temperature greater than the critical one, there is only one pure equilibrium state and the spins are disordered. Below the critical temperature there are two translational invariant pure equilibrium states, one with positive magnetization, the other with negative magnetization.

Generally speaking a pure equilibrium state is any state which is at local equilibrium and whose connected correlation functions go to zero at large distance; the expectation value in the state α will be denoted by $<>_\alpha$. It is clear that the linear combination of two pure equilibrium states will be a mixed equilibrium state. According to the conventional wisdom, a system, which is not at equilibrium at initial time, evolves toward a pure equilibrium state and the time needed to go from one equilibrium state to another equilibrium state is exponentially large.

From the technical point of view a pure equilibrium state is characterized by the clustering property: all the connected correlations functions go to zero at large distance. Equivalently intensive quantities do not fluctuate. In a nutshell a pure equilibrium state corresponds to our intuitive idea of a normal equilibrium state. All equilibrium states have the same free energy density and states with higher free energy density are metastable.

We have already seen that spin glasses may arrange the directions of their spins in many different ways and all these arrangements give similar values of the free energy; in other words there exist many different ground states of spin glasses which are nearly degenerate in free energy. It is natural to assume that an Ising spin glass has many more equilibrium states than a ferromagnetic Ising model[3]. The configurations which have lower free energy correspond to pure equilibrium states while those configurations which have a higher free energy density in the infinite volume limit correspond to metastable states.

We face the new problem of characterizing the ensemble of equilibrium states. The *a priori* probability w_α for the state α to appear in the ensemble is:

$$w_\alpha \propto \exp(-\beta F_\alpha), \tag{24}$$

where F_α is the free energy of the state α.

The second question is to study how all these states differ from one another, more precisely to define a distance between these states. A simple possibility for the distance between the state α and the state β is the following:

$$d^2_{\alpha\beta} = 1/N \; \Sigma_i \; (m_i^\alpha - m_i^\beta)^2, \tag{25}$$

where N is the total number of spins, the sum over i is from 1 to N and m_i^α is the average magnetization of the spin i in the state α, i.e. $m_i^\alpha = \langle\sigma_i\rangle_\alpha$. In the same way we can define the overlap between two states α and β as;

$$q_{\alpha\beta} = 1/N \; \Sigma_i \; m_i^\alpha \, m_i^\beta. \tag{26}$$

If all the states have the same overlap with themselves (as should happen in spin glasses), i.e. $q_{\alpha\alpha} = q_{\beta\beta} \equiv q_{EA}$, the distance is very simply related to the overlap:

$$d^2_{\alpha\beta} = 2 \; (q_{EA} - q_{\alpha\beta}). \tag{27}$$

It is possible to define the probability distribution of overlap P(p)[8]:

$$P(q) = \overline{P_J \; (q)} = \Sigma_{\alpha\beta} \; w_\alpha \; w_\beta \, \delta(q_{\alpha\beta} - q), \tag{28}$$

where the bar denotes the average over different samples with different values of the couplings J.

In other words P(q) is the probability of finding two states with overlap p, weighting each state with its probability of appearing in the ensemble. Only states which have a non-zero probability ($w_\alpha \neq 0$) contribute to eq. (28).

The physical interpretation of replica symmetry breaking is based on the highly non- trivial result[8] that the two functions P(q), defined in eqs. (15) and (28), coincide.

In the ferromagnetic Ising model at non-zero magnetic field P(q) is a delta function because there is only one equilibrium state. Exactly at zero magnetic field the function P(q) contains two delta funtions, one at $q=m^2$, the other at $q=-m^2$, where m is the spontaneous magnetization.

In spin glasses in the mean field approximation the funtion P(q) at high temperatures is a delta function whereas at low temperatures at non-zero magnetic field it has two delta functions with a smooth region in between. In this case, where the function P(q) is not a simple delta function, we say that the replica symmetry is broken. The transition is very smooth from the thermodynamic point of view and only the second derivative of the specific heat is discontinuous.

In the usual approach all possible probability distributions with more than one overlap can be computed from the function P(q). The free energy can be computed as a functional of P(q) and the function P(q) can be found by looking for the maximum of free energy. The first interesting result is that $P_J(q)$ fluctuates from sample to sample[9].

$$\overline{P_J(q_1)\, P_J(q_2)} = 1/2\ P(q_1)\, P(q_2) + 1/2\ \delta(q_1-q_2)\, P(q_1) \quad (29)$$

This result indicates that only a few states give an important contribution to the function $P_J(q)$ and their relative weight and overlap fluctuate from sample to sample.

The most interesting and unexpected result concerns the probability distribution of the overlap of three or more states. One finds that the distance satisfies the ultrametric inequality[9]:

$$d_{\alpha\beta} < \max\ (d_{\alpha\gamma}, d_{\beta\gamma})\ \forall \gamma. \quad (30)$$

As is well known to mathematicians, this inequality implies that the space of states, can be divided into clusters of states; each cluster of states may be further divided into subclusters and so on.....; a cluster can be characterized as the set of states whose distance from a given (but arbitrary) state of the cluster is smaller than a given distance.

A more intuitive definition of an ultrametric distance is that two spheres are identical if they have a point in common. As a consequence, if we consider a random walk in the space of the states of the system and the length of each step is less than or equal to δ, after M steps the distance between the initial point and the final point will also be less than or equal to δ. The whole cluster can be reached in one step and there is nothing else to explore at distances not greater than δ.

This result is extremely important because it may be the basis for the existence of many different scales of relaxation times. Indeed let us consider a system which evolves in time and relaxes toward equilibrium with many different scales $\tau_1 << \tau_2 << \tau_3 << .. << \tau_n$. If the system is in a given configuration at time 0, we call R_k the region of configuration space that the system may explore in time τ_k.

All the points in configuration space which belong to R_k but not to R_{k-1} are said to be at distance k.

It is evident that this definition of distance satisfies the ultrametric inequality eq. (30). Let us suppose the contrary, i.e. there are two configurations (α and β) at distance k and there is another configuration (γ) at distance k-1 from both α and β. Now the times needed to go from α to γ and from γ to β are both equal to τ_{k-1} and the estimated time to go from α to β via γ is about $2\tau_{k-1}$ in contradiction with the inequality $\tau_{k-1} \ll \tau_k$.

This result is not surprising: many different times of relaxation are naturally present if the free energy landscape in configuration space contains many valleys which are separated by high mountains. The time to go from one valley to an other valley is the exponential of the height of the lower saddle between the two valleys.

The division of configuration space in valleys is clearly ultrametric: more precisely if the distance between two configuration is defined as the height of the highest saddle one must cross doing the most convenient trip between the two configurations, this definition of distance satisfies the ultrametric inequality (30).The real surprise is that the space of equilibrium states of a spin glass is ultrametric using the natural definition of the distance.

The dynamics of spin glasses at finite temperature should be the following: at relatively short times the system evolves exploring the configuration space corresponding to one pure equilibrium space, while at much larger times (e.g. for times proportional to $\exp((Nd)^{1/4})$) the system may arrive at other equilibrium states at distance d from the original one. A careful study of the dynamics is very important[10]. Very interesting result have been obtained, in particular it has been stressed that the behaviour of the system may be rather different if the system starts really from equilibrium or from a state which is slightly off equilibrium[11].

Another result which is very interesting is that the specific form of the equilibrium states is very sensitive to external parameters. A small change of the external magnetic field, of order $1/N^{1/2}$, is sufficient to completely upset the microscopic details of the equilibrium states. The equilibrium states at two different magnetic field have very small overlap[3].

All these results certainly hold when fluctuations are neglected. This approximation is good when the dimensions of the space are very large and the prediction of this approach compare well with the results of the numerical simulation in infinite dimensions, i.e. for the Sherrington and Kirkpatrick model[12].

If we decrease the dimension the situation is less clear. The most likely scenario, although infinitely many different possibilities are open, is that the picture does not change for dimensions greater than D_c. At the lower critical dimension D_c, the critical temperature becomes zero and there is no transition to a spin glass phase. However if the coherence length ξ is large and if the relaxation time is proportional to exp $(\xi^{1/4})$, it is possible that the relaxation time becomes very large (seconds, days, years...) and for reasonably long observation times the system seems to be in the spin glass phase.

In three dimensions the situation is confused: many data can be fitted as $(T - T_c)^{-\nu}$ with a reasonable value of T_c. However, the same data can also be fitted as exp (A/T^3) [13]. It is clear that we cannot decide between the two different options by looking at the value of the χ^2 of the fit. It is difficult to find a crucial test.

A very neat way of solving the problem consists of computing the corrections induced by fluctuations in the framework of the present theory and comparing them to the experimental and numerical data. If the agreement is good, the value of the lower critical dimension should be the theoretical one. Unfortunately such computation is very difficult and progress has been made very slowly[14].

References

1. For a nice short experimental review of spin glasses see J. A. Mydosh, in "Disordered Systems and Localization", ed. by C. Castellani, C. Di Castro and L. Peliti, Springer-Verlag (1981).
2. N. Sourlas LPTENS preprint 85/9 (1985).
3. G. Parisi, Phys. Rev. Lett. 43, 1754 (1979), J. Phys. A13, 1101, 1887, L115 (1980). For a recent theoretical review see M. Mezard, G. Parisi, M. Virasoro "Spin Glass Theory and Beyond", World Scientific, Singapore (1987).
4. D. Sherrington and S. Kirkpatrick, Phys. Rev. Lett. 35, 1792 (1975); Phys. Rev. B17, 4384 (1978).
5. S.F. Edwards and P.W. Anderson, J. Phys. F5, 965 (1975); F6, 1927 (1976).
6. J.R.L. de Almeida and D.J. Thouless, J. Phys, A11, 983 (1978).
7. M. Mezard, G. Parisi, M. Virasoro, Eur. Phys. Lett. 1,105 (1986)
8. G. Parisi Phys. Rev. Letters 50, 1946 (1983).
9. M. Mezard, G. Parisi, N. Sourlas, G. Toulouse and M. Virasoro, Phys. Rev. Lett. 52, 1156 (1984); J. Physique 45, 843 (1984)
10. H. Sompolinsky and A. Zippelius, Phys. Rev. Lett. 50, 1297 (1983) and references therein.
11. A. Houghton, S. Jain, A. P. Young, Phys.Rev. B25, 2630 (1983).
12. A. P. Young, Phys. Rev. Lett. 43, 1206 (1983); K. Namoto, Hoikkaido University Preprint (1986).

14. C. De Dominicis and I. Kondor J. Physique Lett. 46, L-1037 (1985) and references therein.

5. On the Emergence of Tree-like Structures in Complex Systems

Giorgio Parisi

5. 1. INTRODUCTION

Physicists have studied for a long time ordered systems trying to find a hidden order; disorder was perceived as a negative feature. Disorder was characterized as absence of order. This attitude has changed in recent years and there is now a strong interest in studying the behavior of disordered systems, focusing on those features which are proper of disorder. The most interesting results have been obtained when it was found that the laws controlling the evolution of the system were themselves disordered, i.e. chosen at random (Balian et al. 1979; Mezard et al. 1987; Souletie et al. 1987; Livi et al. 1988). I stress that when I use the word random, I do not mean that the relevant quantities are absolutely random (i.e. they can be anything), but that they are extracted with a probability distribution from a given class of objects. The goal of such an approach consists in identifying and computing the properties of the system which do not depend on the particular choice of laws, but depend only on the probability distribution used to extract the laws.

A physical example may help to clarify the concept. We can consider the motion of electrons in a metallic glass. A glass is an amorphous material which contains different kinds of atoms. In a glass the atoms do not move and their position has been fixed randomly at the moment of solidification. A glass is essentially a frozen fluid in which the atoms are disordered as in a liquid, but they do not move, as in a solid. The forces affecting the electrons depend in a random way on the position of the electrons. However, these forces do not change if the electron returns after some time to the same point. We are not interested in computing the electric resistance of the sample, starting from the position of the atoms. We know that if the samples are not too small (i.e. they contain many atoms) the conductivity is the same in all samples which have been prepared with the same experimental protocol, although the relative positions of single atoms in the two samples are quite different (Castellani et al. 1981).

In the same spirit we could guess that the brains of two homozygote twins work in similar ways, although the position of the individual neurons are different.

I do not want to discuss here the physical aspects of these recent findings. Instead I would like to explain why theoretical physicists have started to study systems where the laws of time evolution are chosen randomly. A very rich theory has been developed, quite advance mathematical tools have been used, and the final results are very interesting. Completely unexpected mathematical structures have been discovered and these structures have many properties which recall biological systems. Physicists have just started investigations of these disordered systems. It is likely that much more precise, interesting and general results will be obtained in the future.

I believe that the time is ripe to start thinking how these ideas, at least at the descriptive level, can be applied in the biological sciences. The aim of this chapter is to explain these new results using a language as plain as possible and to present some possible applications to biology. I am not fully convinced that the particular applications that I will describe are completely well founded. However I present them to show how the concepts introduced may have concrete applications in biology.

I fear that a biologist may be disturbed by my approach and by what may be considered disregard for the details of the underlying mechanisms. Indeed there is a difference between what a physicist and a biologist consider a satisfactory explanation of a phenomenon. For example the distribution of the number of individuals in a species, as a function of the species. I suspect that a biologist would only be satisfied if he determines the correct nature of the mechanism which produces the experimentally observed distribution. On the other hand a physicist would be happy if he can show that systems which only vaguely resemble a real one, have similar properties. From this he would argue that the observed behavior of the system is proper of a wide class of systems, each one having a somewhat different microscopic mechanism.

Theoretical physicists in recent years have discovered that the collective behavior of a macro-system, i.e. a system composed of many objects, does not change qualitatively when the behavior of single components are modified slightly. There are many universal classes which describe the collective behavior of the system, and each class has its own characteristics; the universal classes do not change when we perturb the system. The most interesting and rewarding work consists in finding these universal classes and in spelling out their properties. Although in many cases simple arguments can be used to find the universal class to which a given system belongs, the fine details of the explanation of why the system belongs to one or another class are much less interesting. Starting from the interatomic potential between molecules of H_2O, after a very long computation, we could arrive at the conclusion that water is still liquid at 99° C. Such a computation is definitely interesting from a technical point, but it is not one of the most exciting results in the theory of phase transitions.

This conception has always been implicit in most investigations and it has become quite explicit during studies done in the last twenty years on second order phase transitions (Wilson & Kogut 1974). The objective, which has been mostly achieved, was to classify all possible types of phase transitions in different universality classes and to compute the parameters that control the behavior of the system near the transition point as a function of the universality class. This point of view is not very different from the one expressed by Thom in the introduction of *Structural Stability and Morphogenesis* (Thom 1975). It differs from Thom's program because there is no *a priori* idea of the mathematical framework which should be used. Indeed Thom considers only a restricted class of models (ordinary differential equations in low dimensional spaces), while we do not have any prejudice regarding which models should be accepted.

One of the most interesting and surprising results obtained by studying complex systems is the possibility of classifying the configurations of the system taxonomically. It is well known that a well founded taxonomy is possible only if the objects we want to classify have some unique properties, i.e. species may be introduced in an objective way only if it is impossible to go continuously from one species to another; in a more mathematical language we say that the objects must have the property of ultrametricity (Rammal et al. 1986).

More precisely it was discovered that there are conditions under which a class of complex systems may only exist in configurations that have the ultrametricity property and consequently they can be classified in a hierarchical way. Indeed, it has been found that only this ultrametricity property is shared by the near optimal solutions of many optimization problems of complex functions, i.e. corrugated landscapes in Kaufman's language (Kaufman & Levin 1987).

These results are derived from the study of spin glass models, but they have wider implications. It is possible that the kind of structures that arise in these cases is present in many other apparently unrelated problems. It is therefore worthwhile to understand in detail the mathematical structures which arise in spin glasses. These structures are already quite complex and display very interesting properties. My aim is to explain these findings and to see some possible applications to biology.

This chapter is organized as follows: In section 5.2 I give a qualitative description of what I believe a complex system is; in section 5.3, I give a simple description of spin glasses, which are archetype of complex optimization problems. I focus my attention on the two most important characteristics of these systems, *randomness* and *frustration*, and I try to spell out some simple consequences. In section 5.4 I introduce other models in very different domains of physics and biology and I show that they have many characteristics in common with spin glasses. In section 5.5 I show how a taxonomic classification of spin glass configurations is possible and I describe the consequent emergence of a tree like structure of solutions of this complex optimization problem. In this section the description will be mostly verbal. Section 5.6 has a context very similar to the previous section, but uses a more mathematical language, and can be skipped by the less mathematically inclined reader.

The consequences of the results of the previous sections on time evolution of the system are discussed in section 5.7 and protein dynamics is presented as a possible example. Section 5.8 and 5.9 contain some speculative applications of these ideas to biology. In section 5.8 I try to compare the tree structures we have found in these models with the real biological classification of living beings. As a concrete example (Epstein & Ruelle 1989). I consider the taxonomic distribution of European mono- and dicotyledons. In section 5.9 I present some considerations on evolution, paying particular attention to the process of convergent evolution. In section 5.10 I present my conclusions.

5. 2. COMPLEX SYSTEMS

Defining a complex system is not an easy task. Practically every researcher has his own definition (Peliti & Vulpiani 1987). It may range from the classical algorithmic complexity to more recent and sophisticated definitions. In the literature a complex system is sometimes defined in very general terms, i.e., a complex system is a complicated system!composed of many parts, whose properties cannot be understood. Consequently it may be useful to explain my own definition (Parisi 1987). It should be clear, however, that any given definition (especially a mathematical one) cannot capture all the complex meanings we associate with the word complexity.

The basic idea is that the more complex the system, the more can be said about it. I am excluding the factual description of the system, which may be very long. I refer only to the global characteristics. A few examples will help clarify this point. If I have a sequence of randomly tossed coins, 50% probability head, I have described the system. The only improvement would be the knowledge of the sequence itself. If on the contrary the sequence of bits represents a book, there is much more information, such as style, choice of words, the plot and so on. It the book is really deep, complex, there are a very large number of things that can be said about it. Sometimes the complexity is related to the existence of different levels of description: one can describe an *Escherichia coli* at the molecular level, at the biochemical level, and at the functional level.

If we move towards a mathematical definition, we must realize that the concept of complexity, like entropy, is of a probabilistic nature and it can be defined more precisely if we try to define the complexity of ensembles of objects of the same category. Of course, if you have only one object which changes with time, you!can study the complexity of the time dependence (or behavior) of this object.

The simplest situation for which a consistent definition of complexity can be given arises when we try to classify an ensemble of sets. Everybody has some experience in classifying. Indeed one of the main activities of the mammal mind consists in finding relations and classifying them among the extremely large amount of sensory information received. For example different images of the same object are usually correctly classified as different images of the same object, in spite of extremely large differences of these images on the retina.

In order to be more precise about what I mean by the word classification, let us consider some examples. If the configurations are a sequence of completely random numbers, no classification is possible and all the configurations belong to the same class. If we consider bottles of wine which may be empty or filled we can classify them into two sets: the empty and the full bottles.

These two examples correspond to rather trivial classifications where no structure emerges. If the configurations are all the living objects on Earth, the situation is much more interesting and it is quite likely that the best classification is the one done by biologists. If we work in History of Art and we have to classify different painters, the task is more difficult. A given painter may belong to the French School, but the way he uses color may be strongly influenced by German painters; the relative influence of one painter on others will also have to play a role in the classification.

In the first three examples the configurations are classified like leaves of a tree (taxonomy), the tree being trivial in the first two cases but quite large in the case of living beings. In the last example the situation is more complex and a simple genealogical tree cannot be established (a given painter may be simultaneously under the influence of many painters). In our mind classification is equivalent to establishing some relations of kinship (or distance) among different configurations and in general the final output is not a tree.

A classification may be quite simple, as the first two examples show or it can be quite complex, as in the last two examples. The meaning of a complex classification is quite clear intuitively: a classification is very complex if there are many levels (i.e. orders, family, genera) and there are many elements in each level. Consequently a reasonable mathematical definition of the complexity of a classification should be possible.

If for the time being we assume we know the complexity of a classification, we could take a behaviorist point of view. We classify all the possible configurations!which a given system assumes and we use the complexity of this classification as a measure of the complexity of the system. This proposal can be condensed in the motto *a system is complex, if its behavior is complex.*

This brief discussion should have clarified the intertwining between the concept of complexity and the problems arising in classifying objects; it may also serve as an introduction to these two themes, complexity and classification, which will play a crucial role in the rest of the chapter.

5.3. SPIN GLASSES

A spin glass is a kind of alloy that contains a small percentage of magnetic material, which is distributed randomly in a non magnetic matrix (Balian et al.

1979; Souletie et al. 1987). Here I am interested mostly in the mathematical structures arising from the study of spin glasses and not in the particular applications of these structures to the physics of real spin glasses. In presenting these results I have decided to skip the description of real spin glasses, and to describe only some more familiar situations which can be modelled mathemati cally as spin glasses (Mezard et al. 1987).

It is a common experience in life to have goals that are not mutually compatible. The only solution to such a dilemma is to give up some of them. In making such a choice we feel frustrated. It is very difficult be friendly to both Mr. White and Mr. Smith who hate each other. We must choose one of them and this also is a frustrating situation.

The situation is more complex when many people are involved. In Shakespeare's tragedies such situations are very frequent. There is fight be tween two fields and the various characters on the scene have personal relations to each other. Some are friends and some are enemies.

Let us consider three characters: A, B, and C. If A and B, and B and C, and C and A have good relations there is no problem and they will be on the same side. If A is a friend of B and C is an enemy of both A and B, A and B will be on the same side and C will be on the other side. If however, A, B, and C hate each other and there are only two sides to choose, someone must stay on the same side with his enemy, a frustrating situation.

This analysis can be formalize by assigning each pair (A and B) a number (J_{AB}) which is 1 if A and B are friends, 1 if they are enemies. The relations between three persons (A, B, and C) are frustrated if

$$J_{AB} \cdot J_{BC} \cdot J_{CA} = 1$$

When the relations of many triplets are frustrated, it is evident that the two situations on the scene are unstable and many rearrangements of the division in two side are possible.

At a given moment of the tragedy it is possible to define the dramatic tension as

(number of frustrated triplets) / (total number of triplets)

Detailed studies have shown that in many of Shakespeare plays the dramatic tension has a small value at the beginning of the play, it reaches a maximum towards the middle and decreases at the end.

Leaving tragedy and returning to real life, we could consider the following mathematical model: we have N persons and the J's which control their relations have been preassigned in some way. Each person can choose the field to which it belongs and will make its choice with the objective of maximizing the number of friends minus the number of enemies in his field.

At the beginning the persons are distributed randomly in the two fields and sequentially each person is asked to change fields (if he or she wishes). After everybody has been asked, the procedure starts again by asking everybody if he or she wishes to change the previous decision due to the shifts done by other people. Many cycles of this procedure are repeated until no one asks to change and a stationary state is reached. This state is called a locally optimized state, because no improvements can be obtained by moving only one person at a time.

There are two possibilities:
(a) the stationary state is reached after a finite number of interactions (normally a few);
(b) the stationary state is never reached.

This second possibility can be realized if the relation of being friends or enemies is not symmetric. In other words, if John loves Mary, but Mary does not like John at all. When it is John's turn to decide, he will move near Mary, and when it is Mary's turn, she will go to the opposite field. The process will never end (unless John or Mary change their minds, a realistic possibility, which is not considered in the model).

On the contrary if love (or hate) is always returned, one can prove that the process will end in a few steps. Only in this case, where both persons are either friends or enemies, there exists a quantity (i.e. the total number of friendly couples minus enemy couples in the same field) that everybody agrees to minimize. The combination of the local optimization process (the decisions made by each individual for his own purpose) go in the direction of global optimization. Of course, if only one person is moved at a time, it is quite possible that the process will not reach the globally optimized situation, which may be reached if we allow groups of friends to change simultaneously the field.

The case in which the whole process tends to optimize a given quantity is far from being the general case. However, it is the simplest one (e.g. limit cycles are absent) and it is the only one for which we can present a complete theory. Many physical systems, in which individual components tend to minimize energy, or free energy, can be represented in this way. In the rest of this chapter I will restrict the discussion to this second case, not because the more general case is less interesting, but for the simple reason that it is the only case that is well understood.

The output of the previously described process clearly depends on J. The following questions are relevant: if we change the initial conditions at fixed J's, does the final configuration change? Does the local optimization process arrive at the global optimum or for some initial configurations does it become stuck in a local (rather than in the global) optimum?

If people can be divided into two groups, men and women, and we suppose that men like men and dislike women, and women like women and dislike men, any starting configuration will produce a final situation where men and women form two separate groups. No frustration is present in such a case.

If we have the opposite situation (men like women and dislike men), if men and women are in equal numbers, two groups will be formed, each having the same number of women and men. In this case many outputs are possible and each of them is as good as the others. The system is frustrated in a regular way.

If choice of the J's is not regular as in these two cases (as in the extreme case where the J's are chosen randomly) many different locally optimal solutions may be reached and some of these solutions will be more satisfactory than others.

In the last case (J chosen randomly according to a given probability law) a random frustration is present and a local algorithm is unable to find the maximum. This is not surprising because it is believed (however it has not been proven) that even the best algorithm will take a time which increases exponentially with N (the number of persons) for large N's.

The aim of the theory of spin glasses is not to determine the best algorithm for finding the optimal solutions but to study the properties of the optimal and of the nearly optimal solutions. We may suppose that, when we find in nature different solutions of the same optimization problem, they have been well optimized by natural evolution; we can study their properties, disregarding the way in which they have been produced.

The simplest question is if nearly optimal configurations differ in a significant way from the optimal one. If they are very similar, there is nothing interesting to classify. If the opposite happens, we could try to classify these near optimal configurations. It is possible that they are all on the same footing; it is also possible that they could be grouped into species, which we can group into families, and so on. In this last case we want to understand the kind of classification we obtain.

Of course the answer to this question will depend crucially on the probability distribution that we use to chose the J's. If all the J's are chosen randomly we have what is called a long-range model. On the contrary, if we assume that the two individuals may be indifferent from each other (this could correspond to $J = 0$), different models may be obtained. For example if only a few people are not indifferent to any given individual (nearly the totality is indifferent), we obtain what is called a short range model. Many different kinds of short-range models may be constructed. For example we can assume that two individuals who have no interest in common are indifferent to each other. In other words you may be a friend or enemy of someone who has the same interest as you. In this case the form of the short range model depends on the set of possible interests of different individuals.

There are two ways of approaching this theoretical problem. The experimental approach consists of considering some instances of the J's and in computing explicitly the near optimal solutions and observing their properties.

The other possibility consists in studying the problem analytically and deriving the properties of the near optimal solutions when the J's are chosen randomly inside a given class. The second approach gives much clearer results. Unfortunately it is too difficult for most problems. We have been lucky enough that the spin glass problem we have described can just barely be studied analytically. Many interesting results have been derived and they will be described in sections 5.5 and 5.6.

It is also possible to consider a more general model in which individuals make errors in their judgements: the errors are less likely the more evident the choice is. In this case the evolution of the system is quite different and depends strongly on the error rate.

A small value of the error rate keeps the system from remaining stuck in a stable, local, but not global minimum. The system will eventually arrive, perhaps after a very large number of steps, near the global minimum. Of course if the error rate is large, the evolution of the system will be much faster, but the system will never arrive near the optimal configuration.

A non zero error rate corresponds in physics to the introduction of a temperature (only at zero temperature is energy really minimized). In this case we can study the time evolution of the system and we identify two different time domains, the non equilibrium region, where the system is approaching the optimal solution and the equilibrium region, where the system moves near the global optimum. The dynamics in both regions are well studied, although they are somewhat less understood than the static properties; some of the most interesting!results will be presented in section 5.7.

Before leaving this section let us lay down in a more mathematical language the ideas which we have just described (Mezard et al. 1987). We could say that we have N variables σ_i, one for each person (i = 1,...,N): σ_i takes the value 1 or -1 depending on which side the ith character stays. For a given set of the $J_{i,k}$ the function we want to optimize is

$$O_J[\sigma] = \sum_{i,k=1,N} J_{i,k}\, \sigma_i\, \sigma_k$$

where we have used the square parenthesis to stress that O is a function of the σ's together. If $J_{i,k}$ is positive the two spins σ_i and σ_k tend to have the same sign, otherwise they tend to have the opposite sign.

The function O is often called (depending on the contest) the objective function (or the fitness). We also could say that it is minus the cost or minus the Hamiltonian (the energy): $O_J[\sigma] = -H_J[\sigma]$.

The dynamics without errors corresponds to examining σ_i sequentially and to set the σ_i at the next step equal to

$$\text{sign}\,(h_i),$$

$$h_i = \sum_{k=1,N} J_{i,k}\,\sigma_k \qquad (1)$$

where the sign function is +1 when the argument is positive and it is -1 when the argument is negative. It is easy to check that each move increases the objective function (indeed the variable σ_i is flipped if and only if this flip increases the energy). After a not too large number of steps the evolution stops (if the J's are symmetric, i.e. if $J_{i,k} = J_{k,i}$) and we reach a local maximum of the function O_J.

It is well known and a difficult mathematical problem, to find a fast algorithm which (for a given instance of the J's) computes the set (or sets) of the σ's which maximize O_J. From the point of view of complexity theory the problem of finding the global maximum of the function O_J is NP complete, which means that it is very likely that there is no algorithm which can find the minimum using a number of steps which increases as a power of N (for N large). It is believed (but it has not been proven) that the time needed by the best algorithm increases as $\exp\{cN\}$, c being an N-independent constant.

One of the simplest results that can be obtained in the analytic approach is the following. Let us consider the simplest case where the J's are chosen to be equal with probability 1 and -1. For very large values of N, the maximum value of $O_J / N^{3/2}$ is approximately the same for almost all realizations of the J's and this value is about -0.7633. In other words if we come back to our problem, where N people know each other and they are randomly friends or enemies and we split them into two groups of N/2 people, trying to keep (as far as possible) friends in the same groups and enemies in the opposite groups, the best score possible is that the average number of friends (of a generic person) minus the enemies (of the same person), which belong to the same group, is $0.7633/2\, N^{1/2}$ when N goes to infinity.

If we want to introduce errors in the optimization process, the most elegant way consists in substituting eq.(1) with

$$\sigma_i = 1 \text{ with probability } p_+(h_i),$$

$$\sigma_i = -1 \text{ with probability } p_-(h_i), \qquad (2)$$

where h_i is defined as in equation 1 and the functions p_+ and p_- are given by

$$p_+(h) = \exp(\beta h) / (\exp(\beta h) + \exp(-\beta h))$$

$$p_-(h) = p_+(-h).$$

Conservation of probability requires $p_-(h) + p_+(h) = 1$, as can be easily checked.

In the limit $\beta \to \infty$ eq.(2) reduces to eq.(1). This way of introducing errors has the advantage that we can compute the probability distribution of the configurations of the σ's at equilibrium. In this case we know that, after a sufficiently long time each configuration of the spins appear with a probability

$$P_J[\sigma] = \exp \{ \beta O_J [\sigma]\} / Z \qquad (3)$$

where the *partition function* Z is such that the total probability is equal to 1:

$$\sum_{\{\sigma\}} P_J [\sigma] = 1; \quad Z = \sum_{\{\sigma\}} \exp \{\beta O_J [\sigma]\} \qquad (4)$$

In statistical mechanics $-O_J \equiv H$ is the energy of the system and eq.(3) is the Gibbs probability distribution and β is given by $1/(kT)$, where k is the Boltzman constant and T is the absolute temperature (Amit 1989).

There are many different dynamics which lead to the probability distribution (3); it is also interesting to consider eq.(3) as a way of weighting the configurations of the σ's which are not optimal. Indeed for $\beta \to \infty$ the probability distribution is concentrated on the optimal solution, while for large but finite β the probability distribution is concentrated on the nearly optimal configurations.

According to the usual conventions, we will denote by $<A>$ the expectation value of a function $A[\sigma]$ according to the probability distribution (3):

$$<A>_J = \sum_{\{\sigma\}} P_J [\sigma] A [\sigma], \qquad (5)$$

(although A depends on J, sometimes we will not explicitly indicate this dependence in order to lighten the notation).

The study of the dependence of β on the expectation values of different quantities in the region of large β give us information on the properties of the nearly optimal configurations.

5. 4. OTHER FRUSTRATED MODELS

The presence of frustration implies that there are many local maxima of the functiōn we want to maximize and therefore a local optimization process cannot find the global maxima in most cases.

The case where the number of local maxima is very high has been denoted, in very effective way, a corrugated (or rough) landscape (Kaufman & Levin 1987). Indeed, if we consider a two dimensional picture of a landscape with many valleys and mountains, there are many local maxima (and many local minima). In a real mountain any one who finds his way just trying to go always up quite likely will reach the top of a small hill not that of the highest mountain. All optimization problems in which there are global constraints which forbid local optimization to produce the most optimized solution are in some sense frustrated. Roughly speaking, if the problem can be solved easily the problem is not frustrated; if the problem is difficult, frustration is present.

A typical frustrated optimization problem is the travelling salesman (Kirkpatrick et al. 1983; Lawler et al. 1985). In order to specify an example of the problem we must have a list of cities and we must know their distances; the task consists in finding the order in which a travelling salesman should visit them if he wants to travel the shortest path. The problem is frustrated. Indeed, in the optimal solution the travelling salesman does not always go to the nearest city he has not visited. If he did, at the end, the length of his path would be greater than that of the optimal solution.

There are many examples of corrugated landscapes at the molecular level. Let us consider the tertiary structure of a protein with a fixed primary structure. Here we know that equilibrium statistical mechanics works and the probability of finding a protein in a given configuration should be given by something like eq.(3), where minus the energy or the free energy plays the role of the objective function. It is well known that a change in one amino acid may strongly affect the folding of the whole protein. Numerical simulations and a careful analysis of experimental data (Fraunfelder et al. 1989) show that the potential has many local minima, as one could imagine, and that the change of one amino acid has the effect of transforming a local minimum into a global one and a global minimum into a local one.

Another example is given by the binding energy of a protein to a given structure as a function of the primary structure (or of the genetic code) of the protein. The situation is similar to the previous example and many local maxima of the binding energy are also expected here. The problem has been investigated most exhaustively in the case where the protein is an antibody (Kauffman et al. 1988).

A more complex example is the survival probability (or fitness) of a living being as a function of its genetic code (Eigen & Schuster 1977; Eigen 1988; Kaufman 1990). If the fitness is not increased by any point mutations we say that the genetic code corresponds to a local maximum. It is quite likely that there exist many local maxima which correspond to cases in which an improvement in the fitness may be obtained only if two (or more) point mutations happen simultaneously (see chapter 6 by Schuster).

In all these cases the functions which we wish to minimize (or maximize) are quite complex. The simplest approach consists in approximating them with randomly chosen functions having a certain number of properties. This requires looking for those properties of proteins which do not depends on the detailed chemistry of amino acids.

I stress again that when I say randomly chosen functions I do not mean totally random, but randomly chosen inside a given class. For the protein the possibility of dividing the amino acids into hydrophobic and hydrophylic may play a crucial role. The main point is that we would be very surprised if some of the qualitative properties of the proteins, e.g. the possibility of existing in different folding states, were to depend on the fact that valine is more hydrophobic than guanine (or *vice-versa*).

In other words we do not want to study the detailed consequences of the potential between these amino acids which happen to be the real ones. We want to understand the consequences of a generic potential which is not too different (in some sense) from the real one. We hope that the real potential does not have any peculiar property which distinguishes it from the others. Using Thom's language we could say that the consideration of randomly chosen potentials implies that we only study structurally stable properties of the proteins, i.e.

properties which do not change when we change slightly the chemical properties of the amino acids.

All these models have not been solved analytically (some have not yet been formulated in a clear mathematical setting). The spin glass model presented in the previous section is a lucky exception in this respect. However one may hope that many of the characteristics of the solution of the spin glass model will survive also in these more complex cases.

5. 5. THE TAXONOMY OF NEARLY OPTIMAL CONFIGURATIONS

In this section we will present the results that have been obtained for spin glasses for the nearly optimal configurations, i.e. those configurations whose fitness (or objective) function has a value quite close to the absolute maximum (Parisi 1984; Mezard et al. 1987; Parisi 1987).

Given two configurations we can define their distance as the percentage of the number of variables σ_i which differ from one configuration to the other (the σ's may take only the value -1 or 1). For example, if we have 100 variables σ's, 90 equal in the two configuration and 10 different, the distance is 0.1. Such a distance belongs to the interval (0-1), and it is zero only if the configurations are equal.

Clear cut results are obtained for the problem of spin glasses presented in section 5.3 only in the limit where the number (N) of σ is very large (mathematically speaking infinitely large). In this case two configurations may stay at a zero distance if the number of differences remains finite when the total length goes to infinity. The percentage of different σ's and not their absolute number is relevant in the definition of distance.

Let us consider now the set of nearly optimized configurations. They are defined to be those configurations whose objective function differs only by a small percentage from the optimal one. Let us start by presenting some general considerations that are valid not only for spin glasses, but in most statistical mechanical problems.

The number of nearly optimized configurations is extremely large when the number of variables N goes to infinity (it is exponentially large) and these configurations may be classified into species.

Species are defined as follows:
(a) Each of the species contains an exponentially large number of configurations.
(b) More crucially if we take two different generic configurations belonging to species A and species B respectively, their distance does not depend on the generic representatives of the species, but only on A and B.
(c) It should also be true that the distance between two configurations of the same species should be strictly smaller than the distance between two configurations of different species. This last property may be written as

$$d_{A,A} < d_{A,B}$$

where $d_{A,B}$ is the distance between species A and B, i.e. the distance between two generic representatives of species A and B.
(d) the classification into species is the finest one which satisfies properties a, b, and c.

We notice that the same results may be obtained also if we consider all the possible individuals, but we extract them with the probability given in eq.(3). In this case the classification into species will depend on the parameter β, i.e. from the temperature of the physical system.

From general hypotheses in statistical mechanics it can be proven that the classification into species is possible. In physical systems species are called phases and normally the classification is quite poor. For usual materials, in the generic case there is only one phase (one species) and the classification is not very interesting. In slightly more interesting cases there may be two species. For example if we consider the configurations of a large number of water molecules at zero degree centigrade, we can classify them as water or ice. Two species are present. In slightly more complex cases, if we choose carefully external parameters like pressure or magnetic field we may have the coexistence of three or more phases (tricritical or multicritical) points.

In all these cases the classification is rather poor and the number of species is quite small. It was really a surprise when it was discovered that in the spin glass model described in the previous section the number of species is very large. It goes to infinity with N (the number of variables) and a very interesting nested classification of species is possible.

Let us now describe the results that have been obtained in the study of spin glasses. First of all there is a kind of democracy between species in the sense that species may differ only in the number of individuals (configurations) which belong to that species and in the distances with other given species. There is no intrinsic difference; in particular the variability inside species A (i.e. $d_{A,A}$) does not depend on A: it may be convenient to call it D_S.

Most interestingly an explicit computation shows that species may be classified in a taxonomic way (Rammal et al. 1985). For example one may take a distance D_G, with $D_G>D_S$, and introduce genera $\alpha\ \beta\ \gamma$... in such a way that all species belonging to the same genus have distances less or equal than D_G, while species belonging to different genera have distances greater than D_G. The value of D_G is free; however if D_G is very close to D_S the genera will be quite similar to species, while if D_G is much greater than D_S the classification into genera will be very coarse. It is remarkable however that independently from the value of D_G the number of species in any genus is infinite (it goes to infinity with N, the number of variables).

In the same manner one can introduce families by choosing a distance $D_F>D_G$ and by grouping together genera at distances smaller or equal than D_F. Families are similar to genera (they are a coarser classification). In particular each family contains a very large number of genera.

It should be evident to the reader that if we are not satisfied with the classification into species, genera, and families, we can introduce orders by choosing a distance $D_O>D_F$ and by grouping together families at distances smaller or equal than D_O. Orders too contain a very large number of families.

We could go on for a while by introducing more levels of classification. Let us stop here and try to clarify some crucial points. The number of levels and the distances at which levels are defined (D_G, D_F, D_O and so on) are obviously arbitrary: we can introduce a classification which is coarser than genera, but finer than families. The situation may be similar to the one described in fig. 5-1 where we represent the classification as a tree. This tree may recall the phylogenetic tree, but it definitely is not a phylogenetic tree, as there is no

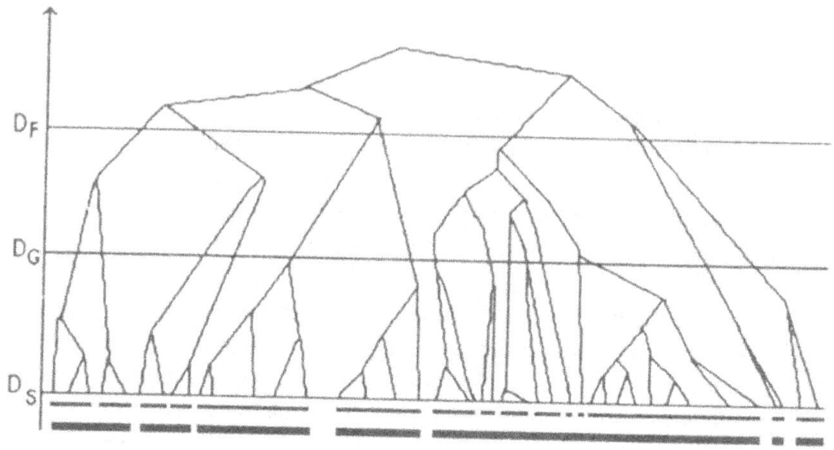

Fig. 5-1. An example of a tree with the associate classification.

evolution in this model. We are only considering the set of configurations with nearly maximal fitness.

In fig. 5.1 the leaves of the tree are the species and the distance between different species is represented by the level we have to reach to find a path joining the two species. If the quantities D_G and D_F are the ones represented by the two thick horizontal lines, the classification into genera and families is represented by the large horizontal lines below the tree. The 41 species are classified into 15 genera and 7 families. It is evident that the levels which we have chosen to define genera and families are arbitrary.

The tree of species in spin glasses is much more dense and branched than the one in fig. 5-1. It is indeed an infinitely branched tree. Branches are presented at any level and there is an infinite number of branches in any interval of distance. The very mathematical existence of such an object is far from being trivial and it has been proven only recently by Ruelle (1988).

Many more results are known if we consider also the populations inside each given species. Let us call W_A the frequency of individuals of species A; the sum over all the species of W_A is obviously one; W_A is the population in species A divided by the total population.

If one makes a histogram of the number of species with W_A greater than p (which we call $N_S(p)$), one finds that for small p it goes like

$$N_S(p) \approx p^{(-y_S)} \quad (6)$$

where y_S is a given number, which characterizes the tree. In a similar manner

$$N_G(p) \approx p^{(-y_G)} \quad (7)$$

$$N_F(p) \approx p^{(-y_F)}$$

$$N_O(p) \approx p^{(-y_O)}$$

where $N_G(p)$ is the number of genera having a percentage of population greater than p, the population of a genus being defined as the sum of the populations of the species that belong to that genus. A similar definition holds for families and orders. Mathematical consistency requires

$$y_S < y_G < y_F < y_O.$$

The full characterization of such a tree is given by a function y(D) such that

$$y_G = y(D_G).$$

which tells how the exponent y changes when one changes the definition of genera (the same function plays the equivalent role for families and orders).

The power law distribution of frequencies is a quite common phenomenon in nature and it is quite gratifying that we have found that it appears in this more abstract setting.

Different trees have different function y(D) as they depend on the precise nature of the optimization process. In simple cases y(D) can be computed. For example this has been done in the spin glass model and it has been found that y(D) depends on temperature.

This infinitely branched tree, in spite of its apparent complications is the simplest mathematical structure for an infinitely branched classification and it is rather likely to be present (with different forms of the function y(D)) in many physical problems. We would like therefore to see if such a structure may be identified in biological systems. Before doing this, we present in the next section a slightly more technical exposition of the ideas just presented.

5.6. MORE INFORMATION ON THE TREE

In this section we have two goals:
(a) give a more precise definition of species, and
(b) specify details about the tree.

Let us start with the first goal. We suppose that the configurations of our system are weighted according to a probability/distribution P, like eq.(3) and we indicate the expectation value of a quantity A by A as in eq.(5). The fluctuation of the quantity A is defined as

$$< A^2 > - < A >^2 = < (A - <A>)^2>$$

Intensive quantities may be defined as

$$1/N \sum_{i=1,N} < A_i (\sigma_i) , >$$

where the functions A_i depend only on the value σ_i (they may depend on the site i).

The crucial question is *do intensive quantities fluctuate?* Intuitively, we would like a negative answer, since intensive quantities are averages over the whole system. However, this is not always the case. To illustrate this point it is convenient to consider a simple example.

We start with the model of section 5.3 where only men are present and they like each other; we also suppose that the error rate is quite small (Physicists call this situation a ferromagnetic system). In such a situation nearly all men will belong to one field (a few will form a minority by mistake). However two possibilities are open: the majority in one field and the minority in the other or the reverse situation, and they have the same probability. The percentage of people in each given field is 1/2, but it will be the average of configurations with

99% (let us say) and 1%. In this case the percentage fluctuates from configuration to configuration.

In this case it is possible to classify configurations according to which field the majority belongs to and to define an average restricted to this kind of configurations. If we define these restricted averages the percentage will not fluctuate.- More precisely if

$$<\sigma_i> = \frac{\sum_{[\sigma]} \sigma_i \exp(-\beta H)}{\sum_{[\sigma]} \exp(-\beta H)}$$

we have that

$$<\sigma> = 0$$

By classifying the configurations into two sets we can define restricted averages $< >_+$ $< >_-$ such that

$$1/2 < A >_+ + 1/2 < A >_- = < A >$$

$$< \sigma >_+ = m; \ < \sigma >_- = -m$$

It is possible to prove that intensity quantities do not fluctuate in $< >_+$ and in $< >_-$. The decomposition of a probability distribution in which intensity quantities fluctuate into the linear combination of restricted probability distributions in which the intensity quantities do not fluctuate can be performed in many cases in statistical mechanics. These restricted probability distributions correspond to different species!and the species will be identified by the expectation value of intensity quantities (in physics species are also called phases or pure states). Generally speaking we can thus write the decomposition in species as

$$< A > = \sum_\alpha W_\alpha < A >_\alpha ,$$

where α labels the species and W_α is the frequency of species α.

We must now specify the probability distribution of the W's and how different states differ from others. This will depend on the model. We now describe the results for the infinite-range spin glass. To this end it is convenient to introduce the distance (d) of two states:

$$d_{\alpha\gamma} \equiv \frac{1}{(4N)} \sum_{i=1,N} (m_i^{\alpha} - m_i^{\gamma})^2$$

$$m_i^{\alpha} \equiv < \sigma_i >_{\alpha}$$

At zero temperature, where all the m's are equal to ± 1, d becomes proportional to the usual Hamming distance we introduced in section 5.3.

In spin glasses species do not have microscopically different properties and

$$d_{\alpha,\alpha} = d_{\gamma,\gamma} = D_S$$

After a long computation it turns out that in spin glasses the w's are random (not independent) variables whose probability law is defined as follows (Mezard et al 1987; Parisi 1988).

First we introduce the variable f_α for each species α which we may call the free energy of the species α. The W's can be computed from the f's as follows:

$$W_\alpha = \exp(-f_\alpha) / \sum_\gamma \exp(-f_\gamma).$$

It is evident that if we shift all the f's by the same amount, the W's are left invariant. It turns out that the f's are independent random variables and that the probability of finding a state with a free energy in the interval f, f + df is given by

$$\exp[^{y}s(f-\tilde{f}]df \qquad (8)$$

where $\tilde{f}$ is a reference free energy (roughly speaking $\tilde{f}$ is the free energy of the system) and ys is a positive parameter (we also assume that $y<\beta$). Eq. (8) yields the probability distribution of the frequency of species and it is essentially equivalent to eq.(6).

We must now state which is the probability distribution of the d's and which are the correlations between the d's and the W's. Moreover we would like to know if the species may be classified in some way. It turns out that the following inequality is satisfied

$$d_{\alpha,\gamma} \leq \max\,(d_{\alpha\beta}, d_{\gamma\beta}) \quad \text{for any}\ \beta. \tag{9}$$

Eq.(9) tells us that the space of species is ultrametric; it amounts to saying that, for any choice of the number D, species may be classified into non overlapping clusters (of diameter D) such that α and γ belong to the same cluster if and only if $d_{\alpha,\gamma} \leq D$. Clusters may be decomposed into subclusters of smaller diameter and so on. The whole distribution of states is characterized if we specify the clusters and their overlaps.

Ultrametricity is the crucial property that allows a taxonomic classification. A space is ultrametric if two spheres of the same diameter coincide if they have at least one point in common: if α belongs to the same family as β ($d_{\alpha\beta} < D_F$) and β belongs to the same family as γ ($d_{\beta,\gamma} < D_F$), then α belongs to the same family as γ ($d_{\alpha,\gamma} < D_F$).

The description of the combined probability distribution of distances and frequencies can be obtained first in the simplest case where we suppose that the distances may assume a finite number of values, which we indicate by d_j, $j=0,1,\ldots,k$, (for simplicity we suppose that the d's are an increasing sequence). For sake of space we will write down the formulae for $k=3$; the generalization to arbitrary k being evident. At the end we will perform the limit $k\to\infty$, in this case the distance will take a continuous range of values.

If $k=3$ the distance among the arbitrary pair of species is at maximum D_F. States at smaller distance belong to the same family. If the distance is D_G (not D_F) they belong to the same genus; their distance may be $D_S \equiv 0$ only if they coincide.

We now describe the frequency distribution of species and of genera inside the families and of the species inside the genera. The rules are so simple that they can be easily generalized to an infinite hierarchy.

The number of families goes to infinity with N. We label them with an index α_3 and we associate to each of them a "free energy" $f_{\alpha 3}$. The f's are random independent variables, whose probability distribution is supposed to be:

$\exp [y_F (f_{a3} - f)]\, df_{\alpha 3},$

In each of the families there are infinitely many genera of diameter D_G. We label them with a pair of indices α_3,α_2 and we associate to each of them a "free energy" $f_{a3,\alpha 2}$. The f's are random independent variables, whose probability distribution is!supposed to be:

$$\exp [y_G(f_{\alpha 3,\alpha 2} - f_{\alpha 3})]\, df_{\alpha 3,\alpha 2}$$

the parameter y_G satisfies the inequality $y_G < y_F$.

Finally the states are labelled with three indices $(\alpha_3,\alpha_2,\alpha_1)$ and the probability distribution of the associated free energies (which are as usual random independent variables) it is assumed to be

$$\exp [y_S(f_{\alpha 3,\alpha 2,\alpha 1} - f_{\alpha 3,\alpha 2})]\, df_{\alpha 3,\alpha 2,\alpha 1}.$$

where $y_G < y_F < 1$. Integrating over the free energies of the families and of the genera one finds that the probability distribution of the free energies of the states is given by equation 8. It can be proved that these last equations implies eqs (6-7) of the previous section.

This construction may be generalized to any value of k. In this way we obtain a distribution of the probability of the states and of their distances, the only parameters being the D's and the y's. We finally send k to infinity in such a way that the D's become dense in the interval D_S, D_M. In this way we obtain a monotonous decreasing function y[D], which is defined in the interval D_S, D_M: the distribution of phases and of their distances depends only on this function y[D]. It is not evident whether this construction survives in the limit k going to infinity, where an infinite number of levels is present. Fortunately a detailed analysis by Ruelle shows that this construction really defines an infinitely branched tree.

Many more questions may be posed and we know the answers to some of them. We will not discuss them for lack of space. (for further information see Parisi 1984, 1988; Mezard et al. 1987).

5.7. DYNAMICS WITH AN APPLICATION TO PROTEINS

In an abstract optimization problem no time direction is present. However in many cases the system depends on time and the laws of motion are such that the system evolves in such a way as to optimize the objective function.

In a typical physics example the objective function is given by minus the energy. The time evolution of a physical system at low temperatures is such to minimize the energy, if we neglect small thermal fluctuations. We have already noticed that a non zero temperature is equivalent to introducing errors in the optimization process, as described in equation 2. If we consider the free energy, we also include the effect of thermal fluctuations and we arrive at a sharper statement: the system evolves in such a way as to minimize the free energy (which is given by the internal energy minus the temperature times the entropy).

In many cases after a sufficient large time the system reaches a configuration near the minimum energy (or free energy), independent of the initial state. In these cases, after a transient during which the system is not at equilibrium, the system reaches the equilibrium configuration which is unique. The probability distribution at equilibrium is given by the Boltzman distribution (eq.4).

A spin glass behaves differently. Let us consider the time evolution of a system whose states have the distribution described in the previous sections. In this case we must distinguish three time ranges. For short time the system remembers the initial state and its probability distribution is not given by eq.(4). Only after a large time will the system be nearly equilibrated in a given phase, and the probability distribution will be given by eq.(4) becomes fully satisfied. For simplicity we will now discuss only what happens in this last region, where equilibrium thermodynamics may be applied without restrictions.

If we look at the system at a given moment at very large times, the probability of finding it in the phase (or species) α is given by W_α. What happens if we look at it at a later time? In order to go from the phase α to the phase γ the system passes through an intermediate configuration between the two and we have already excluded that such an intermediate configuration exists with non negligible probability. We are led therefore to conclude that the system will remain frozen in the phase α forever.

This conclusion is too hasty, because the probability of finding the system in an intermediate state is small, but it is not strictly zero. Indeed those configurations, which do not satisfy the ultrametricity inequality and are intermediate between α and γ, have a much higher free energy than normal configurations (let us denote by Δf the difference in free energy). They are present with small probability, proportional to

$$\delta \equiv \exp(-\Delta f/(kT)).$$

We finally arrive at the conclusion that the system will remain in the phase α for a long time (proportional to $1/\delta$) and it will make a fast transition (in a relatively short time) to the phase δ. In other words *punctuated equilibria* are present. If we look at the system at a given time it is extremely unlikely that we will observe a transition between two phases.

The quantity that controls the dynamics at very large times is Δf, which is unfortunately a hard quantity to compute. In the infinite range spin glass model it is expected that Δf goes to infinity when N goes to infinity. More detailed estimates suggest that Δf increases as a power of N, may be like $N^{1/4}$, N being the total number of elements of the system (Vertecchi & Virasoro 1989).

It is also believed that Δf depends on the distance $d_{\alpha,\gamma}$ between states α and γ (due to ultrametricity after many jumps at a distance d, one does not arrive at a distance larger than d) and we thus define a function $\Delta f(d)$. One finally finds that, starting from the state α, after a very large time t the system has visited nearly all states within a distance $d_{\alpha,\gamma} \leq d$, where we have

$$\Delta f(d) \approx kT \log(t)$$

Indeed the characteristic time for the system to go from a state to another at distance d is given by $\exp(\Delta f(d)/(kT))$. Unfortunately in most cases we are not able to compute the function $\Delta f(d)$ from first principles.

The final conclusion is that the system has many large time scales and a non trivial dynamics may be observed up to times of the order of $t_M \equiv \exp(\Delta f(d_M)/(kT))$ (Sompolinsky and Zippelius 1982). In many cases t_M may be extremely large (practically infinite), so that the system evolves at all times and never reaches complete equilibrium.

The very large value of t_M implies that we may easily study experimentally the behavior of the system before it reaches real equilibrium. In many cases only the region of time $t << t_M$ may be observed. Unfortunately the theoretical understanding of a system, when it is not in equilibrium, is not very developed, and it is difficult to make precise predictions. We hope that this situation will improve in the future.

These ideas have been applied recently to protein folding. It has been suggested that all random heteropolymers (i.e. polymers composed of different monomers in a random order), and in particular proteins, may have properties in common (Stein 1983; Garel and Orland 1988; Shaknovich and Gutin 1989).

When the heteropolymers are in the globular phase, it is crucial to study the behavior of the following quantity:

$$d^2_{\alpha\beta} = 1/N \sum_{i=1,N} (x_i - y_i)^2,$$

where α and β are two equilibrium configurations of the hetero-polymer with corresponding coordinates x and y (i labels the monomers whose number is N). We notice that in order to avoid trivial results the relative position and orientation of the two heteropolymers have to be adjusted in order to minimize d.

Depending on the probability distribution of d, we can distinguish three situations in the collapsed phase:
(a) d does not depend on the choice of α and β and it is of order R, i.e. the radius of the globule. We can say that the heteropolymer is randomly folded.
(b) d does depend on the choice of α and β and it is much smaller than R. In this case the heteropolymer is folded in only one way.
(c) d depends on the choice of α and β and it can take values which range from very small up to R. In this case many folding possibilities are present.

On the basis of a sophisticated theoretical analysis it has been suggested that these properties of the heteropolymers may be investigated with techniques very similar to those used in spin glasses. In particular it has been suggested that the situation (b) is not present. As soon as one preferential folding is possible, many different foldings are available (Shaknovich and Gutin 1989).

In some cases the hierarchical classification described in the previous sections can be used and we can apply the theory developed for spin glasses. If this happens, the previous arguments imply that many time scales should be present in the dynamics.

It is very interesting to see if these results are confirmed by a more careful analytic study and by numerical simulations. Real proteins have characteristic time scales for different kinds of motions which go from the picosecond to the second; the same protein may exist in different foldings and sharp protein-quakes characterize the transition from one folding to another. The time dependence of the response of proteins to external perturbations is quite similar to that of spin glasses and a strong experimental effort has been devoted to finding out the similarities and the differences among proteins and spin glasses (Ansari et al. 1985; Iben et al. 1988; Frauenfelder et al. 1989).

It would be extremely interesting to find out that some of the experimentally observed properties are not peculiar of real proteins, but they are shared with all heteropolymers. In this case they could be understood from general principles without having to look into the details of the chemistry of amino acids.

The theoretical analysis of protein folding from the point of view of statistical mechanics is still in its infancy. Before drawing definite conclusions we must wait for the development and refinement of a general theory of heteropolymer folding.

5. 8. A COMPARISON WITH REAL TAXONOMY

In this section I will describe an attempt made by Epstein and Ruelle (1989) to apply some of the ideas presented here to plant taxonomy. Their aim was to understand if the taxonomic tree found in nature, has some properties in common with the one described in the previous sections. They explored whether this happens independently of the evolutionary processes that lead to the appearance of this property. The tree described in the previous section is a well defined and relatively simple structure, rather natural from the mathematical point of view, and it is likely that many different processes may produce such a structure.

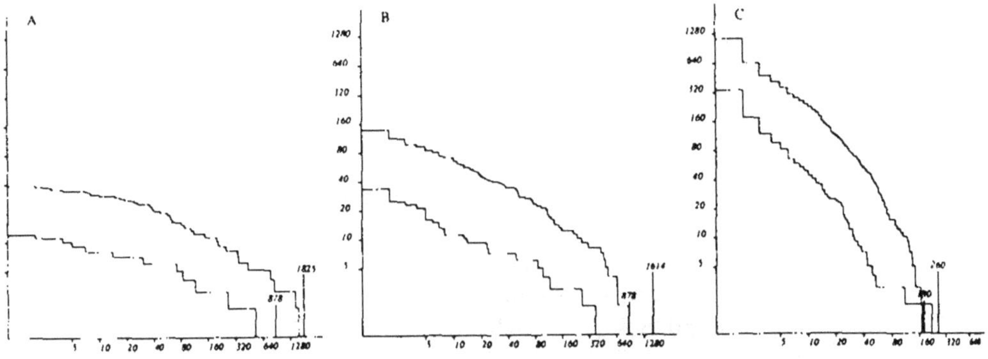

Fig. 5-2. Figures A, B, C represent respectively the functions N_O (k), and N_F (k), i.e., the numbers of orders, families, and genera with more than k species in *Flora Europea* 1980. In each case the upper curve is for dicotyledones and the lower curve is for monocotyledons. The vertical bar is at the log of the largest number of species in an order, family and genus, i.e., where the functions N(k) become zero. (From Epstein and Ruelle 1989, by permission).

In their analysis Epstein and Ruelle concentrated their attention on the species, genera, families, and orders of European dicotyledons and monocotyledons in order to check the validity of eqs. 6-7.

The first step consists in assigning a weight to each species. A natural possibility is to give a weight which is proportional to the number of individual members of that species. However in such a case small species which tend to have many individuals may be favored. Another possibility is to weigh according to the biomass of the species. Both possibilities represent a measure of the success of the species. Unfortunately it is impossible to obtain either of these numbers.

The simplest solution is to give each species the same weight. In this case each order, family, or genus has a weight which is proportional to the number of species which belong to that order, family or genus. With this choice the formulae in eq.(7) become

$$N_G(k) \approx k^{(-y_G)} \qquad (10)$$

$$N_F(k) \approx k^{(-y_F)}$$

$$N_O(k) \approx k^{(-y_O)},$$

where $N_G(k)$ is the number of genera having more than k species. A similar definition holds for families and orders.

The data both for monocotyledons and dicotyledons are shown in fig. 5-2 in a log-log scale and look quite similar in both cases. The power law in eq.(10) corresponds to a straight line on the plot and it is a reasonably good approximation. If we make a fit, we find approximately

$$y_G \approx 0.8; \; y_F \approx 0.35; \; y_O \approx 0.12$$

However the fit of a straight line is not perfect. The most evident deviation is a convexity of the experimental data. For examples orders with large weight are under-represented with respect to the theoretical prediction. It is not clear to me if the effect is real (i.e. whether it is produced by the law that controls evolution) or it is an artifact (for example there could be a tendency among taxonomists to divide large genera).

It is not surprising to find that $N_G(k)$ obeys a power law. It has been discussed in the literature and has been verified repeatedly, with systematic deviations corresponding to under-representation of large genera (Yule 1924; Williams 1964; Engen 1978). It is surprising however that similar results hold not only for genera, but also for families and orders. The value of the exponent y being quite different in the last two cases.

It would be interesting to find out if these conclusions are valld also for the classification of other kinds of organisms.

5. 9. ON EVOLUTION

Evolution will be described more extensively by Schuster (chapter 6). Here we only present some preliminary remarks. We will consider only the evolution of single species. Under this restriction it may be reasonable to assume that evolution is a random process which is biased toward the maximization of fitness (Smith 1970; Gillespie 1984). This is a very strong assumption, because it neglects the interaction and competition among species, but it could be a starting point.

We could boldly construct a model of natural evolution in which fitness is a rather complex function of the genome and try to use the considerations of the previous sections to study its consequences. Before attempting any computation a few considerations are in order which outline the difficulties of such an enterprise.

We have seen that it is crucial to distinguish between an equilibrium and a non equilibrium situation. If we compare nature with the system described in the previous sections, we will observe that natural evolution is still going on. In other words the various species have not yet reached the optimal fitness; the time interval has been too short.

This observation implies that in our hypothetical model we are far from the equilibrium situation and the Boltzman distribution eq.(4) is not valid when we take for the Hamiltonian the fitness function. Therefore the analysis of sections 5.5 and 5.6 cannot be applied directly. Unfortunately the behavior of the system away from equilibrium is little understood and for the moment we have no general and fully developed theory to help us make sharp predictions in this region.

In modelling evolution it would be quite important to understand the frequency of mutations that decreases fitness, followed by a second mutation resulting in a net increase of fitness. If these double steps are too rare to be important in evolution only neutral or fitness increasing mutations must be considered and we have to study what could be called an adaptive walk on a rugged landscape. Although this is not a simple problem, it is within the reach of present day theoretical techniques and some progress is slowly being made (Amitrano et al. 1989; Kauffman 1990, Kauffman et al. 1988; Macken and Perelson 1989; Schuster 1988) (see also chapter 6 by Schuster).

It must be stressed that there is no need to consider only natural evolution at large. There are many simpler and smaller systems where evolution is important. The formulation of theoretical predictions (and their experimental tests) for such systems may play a crucial role in the construction of a more general theory of evolution. Here for lack of space I mention only the phenomenon of somatic hypermutation of B lymphocytes in the presence of external antigen, where a rate of 10^{-3} mutations per base pair per generation happens *in vivo* (Various authors 1987). In this very interesting case some theoretical analysis is already available (Kauffman et al. 1988; Macken and Perelson 1989).

It is also possible that more interesting results could be obtained if we study the evolution at the functional level rather than at the level of the genome. A striking phenomenon is convergent evolution, i.e. different species developing independently similar organs (e.g. wings in bats and birds). Convergent evolution of the genome is extremely unlikely on account of the large number of nearly neutral mutations. On the contrary it seems that the laws of physics imply that there are not so many different ways in which a vertebrate can fly and consequently the number of solutions is rather small. The fact that all flying species have developed wings independently from one another strongly hints that wings are mandatory for flying. Similarly we could argue that a visual system based on an eye (like the one of vertebrates and octopus) is necessary to have good resolution.

Seen this way convergent evolution reflects the small number of possible viable alternatives. The smallness of the phase space implies that the time needed to reach equilibrium is much shorter if we consider only the functional level. The following analogy may be helpful. If we study the similarities among computer programs for word processing written by different people they look completely different if we compare the machine code bit by bit. They look more similar if we consider only what is seen by the users of these programs. A classification based on the similarities of the machine code is nonsense (unless we want to discover if somebody has copied someone else's programs) and the only reasonable classification is the one based on the functions of the programs.

It may be possible that natural evolution is closer to equilibrium if we consider fitness as a function of the form (in a broad sense) of the organism and concentrate our attention in the functions of the organs, neglecting the molecular level. Unfortunately it is difficult to give a more precise definition of the form of an organism and it is not a simple task to put these last ideas in a clean mathematical form and to test them.

The last problem in which these concepts may be relevant is prebiotic evolution (Anderson 1983). In spite of the effort and of the ingenuity of the people involved, the practical absence of experimental data makes the study of this field very difficult.

5.10. CONCLUSIONS AND GENERAL OUTLOOK

In this paper I have tried to summarize the state of art in our theoretical understanding of the emergence of tree-like structures in complex systems and to outline possible applications to biology. Before ending I must mention that the findings presented in this chapter play a crucial role also in the study of neural networks, starting from the first applications of statistical mechanics (Hopfield 1982; Bienenstock et al. 1986; Mezard et al. 1987; Amit 1989).

Generally speaking a network is composed of many parts which mutually interact. The models described in section 5.3 may be also considered as examples of networks. In neural networks the neurons at a given time may be active or inactive. Active elements may have excitatory or inhibitory effects on other components. The time evolution is not constrained to be an optimization process and the equivalent of the Hamiltonian (or the objective function) does not exist. These systems display a behavior which is much richer than Hamiltonian systems and they are consequently harder to analyze.

A similar structure arises in many other networks quite different from neural nets. They range from the idiotypic network to antibodies (Parisi 1990) to the interaction of different metabolic components such as proteins in a cell or hormones in an organism (Kauffman 1969, 1990). Tree like structures may be very important for understanding the qualitative behavior of these systems (Mezard and Patarnello 1990).

Networks are likely to be the field in which the ideas presented here will find the most interesting applications and there exists a very large amount of work which lack of space prevents me from reviewing.

Acknowledgements

I am grateful to E. Marinari and to A. Sagnotti for useful suggestions and a critical reading of the manuscript.

REFERENCES

Amit, D.J. 1989. Modeling brain functions. Cambridge: Cambridge Univ. Press

Amitrano, C., L. Peliti and M. Saber. 1989. Population dynamics in a spin-glass model of chemical evolution. J. Mol. Evol. 29: 513-525.

Anderson, P.W. 1983. Suggested models for prebiotic evolution: The use of chaos. Proc. Natl. Acad. Sci. USA 80: 3386-3390.

Ansari, A., J. Berendzen, S.F. Bowne, H. Frauenfelder, I.E.T. Iben, T.B. Sauke, E. Shyamsunder and R.D. Young. 1985. Protein states and proteinquakes. Proc. Natl. Acad. Sci. USA 82: 5000-5004.

Balian, R., R. Maynard and G. Toulouse. 1979. Ill-condensed Matter. Amsterdam: North-Holland.

Bienenstock, E., F. Fogelmman and G. Weisbuch. 1986. Disordered Systems and Biological Organization. Berlin: Springer-Verlag.

Castellani, C., C. Di Castro and L. Peliti. 1981. Disordered Systems and Localization. Berlin: Springer-Verlag.

Derrida, B. 1985. A generalization of the random energy model which includes correlations between energies. J. Phys. (Paris) Lett. 46: 401-407.

Eigen, M. 1985. Macromolecular evolution: Dynamical ordering in sequence space. In D. Pines (ed.) "Emerging Syntheses in Science," pp. 25-69. Santa Fe, New Mexico: Santa Fe Institute.

Eigen, M., and P. Schuster. 1977. The hypercycle: A principle of natural self-organization. Naturwissenschaften 64: 541-565

Engen, S. 1978. Stochastic Abundance Models. London: Chapman and Hall.

Epstein, H. and D. Ruelle. 1989. Test of a probablistic model of evolutionary success. Phys. Rep. 184: 289-292.

Frauenfelder, H., P.J. Steinbach and R.D. Young. 1989. Conformational relaxation in proteins. Chemica Scripta 29A: 145-150.

Garel, T. and H. Orland. 1988. Mean-field model for protein folding. Europhys. Lett. 6:307-310.

Garey, M.R. and D.S. Johnson. 1969. Computers and Intractability: A Guide to the Theory of NP-completeness. New York: Freeman.

Gillespie, J.H. 1984. Molecular evolution over the mutational landscape. Evolution 38: 1116-1129.

Hopfield, J.J. 1982. Neural networks and physical systems with emergent collective computational abilities. Proc. Nat. Acad. Sci. USA 79:2554-2558.

Iben, I.E.T., D. Braunstein, W. Doster, H. Frauenfelder, M.K. Hong, J.B. Johnson, S.Luck, P. Ormos, A. Schulte, P.J. Steinbach, A.H. Xie and R.D. Young. 1989. Glassy Behavior of a protein. Phys. Rev. Lett. 62: 1916-1919.

Kauffman, S. 1969. Metabolic stability and epigenesis in randomly constructed genetic nets. J. Theor. Biol. 22: 437-467.

Kauffman, S. and S. Levin. 1987. Towards a general theory of adaptive walks on rugged landscapes. J. Theor. Biol. 128: 11-45.

Kauffman, S.A. 1990. Origins of Order. Self Organization and Selection in Evolution. Oxford: Oxford University Press.

Kauffman, S.A., E.D. Weiberger, and A.S. Perelson. 1988. In A.S. Perelson (ed.) "Theoretical immunology: Part One". Redwood City: Addison-Welsey.

Kirkpatrick, S., C.D. Gelatt and M.P. Vecchi. 1983. Optimization by simulating annealing. Science 220: 671-680.

Lawler, E.L., J.K. Lenstra, A.H.G. Rinnooy Kan, D.B. Shmoys (eds.). 1985. The Traveling Salesman Problem: A Guided Tour of Combinatorial Optimization. Chichester: Wiley.

Livi, R., S. Ruffo, S. Ciliberto and M. Buiatti. 1988. Chaos and Complexity. Singapore: World Scientific.

Macken, C.A., and A.P. Perelson. 1989. Protein evolution on rugged landscape. Proc. Nat. Acad. Sci. USA 86: 6191-6195.

Mezard, M., G. Parisi, and M.A. Virasoro. 1987. Spin Glass Theory and Beyond. Singapore: World Scientific.

Mezard, M. and Patarnello. 1990. in preparation.

Mezard, M., G. Parisi and N. Sourlas, G. Toulouse and M. Virasoro. 1985. Replica symmetry breaking and the nature of the spin glass phase. J. Phys. (Paris) 45: 843-854.

Parisi, G. 1984. An introduction to the statistical mechanics of amorphous systems. In J.B. Zuber and R. Stora (eds.) "Field Theories and Statistical Mechanics," pp. 473-523. Amsterdam: North-Holland.

Parisi, G. 1987. Facing complexity. Physica Scipta 35: 123-124.

Parisi, G. 1987. Spin glasses and optimization problems without replicas. In J. Souletie, J. Vannimeneus, and R. Stora (eds.) "Chance and Matter," pp. 525-552. Amsterdam; North-Holland.

Parisi, G. 1990. A simple model for the immune network. Proc. Nat. Acad. Sci. USA 87: 429-433.

Peliti, L. and A. Vulpiani. 1988. Measures of Complexity. Berlin: Springer.

Rammal, R., G. Toulouse and M.A. Virasoro. 1986. Ultrametricity for physicists. Rev. Mod. Phys. 58: 765-788

Ruelle, D. 1969. Statistical Mechanics: Rigorous Results. New York: Benjamin.

Ruelle, D. 1987. A mathematical reformulation of Derrida's REM and GREM. Commun. Math. Phys. 108: 225-239.

Shakhnovich, E.I. and A.M. Gutin. 1989. The nonergodic (spin-glass-like) phase of heteropolymer with quenched disordered sequence of links. Europhys. Lett. 8: 327-332.

Schuster, P. 1988. Potential functions and molecular evolution. In M. Markus, S.C. Mueller and G. Nicolis (eds.) "From Chemical to Biological Organization," pp. 149-165. Berlin: Springer-Verlag.

Smith, J.M. 1970. Natural selection and the concept of a protein space. Nature 225: 563-564.

Souletie, J., J. Vannimenus and R. Stora. 1988. Chance and Matter. Amsterdam: North-Holland.

Sompolinsky, H. and A. Zippelius. 1982. Relaxational dynamics of the Edward-Anderson model and the mean-field theory of spin glasses. Phys. Rev. B 25: 6860-6875.

Stein, D.L. 1985. A model of protein conformational substates. Proc. Nat. Acad. Sci. USA 82: 3670-3672.

Tutin, T.G., V.H. Heywood, N.A. Burgers, D.M. Moore, D.H. Valentine, S.M. Walters and D.A. Webb. 1980. Flora Europaea. Cambridge: Cambridge University Press.

Various Authors. 1987. Role of somatic mutation in the generation of lymphocyte diversity. Immun. Rev. 96: 5-162.

Vertecchi, D. and M.A. Virasoro. 1989. Energy bariers in SK spin-glass model. J. Phys (Paris) 50: 2325-2332.

Thom, R. 1975. Structural Stability and Morphogenesis. Reading, Mass.: Benjamin.

Williams, C.B. 1964. Patterns in the Balance of Nature and Related Problems in Quantitative Ecology. London: Academic Press.

Wilson, K. and J. Kogut. 1974. Renormalization group and the ε expansion. Phys. Rep. 12C: 75-200.

Yule, G.U. 1924. A mathematical theory of evolution based on the conclusions of Dr. J.C. Wills. Phil. Trans. B 213: 21-87.

J. Phys. A: Math. Gen. 17 (1984) 3521-3531. Printed in Great Britain

On the multifractal nature of fully developed turbulence and chaotic systems

Roberto Benzi†, Giovanni Paladin‡, Giorgio Parisi§ and Angelo Vulpiani‡

† Centro scientifico IBM di Roma, via Giorgione 129, 00100, Roma, Italy
‡ Università di Roma 'La Sapienza', Dipartimento di Fisica, and GNSM-CNR unità di Roma, P.le Aldo Moro 2, 00100, Roma, Italy
§ Università di Roma II, 'Tor Vergata', Dipartimento di Fisica, 00173, Roma and Laboratori Nazionali INFN, Frascati, Italy

Received 1 June 1984

Abstract. It is generally argued that the energy dissipation of three-dimensional turbulent flow is concentrated on a set with non-integer Hausdorff dimension. Recently, in order to explain experimental data, it has been proposed that this set does not possess a global dilatation invariance: it can be considered to be a multifractal set. In this paper we review the concept of multifractal sets in both turbulent flows and dynamical systems using a generalisation of the β-model.

1. Introduction

One of the most tested hypotheses in the theory of fully developed turbulence is that the small-scale statistics of turbulent flows obeys universal scaling properties. This is the celebrated Kolmogorov theory (1941, hereafter K41) which is still the only prediction made on the statistical properties of turbulence. The K41 theory is based upon the concept of self-similarity of the inertial range and upon the dependence of the probability distribution on the energy dissipation. Deviations of the K41 theory are commonly argued to be given by intermittency in the flow as first pointed out by Landau (Landau and Lifshitz 1971). Since Landau's remark there has been a great effort in generalising the K41 theory to include the intermittency correction, however no definite theoretical framework has been found to deal with the problem of intermittency. It has been argued that intermittency is due to the singularity of the Navier–Stokes equations in the small viscosity limit. It has been proposed by Mandelbrot (1974) that singularities are concentrated on a set $A \subset R^3$ with non-integer Hausdorff dimension. In § 2 we shall see how this is related to the scale invariance of the Navier–Stokes equations.

It is important to understand how the dynamical properties of nonlinear energy transfer among the various scales of motions determine the geometrical properties of the set A. Frisch *et al* (1978, hereafter FNS) have clarified this point introducing the well known β-model, also reviewed in § 2. The fundamental ingredients of these models is that there is a detailed balance of energy transfer in the inertial range. For detailed balance of energy transfer we mean that there is an exact balance in any shell k, $k+dk$ (k = wavenumber) between input/output energy. If we relax this constraint, we must then introduce a hierarchy of sets A_i on which singularities with different scaling properties are concentrated. The mechanism of energy transfer has to be constrained, not in detailed balance between local interaction in the k-space, but on the average. In other words we look at the history of the nonlinear interactions both in space and in time. It turns out that to this aim we need to introduce the concept of multifractal sets as defined in Frisch and Parisi (1983) and reviewed in § 2. Whether turbulence can be described by ordinary fractal sets or multifractal sets can be inferred only from experimental data at this stage. Within the present experimental results we show that there are some indications that multifractal sets are indeed necessary to describe the properties of structure functions in turbulent flows. In § 3 we propose a simple model, the random β-model, which generalises the results of FNS to take into account the multifractal structure of turbulence.

Multifractal sets, which originally were embedded in the weight curdling of Mandelbrot (1974), have been recently invoked by Paladin and Vulpiani (1984) to study the attractor sets in deterministic chaotic systems. It is the aim of this paper to review the concept of multifractal sets not only for what concerns the turbulence theory but also in connection with strange attractors of dynamical systems. This is done in § 4 together with some numerical analysis of simple attractors.

2. Turbulence and multifractal sets

Let us consider the Navier–Stokes equations:

$$\partial_t V + (V \cdot \nabla) V = -\nabla p/\rho + \nu \Delta V. \tag{2.1}$$

In the limit $\nu \to 0$ the equations (2.1) are formally invariant under the scaling transformations (see Frisch 1983):

$$r \to \lambda r, \qquad V \to \lambda^h V \qquad \lambda > 0, \qquad t \to \lambda^{1-h} t. \tag{2.2}$$

For finite ν we can still ensure invariance of equations (2.1) if

$$\nu \to \lambda^{h+1} \nu. \tag{2.3}$$

Note that the Reynolds number VL/ν is invariant under transformations (2.2) and (2.3). Assuming that small-scale turbulence is statistically invariant under the above scaling law, one can select h by using physical arguments. Kolmogorov (1941) proposed that the scaling laws of turbulence preserve energy transfer, assuming that nonlinear interactions are local in the k-space. This assumption implies that energy dissipation ε is invariant under the scaling laws (2.3). By definition $\varepsilon = \nu\langle(\nabla V)^2\rangle$ where $\langle\ldots\rangle$ denotes ensemble average. It follows that:

$$\varepsilon \to \lambda^{3h-1} \varepsilon. \tag{2.4}$$

The invariance of ε implies $h = 1/3$. The K41 theory has strong implications for the velocity gradients ∇V. Let us consider the quantity

$$\Delta V/\Delta x^{1/3} = [V(x) - V(y)]/(x-y)^{1/3}.$$

Scaling laws (2.3) with $h = 1/3$ imply

$$\lim_{x \to y} \Delta V/\Delta x^{1/3} \neq 0. \tag{2.5}$$

Thus the velocity gradient is singular. However the above considerations do not imply that the set of singular points of the Navier-Stokes equations are space-filling as originally assumed by Kolmogorov. Following Mandelbrot (1974), we can define the Hausdorff dimension D of the set of singular points (2.5). If $D<3$ then the probability for a point to be singular behaves as λ^{3-D}. It follows that energy dissipation is a fluctuating quantity in space. Its probability distribution determines the value of D. This is the physical basis of Landau's remark.

A clarifying picture of the mechanism underlying the above ideas is obtained following FNS. Let us consider the scales:

$$l_n = l_0 2^{-n}$$

where l_0 is the scale on which energy is injected. Nonlinear interactions produce on the average N eddies of scale l_{n+1} i.e. energy is transferred from scale l_n to the N eddies of scale l_{n+1}. These N eddies occupy a fraction β $(0 \leqslant \beta \leqslant 1)$ of the l_n-eddy volume. After n generations the volume occupied by 'active' eddies is β^n. If V_n is the velocity difference of an active l_n-eddy, then the energy per unit mass on scale l_n is

$$E_n \sim \beta^n V_n^2,$$

and the energy transfer is

$$\varepsilon_n \sim \beta^n V_n^3 / l_n.$$

Assuming that

$$\varepsilon_n \sim \varepsilon = \text{constant}, \tag{2.6}$$

we obtain $V_n \sim \varepsilon^{1/3} l_n^{1/3} P^{-1/3}$ where $P \sim (l_n/l_0)^{3-D}$ and $\beta = 2^{D-3}$. Benzi and Vulpiani (1980) have estimated $D \simeq 3-2/3$ in good agreement with known data on the correlation functions of ε. Either the β-model and the scaling law (2.3) with $h=1/3$ predicts a linear behaviour of the coefficients

$$\langle \Delta V^p \rangle \sim r^{\zeta_p}, \qquad \zeta_p = hp + 3 - D_F, \qquad h = (D_F - 2)/3.$$

In figure 1 we report data on ζ_p for various experimental tests (Anselmet *et al* 1983). Although a linear fit is not inconsistent with the experimental accuracy, there is a

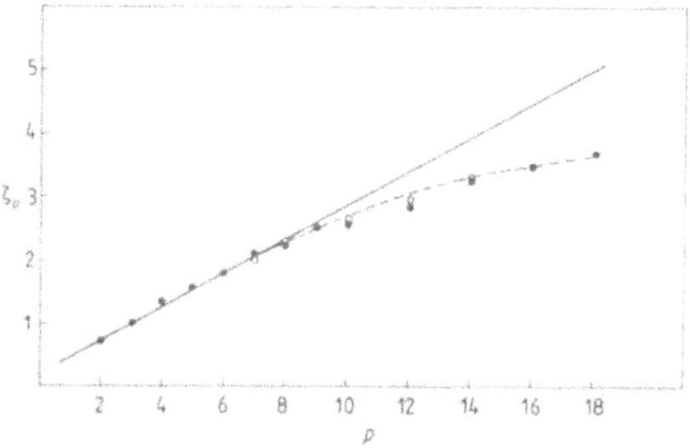

Figure 1. ζ_p against p. Dots and circles represent experimental data by Anselmet *et al* (1983). Full line is the β-model of FNS with $D_F = 2.83$. The broken line refers to equations (3.8) and (3.9) with $x = 0.125$.

tendency for ζ_p to behave in a nonlinear way. If this is assumed to be the case, it follows that neither equation (2.4) nor equation (2.6) is valid. In other words the rate of energy transfer is not constant among the various scales of motion but fluctuates both in space and in time. This idea has been originally proposed by Mandelbrot (1974). A simple way to include fluctuations in the energy dissipation within the scaling laws (2.2), (2.3) has been recently proposed by Frisch and Parisi (1983, hereafter FP). Let us define $S(h)$ the set of points for which

$$\lim_{x \to y} [V(x) - V(y)]/(x-y)^h \neq 0$$

and let us denote by $d(h)$ the Hausdorff dimension of $S(h)$. The Kolmogorov theory simply implies that $d(h) = 3\theta(h-1/3)$ while the β-model gives $d(h) = D\theta[h-(D-2)/3]$. Existence of singularities with arbitrary exponents h is consistent with the Navier–Stokes equations in the limit $\nu \to 0$. Thus the possibility that $d(h)$ is not a step function of h could be embedded into the scaling law of the velocity field. Generalising the considerations done at the beginning of this section, we can assume that the probability for a point to belong to $S(h)$ is proportional to $\lambda^{[3-d(h)]}$. It follows that

$$\langle \Delta V^p(r) \rangle \propto \int d\mu(h) r^{[ph+3-d(h)]} \tag{2.7}$$

where $\mu(h)$ is a measure concentrated on the region where $d(h) > 0$. The RHS of equation (2.7) can be estimated using a saddle-point technique. We obtain

$$\zeta_p = \min_h [ph + 3 - d(h)]. \tag{2.8}$$

For a proper choice of $d(h)$, equation (2.8) might fit the experimental data of figure 1. Physically, equation (2.8) implies that for a given value of p, ζ_p is dependent on a particular value of h, i.e. a particular value of $S(h)$. Hence the kind of instabilities needed to set up the sets $S(h)$ are picked out by the moments of the velocity differences. We mean, for example, instabilities which lead to vortex sheets or vortex tubes (for a review see Monin and Yaglom (1975)). Figure 1 can then be interpreted as the evidence of different mechanisms acting on the flow to select the probability distribution of energy transfer and dissipation. It is therefore clear that the β-model of FNS is not able to take into account the different nature of energy transfer. In the next section we discuss how to improve the β-model to deal with different sets of singularities. The idea that different moments of strongly fluctuating distribution can be dominated by different singularities has been explored, in a different context, by Berry (1977 and 1982).

3. A random β-model

Let us assume that, in the spirit of Mandelbrot's weight curdling, the contraction factors β are independent random variables, which can take different values in each eddy i at the step n of the energy cascade. The $\beta_n(i)$'s are distributed according to a given probability distribution $P(\beta)$. In this way the geometrical structure of intermittency does not possess a global dilatation invariance. The number of active eddies which are generated in a step is not fixed by a parameter as for ordinary 'homogeneous' fractals. The rules which generate multifractal inhomogeneous sets are drawn at random at each step in length scale. Figure 2 shows two different naive pictures of intermittency according to either the deterministic or the random β-model. It is interesting to compute the fractal dimension of the multifractal set which is defined by

$$\langle N_n \rangle \sim l_n^{-D_F} \tag{3.1}$$

where N_n is the number of active eddies at the nth step and $\langle\ \rangle$ indicates a space average, see Mandelbrot (1982, p 211 where D_F is called the similarity dimension). We shall also use the average $\{\ \}$ on the distribution $P(\beta)$

$$\{f\} \equiv \int d\beta\, P(\beta) f(\beta).$$

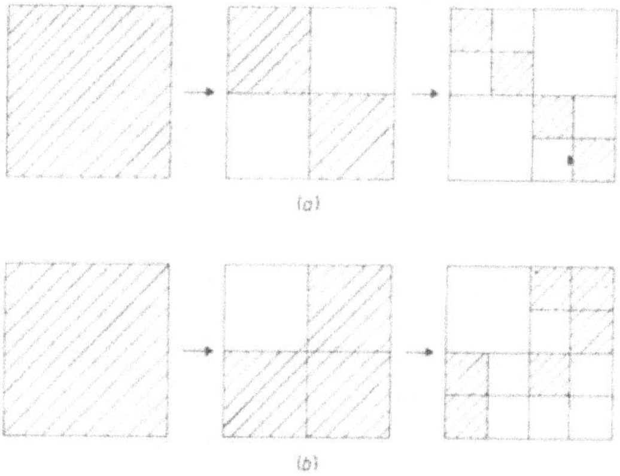

Figure 2. (*a*) Schematic view of the β-model and (*b*) compared with the random β-model. The shaded areas are the zones active during the fragmentation process.

An eddy of size $l_{j+1} = l_j/2$ covers a fraction $B_j = \Sigma_i \beta_j(i)/N_j$ of the volume occupied by its hypercube 'father'. It follows that the number of active eddies after n steps is given by

$$N_n = 2^{3n} \prod_{j=1,n} B_j. \tag{3.2}$$

We can easily average (3.2), noting that the β's are independent random variables:

$$\langle N_n \rangle = 2^{3n} \left\langle \prod_{j\ 1,n} B_j \right\rangle \sim l_n^{-(3+\log_2\{\beta\})}.$$

We have transformed the space average in β-average. By definition (3.1) we can obtain

$$D_F = 3 + \log_2\{\beta\}. \tag{3.4}$$

D_F cannot completely characterise an 'inhomogeneous' fractal because it does not give complete information on the probability distribution of β. The main point of this section is to show that this information is provided by the exponents ζ_p.

Let us denote by $l_n(k)$, $k = 1, \ldots, N_n$, the N_n active eddies at the nth step. Each $l_n(k)$ generates eddies of size $l_{n+1}(k)$ where k indicates its origin. The rate of energy transfer is constant among $l_n(k)$ and $l_{n-1}(k)$:

$$V_n^3(k)/l_n(k) = \beta_{n+1}(k) V_{n+1}^3(k)/l_{n-1}(k). \tag{3.5}$$

This relation implies that the velocity difference $\delta V(l_n) = V_n$ in an eddy generated by a particular set of fragmentations $\beta_1, \beta_2, \ldots, \beta_n$, is

$$V_n \sim l_n^{+1/3} \left(\prod_{i=1,n} \beta_i \right)^{-1/3}. \tag{3.6}$$

The structure functions are then

$$\langle |\delta V(l_n)|^p \rangle = \int \prod_{i=1,n} \mathrm{d}\beta_i P(\beta_1, \ldots, \beta_n) \beta_i | V_n|^p. \tag{3.7}$$

Because we assumed that there are no correlations among different steps of the fragmentation process, it follows that

$$\prod_{i=1,n} \mathrm{d}\beta_i P(\beta_1, \ldots, \beta_n) = \prod_{i=1,n} P(\beta_i)\, \mathrm{d}\beta_i.$$

From (3.7) we can compute the exponents ζ_p:

$$\zeta_p = p/3 - \log_2 \{\beta^{(1-p/3)}\}. \tag{3.8}$$

The probability distribution $P(\beta)$ is known in principle from the knowledge of all the β moments, i.e. of all the exponents ζ_p. Figure 1 shows that the simple form

$$P(\beta) = x\delta(\beta - 0.5) + (1 - x)\delta(\beta - 1) \tag{3.9}$$

leads (with $x = 0.125$) to a good fit to the available experimental data, x being the only free parameter. There is no good reason to choose a two-step probability distribution for β, of course. We have assumed that an active eddy can generate either velocity sheets ($\beta = 0.5$) or space-filling Kolmogorov-like eddies ($\beta = 1$) (see Saffman 1968). We see, by comparing relation (3.8) and (3.4), that in our model the fractal dimension is

$$D_F = 3 - \zeta_0, \tag{3.10}$$

nevertheless D_F is often computed by the energy dissipation correlation

$$\langle \varepsilon(x + r)\varepsilon(x) \rangle \sim r^{-\mu}. \tag{3.11}$$

$D^* \equiv 3 - \mu$ is considered equal to D_F. FNS have shown that under general conditions

$$D^* = 1 + \zeta_6. \tag{3.12}$$

This relation is still valid in our model, with the further inequality

$$D_F = 3 + \log_2 \{\beta\} \geq D^* = 3 - \log_2 \{\beta^{-1}\}. \tag{3.13}$$

The equality $D_F = D^*$ holds only in the deterministic β-model. From the fit of the data of the structure functions given in figure 1, we have obtained

$$D_F = 2.91, \qquad D^* = 2.83. \tag{3.14}$$

This result is an indirect check of the multifractal nature of fully developed turbulence.

4. Fractal structure of strange attractors

The scenario of the random β-model is quite general and can be extended to the analysis of dynamical systems. Indeed the chaotic motions often lie on complicated manifolds of the phase space, called strange attractors, which can have an intricate multifractal structure. The fractal dimension D_F cannot fully characterise an attractor, and further the FNS β-model does not describe the intermittency in a satisfactory way. We shall therefore introduce a set of easily computable exponents generalising the fractal dimension which are defined in terms of a 'local density' $n(r)$.

Let us consider a time series of points $X_i \equiv X(i\Delta t)$, $(i = 1, 2, \ldots N)$ of the dynamical system

$$dX/dt = f(X), \qquad \text{where } X \text{ belongs to } R^d. \tag{4.1}$$

The fraction of points which are contained in a hypersphere of radius r and centre X_i is

$$n_i(r) = \sum_{j \neq i} \theta(r - |X_i - X_j|)/(N-1). \tag{4.2}$$

The moments of such a local density are (via a space average)

$$\langle n(r)^q \rangle = \lim_{N \to \infty} \sum_{i=1,N} n_i(r)^q / N. \tag{4.3}$$

We define a new set of exponents $\phi(q)$ by the relation:

$$\lim_{r \to 0} \langle n(r)^q \rangle \sim r^{\phi(q)}. \tag{4.4}$$

In a homogeneous fractal, (see figure 2(a)), $n(\lambda r)$ has the same statistical properties as $n(r)\lambda^{D_F}$ and this implies:

$$\phi(q) = D_F q. \tag{4.5}$$

On the contrary, attractors do not possess a global dilatation invariance and it is only possible to show that $\phi(q)$ is convex in q by a general theorem of probability (Feller 1971). It is worth pointing out that $\phi(1)$ is the exponent ν proposed by Grassberger and Procaccia (1983) to estimate the fractal dimension of attractors. $\phi(q)$ is plotted versus q for the Lorenz model and the Hénon map in figure 3. One sees that $\phi(q)$ is nearly linear at $|q| \leq 1$ but deviations from the line $D_F q$ appear at larger values of $|q|$. It is evident that the $\phi(q)$ are analogous to the exponents ζ_p for the velocity fluctuations in a turbulent flow: the ζ_p are linear in p in the FNS scheme where the energy dissipation is concentrated in a homogeneous fractal set. Grassberger (1983) has recently introduced some exponents essentially equivalent to ours. We note that the $\phi(q)$ as defined in Paladin and Vulpiani (1984) are easier to compute than the Grassberger ones. Indeed we do not need to use the box-counting method which is not easily handled for topological dimension $d > 3$. We shall see that our approach shows a connection between the structure of the attractor and the dynamics of the system.

We have to relate the exponent ϕ to the dynamical properties of the system (4.1) by an adaptation of the model proposed in § 3. We shall assume therefore that the same statistics are obtained by considering the positions in the phase space, at large times, of N points which are uniformly distributed in a hypercube of size l_0 at the initial time, instead of the N position at times $i\Delta t$ of the evolution of one point. This

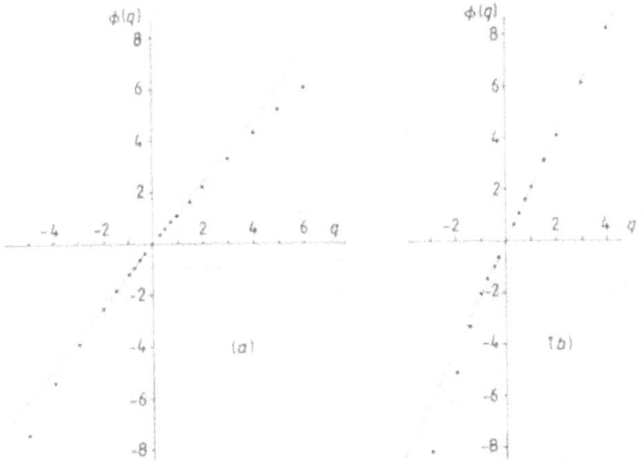

Figure 3. Variation of ϕ against q. Broken lines correspond to the line $\phi = qD_F$. Dots are $\phi(q)$ computed with $N = 10^4$ for: (a) Hénon map ($a = 1.2$, $b = 0.3$) (Hénon 1976); and (b) Lorenz model ($r = 28$) (Lorenz 1963).

possible because of the mixing property which is satisfied for chaotic systems. Let us describe the dynamics by a fragmentation mechanism which allows us to apply the formalism of the preceding section. At time Δt_1 after the initial time, our N points in the phase space of the system (4.1) will be distributed in $N_1 = a^d\beta_1$ hypercubes of size $l_1 = l_0/a$ ($a > 1$) where β_1 is a parameter given by the dynamics. Each new hypercube generates $N_2 = a^d\beta_2$ hypercubes of size $l_2 = l_1/a$ after a time Δt_2. The β_2's depend, as before, on their particular hypercube 'father'. We can iterate this process being careful to choose the breaking times Δt_i so that $l_{i+1} = l_i/a$. In this way the initial hypercube of size l_0 is reduced to a large number of hypercubes of size $l_n = a^{-n}l_0$ after n steps. Each hypercube has its own history determined by a succession $\beta_1(\alpha), \ldots, \beta_n(\alpha)$ where α indicates the history which is considered. We have now to impose conservation of the number of points at each step. We have

$$l_{j+1}^d\rho_{j+1}(\alpha) = [(l_{j+1}/l_j)^d\beta_j(\alpha)]l_j^d\rho_j(\alpha) \tag{4.6}$$

where $\rho_j(\alpha)$ is the point density of a hypercube of size l_j obtained by a fragmentation history α. It follows from (4.6) that

$$\rho_j(\alpha) \sim \left[\prod_{i=1}^{j} \beta_i(\alpha)\right]^{-1}. \tag{4.7}$$

We can estimate (4.4) putting $r = l_n = l_0a^{-n}$:

$$\langle n(l_n)^q\rangle \sim \langle (l_n^d\rho_n)^q\rangle = l_n^{dq}\langle\rho_n^q\rangle \sim a^{-ndq}\left\langle\left[\prod_{i=1,n} \beta_i(\alpha)\right]^{-q}\right\rangle, \tag{4.8}$$

where the β_i's can be assumed to be independent random variables. We can therefore perform the average (4.8) as an average on the probability distribution for the β of a single fragmentation:

$$\langle n(r)^q\rangle \sim a^{-ndq}\langle\beta^{-q}\rangle^n \sim a^{-n[dq - \log_a\langle\beta^{-q}\rangle]}. \tag{4.9}$$

The exponents ϕ are thus

$$\phi(q) = dq - \log_a\{\beta^{-q}\}. \tag{4.10}$$

Relation (4.10) is quite analogous to the relation (3.8) in § 3. One has in the case of a homogeneous fractal:

$$\beta_i(\alpha) = a^{(D_F - d)}, \qquad \text{for any } \alpha \text{ and } i \tag{4.11}$$

and (4.5) trivially follows from (4.10). We use now the definition (3.1) to compute the fractal dimension of an attractor in our model. The number $N(r)$ of hypercubes of size $r(=l_n)$ necessary to cover an attractor, in the limit $r \to 0 (n \to \infty)$, is

$$N(l_n) \sim a^{nd} \prod_{i=1,n} B_i, \tag{4.12}$$

where B_i has been defined in § 3 as

$$B_i = \sum_{i=1,N_i} \beta_j(i)/N_j.$$

Averaging (4.12) one obtains

$$\langle N(l_n)\rangle \sim a^{n[d+\log_a\{\beta\}]}. \tag{4.13}$$

The relation

$$D_F = -\phi(-1) \tag{4.14}$$

follows from the definition of the ϕ's in (4.10), the estimate (4.14) is in good agreement with the results obtained by the box-counting methods (Russell *et al* 1980):

Hénon($a = 1.2, b = 0.3$)	$-\phi(-1) = 1.20 \pm 0.01$,
Lorenz($r = 28$)	$-\phi(-1) = 2.06 \pm 0.01$.

The inequality $\{\beta\}^{-1} \leq \{\beta^{-1}\}$ implies $\nu \leq D_F$. ν gives therefore a higher weight to the lower density regions of an attractor. $\{\beta^{-1}\}$ can however be a more interesting parameter than $\{\beta\}$ as the average value and the most probable one can differ. It is amusing to note that the estimate of D_F by ν corresponds to the estimate of D_F by D^* in fully developed turbulence. Both $\phi(1)$ and ζ_6 are determined by $\{\beta^{-1}\}$ while D_F is related to $\{\beta\}$.

We want finally to discuss, in terms of our model, Mori's (1980) estimate of the fractal dimension from the Lyapunov exponents (say $\lambda_1 \geq \lambda_2 \ldots \geq \lambda_n$) of a dynamical system:

$$D_F = d_+ + d_0 + d_-\left[\sum_{i=1,d_+} \lambda_i \Big/ \left|\sum_{j=1+d_0+d_+,d} \lambda_j\right|\right], \tag{4.15}$$

where d_+, d_0 and d_- are the number of λ_i which are respectively greater than, equal to or less than zero.

Let us define A_{ij}, the matrix that describes the linearised evolution of the system around the time t_n:

$$A_{ij} = \partial f_i/\partial x_j|_{t=t_n},$$

where $t_n = \Sigma_{i=1\ldots n} \Delta t_i$. Let $K_n(i)$ be the eigenvalues of A_{ij}. It is possible to write B_n

as a function of $K_n(i)$:

$$B_n = \sum_{j=1,N(l_{n-1})} \exp\left(\sum_{i=1,d} K_n^j(i)\Delta t^j_n \right) \Big/ N(l_{n-1}) \tag{4.17}$$

$K_n^j(i)$ is the ith eigenvalue of the matrix A_{ij} computed in the centre of the jth hypercube ($i = 1, d$; $j = 1, N(l_n - 1)$), Δt^j_n is the time needed to reduce the size of the jth hypercube from l_{n-1} to l_n i.e. the time needed to have the fragmentation. Δt^j_n is determined by the eigenvalues $K^j_n(i) < 0$:

$$\Delta t^j_n = \ln a / |\kappa_n(j)| \tag{4.18}$$

where $\kappa_n(j)$ is the local contraction rate at the time t_n around the centre of the jth hypercube.

An estimate of $\kappa_n(j)$ is not trivial. Mori assumed for $\kappa_n(j)$ an average over the negative eigenvalues of the matrix A_{ij}:

$$\kappa_n(j) = \sum_{i=1+d_0+d_+,d} |K^j_n(i)|/d_- . \tag{4.19}$$

By equations (4.12), (4.17) and (4.18) we have

$$N(l_n)/N(l_{n-1}) = \sum_{j=1,N(l_{n-1})} (l_{n-1}/l_n)^{[d+\sum_{i=1,d}(K^j_n(i)/\kappa_n(j))]} / N(l_{n-1}). \tag{4.20}$$

As $\langle N(l_n)\rangle \sim l_n^{-D_F}$ the fractal dimension D_F becomes

$$D_F = d + \ln_a \langle a^{[\sum_{i=1,d} K^j_n(i)/\kappa_n(j)]} \rangle. \tag{4.21}$$

By (4.21) it is possible to see that D_F is not related to any quantity easily estimated from the Lyapunov numbers. Mori's estimate (4.16) follows from (4.21) by assuming only (4.19) and no fluctuations of the eigenvalues of the matrix A_{ij}.

Acknowledgment

We thank U Frisch for fruitful discussions.

References

Anselmet F, Gagne Y, Hopfinger E J and Antonia R A 1983 *Preprint, Institute de Mecanique de Grenoble*

Benzi R and Vulpiani A 1980 *J. Phys. A: Math. Gen.* **13** 3319

Berry M V 1977 *J. Phys. A: Math. Gen.* **10** 2061

—— 1982 *J. Phys. A: Math. Gen.* **15** 2735

Feller W 1971 *An Introduction to Probability Theory and its Applications* vol 2 (New York: Wiley)

Frisch U 1983 *Turbulence and Predictability of Geophysical Flows and Climate Dynamics, Varenna Summer School* LXXXVIII

Frisch U and Parisi G 1983 *Turbulence and Predictability of Geophysical Flows and Climate Dynamics, Varenna Summer School* LXXXVIII

Frisch U, Sulem P and Nelkin M 1978 *J. Fluid Mech.* **87** 719

Grassberger P 1983 *Phys. Lett.* **97A** 227

Grassberger P and Procaccia I 1983 *Phys. Rev. Lett.* **50** 347

Hénon M 1976 *Commun. Math. Phys.* **53** 69

Kolmogorov A N 1941 *C.R. Acad. Sci., USSR* **30** 301

Landau L D and Lifshitz E M 1971 *Mecanique des Fluides* (Moscow Mir ed)
Lorenz E N 1963 *J. Atmos. Sci.* **20** 130
Mandelbrot B 1974 *J. Fluid Mech.* **62** 331
—— 1982 *The Fractal Geometry of Nature* (San Francisco: Freeman)
Monin A S and Yaglom A M 1975 *Statistical Fluid Mechanics* vol 2 (Cambridge, Mass: MIT Press)
Mori H 1980 *Prog. Theor. Phys.* **63** 1044
Paladin G and Vulpiani A 1984 *Lett. Nuovo Cimento* **41** 82
Russell D A, Hanson J D and Ott E 1980 *Phys. Rev. Lett.* **45** 1175
Saffman P 1968 *Topics in Non Linear Physics* ed N Zabusky (Berlin: Springer) p 485

Part III. Simulations

RECENT PROGRESSES IN GAUGE THEORIES

G. Parisi
INFN, Laboratori Nazionali di Frascati, 00044 Frascati, Italy

1. - INTRODUCTION

In the last years we have seen many developments in our understanding of gauge theories, expecially toward the construction of new tools for doing reliable computations in the non-perturbative region. The motivations are clear: we believe that quantum cromodynamics (QCD) is the true fundamental theory for strong interactions; this interaction is characterized by an effective running coupling constant $\alpha(q^2)$ which goes to zero when $q^2 \to \infty$, i.e. at short distances (this property of the coupling constant is called asymptotic freedom)(1). More precisely, if the number of quark flavours is four, we find:

$$\alpha(q^2) \to \frac{12}{25 \ln (q^2/\Lambda^2)} \qquad q^2 \to \infty \ . \tag{1.1}$$

The parameter Λ^2 is experimentally measured (e.g. in deep inelastic scattering) and it should be in the range 0.2-0.5 GeV^2. QCD is a compleate theory of strong interactions: using Λ^2 and the quark masses as input, we should be able to compute all the physical quantities, in particular the mass spectrum of hadrons. However perturbation theory can be used only to compute hard processes (the coupling constant being small) and non-perturbative techniques are badly needed in the soft region.

Many interesting results have been obtained; for lack of time it is impossible to mention all of them: in this talk I will speak only those ideas which are more familiar to me.

For the time being most of the theoretical effort has been concentrated on the study of pure gauge theories without fermions, where only double coloured gluons interact; fermion should be included pertubatively at a later stage. In such a simplyfied theory Λ is the only parameter: we want to comput the static potential between quarks (V(r)) and the glueball spectrum. Other quantities, like the energy dependance of the total cross section for glueball scattering, are more difficult to obtain(5). In this talk I will try to give you a rough idea of how to carry on these non-perturbative calculations. In Section 2, I present the formalism of Euclidean quantum field theory, which is essential to master most of the new developments. In Section 3 I discuss some qualitative ideas on confinement. Most of the efforts to obtain quantitative results can be divided into two categories: "brute force" computations and analytic computation. Brute force computations are usually very long and the results can be obtained only after spending a lot of human or computer time. They are mainly based on lattice gauge theories (Section 4) and can be divided into two large groups: computer simulations (Section 5) and high temperature expansions (Section 6). Some of the difficulties to deal with the high temperature expansion are connected with the possible existence of an elusive roughening transition (Section 7).

The most interesting analytic approach is based on the idea of writing equations for W(C), the vacuum expectation value of the Wilson loop:

$$\overline{W}(C) = \langle W(C) \rangle \; ; \qquad W(C) = \mathrm{Tr}\left\{P\left[\exp\left(i\oint_C A_\mu(x)\, dx_\mu\right)\right]\right\}. \qquad (1.2)$$

One obtains equations which are rather difficult to be solved, however it is known that rather impressive simplifications are present in the limit in which the number of colours become infinite. This fact enable us to write simple closed equations for an SU(N) in the limit $N \to \infty$ (Section 8). Although these equations seem formidable there is some hope that they can be solved, at least approximatively.

2. - EUCLIDEAN FIELD THEORY

After the first works by Schwinger[6], it has strongly enphasized by Symanzik[7] that there are deep similarities between quantum field theory and classical statistical mechanics: indeed a field theory defined on a D-dimensional Minkowski space (D-1 space directions, one time direction) is connected to the corresponding field theory on a D dimensional Euclidean space by an analytic continuation (Wick rotation): the Minkowski metric $x^2 = \sum_1^3{}_i x_i^2 - x_o^2$ become the Euclidean metric $x^2 = \sum_1^4{}_i x_i^2$ if we set $x_4 = i x_o$.

The equivalence of the two theories has very deep consequences: if we quantize an Euclidean field theory using the Feynman path integral formulation, we obtain a special kind of classical statistical mechanics. Now classical statistical mechanics is a much older discipline than relativistic quantum field theory and we have a much better physical intuition of it: statistical mechanics deals with probabilities, not with amplitudes as quantum mechanics and the variety of statistical systems is large also in everyday life.

Only the beginning of the seventies Symanzik ideas became popular and they started to be applied in rather different fields: after the key works of Nelson[8] and of Osterwalder and Schrader[9] and the beautiful results of Guerra, Rosen and Simon[10] Euclidean field theory became an essential tool in the rigorous approach to the construction of an interacting quantum field theory[11].

In a different contest it was enphasized by Migdal and Poliakov[13] that the problem of computing the critical exponents for second order phase transitions is connected to the control of infrared divergencies in a theory with masless particles. Somewhat later in a beautiful series of paper Wilson[14] succeded to compute the critical exponents: his approach was a combination of the block spin picture of Kadanoff[14] and the renormalization group which was used by Gellman and Low[15] to study the high energy limit of QED[16].

In these last years we have seen a very fruitful cross-fertilization[18] of statistical mechanics and quantum field theory[19]; from a conceptual point of view quantum field theory has started to be adsorbed in the general framework of classical statistical mechanics: this process is arrived to such a stage that it could be said (although it is not quite true) that quantum field theory is an high specialized branch of statistical mechanics.

To help the reader to orient himself in these recent developments I have inserted a table showing the relations between the main quantities in quantum field theory and the corresponding quantities in statistical mechanics.

In Table I the bracket indicates as usually the statistical expectation value, e. g.

$$\langle\varphi(x)\varphi(0)\rangle = \int d[\varphi]\,\varphi(x)\varphi(0)\exp(-\beta H)/\int d[\varphi]\exp(-\beta H)\,. \qquad (2.1)$$

Let us see an example in details: we suppose that the field creates from the vacuum an infinite number of particles of mass m_n. In momentum space we can write:

$$G(p) = \Sigma_n\, C_n/(p^2+m_n^2)\,. \qquad (2.2)$$

In position space we obtain (to avoid Bessel functions let us consider the case D=3):

$$C(x) = \langle\varphi(x)\varphi(0)\rangle = \Sigma_n\, C_n \exp(-m_n|x|)/|x|\,. \qquad (2.3)$$

If we know the function C(x) analytically, it is trivial to compute all the m_n, however if C(x) is only known numerically with some er-

TABLE I

Quantum field theory	Classical statistical mechanics
Minkowski space	Euclidean space
$\hbar$	$\beta^{-1} = kT$
$\mathscr{L}$	H
Feynman factor for amplitudes : $\exp(-i\mathscr{L}/h)$	Boltzman factor for probabilities : $\exp(-\beta H)$
Sum of all vacuum to vacuum diagrams : $\int d[\varphi]\exp(-i\mathscr{L}/h)$	Partition function : $\int d[\varphi]\exp(-\beta H)$
Vacuum energy	Free energy
Vacuum expectation value	Statistical expectation value
$\langle 0 \| A \| 0\rangle$	$\langle A\rangle$
Quantum fluctuations	Statistical fluctuations
Time ordered products	Simple products
Existence of a mass gap	Exponential decrease of correlations
Mass	(Correlation length)$^{-1}$
Green functions :	Correlation functions :
$\langle 0 \| T[\varphi(x)\varphi(0)] \| 0\rangle \sim$ $\sim \exp i m \|t\|$	$\langle\varphi(x)\varphi(0)\rangle \sim \exp - m\|x\|$
Changement of vacuum	Phase transition
Goldstone bosons	Spin waves
Decrease to zero the mass	Approach a second order phase transition
Free scalar bosons	Random walk (Free curves)
Scalar bosons with repulsive interaction	Self avoiding walk (Interacting curves)
Gauge theories	Interacting surfaces ?
Wightman axioms	Osterwalder and Schrader axioms
Cutoff	(Lattice spacing)$^{-1}$
Hamiltonian	Transfer matrix
Instantons	Defects (vertices, dislocations)

rors, it may be not so simple to extract m_1 and the numerical evaluation of m_2 may easyly present serious difficulties ; however this is a pratical problem in numerical computations and it does not involve questions of principles.

The Euclidean formulation of gauge theories does not present any special difficulty. In the SU(N) case (N = 3 is the physical one) the gluon field is a doubly coloured vector:

$$A_\mu^{a,b} , \qquad\qquad a, b = 1, N$$

and b are the colour indeces.

The lagrangian which describes a pure gluonic wordl is:

$$\mathscr{L} = \frac{1}{g^2} \int d^D x \, F_{\mu\nu}^2 \tag{2.4}$$

where g is the chromatic charge ($\alpha = g^2/4\pi$) and $F_{\mu\nu} = \partial_\mu A_\nu - \partial_\nu A_\mu + [A_\mu, A_\nu]$(21). The presence of the commutator in the definition of $F_{\mu\nu}$ is the origine of the interaction among gluons. The factor $1/g^2$ seems to be unusual: using the same convenction in electromagnetism, one would find:

$$\text{div } E/e^2 = \varrho . \tag{2.5}$$

ϱ being the electron density. In other words we have set the electron charge equal to one and we have rescaled the electric field ; the conventional electron field E_c can be easily obtained:

$$E_c = E/e . \tag{2.6}$$

In order to define the theory in 4 dimension we must renormalize the coupling constant to avoid ultraviolet divergences ; one finally obtains an effective running coupling constant which in the large q^2 region behaves as:

$$\alpha(q^2) \sim 12\pi/(11 N \ln q^2/\Lambda^2) . \tag{2.7}$$

The only parameter of the theory is Λ; dimensionless quantities are therefore fixed and cannot be changed by changing the coupling constant.

Another important feature of the theorv is gauge invariance: in the electromagnetic case only the fileds E and H and not the potentials A are well defined; the value of the potential in one point is arbitrary: it can be changed by a gauge transformation:

$$A_\mu^{(x)} \rightarrow A_\mu^{(x)} + \partial_\mu \lambda(x) \quad . \tag{2.8}$$

Although in perturbation theory one usually remove this ambiguity by fixing the gauge(22), only gauge invariant quantities have a clear physical meaning.

All these features of the theory in the usual Minkowski formulation are also true in the Euclidean version of the theory.

As we said in the introduction our aim is to compute the glueball m_G mass and the static potential between quarks. It is easy to prove that in Euclidean space:

$$m_G = - \lim_{r \to \infty} \frac{1}{r} \ln \langle F^2(r)\, F^2(0) \rangle \; ,$$

$$V(L) = - \lim_{T \to \infty} \frac{1}{T} \ln \langle W(C) \rangle \; , \qquad C = T \times L \tag{2.9}$$

where $C = T \times L$ indicates that the circuit C is a rectangle of sizes L and T.

As we see it is rather clear how to extract the most physically interesting information from the Euclidean version of QCD.

3. - CONFINEMENT

It is clear that the Wilson loop $W(C)$ plays a very important role in the study of gauge theories: it is the most natural gauge invariant observable. Indeed in the abelian case $\Phi(C) = i \ln W(C)$ (i.e. $W(C) = \exp - i\Phi(C)$) is the flux concatenated with the circuit C.

The quantity $A(C) = \ln \langle W(C) \rangle$ (24) is the contribution of the chromodynamic field to the action of a coloured particle having C as trajectory. $A(C)$ is well defined in perturbation theory(25) apart from a linear divergence proportional to the length of the circuit C, which corresponds to the classical self energy of the electron.

At the first order in perturbation theory (or in the abelian case), i.e. neglecting gluons self couplings, $A(C)$ is the standard electromagnetic self induction of the circuit C:

$$A(C) = e^2 \oint_C dx_\mu \oint_C dy_\mu \; 1/(x-y)^2 \; . \tag{3.1}$$

Perturbation theory will obviously break down at large distances. If the potential $V(L)$ increase at infinity like σL (confinement), we find that:

$$\overline{W}(C) = \langle W(C) \rangle \sim \exp -\sigma L \cdot T = \exp -\sigma S . \qquad (3.2)$$

The decrease of the expectation value of the Wilson loop like the surface may be considered as a criterion for confinement. In such a situation we expect the formation of a physical string between the two quarks where the energy is concentrated. In the time evolution the string will describe a surface: in the Euclidean space we expect that there will be a region of space on which the increase in action will be concentrated. We can introduce the parameter a(x, C) defined by(26-28):

$$a(x,C) = \langle F^2(x)\, W(C) \rangle / \langle W(C) \rangle . \qquad (3.3)$$

Intuitively a(x,C) has the meaning of the increase of the action at the point x as effect of the Wilson loop: more precisely:

$$d/d(1/g^2)\, W(C) = \int d^4x\, W(C) \cdot a(x,C) . \qquad (3.4)$$

If the theory confines we expect that there will be a region in which a(x,C) is substantially different from zero this region become a surface of some thickness(29) having C as boundary in the limit in which the loop C becomes very large.

Perturbation theory tell us that at short distances V(L) defined in eq. (2.9) behaves as $\alpha(1/L^2)/L$ and it does give us no informations on the large distance behaviour.

"Has anybody proved confinement in QCD ? " is the standard question to the expert. However this is not the most important question; there are general arguments showing that or QCD confines, or gluons take mass like in the Higgs mechanism, or there are long range forces(30). This statement is very similar in spirit to the sentence "a material is solid or liquid or gas" ; we need an explicit computation to find if Helium is liquid at zero temperature and it would very difficult to decide the issue using general theorems. Here the situation is the same: what we need in QCD, is efficient way to do computations in the low energy region: the output should be the whole mass spectrum.

Let me present for compleatness a simplified argument which shows how confined may be realized. I will neglect for semplicity the non abelian character of the theory. We consider a flat surface of area S. We divide it in N smaller surfaces of area S/N. We obtain:

$$W(C) = \exp i\Phi(C) = \exp\left(i \sum_{1}^{N}{}_i \Phi_i\right) = \prod_{1}^{N}{}_i \exp(i\Phi_i) , \qquad (3.5)$$

where Φ_i is the flux going through each of the smaller surfaces. We have to compute the statistical average of W(C). Let us assume that the Φ_i^2 are statistically independent (so happens in two dimensional theories):

$$\langle \Phi_i \rangle = 0 , \quad \langle \Phi_i^2 \rangle = f , \quad \langle \Phi_i \Phi_j \rangle = 0 , \quad i \neq j \quad (3.6)$$

and let us pospone the discussion on the origine of (3.6).

Now confinement is trivial; indeed:

$$\langle \exp - i \Phi_i \rangle \simeq \exp(-f/2) ,$$
$$\langle W(C) \rangle = \prod_{1}^{N}{}_i \langle \exp i \Phi_i \rangle \simeq \exp - \frac{Nf}{2} . \quad (3.7)$$

In other words the total flux $\Phi(C)$ is the sum of N statistically uncorrelated variables of zero mean and fixed variance; the central limit theorem tell us that the probability distribution of Φ is a gaussian with variance proportional to S:

$$P(\Phi) = \frac{1}{(2\pi f)^{1/2}} \exp(- \frac{N \Phi^2}{2f}) . \quad (3.8)$$

We finally get:

$$\langle W(C) \rangle = \int d\Phi \, P(\Phi) \exp i\Phi = \exp - \frac{fN}{2} . \quad (3.9)$$

Confinement is a simple consequence of the large fluctuations of the flux concatenated to large circuits and the statistical independence hypothesis naturally lead to (3.8-9). Of course (3.6) is not true in perturbation theory, where the conservation law for $\nabla_\mu F_{\mu\nu} = 0$ gives strong costraints; beyond perturbation theory everything is possible, as it has been advocated by many authors[31, 32], the main difference being in mechanism producing eq. (3.6) chromomagnetic; monopoles, dense instantons, merons, condensations of flux tubes have been suggested; all these approaches share have one common point: the practically impossibility of using them to obtain reliable quantitative answers[33].

In the next Section we shall see other approach which should be able to give quantitative predictions.

4. - LATTICE GAUGE THEORIES

In the standard formulation of the theory ultraviolet divergences are present; although these ultraviolet divergences can be removed in perturbation theory, in order to give a non perturbative definition of the theory it is better not to introduce them from the beginning. This can be easily done by using a cutoff M and send M to infinity only at the end (momenta greater than M are disregarded). This can be done by discretizing the Euclidean space introducing a lattice, the fields will be defined only on the points or links of the lattice. In any computation the momenta will be bounded inside the first Brilloin zone; if we consider an hypercubic lattice of spacing a, each component p will belong to the interval

$$\left[-\frac{\pi}{a}, \frac{\pi}{a}\right], \qquad \text{M being equal to } \frac{\pi}{a}.$$

In principle it is also possible to work in the real Minkowski space and to discretize only the space and not the time. This approach has been suggested long time ago, but only recently it has been strongly developed. The introduction of a lattice is a device for dealing only with a finite number of degrees of freedom: it is very similar to the introduction of a mesh of points for solving differential equations[34].

There are many ways in which one can write a field theory on the lattice, however it is better to conserve the symmetries of the original problem as far as possible. The symmetry we want to preserve here is gauge invariance. As a first step we must define the gauge fields A_μ and the Wilson loop W(C). We associate to each link (i, k) of the lattice[35] a variable U_{ik} belonging to the group[36,37]: it is the lattice equivalent of $\exp(i \int_i^k A_\mu dx_\mu)$[38]. The Wilson loop is simple given by:

$$W(C) = \mathrm{Tr} \prod_{(i,k) \in C} U_{ik}, \tag{4.1}$$

where the product runs over all the links belonging to the path C.

We notice that for a small loop $C_{\mu\nu}$ of area a^2 laying in the $\mu\nu$ plane we have in the continuum case:

$$W(C_{\mu\nu}) = 1 + i a^2 \mathrm{Tr} F_{\mu\nu} - \frac{a^4}{2} \mathrm{Tr} F_{\mu\nu}^2 + O(a^6). \tag{4.2}$$

We find:

$$F^2_{\mu\nu} \simeq \frac{1}{a^4}\left[1 - W(C_{\mu\nu}) + h.c.\right] ,$$

$$\int \left(\sum_{\mu,\nu} F^2_{\mu\nu}\right) d^D x = a^{D-4} \sum_P \left[1 - W(P) + h.c.\right] , \tag{4.3}$$

where the sum runs over all the plaquettes P (faces of the cubes) of the lattice (W(P) is the Wilson loop associated to the plaquette P).

The final expression for the Partition function and the Wilson loop are :

$$Z = \int dU \exp(-H/g_B^2) ,$$

$$\langle W(C) \rangle = \int dU \exp(-H/g_B^2) W(C)/Z , \tag{4.4}$$

$$H = a^{D-4} \sum_P \left[1 - W(P) + h.c.\right] .$$

Formally g_B^2 and $1/g_B^2$ play the role of the temperature and β respectively. The low coupling expansion (i.e. in powers of g_B) can be done using the saddle point method: Feynman rules can be derived(40, 41); they are more complex than the usual ones : the interaction is non polynomial but no ultraviolet divergences are present, all momenta being bounded. In the limit a going to zero, one recovers the standard Feynman rules.

The bare coupling constant ($\alpha_B = g_B^2\ 4\pi$) is approximately the running coupling constant evaluated at $q^2 = \pi^2/a^2$. More precisely one obtains for the SU(2) group

$$\alpha(M^2) = \alpha(\pi^2/a^2) = \alpha_B + H\alpha_B^2 + O(\alpha_B^3) ,$$

$$\frac{11}{6\pi}\alpha(q^2) = \frac{1}{\ln q^2/\Lambda^2 + \frac{102}{121}\ln\frac{11}{6\pi}\alpha(q^2)} , \tag{4.5}$$

where the running coupling constant is defined in the momentum subtraction scheeme and the constant H has been computed by Hasenfratz and Hasenfratz(40) ; ($H \simeq 3.39$).

From eq. (4.5) we trivially get :

$$\Lambda = M\left(\frac{11}{6\pi}\alpha(M^2)\right)^{-\frac{51}{121}} \exp\left(-\frac{3\pi}{11\alpha(M^2)}\right) \simeq$$

$$\simeq M\left(\frac{11}{6\pi}\alpha_B\right)^{-\frac{51}{121}} \exp\left[-\frac{3\pi}{11}\left(\frac{1}{\alpha_B} - H\right)\right]. \tag{4.6}$$

It is clear that the continuum approximation can be good only in the region $\alpha_B H \ll 1$. We need of such a small value of α_B in order to apply perturbation theory: indeed a simple computation(42) shows that the mean value of the plaquette (U) (normalized to 1)(43) is equal to

$$U = 1 - \frac{3\pi}{4}\alpha_B . \tag{4.7}$$

For α_B as small as 0.2, U is equal to only 60% of its free value. The presence of terms proportional to $\pi\alpha_B$ is typical of lattice gauge theories: the relevant expansion parameter is g_B^2 and not α_B/π.

H is a fundamental constant in the comparison of the results of the lattice theory with the continuum version; let us present a rough qualitative computation of H. We first notice that the renormalized charge is different from the free one also in the pure electromagnetic case, the origine of this difference is the non linearity of the lattice action. If α is not zero, the thermal fluctuations renormalize the charge; let us try to estimate this effect. In order to compute the renormalized charge, we must know the variation of the action with respect to an external perturbation; let us decompose the field A as $A_f + A_e$: A_f is the fluctuating part and A_e is the external field; $F_{\mu\nu}$ is essentially given by :

$$\frac{1}{g_B^2}\cos A = \frac{1}{g_B^2}(\cos A_f \cos A_e - \sin A_f \sin A_e). \tag{4.8}$$

If we do the mean over the fluctuating field A_f we get

$$\frac{1}{g_B^2}\cos A \sim \frac{\langle \cos A_f \rangle}{g_B^2}\cos A_e = \frac{\cos A_e}{\tilde{g}_B^2}, \tag{4.9}$$

$$\tilde{g}_B^2 = g_B^2/U .$$

In other words the more appropriate expansion parameter should be :

$$\tilde{\alpha}_B = \alpha_B/U = \alpha_B + 0.75\pi\alpha_B^2 + \cdots . \tag{4.10}$$

This elementary computation give an estimate of H which is 70% of the correct value. Using the new variable $\tilde{\alpha}_B$ one gets:

$$\alpha(M^2) = \tilde{\alpha}_B + 1.03\tilde{\alpha}_B^2 + O(\tilde{\alpha}_B^3) ,$$

$$\Lambda = 2.4 M \left(\frac{6\pi}{11\tilde{\alpha}_B}\right)^{-\frac{51}{121}} \exp\left(- \frac{3\pi}{11\alpha_B}\right) . \qquad (4.11)$$

A good choice of the expansion parameter is very important: the two expression for Λ in eq. (4.6); which are equivalent in the limit $\alpha_B \to 0$, differ by a factor 10 for $\alpha_B \simeq 0.15$. Although we need a formidable two loop computation to have reliable results, it may be useful to investigate the problem in an abelian theory where two loops computations are much simpler than in the non abelian case (in other words we should compute those diagrams giving contributions proportional to powers of $\frac{N^2-1}{N}\alpha_B$)[43].

5. - COMPUTER SIMULATION

In the last years the most spectacular results have been obtained doing computer simulations using the Montecarlo technique. Although the Montecarlo technique is time honoured in the framework of statistical mechanics[44], only recently Wilson[45] suggested to apply it to the study of gauge theories; let me spend some time to give a physical picture of the method.

Suppose that we consider a finite piece of the lattice: a cube of size d ($d \gg \Lambda$). In this situation we have pratically reached the thermodynamic limit $d \to \infty$; if periodic boundary conditions are used the corrections to the thermodynamic limit should be small as $\exp(-d/\Lambda)$[46].

Using the integral representation (4.4) the expectation value of the Wilson loop can be written as the integral over all the configurations of the fields in the cube. The number of fields N is of order $(d/a)^4$, so that also for small values of d the evaluation of the integral using the Simpson rule is practically impossible (i.e. the computer time needed is greater than the age of the universe). We need a method such that the time computer increase like N (or a small power of N) and not as exp(N).

In order to find the method we must go back in time and undo what Boltzmann and Maxwell did. Equilibrium statistical mechanics was introduced to study the large time behaviour of the system. Let us consider a classical example: we study the time behaviour of N particle in a box, whose trajectories $x_i(t)$ $(i = 1 - N)$ satisfy the Newton law:

$$m\ddot{x}_i = -\frac{\partial U}{\partial x_i} = F_i \qquad (5.1)$$

where $U(x_1 \ldots x_N) = U[x]$ is the interparticle potential.

Standard arguments based on ergodic theorems, tell us that in most of the cases after enough time the system will reach equilibrium; for large N the microcanonical distribution will be equivalent to the canonical distribution, given by the Boltzman factor. We finally find the highly non trivial result:

$$\lim_{\tau\to\infty} \frac{1}{\tau} \int_0^\tau f[x(t)]\,dt = \overline{f[x]} = \frac{\int dx_i \exp - \beta U[x]\, f[x]}{\int dx_i \exp - \beta[U]} . \qquad (5.2)$$

The temperature T $(\beta = 1/kT)$ can be computed as function of the energy E (which is a conserved quantity) using a thermometer; in this case the momenta themselves may play the role of the thermometer; indeed we know that:

$$\lim_{\tau\to\infty} \frac{1}{\tau} \int_0^\tau \frac{p_i^2(t)}{2m}\,dt = \overline{\frac{p_i^2}{2m}} = \frac{3}{2}kT . \qquad (5.3)$$

Now the point of view of a computer is the opposite of Boltzman; the number of steps needed to solve the coupled Newton equations is of order N x T, i.e. it increase linearly with N: the right hand side of eq. (5.2) is much easier to compute then the right hand side[48].

What happens for finite times? Also at equilibrium random thermodynamic fluctuations are present; we finally obtain:

$$\frac{1}{T} \int_0^\tau f(x)\,dt = \langle f \rangle + O(1/\tau^{1/2}) . \qquad (5.4)$$

The $\tau^{-1/2}$ law come from the mean of independent random fluctuations (the practically random behaviour of the deterministic system (5.1) is the basis of thermodynamics).

Of course we must understand how the time τ_e for which equilibrium is reached (i.e. the time for which $\tau^{-1/2}$ corrections are small), depends on N. The physical intuation tell us that if N is increased at fixed density and the potential is not pathological, equlibrium is reached locally in a time which is independent from N[49].

Near a second order phase transition (when long range correlations, i.e. zero mass particles,are present) the $\tau^{-1/2}$ law is no more valid and the pace of the approach to equlibrium is much lower, this phenomenum being called critical slowing down[50]. Large times are also needed to reach the equilibrium near a first order transition; the system may be locked into a metastable state untill a fluctuation grater than a critical size, is formed and becomes the germ of the condensation[51]; this difficulty may be avoided by fixing the initial conditions in such a way that in half of the box there is one phase (gas) and in the other half there is the other phase (liquid) and studying the movement of the interphase boundary[52].

In other words, instead of using the Boltzmann integral representation, it is more convenient to introduce a fictitious time t, to write appropriate equations of motion and to study the large time behaviour of the system.

The equation of motion can be freely chosen, provided that thermodynamic equilibrium is asymptotically reached. For example we can consider the same particles an in eq. (5.1) moving in highly viscous liquid at temperature T. One finds the Langevin equation[50, 53]:

$$\eta \dot{x}_i(t) = F_i\left[x(t)\right] + b_i(t) \tag{5.6}$$

$$\langle b_i(t) \rangle = 0\,, \quad \langle b_i^{\prime}(t)\, b_j^{\prime}(t) \rangle = \delta_{ik}\,\delta(t-t')\,B\,,$$

$$B = 2KT\eta\,,$$

where η is the viscosity, $b_i(t)$ are random gaussian variables, uncorrelated in time and represent the effect of the Brownian motion. The relation among the viscosity, the temperature and B dates back to Einstein[54].

It is intuitive that at large time the particles must go to thermal equilibrium, they are in contact with the liquid that plays the role of heat reservoir. The formal proof of this statement can be done using the Fokker Plank equation[53, 55].

It is possible to do computer simulations based on the Langevin equation, however for practical purpose we must discretize the time interval and some errors are introduced. It is also possible to write random equation of motion for discreate times whose solution goes to equilibrium at large times: in this way we obtain the Montecarlo technique; in the limit of small steps the Montecarlo technique reduces to the Langevin equation: it can be considered as a discretized form (discretization in time) of the Langevin equation, such to preserve the asymptotic limit. From the physical point of view there is no substantial difference between the Langevin equation and the Montecarlo procedure(57): My impression is that the Montecarlo method seems to be faster while the Langevin equation have the advantage of having a simpler analytic form(59).

Now in the last year many computer simulations have been done and many very interesting results have been obtained; we start to have a good understanding of the physics with finite groups and of the problem of approximating a continuous group with a discreate subgroup(61). For lack of time I will not discuss here this very interesting problematics, and I will present the results for the continuous groups (mainly SU(2) and U(1)).

Let me describe a typical computer simulation: we start with a lattice having cubic shape(62), the length of the cube range from 4 to 16 lattice spacings in most of the computations.

The first thing to do is to look for phase transitions by studying the temperature dependence of the internal energy (the expectation value of the Wilson loop around one plaquette) on the temperature(63). Phase transitions can be divided into two groups; first order: U is discontinuous but it is infinitely differentiable from both sides; it is believed that U can be approximatively analitically continued from each side beyond the transition point(64), giving the results for the metastable phase(64). If a second order phase transition the internal energy is continuous, there is no metastability, but only a "critical slowing down", and the internal energy has a singularity proportional to $|T - T_c|^{-\alpha+1}$ $\alpha = 2 - D\nu$(65), the energy connected correlation function (in the case of gauge theories the connected correlation functions of two plaquettes) decrease to zero like $r^{-2D+2/\nu}$.
If α is negative it is very easy to see the transition by plotting U against T. For α strongly positive it is not so easy; it is also difficult to distinguish an higher power decrease of the correlation function(66) from an exponential decrease; it is therefore possible to miss a second order transition.

Let see some of the results: in Fig. 1 we have the internal energy of 5 dimensional SU(2) theory(42), some of the points are obtained decreasing the temperature, other by increazing it, so that we see an hysteresis loop which can be interpreted as a first order phase transition(67). The high temperature expansion (described in the next Section) tell us that the theory is confined at

Fig. 1. The expectation value of a single plaquette P (defined in the text as 1-U) as function of β in the 5-dimensional SU(2) gauge theory(42). Crosses heating; circles cooling.

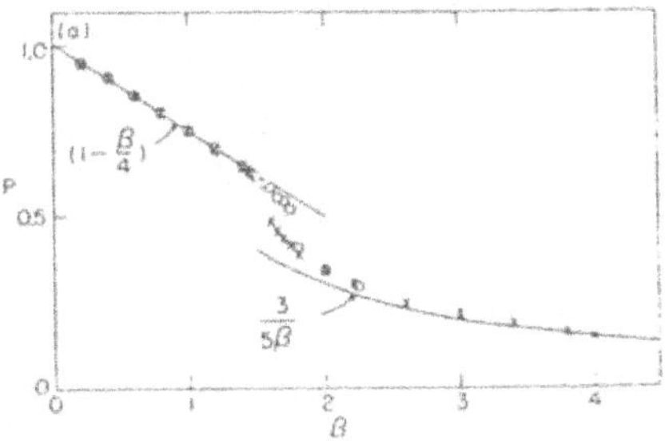

high temperature, the absence of infrared divergences in the low perturbative expansion (the standard perturbative expansion) suggest us that we see the transition from the confined and the unconfined phase.

In Fig. 2 we have shown the same plot for the 4 dimensional U(1) theory: we still have a transition between the confined and the unconfined phase; a carefull analysis shows that the hysteresis loop is not due to metastability, but to critical slowing down: the transition is a second order one $\nu \sim 1/3$(69, 70).

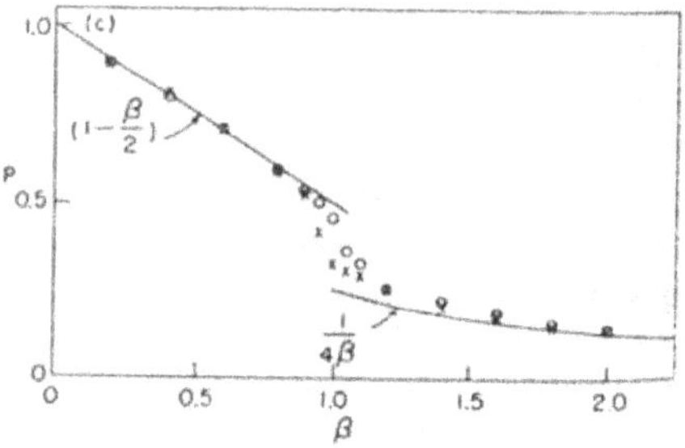

Fig. 2. The expectation value of a single plaquette P (defined in the text as 1-U) as function of β in the 4-dimensional U(1) theory(42). Crosses heating; circles cooling.

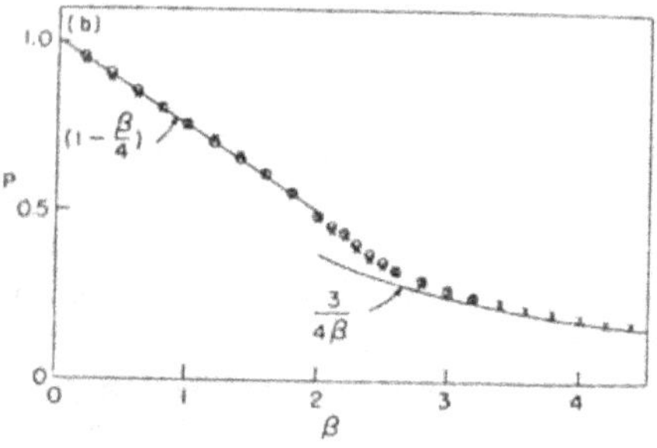

Fig. 3. The expectation value of a single plaquette P (defined in the text as 1-U) as function of β in the 4-dimensional SU(2) theory(42). Crosses heating; circles cooling; no evident hysteresis loop is appreciable.

In Fig. 3 we see the internal energy for the more interesting case: the 4 dimensional SU(2) theory; there is no evidence for a phase transition, although the glitch around $\beta = 2$ ($\alpha = 1/2$) may suggest a second order transition with a strong negative value of α.

Let us see what happens to the Wilson loop in order to decide if the theory is confined also for $\beta > 2$. Let me recall what we expect in the confined phases for large r: for the static potential V(r) and for the expectation value of a Wilson loop of size L x T

$$V(r) \simeq \sigma r + \mu + \frac{\lambda}{r} ,$$

$$W(L,T) \simeq LT\sigma + \mu(L+T) + \ln(L+T) + f_c(L/T) + \cdots \tag{5.7}$$

for $r \gg \Lambda^{-1}$. In the perturbative region a $r \ll \Lambda^{-1}$ we get:

$$V(r) \sim \mu + \frac{a}{r} ,$$

$$W(L,T) \sim \mu(L+T) + a f_p(L/T) . \tag{5.8}$$

The presence of the last term in the confined case has been suggested in a beautiful fundamental paper of Lüscher, Symanzik and Weitsz[71] and it is connected to the presence of oscillations of the string (as we shall discuss in the Section 7).

A careful analysis is needed to separate the term proportional to the surface from the other ones[72]. This has been done by Rebbi[39] in the case of a discrete subgroup of SU(2) and by Creutz for the group SU(2)[73]. According to our policy of not discussing the results for finite groups for lack of space (although the analysis in more accurate) we present only the results for the SU(2) group.

Let us consider the quantity:

$$R(L,a) = \ln\left[W(L,L)W(L-1,L-1)/W^2(L,L-1)\right]^2 . \tag{5.9}$$

The contribution linear with L has been eliminated we expect that

$$R(L,a) \sim (a^2\sigma + \lambda^2/L^2) , \qquad L \gg \Lambda^{-1}(a) ,$$

$$R(L,a) \sim a(\pi^2/L^2) , \qquad L \ll \Lambda^{-1}(a) . \tag{5.10}$$

The results are shown in Fig. 4. The straight line is a rough approximation to Λ. The value of σ can be reasonably estimated in the region $\beta < 2.3$.

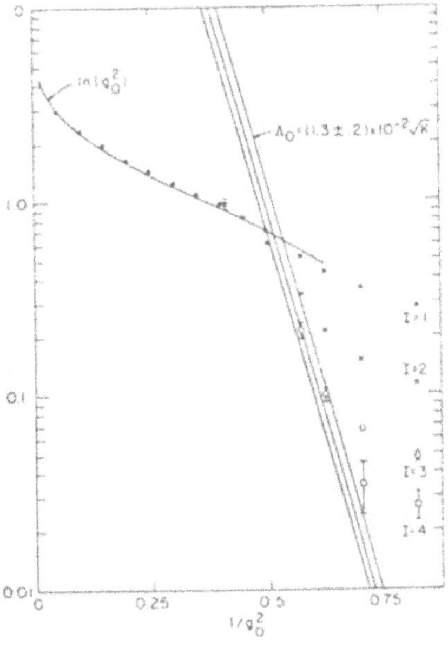

Fig. 4. The quantity $R(I,a)$ defined in the text agains $1/g_B^2 = \beta/4$[73] for the 4-dimensional SU(2) theory.

We notice that in the region of small $\alpha_B = 1/\beta\pi$ and large L, the function R(L, α) should depend only on a renormalization invariant quantity[27]:

$$\frac{L^2}{\alpha} R(L,\alpha) = f(z) ,$$
$$z = \frac{L^2}{\pi^2} \left(\frac{6\pi}{11} \alpha(M^2)\right) \exp\left(\frac{6\pi}{11\alpha(M^2)}\right) \tag{5.11}$$

Standard perturbation theory allow us to compute the coefficients of the expansion in power of log(z) around z = 0, while if eq. holds, in the large z region we have:

$$f(z) \rightarrow z + \text{Const} . \tag{5.12}$$

The function f(z) controll the cross over from perturbation theory to confinement: it would be rather interesting to extract it from the Montecarlo data[74] and to see if Const $\neq$ 0.

It is rather unfortunately that the smallness of the lattice (8^4) prevents us from studying Wilson loop greater than 4 without feeling the effects of the periodic boundary conditions[75]. However from these relatively small lattices we can definitely say that the SU(2) theory confines also in the small coupling region as it was predicted by the Migdal Kadanof recurrence equation[76]: the exponential decrease of σ with a slope similar to the one suggested by the renormalization group is a greater succes of the theory. The drastic changement in the behaviour around $\beta = 2$ ($\alpha = 1/2\pi$) can be easyly understood: the quadratic terms dominates over the linear ones in eq. (4.5) and pertubation theory breaks down. The ratio σ/Λ^2 is found to be of order 1, although higher values are not excluded, given our ignorance on the precise form of Λ.

The glue-ball mass could be extracted by studying the decrease of the gauge invariant plaquette-plaquette connected correlation function

$$C(r) = \langle W(P(0)\,W(P(r)) \rangle - \langle W(P) \rangle^2 . \tag{5.13}$$

There are two difficulties. The connected term is defined as the difference of two terms of order 1 and it is very difficult to measure at distances greater than one[77]; at high distances one sees no signal, only noise, unless one waits a very large time. Perturbation theory tell us that this correlation function decrease like α^2/r^8 at small distances[78]; one should see a cross over from $1/r^8$ to $\exp(-M_G r)/r^{3/2}$. This cross over is invisible unless one knows the function with high accuracy.

Before changing argument let me mention a very interesting computation[79]. It was suggested[80, 81] that QCD in Minkowski space has a phase transition from a confined to an unconfined phase when the temperature become greater than T_c (Here the temperatu re is really the temperature, measured in Kelvin). In the interpre- tation of ref. (80) this phenomenum is connected to the exponential increase of the hadronic mass spectrum and the Hagerdon limiting temperature should be interpreted as a phase transition[82].

Finite temperature field theories can also simulated on the computer using the same Montecarlo approach. This was done in ref. (79) where some extimate of T_c are presented ; at this preliminary stage it is unclear if the transition is first or second : a more detailed analysis and more computer time is needed ; the results are however very promising.

6. - THE HIGH TEMPERATURE EXPANSION

The high temperature expansion is a familiar technique is statistical mechanics[83]: at infinite temperature the theory is tri- vial : entropy dominates over energy and no correlation is present between different variables, i. e. all connected correlation functions are zero. If the lattice hamiltonian is enough simple (nearest neigh- nour interaction is the ideal) is easy do develop any physical quanti- ty (e. g. $f(\beta)$) in powers of β.

$$f(\beta) = \sum_{0}^{\infty}{}_k\, f_k \beta^k \, . \tag{6.1}$$

Under general conditions it is possible to prove that this expansion has a finite radius of convergence (that is not true with the usual per turbative expansion), i. e. the function $f(\beta)$ is analytic around $\beta = 0$[84]. A well known theorem tell us that the knowledge of the func- tion $f(\beta)$ around $\beta = 0$ (i. e. all the f_k) is sufficient to fix the func- tion in the whole analyticity domain. The boundaries of the domain of analyticity are phase transitions for real temperatures and other phy sical uninteresting singularities for complex temperatures. If no phase transitions are present for real positive temperature[85], the behaviour at low temperatures can be extracted from the high tempe rature expansion ![86].

So long with the theorems ; let us come back with the reality. One could think that it would be enough to compute some of the f_k, construct the Padé approximations to $f(\beta)$ and look to the computer output. Although this procedure is convergent it is highly inefficient; it is a general rule that it is difficult to extract "quantitative" infor- mations from the high temperature expansion without a "qualitative" input[88]. Let me give classical examples : we consider the three di mensional Ising model ; the magnetic susceptibility $\chi(\beta)$ is believed to have a singularity of the form $(\beta_c - \beta)^{-\gamma}$. It is pure nonsence to use the first 10 or 20 χ_k known to construct Padé approximant to χ and to fit the output in order to find β_c and γ. A Padé approximant has only simple poles and it does not approximate well a cut near the tip of the cut. It is much wiser to construct the logarithmic derivative

$$\mathrm{ld}(\beta) = \frac{d}{d\beta} \ln \chi(\beta) . \tag{6.2}$$

$\mathrm{ld}(\beta)$ has a simple pole at β_c (plus a subdominant cut) and the residuum is the critical exponent γ.

This example shows that it is much better to use approximants[89] which have automatically the correct singularity structure. The informations on the nature of the singularities should be deduced from physical arguments. If we do not have this informations, we should try to extract it from the k dependence of the f_k. The following Appel comparison theorem is usual very useful.

If the nearest singularity to the origine has the form $(\beta_c - \beta)^{-\gamma}$, the asymptotic behaviour of the f_k is given by:

$$f_k \propto k^{\gamma - 1} |\beta_c|^{-k} \exp - i n \theta , \qquad \beta_c = |\beta_c| \exp i \theta . \tag{6.3}$$

If the function is real and β_c is complex we must have a pair of complex coniugate singularities and $\exp - i n \theta$ is substituted by $\cos(n\theta + \varphi_0)$. If all terms are positive the nearest singularity is on the positive real axis. We can therefore use the ratio test:

$$\beta_c = \lim_{k \to \infty} R_k , \qquad \gamma = \lim_{k \to \infty} \left[1 + k\left(\frac{R_k}{\beta_c} - 1\right) \right] ,$$
$$R_k = \frac{f_k}{f_{k-1}} . \tag{6.4}$$

We must now extrapolate the values for R_k and γ_k we have computed,up to $k = \infty$. If the sequence is smooth a fit with inverse powers of k normally give the correct result[90], if the sequence is not smooth, you are in trouble.

Without a preliminary estimation of the effective radius of convergence of the series there is the danger of using it in the region where they are not convergent or slowly convergent. Unfortunately in many paper I have seen on the high temperature expansion for lattice gauge theories this elementary precaution has not been taken and unreliable results have been obtained.

Life is not always so easy. Sometimes the nearest singularity is on the negative axis: however a simple conformal mapping like $\beta = \frac{z}{b+z}$ (91) may map far away the singularity on the negative axis and the physical singularity on the real positive axis become the nearest to the origine.

We must also add a word of caution; first order transitions are normally invisible in the high temperature expansion, one obtains automatically the analytic continuation of the free energy in the metastable region. Of course the combined use of the high and the low temperature expansion is very efficient to locate first order transitions.

Let us come back to gauge theories. At infinite temperature there is no correlation among variables at different points. The mass of the glue-ball (the inverse of the correlation length) is infinite in this limit. Condition (3.6) is clearly satisfacted and the theory is confined. The construction of the high temperature expansion is straightforward: in order to compute the term proportional to β^k in the free energy, we must count how many closed surfaces of k plaquette can be imbedded on the lattice and weight each surphace with a group theoretical factor depending on the topology of the surphace[92].

Before analyzing real interesting series for 4 dimensional models it is wise to study what happens on an hypercubic lattice when the dimension of the space (D) goes to infinity. In this situation the high temperature expansion can be exactly summed up[93]; in the SU(2) case (and also in the Z2 case) one finds[94]

$$U(\beta) = \beta\left[u_0(\beta^2/D^{1/2}) + \frac{1}{D^{1/2}}\, u_1(\beta^2/D^{1/2}) + \cdots\right] \qquad (6.5)$$

The functions u_0, u_1, etc. can be explicitely computed. When D goes to infinity one finds two singularity $\beta^2 = \pm 1/D^{1/2}$; the 1/D corrections show that the negative β^2 singularity is the nearest to the origine. As explained in details in ref. (93), near the transition point, the surface, on which is concentrated the energy in presence of a large Wilson loop, is no more flat, there are many tree like deformations, which form a branched polymer of cubes. At the transition point, the lenght of this tubes arrives to infinity, their thickness remaining fixed; we can call this transition a "local roughening" transition, because the deformation can start from a small size of the surface. The critical exponent α is 1/2.

A more careful analysis shows that there is a first order phase transition at $\beta = 1/D$ which separate the confined and the unconfined phases[39, 93]. The second order transition is in the metastable phase and it is a virtual transition not directly relevant for the physics.

Descreasing the dimensions the two transitions become rather near ; the analysis is very clear in the 4 dimensional Z2 theory : there is a first order transition at the self dual point[37, 95, 96] and there are two virtual second order transition very near : about $\beta_{II} \simeq \simeq 0.49$ and 0.40 respectively, as can be easily seen using the ratio test and conformal mapping[97].

I have analyzed the 4-dimensional high temperature expansion for $U(\beta)$ in SU(2) up to β^{15}[39]. A first sight one finds that a transition at $\beta \simeq 2.15$ with $\alpha \simeq 1/2$ is strongly suggested ; however this result is in strong contrast with Montecarlo : if we resume the high temperature expansion according to this hypothesis the expression for $U(\beta)$ violently disagree with the Montecarlo results in the region near the critical temperature. In order to reach compatibility of the high temperature expansion with computer simulations we must decrease to 2 and to $-1/2$[98]. This drastic solutions still compatible with the first fifteen orders of ref. (37), however a preliminary analysis of the 21 orders of Wilson shows that the higher orders have the tendency of preferring higher still values of β and α[100]. Indeed a simple minded analysis would suggest $\beta_c \sim 3$ and $\alpha = 2$ which is a pure nonsense. The only deceiving conclusion is that it is impossible to use the high temperature expansion to extrapolate beyond $\beta \simeq 2.1$ unless 50 orders are computed. Similar conclusions can be extracted from the high temperature expansion for σ[101] and from $a(x)$[28] defined in eq. (3.3).

The reason for this debacle is clear. In the mean field theory the deformations of the surface have Hausdorf dimensions 4[93, 102], as ordinary branched polimers, in other words there radius increase like $N^{1/4}$, N being the number of steps[103]. This means that these polimers of cubes, if self-repulsion is neglected, would be strongly overlapping. The self-repulsion, due to the non linear superposition of fluxes, decrease the phase space allowed to these deformation and probably forbids the transition[93]. In the high temperature expansion this effect comes from excluded volume effects : however to construct a diagram of a tube which bends back whith the two ending point touching we must go to the 17th order for $U(\beta)$. In other words the effect that is potentially able to stop the transition appears only at rather high orders and low order computations do not contain enough informations on what happens beyond the would be transition point.

It may be possible that better results are obtained if we use all the informations we have on the high temperature expansion at arbitrary non integer dimensions ; e. g. we set $D = 2 + \frac{2\beta^2}{1+\beta^2}$ and we extrapolate at $D = 4$ to the point $\beta^2 = \infty$ by changing D together with β^2. This procedure avoid us to pass near the region $D = 4$, $\beta \sim 2.1$ which is not smooth. We can also try more fancy parametrizations like $D = 4 - \frac{8\beta^2}{16+\beta^4}$ to outflank the non existing local roughening transition.

The impossibility of extracting informations for the behaviour in the low coupling region from the high temperature expansion, should dissuate from using shorter series in the same region where longer series for $U(\beta)$ do not give the correct result.

As usually happens, it is possible that using unjustified procedures one gets the correct order of magnitude, only because it is very difficult to make errors in the order of magnitude ; however I want to recall the attention of the reader on ref. (104, 105) : in these two papers the same high temperature expansion is used to extrapolate at zero temperature and the results for a quantity like σ/Λ^2 differ of a few orders of magnitude. A blind use of the matching conditions may give rather serious errors.

7. - THE ROUGHENING TRANSITION

In the strong coupling limit the surface associated to the Wilson loop is very rigid, the surface tension being very high ($\sigma \sim -\ln\beta$); it coincides with the minimal area surface and no fluctuations are present. If we consider a Wilson loop on the x_1, x_2 plane the parameter $a(x)$ introduced in eq. (3.3) will be different from zero only on this plane:

$$a(x_1, x_2) = \delta(x_\perp) . \qquad (7.1)$$

By decreasing the temperature, the surface tension decreases and the surface is no more flat. The simplest defect consistes of shifting one lattice element in one of the 2(D-2) directions and adding four more plaquettes. The probability for this deformation is proportional to $2(D-2)\beta^4 \doteq \lambda$. When the parameter λ becomes of order 1 the surface full of defects; each of them corresponding to placing a three dimensional cube on the surface in one of the 2(D-2) allowed directions(106).

If two near by cubes are parallel they gain a factor β^2 in energy but they loose a factor 2(D-2) in entropy(96). The probability of being parallel and vanish when the dimension go to infinity(107) at fixed λ, for lower dimensions it is possible that the cube organize themselves and their directions start to be correlated on large scale(108). In this case we can speak of "global roughening" while if the direction of the cubes are not correlated we have a "local roughening".

A local roughening transition happens when adding cubes on the top of cubes the deformation may arrive to infinity (see Fig. 5). At the transition point one would find for Wilson loops of any size :

$$a(x) = \delta(x_\perp) + \tilde{a}(x) , \tag{7.2}$$

$\tilde{a}(x)$ going slowing to zero at infinity.

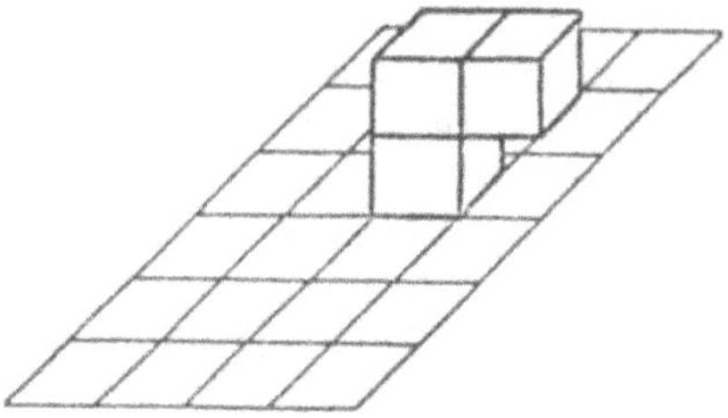

Fig. 5. A typical deformation of a surface (hangover) near a local roughening transition.

For a local roughening transition all thermodynamic quantities, in particular the expectation value of Wilson loops of any size, are singular.

A global roughening transition(28, 30) can be defined only in the limit of infinite size of the Wilson loop: the surface tension is singular but no singularity is present in the expectation value of the Wilson loop of finite size. In the case of a Wilson loop of size $L \times L$, the function $a(x)$ should be substantially different from zero inside a region having transversal dimensions $R(L)$, $R(L)$ going to infinity with L.

Both transitions have their correspective in dual models. The local roughening transition correspond to the tachion of the conventional dual models(93) and the global roughening transition is responsible of the last term in eq. (5.7) which was found in ref. (71) using a new solution of the equations of motions of the string, which should not suffer of the drawbacks of the conventional solution; according to this analysis the function $R(L)$ should behave like $\ln L$(30).

In 4-dimensional SU(2) theories the high temperature expansion suggest the presence of a local roughening transition at $\beta \simeq 2.15$, however as we said in the previous Section, this indication cannot be taken seriously. It is also possible that the self-repulsion effect transform an incipient local roughening transition in global one(110). This issue may decided only if good quality compute simulations for the quantity $a(x)$ are available.

In 3-dimensional SU(2) theories the situation is much clearer. No singularity is seen in the high temperature expansion for the internal energy(111), although the extrapolation at zero temperature is problematic, while the surface tension clearly shows a singularity at about $\beta = 1.4$(112). Summarizing, we have a good evidence for a global roughening transition in $D = 3$, a similar transition is likely present in the 4-dimensional case, but there is no serious evidence point.

8. - THE LARGE N EXPANSION

Many progresses have been done using the large N expansion. This technique has been introduced long time ago[113] in statistical mechanics : the main idea is that if the number of colours (N) goes to infinity, it is possible to use statistical theorems also in colour space. *In the simplest situation fields have only one colour index* and the problem can be exactly solved[114] : field theory can be reduced to an integral equation (the Hartree-Fock approximation is correct). This large N expansion is a wonderful laboratory to study the formal properties of field theory such as infrared finiteness[115], existence of non renormalizable interactions[116], analiticity in the Borel plane[117], etc.

It was remarked by t'Hooft[118] if the field is double coloured (as the gluon)[119], the theory can still be defined in the infinite N limit : remarkable semplifications are present, only planar diagrams survive[120]. Although in this case the theory cannot be solved (exception done for some notable cases), it has been argued that in this limit only zero with resonances are present[121]. In other words if we write formally :

$$SU(N) = SU(\infty) + A_1/N^2 + A_2/N^4$$

the mass spectrum is contained in the leading term $(SU(\infty))$ the whith of the resonances in the first correction (A_1/N^2) and the third term (A_2/N^2) gives (for $N = 3$) 1% corrections to the mass spectrum, which hopefully can be neglected[122].

If we want to solve the $SU(\infty)$ theory, we must fint something better than to sum all the planar diagrams ; we can take advantage of the fact that statistical (quantum) fluctuations are strongly depressed in the limit $N \to \infty$. Indeed if we consider two quantities having a finite expectation value when N goes to infinity, we find the factorization property[125] :

$$\langle AB \rangle = \langle A \rangle \langle B \rangle + O(\frac{1}{N^2}) \ . \tag{8.1}$$

No fluctuations are present (in other words the commutator A, B in the standard operatorial approach can be neglected). Although we are inclined to think that the saddle point method (or the classical equations of motion) must give the correct result, the situation is more complex. Let me spend some time to present the results for the quantum mechanics which, together with 2-dimensional QCD, is the better understood model.

In the first case we consider the hamiltonian

$$H = \frac{1}{2}P^2 + V(X^2) , \quad P^2 = \frac{1}{N} \Sigma_i p_i^2 , \quad X^2 = \frac{1}{N} \Sigma_i x_i^2 \tag{8.2}$$

i.e. central potential for a particle moving in N dimensions.

As can be checked in the harmonic oscillator, the variables are well defined in the limit $N \to \infty$ however the commutator P^2, X^2 vanishes when N goes to infinity.

The ground state energy is given by(123):

$$E_o = \min \left(\frac{1}{x^2} + V(x^2)\right) , \tag{8.3}$$

i.e. the classical result, plus the centrifugal term for a N dimensional s wave.

In the second case the variables X and P are $N \times N$ matrices: the Hamiltonian is:

$$H = \frac{1}{N} \mathrm{Tr} \frac{P^2}{2} + \mathrm{Tr}\left[V(X)\right] . \tag{8.4}$$

The ground state energy is given by

$$E = \frac{1}{2\pi} \int h(p,x)\, \theta(\varepsilon - h(p,x)) , \quad h(p,x) = \frac{p^2}{2} + V(x), \tag{8.5}$$

where ε is fixed by the condition:

$$\frac{1}{2\pi} \int \theta(\varepsilon - h(p,x)) = 1 . \tag{8.6}$$

Eqs. (8.5) and (8.6) can be considered as a variation of the conventional WKB-Thomas-Fermi approximation.

Quantum effects do not desappear in the limit $N \to \infty$ but the can be easyly computed. A better understanding of the physical origine of eqs. (8.5) and (8.6) (it is not self evident that they describe the sum of all the planar diagrams), would be very useful, expecially for extending these results to real scalar field theories in higher dimensions(126).

Fortunately the simple geometrical interpretation of gauge theories allow us to write directly useful equation in the limit $N \to \infty$. The fundamental variable are the expectation values of the Wilson loops associated to an arbitrary path C. It is convenient to introduce the functional derivative $\frac{\delta}{\delta \sigma_{\mu\nu}(x)} W(C)$, which quantifies the variation of W(C) under an infinitesimal deformation of the path C around the point x[127]. It was shown by Mandelstam[131] that:

$$\frac{\delta W}{\delta\sigma_{\mu\nu}(x)} = \mathrm{Tr}\Big[P(F_{\mu\nu}(x)\,\exp i\oint_x^x A_\mu(z)\,dz_\mu)\Big]. \qquad (8.7)$$

We can now transcribe the "Maxwell" equations ($D_\mu F_{\mu\nu} = J_\nu$, $D_\mu \tilde{F}_{\mu\nu} = 0$) in functional equations for W(C)[132]; we find in the limit $N \to \infty$[128]:

$$\partial_\mu \frac{\delta}{\delta\sigma_{\mu\nu}(x)}\overline{W}(C) = \int dy_\nu\,\delta(x-y)\,W(C_{xy})\,W(C_{yx})$$

$$\partial_\varrho \frac{\delta W}{\delta\sigma_{\mu\nu}}\,\varepsilon_{\varrho\mu\nu\lambda} = 0. \qquad (8.8)$$

The δ-function implies that the points x and y coincides: the loop C looks like an eight, C_{xy} and C_{yx} are the two smaller loops into which the eight may be decomposed.

It is very important to understand the physical meaning of eq. (8.8).

A serious step in this direction has been done by Foester and by Migdal and Makeenko. These last two authors have been shown that an approximate solution of the eqs. (8.8) satisfy the integral equation[134]:

$$\frac{\delta}{\delta\sigma_{\mu\nu}(x)}\ln\overline{W}(C) = \int dy_\nu\,dt\,dP^t_{xy}[\omega]\,\omega_\mu(t)\,\cdot$$

$$\cdot\,\frac{\overline{W}(C_{xy\omega})\,\overline{W}(C_{\omega xy})}{\overline{W}(C)} - (\mu \leftrightarrow \nu), \qquad (8.9)$$

where $dP^t_{xy}[\omega]$ stands for the sum over all the paths $\omega(t')$ going from x to y in "time" t (Wiener measure); the point x and y divide the closed path C into two open paths C_1 and C_2, $C_{xy\omega}$ and $C_{\omega xy}$ denote the two closed path obtained by adding ω to C_1 and C_2.

The meaning of eq. (8.9) is clear[135]; $\delta\ln\overline{W}/\delta\sigma_{\mu\nu}$ is roughly speaking the field induced at the point x by the Wilson loop: gluons coming from any point (y) of the loop contribute to it; in the abelian case, the factor $\overline{W}(C_{xy\omega})\overline{W}(C_{\omega xy})/\overline{W}(C)$ is substituted by 1 and the photon trajectories are free one, in the non abelian case the trajectories of the gluons are strongly influenced by the presence of the Wilson loop.

The most interesting result is contained in the last paper of Migdal[136]: he shows that the solutinn of eq. (8.8) can be written as a two dimensional field theory involving both bosons and fermions[137]. For lack of space I cannot enter in the details and the reader is strongly recomended to read the original literature on the subject.

I believe that this field is still in its infancy and that many very spectacular results are waiting us in the next future.

DISCUSSION

Q1 : Frampton, Harvard: In the relationship between the roughening transition in 4-dimensional QCD and the condensation of tachyons in dual models, how explicit can this relationship be because the dimension in dual models is fixed and cannot be varied continously, unless there is something completely new in dual models ?

A1 : Well, personally I an inclined to think that dualist ought to find something completely new in dual model, the impossibility of varying the dimensions being a spourious effect (Anyhow the high temperature expansion shows that gauge theories have a local roughening transition also at higher dimensions, at least in the metastable phases).
In my talk I wanted to underline the fact that also in conventionally defined dual models the surface of the string is not smooth, but locally rough.

Q2 : Dolan, Rockfeller : How relevant is the existence of the roughening transition for the continuum limit ?

A2 : This has no effect on the continuum limit. It is only relevant to know if the strong coupling calculations on the lattice should be pursued, or how to modify them. For example, let us suppose that we have rather long series and we discover that there is a global roughening transition. In this case we should study not $\sigma(\beta)$, which has a singularity, but W(C) for finite loop and extrapolate at infinitely large loops only when β goes to infinity.

Q3 : Kawamoto, Amsterdam : Could you comment about the introduction of fermions in the Montecarlo method ?

A3 : Work is in progress on this subject. I believe it is technically possible and in a year we will have some results.

A3 : Nahm, CERN : Fermions are definitely very important. They dominate the high energy behavior of scattering (see the work of A. White). They may also suppress a roughening transition, if the analogy with tachyonic strings works, as tachyons are ab sent in the supersymmetric string model in 10 dimensions.

Q4: Brodsky, SLAC: Does the infrared singularity described by Frankel and Taylor for inelastic scattering in perturbation theo ry in non-Abelian gauge theories give any clue as to which theo ries confine?

A4: Frankel, Pennsylvania: We have to wait and see if results to all orders in perturbation theory support the existence of this singu larity.

A4: I don't believe confinement can be seen in perturbation theory.

Q5: Pasupathy, Bangalore: How relevant are nontrivial topological configurations for confinement?

A5: Physicists are divided on this point. I believe since instantons are not apparent in the large N equations and since the large N limit appears to confine, instantons are not relevant, neither for confinement, nor to solve the U(1) problem: the use of topological classification of configurations seems to be rather doubtful in an asymtotic free field theory, where fields strong ly fluctuate at large distances and are very far from a pure gauge.

FOOTNOTES AND REFERENCES

(1) - Long time ago, Landau(2), using general arguments, suggested that strong interaction should be asymptotically free: however the only asymptotically free theory known at that time was the $g\phi^4$ with negative coupling (Landau's arguments were too advanced for the time and they have been forgotten; in modern times the first example of an asymptotic free theory was given by Symanzik(3): he also stressed the importance of being asymtotically free). A modern discussione of Landau philosophy can be found in ref. (4).

(2) - L. D. Landau, A. A. Abrikosv and I. M. Khalatnikov, Dokl. Akad. Nauk USSR 95, 773, 1177 (1954); 96, 261 (1954); L. D. Landau and I. Pomeranchuk, Dokl. Akad. Nauk USSR 102, 489 (1955); L. D. Landau, Niels Bohr and the Development of Physics, ed. by W. Pauli (Pergamon, 1955).

(3) - K. Symanzik, Lett. Nuovo Cimento 6, 420 (1973).

(4) - G. Parisi, Phys. Rev. D11, 909 (1975); L. Maiani, G. Parisi and R. Petronzio, Nuclear Phys. B136, 115 (1978); N. Cabibbo, L. Maiani, G. Parisi and R. Petronzio, Nuclear Phys. B158, 295 (1979).

(5) - V. Alessandrini and A. Krzywicki, Orsay Preprint LPTHE 80/24 (1980).

(6) - J. Schwinger, Proc. Nat. Acad. Sci. (U. S.) 44, 956 (1958).

(7) - K. Symanzik, J. Math. Phys. 7, 510 (1966); Rendiconti della Scuola Internazionale di Fisica "E. Fermi", Corso 45, Varenna 1968, ed. by R. Jost (Academic Press, 1969), pag. 152.

(8) - E. Nelson, in "ConstructiveQuantum Field Theory", ed. by K. Velo and A. S. Wighman (Springer, 1973); See also: F. Guerra, Phys. Rev. Letters 28, 1213 (1972), and F. Guerra and P. Ruggero, Phys. Rev. Letters 31, 1022 (1973).

(9) - K. Osterwalder and J. Schrader, Comm. Math. Phys. 31, 83 (1973).

(10) - F. Guerra, L. Rosen and B. Simon, Comm. Math. Phys. 27, 10 (1972).

(11) - Technically speaking the fact that probabilities (not amplitudes) are positive definite allow us to write very powerful inequalities (the use of inequalities is a very common procedure in statistical mechanics), e. g. the intuitive absence of a two particle bound state in presence of a repulsive interaction turns out to be a rigorous consequences of the Lebowitz inequalities(12) for Ising spin system.

(12) - For a review see: J. Glimm and A. Faffe, Cargese Summer School 1976 (Plenum Press, in press).

(13) - A. M. Poliakov, Soviet Phys. -JEPT 28, 533 (1969); A. A. Migdal, Soviet Phys. -JEPT 22, 1036 (1969).

(14) - K. G. Wilson and M. E. Fisher, Phys. Rev. Letters 28, 234 (1972); K. G. Wilson, Phys. Rev. Letters 28, 548 (1972); K. G. Wilson and J. Kogut, Phys. Rep. 12C, 77 (1974); L. P. Kadanoff, W. Götze, D. Hamblen, R. Hecht, E. A. S. Lewis, V. V. Palcianskas, M. Rayl, J. Swift, D. Aspnes and J. Kane, Rev. Mod. Phys. 39, 395 (1967).

(15) - M. Gellman and F. Low, Phys. Rev. 95, 1300 (1954).

(16) - The relevance of the renormalization group in the study of second order phase transitions was also stressed by Di Castro and Jona-Lasinio[17].

(17) - C. Di Castro and G. Jona-Lasinio, Phys. Letters 29A, 332 (1969); F. de Pasquale, C. Di Castro and G. Jona, Rendiconti della Scuola Internazionale di Fisica "E. Fermi", Corso 51, Varenna 1970, ed. by M. S. Green (Academic Press, 1971), pag. 123.

(18) - D. Amit, The role of statistical mechanics in contemporary physics, presented at Camerino (1979), unpublished.

(19) - For a more detailed discussions of the relations between statistical mechanics and quantum field theory at the beginning of the seventies see refs. (18, 20).

(20) - A. Baracca, G. Parisi, L. Peliti, M. Rasetti e M. Valdacchino, Le transizioni di fase e i problemi attuali della fisica delle particelle elementari, in: A. Baracca: Manuale Critico di Meccanica Statistica (CULP, Catania), in press.

(21) - I will use a matrix notation and I will not write the colour indices in most of the cases; here $[\ ,\ ]$ indicates the commutator.

(22) - The Langevin equation formulation of field theory (see Sect. 5) may be used to construct a perturbative diagrammatic approach to gauge theories in which no gauge fixing is needed[23].

(23) - G. Parisi and Wu Yong-shi, Preprint of the Institute of Theoretical Physics of the Academia Sinica ASITP-80-004 (1980); Scientia Sinica, to be published.

(24) - Please notice that $\ln\langle W(C)\rangle$ is very different from $\langle \ln W(C)\rangle$.

(25) - J. L. Gervais and A. Neveau, Nuclear Phys. B163, 189 (1980); Cargese Summer School 1979 (Plenum Press, in press); Phys. Rep., to be published.

(26) - J. Groenveld, J. Jurkiewicz and C. P. Korthals Altes, Phys. Letters 92B, 312 (1980).

(27) - G. Mack and E. Pietarinen, Phys. Letters 94B, 397 (1980).

(28) - A. Hasenfratz, E. Hasenfratz and P. Hasenfratz, CERN Preprint TH 2890 (1980); C. Itzykson, M. Peskin and J. B. Zuber, Saclay Preprint, to be published; M. Lüsher, G. Münster and P. Weisz, DESY Preprint 80/63 (1980); G. Münster and P. Weisz, DESY Preprint 80/74 (1980).

(29) - As we shall see in the Sect. 7 we have two options: the thickness of the surphace may go to a limit when the diameter L of the circuit goes to infinity, or it may go to infinity with L.

(30) - G. 't Hooft, Nuclear Phys. B138, 1 (1978); B153, 141 (1979).

(31) - C. Callan, R. Dashen and D. Gross, Phys. Rev. D17, 2717 (1978); A. Mandelstam, Phys. Rep. 23C, 237 (1977); A. M. Poliakov, Nuclear Phys. B120, 429 (1977); T. Banks, R. Myerson and J. Kogut, Nuclear Phys. B129, 493 (1977); G. Parisi, Frascati Preprint LNF-76/15 (1976); B. Glimm and A. Jaffe, Phys. Letters 73B, 167 (1978); A. Jaffe, Introduction to Gauge Theories, presented at Helsinki Conference on Mathematics on Physics, 1978; Harward Preprint, unpublished; M. Stone and P. R. Thomas, Phys. Rev. Letters 41, 351 (1979); D. Foerster, Phys. Letters 76B, 597 (1978); T. Yoneya, Nuclear Phys. B144, 195 (1978).

(32) - N. K. Nielsen and P. Olsen, Nuclear Phys. B144, 376 (1978); Phys. Letters 79B, 304 (1978); J. Ambjørn, N. K. Nielsen and P. Olesen, Nuclear Phys. B152, 75 (1979); H. B. Nielsen and N. Ninomiya, Nuclear Phys. B156, 1 (1979); H. B. Nielsen and P. Olesen, Nuclear Phys. B160, 380 (1979); J. Ambjørn and P. Olesen, Nuclear Phys. B170 (FS1), 60 (1980).

(33) - The most serious effort to transform this qualitative model into a quantitative one has been done by the Copenhagen School (see refs.(32)).

(34) - In the study of partial differential equations one approximates the derivative with a finite difference operator, however in many cases the most efficient way for solving differential equations is the finite element method; it would be nice to see if and how this method can be transfered tc the field theory framework.

(35) - i and k are two next neighbour points of the lattice.

(36) - K. G. Wilson, Phys. Rev. D10, 2445 (1974).

(37) - R. Balian, J. M. Drouffe and C. Itzykson, Phys. Rev. D10, 3376 (1974); D11, 2098 (1975); D11, 2104 (1975); D19, 2514 (1979).

(38) - The variables A belong to the Lie algebra, they are the generators of the Lie group; their exponent belongs to the group. Notice that in the lattice formulation of the gauge theory all variable belong to the group: it is possible to construct theories based on discreate groups for which no continuum formulation is possible (e.g. the group of rotations of a cube)[39]. Although in many cases no theory is obtained in the continuum limit (the mass gap is always proportional to the cutoff), a notable exception is the three dimensional Z_2 theory[37].

(39) - C. Rebbi, Brookhaven Preprint BNL-27203 (1980); Phys. Rep., to be published; D. Petcher and D. H. Weingarten, Indiana Preprint (1980).

(40) - A. Hasenfratz and P. Hasenfratz, CERN Preprint TH 2827 (1980).

(41) - V. F. Müller and W. Rühl, Kaiserlautern Preprint (1980).

(42) - M. Creutz, Phys. Rev. Letters 43, 553 (1979).

(43) - We notice that for general N, $H = \frac{N^2-1}{2N}\pi + 0.51\,N$ and the pre factor 2.4 in eq. (4.11) is N independent.

(44) - N. Metropolis, A.W. Rosenbluth, A.H. Teller and E. Teller, J. Chem. Phys. 21, 1087 (1953); for recent applications see for example: K. Binder, in Phse Transition and Critical Phenomena, ed. by C. Domb and M. Green (Academic Press, 1976), vol. 5B; Montecarlo Methods, ed. by K. Binder (Springer, 1979); S. Kirpratick, Les Houches 1978, ed. by R. Balian, J. Meynard and G. Toulouse (North Holland, 1980).

(45) - K. Wilson, talk given at the Crete Summer School 1977.

(46) - More precisely they should be of order $\exp(-d/M_G)$ where M_G is the glueball mass. If we use the experimental information that M_G is at least 2-3 Λ (the value M_G = 1500 MeV was suggested in ref. (47), finite volume effects should be rather small for $d = \Lambda$).

(47) - G. Parisi and R. Petronzio, Phys. Letters 94B, 51 (1980).

(48) - This procedure is currently used and it is called molecular dynamics.

(49) - That is true if we do not store energy in coherent motion of the particles (i. e. macroscopic motion): it would take some time to dissipate it into heat via turbolence.

(50) - L. Van Hove, Phys. Rev. 93, 1374 (1954); See for a review: B. Halperin and P.C. Hoemberg, Rev. Mod. Phys. 49, 435 (1977).

(51) - J.S. Langer, Ann. of Phys. 41, 108 (1967).

(52) - Notice that the speed of the interphase boundary goes to zero, near the critical temperature T_c, like a power of $|T - T_c|$.

(53) - P. Langevin, Comptes Rendus 146, 530 (1908); A.D. Fokker, Ann. d. Physik 43, 812 (1914); M. Planck, Sitz. der Preuss. Akad. 324 (1917); For more recent references see for example: Noise and Stochastic Processes, ed. by N. Wax (Dover, 1954); R.F. Fox, Phys. Rep. 48C, 179 (1978).

(54) - A. Einstein, Ann. d. Physik 17, 549 (1905); 19, 371 (1906).

(55) - I want to profit of my position to suggest to the constructivists that stochastic differential equations (see in this respect ref. (56)) may be an alternative technique to construct an interacting field theory in a mathematically rigorous way.

(56) - G. Parisi and N. Sorlas, Phys. Rev. Letters 43, 744 (1979).

(57) - A detailed description of the Montecarlo procedure technique can be found in refs. (44, 58).

(58) - K. Wilson, Cargese Summer School 1979 (Plenum Press, in press).

(59) - It has been shown in ref. (60) that the computation of correlation functions cen be done with much higher accuracy using the Langevin equation.

(60) - G. Parisi, Correlation functions and computer simulations, Frascati Preprint LNF-80/54 (1980).

(61) - For example if we substitute to the group of rotation in the space O(3) the 60 elements group of rotations of the icosaedrum, there will be a first order transition at a temperature T_c, i. e. at a coupling $\alpha = \alpha_i$; at high temperature ($\alpha > \alpha_c$) the two theories will be very similar, while at low temperature ($\alpha > \alpha_c$) the two theories will behave very differently(37).

(62) - However for special problems it would be more convenient to use rectangular lattices.

(63) - We will use the following definitions $T = g_B^2/4 = \pi\alpha_B$; $\beta = T^{-1}$.

(64) - Some care must be used in doing the analytic continuation; in reality the transition point is an essential singularity and the analytic continuation of U has an exponentially small immaginary part, proportional to the inverse of the mean life of the metastable state(51).

(65) - According to the original Erenfest classification second order transition are characterized by $\alpha \gtrsim 0$, if $0 > \alpha \gtrsim 1$ the transition is third order, the general rule if that the transition is of order k is the k'th derivative of the free energy is discontinuous, the $(k-1)^{th}$ being continous. However one often use the words "second order phase transitions" to indicate any transition of order higher than the first.

(66) - ν must satisfly the bounds $1/D < \nu \lesssim \infty$

(67) - Let me explain this misterious sentence. In order to fasten the approach to equilibrium in Montecarlo simulations it is usual to start at an high temperature, reach the equlibrium, change, the temperature and restart the Montecarlo simulation using as starting point an equlibrium configuration of the previous temperature. It is also possible to go in the opposite direction, i. e. to start from low temperature and gradually increasing the temperature. In special cases it is convenient to have some oscillations in the temperature in order to annehale the defects(68) (a well known procedure in metallurgy).

(68) - S. Kirpatrick and D. Sherrington, Phys. Rev. 17B, 4385 (1978).

(69) - B. Lautrup and M. Nauenberger, CERN Preprint TH 2873 (1978).

(70) - Although Figs. 1 and 2 looks similar, longer computer simulations show a difference between the two cases(42, 69).

(71) - M. Lüscher, K. Symanzik and P. Weisz, DESY Preprint 80/31 (1980). It would be interesting to compare the results of this paper with D. J. Wallace and R. K. Zia, Phys. Rev. Letters 43, 808 (1971) and M. J. Lowe and D. Wallace, Edinburgh Preprint 80/110 (1980).

(72) - For small α, only σ is exponentially small.

(73) - M. Creutz, Phys. Rev. Letters 45, 313 (1980).

(74) - It is also rather interesting to compute at least the first terms in the development of f(z) in powers of ln z and to compare it with the Montecarlo data.

(75) - I am convinced that asymmetric lattices like 6 x 6 x 12 x 12 or 4 x 4 x 12 x 12 are more appropriate to obtain informations on large Wilson loops. Moreover if one finds an observable dependence of the Wilson loop on the perpendicular boundary conditions, it would be rather interesting to analyze it.

(76) - A.A. Migdal, Z. Eksper. Theoret. Fiz. 69, 810 (1975); 69, 1457 (1975); L.P. Kadanoff, Ann. Phys. 100, 359 (1976); S. Caracciolo and P. Menotti, Ann. Phys. 122, 74 (1979); CERN Preprint TH 2899 (1980); G. Martinelli and G. Parisi, CERN Preprint TH 2882 (1980).

(77) - K. Wilson, C. Rebbi and M. Creutz, private communication.

(78) - It would be rather interesting to compute the perturbative expansion also for this function.

(79) - J. Polonyi, K. Szlachanyi and J. Kuti, Talk presented at this Conference.

(80) - N. Cabibbo and G. Parisi, Phys. Letters 59B, 67 (1975).

(81) - A. Poliakov, ICTP Lectures Notes taken by E. Gava, Report IC/78/4 (1978); W. Fischler, J. Kogut and L. Susskind, Phys. Rev. D19, 1188 (1979); T. Banks and E. Rabinovici, Nuclear Phys. B160, 349 (1979); E. Gava, R. Jengo and C. Omero, ICTP Preprint IC/80/77 (1980).

(82) - To study the developments of the limiting temperature hypothesis would be a nice subject for an historical reconstruction of the attitude toward physical laws which was shared by many physicists at that time.

(83) - For a review see: Phase Transitions and Critical Phenomena, ed. by C. Domb and M. Green (Academic Press, 1972), vol. 3.

(84) - See for example: D. Ruelle, Statistical Mechanics (Benjamin, 1972).

(85) - Negative temperature are well defined for bounded hamiltonian when the "field" variables are bounded: they are interesting because they correspond to antiordering.

(86) - This approach has been advocated by ref. (87).

(87) - L.I. Shiff, Phys. Rev. 92, 766 (1952); J.B. Kogut and L. Susskind, Phys. Rev. D11, 395 (1975); C.J. Hamer, J.B. Kogut and L. Susskind, Phys. Rev. Letters 41, 1337 (1978); T. Banks, S. Rabi, L. Susskind, J. Kogut, D.R.T. Jones, P.N. Sharback and D.H. Sinclair, Phys. Rev. D15, 1111 (1977).

(88) - The Lord, whose oracle is at Delphes, neither says nor hides, but hints (Heracleitus).

(89) - An approximant of order k $(f_k(\beta))$ is the only function, inside a given class, such that $f_k(\beta) - f(\beta) = 0(\beta^{k+1})$. If the class of function is given by the polynomial of order k, we have the standard Taylor expansion; if it is given by rational functions, we have the Padé approximants.

(90) - For a simple and efficient procedure to extrapolate see: J. Zinn Justin, Journ. de Physique 40, 969 (1979).

(91) - The value of b can be chosen such to fasten the convergence. The choice of the appropriate conformal mapping is very important.

(92) - J. M. Drouffe, Phys. Rev. D18, 1174 (1978).

(93) - J. M. Drouffe, G. Parisi and N. Sourlas, Nuclear Phys. B161, 397 (1979); G. Parisi, in Proc. Third J. Hopkins Workshop on Current Problems in High Energy Particle Theory, Florence 1979, ed. by R. Casalbuoni et al. (J. Hopkins Univ., 1979), pag. 179; Cargese Summer School 1979 (Plenum Press, in press); J. M. Drouffe, Saclay Preprint (1980).

(94) - In any dimensions $U(\beta)$ is an odd function of β for these two groups.

(95) - M. Creutz, L. Jacobs and C. Rebbi, Phys. Rev. Letters 42, 1390 (1979); Phys. Rev. D20, 1915 (1979).

(96) - J. M. Drouffe, Nuclear Phys. B170 (FS1), 79 (1980).

(97) - I have done this computations using a pocket calculator. Using Padé approximants the higher singularity has been extimated at in the same region $\beta_{II} \simeq 0.55$[96] and $\beta_{II} \simeq 0.46$[98].

(98) - N. Kimura, Hokkaido Univ. Preprint (1980).

(99) - As clearly shown by eq. (6.4) the values of β_c and α are strongly correlated.

(100) - I am grateful to K. Wilson for having comunicated to me his results prior to publication. I believe that he has reached the same conclusions as me; however the responsability of any errors is only mine.

(101) - G. Münster, DESY Preprint 80/44 (1980).

(102) - P. G. De Gennes, Academie des Sciences (Paris) Preprint (1980); G. Parisi, Phys. Letters 81B, 327 (1979).

(103) - B. Zimm and W. Stockmayer, J. Chem. Phys. 17, 301 (1949); T. Lubensky and J. Isaacson, Phys. Rev. Letters 41, 829 (1978).

(104) - D. S. Fisher and D. R. Nelson, Phys. Rev. B16, 2300 (1977).

(105) - G. Parisi, Phys. Letters 90B, 111 (1980).

(106) - We can also add a cube on the top of an other cube, but let us neglect this possibility for the time being.

(107) - In high dimensions this discussion is purely academic because λ is of order 1 only in the metastable phase.

(108) - The argument on the infinite dimensional case should be more refined. The system of cubes is equivalent to a 2 dimensional Pott model with $q = 2(D-2)$ states and it is known that for q greater than 4 a first order transition is present[109].

(109) - R. G. Baxter, J. Phys. C6, L445 (1973); J. B. Kogut, Illinois Preprint (1980), to appear in Phys. Rev.; P. Ginsparg, Y. Y. Goldschmidt and J. B. Zuber, Saclay Preprint DPh-T/63/80 (1980).

(110) - The analogy with the Potts model suggests that the spontaneous orientation of cube-like deformations of the surface is a first order transition for $D > 4$ and it is a second order transition for $D \leq 4$.

(111) - A. Duncan and H. Waidya, Phys. Rev. D20, 903 (1979).
(112) - In this case after conformal mapping the ratio test gives good results in agreement with Padé approximants.
(113) - A. Stanley, Phys. Rev. 176, 718 (1968); R. Abe, Progr. Theor. Phys. 48, 1414 (1972); G. Parisi and L. Peliti, Phys. Letters 41A, 331 (1972); M. Suzuki, Phys. Letters 45A, 5 (1972); E. Brezin and D. J. Wallace, Phys. Rev. B7, 1967 (1973); R. A. Ferrel and D. J. Scalapino, Phys. Rev. Letters 29, 413 (1972); S. Ma, Phys. Rev. Letters 29, 1361 (1972); K. Wilson, Phys. Rev. D4, 2911 (1973).
(114) - The field transform as the fundamental representation of the O(N) or SU(N) group.
(115) - A. Jevick, Phys. Letters 71B, 327 (1977).
(116) - G. Parisi, Nuclear Phys. B100, 368 (1975); K. Symanzik, DESY Preprint 77/05 (1977); Y. Araf'eva, Theor. Math. Phys. 31, 279 (1977).
(117) - G. Parisi, Phys. Letters 76B, 65 (1978); Phys. Rep. 49C, 215 (1978); Nuclear Phys. B150, 153 (1979).
(118) - G. t'Hooft, Nuclear Phys. B72, 461 (1974).
(119) - The field transform as the adjoint representation of the O(N) or SU(N) group.
(120) - J. Koplizk, A. Neveau and S. Nussinov, Nuclear Phys. B123, 109 (1977).
(121) - G. Veneziano, Nuclear Phys. B117, 719 (1976); A. A. Migdal, Ann. of Phys. 109, 365 (1977).
(122) - In the quantum mechanics case studied in ref. (123), one find that the expansion parameter is about $1/N^2\pi^2$ (124). The first two terms give the value of the ground state energy with a relative error of 10^{-4}.
(123) - E. Brezin, C. Itzykson, G. Parisi and J. Zuber, Comm. Math. Phys. 59, 35 (1978).
(124) - E. Brezin, Phys. Rep. 49C, 221 (1978).
(125) - E. Witten, Nuclear Phys. B160, 57 (1979).
(126) - A. Jevicki and B. Sakita, Nuclear Phys. B165, 511 (1980); A. Jevicki and H. Levine, Brown Preprints HET-418 and HET-419 (1980); T. Yoneya, Tokyo Preprint (1980).
(127) - The precise definition of $\delta W/\delta\sigma_{\mu\nu}(x)$ can be found in refs. (128-130): the whole analysis become simpler on a lattice.
(128) - Yu. M. Makeenko and A. A. Migdal, Phys. Letters 88B, 135 (1979); A. A. Migdal, Chernolovka Preprint 27 (1979); Yu. M. Mekeenko, ITEP Preprint 141 (1979); Yu. M. Makeenko and A. A. Migdal, ITEP Preprint 23 (1980).
(129) - D. Foerster, Phys. Letters 87B, 87 (1979); T. Eguchi, Phys. Letters 87B, 91 (1979).
(130) - V. Volterra, Rend. Lincei 4a, III, 274 (1887), and 4a, VI, 127 (1890); A. T. Ogielski, Brookhaven Preprint (1980).
(131) - S. Mandelstam, Phys. Rev. 175, 1580 (1968).
(132) - This program was started in ref. (133).

(133) - G. De Angelis, D. De Falco and F. Guerra, Lett. Nuovo Cimento 19, 55 (1977); Y. Nambu, Phys. Letters 80B, 372 (1978); A. M. Poliakov, Phys. Letters 82B, 247 (1978); J. L. Gervais and A. Neveau, Phys. Letters 80B, 255 (1979); E. Corrigan and H. Hasslacher, Phys. Letters 81B, 181 (1979).

(134) - Yu. M. Makeenko and A. A. Migdal, ITEP Preprint 23, Sections 6-8 (1980), private comunication from A. A. Migdal.

(135) - I am trying to reproduce here what Migdal kindly explained to me more than one year ago at the Alustha Conference (1979).

(136) - A. A. Migdal, The Elf af the String, to be submitted to Phys. Letters, private comunication.

(137) - The need of introducing Fermionic degrees of freedom in the description of the string in QCD was stressed by D. Foerster, Nuclear Phys. 170B, 107 (1980).

PHYSICS REPORTS (Review Section of Physics Letters) 103, Nos. 1–4 (1984) 203–211. North-Holland, Amsterdam

The Strategy for Computing the Hadronic Mass Spectrum

Giorgio PARISI

Dipartimento di Fisica, Università di Roma II, "Tor Vergata", Roma 00173, Italy

and

Laboratori Nazionali INFN, Frascati, Italy

1. Introduction

In the last years many papers have been published in which spectrum and the static properties of the low lying hadrons have been computed for lattice QCD. The goal is to prove or to disprove QCD by a direct comparison of the theory with the "low" energy data. This goal can be reached only if statistical and systematic errors are under control. In this first generation of papers [1] (as it was clearly stated) the control of the systematic effects was lacking because of the relative short CPU time used, and the consequential impossibility of going to larger lattices. At the present moment a second generation of computations is starting [2], and systematic effects will be carefully studied.

It is clear that it is possible to plan a computer simulation in a rational way only in presence of a priori estimates of the size of the errors: systematic errors may be removed only if we understand their origin and we know the dependance of the systematic errors on the various control parameter of the simulation.

In this note we address to these problems in the framework of the quenched approximation [3], i.e. no fermion loops. There are no difficulties to remove the quenching and to introduce fermionic loops, the corresponding increase in CPU time of the computation should range from 3 to 30, for more details the reader may look to refs. [3–4].

In section 2 we discuss the estimates of the statistical errors, while in sections 3, 4, 5 and 6 we discuss the different sources of systematic errors: finite lattice spacing, finite Euclidean time separation, finite volume effects and finite quark masses respectively.

2. Statistical errors

The basic formula needed to estimate statistical errors is rather simple: if A is an observable and $\bar{A}^{(N)}$ is its average after N measurements (which for simplicity we suppose to be statistically uncorrelated) we know that for large N

$$\bar{A}^{(N)} = \langle A\rangle + r(\langle A^2\rangle_c/N)^{1/2}$$
$$\langle A^2\rangle_c = \langle A^2\rangle - \langle A\rangle^2 \qquad (1)$$

where $\langle A\rangle$ is the exact expectation value and r is a Gaussian random number with unit variance. Eq. (1) is essentially the central limit theorem.

It is clear that we would like to have statistical errors as small as possible without having to increase N too much. This can be done if we find an observable B such that

$$\langle A\rangle = \langle B\rangle\,, \qquad \langle A^2\rangle > \langle B^2\rangle\,. \tag{2}$$

An elementary example of this procedure can be found in the Ising model where the Hamiltonian is

$$H = \frac{1}{2}\sum_{i,k} J_{ik}\sigma_i\sigma_k \tag{3}$$

and the σ's are variables which take only the values ± 1.

The Callen identity [5] (which is a special case of the DLR equations [6]) tell us that:

$$\langle\sigma_i\sigma_l\rangle = \left\langle \mathrm{th}\Big(\beta\sum_k J_{ik}\sigma_k\Big)\,\mathrm{th}\Big(\beta\sum_j J_{lj}\sigma_j\Big)\right\rangle \qquad \text{if } J_{il} = 0\,. \tag{4}$$

Now it is quite evident that for not too large β the r.h.s. fluctuates much less than the l.h.s.

This technique may be applied to the study of the expectation value of the thermal Wilson loop with a gain in computer time which may be very high ($O(10^2)$) [7]. Similar DLR equations can also be used to decrease the statistical errors in different simulations, e.g. in the evaluation of the average value of the induced currents in the vacuum in the pseudofermions approach to fermions loops.

Let us now estimate the errors on the hadronic masses [4, 8].

In the quenched approximation for QCD the meson propagator is [1]

$$\begin{aligned} G_\Gamma(t) &= \int \mathrm{d}^3x\, G_\Gamma(\boldsymbol{x}, t) \propto \langle O_\Gamma(0)\, O_\Gamma(t)\rangle\,, \qquad O_\Gamma(t) = \int \mathrm{d}^3x\, (\bar{q}\Gamma q)(\boldsymbol{x}, t)\,, \\ G_\Gamma(\boldsymbol{x}, t) &= \int \mathrm{d}\mu[A]\, \mathrm{Tr}[\Gamma G_q(\mathbf{0}, 0; \boldsymbol{x}, t|A)\, \Gamma\gamma_5 G_q^*(\mathbf{0}, 0; \boldsymbol{x}, t|A)\,\gamma_5] \end{aligned} \tag{5}$$

where $\mathrm{d}\mu[A]$ is the probability measure of pure gauge theory for the gauge field A. $G_q(\cdots|A)$ is the quark propagator in the background field A and Γ is the generic γ matrix. For the time being we suppose that we stay in a box of size L in three dimensions and infinite in the fourth direction (which we call time).

The energy of the low lying states is connected to the behaviour of $G_\Gamma(t)$ at large t, so that the dangerous errors are those which grow with t. Let us concentrate our attention on this region.

For large t we have [4]:

$$\begin{aligned} G_\Gamma(t) &\equiv \int \mathrm{d}\mu[A]\, G_\Gamma(t|A) \equiv \langle G_\Gamma(t|A)\rangle \sim \exp(-E_\Gamma t) \\ G_\Gamma^{(2)}(t) &= \langle G_\Gamma^2(t|A)\rangle \sim \exp(-E_{2\Gamma} t) \end{aligned} \tag{6}$$

where E_Γ and $E_{2\Gamma}$ are respectively the energy of the lowest state created from the vacuum by the action of O_Γ and $O_{2\Gamma}$ where in the second case it is understood that the two quarks and two antiquarks fields have different flavours (quarks interchange and quarks annihilation diagrams are forbidden).

There are two scenarios in the infinite volume limit: (a) $E_{2\Gamma} < 2E_\Gamma$ because of the existence of bound states or of a continuum whose energy is smaller than $2E_\Gamma$. (b) $E_{2\Gamma} = 2E_\Gamma$.

In the last case the analogy with first-order perturbation theory suggests that:

$$E_{2\Gamma} \xrightarrow[t \to \infty]{} 2E_\Gamma + C \frac{l_{ZV}}{L^3 E_\Gamma} \tag{7}$$

C being a constant and l_{ZV} being the Zweig violating scattering length.

In order to understand the meaning of this result let us see what should happen in the real world by considering the cases of the pion, the rho and the delta. Generally speaking the fact that fluctuations come from Zweig violating diagrams suggests that the statistical errors are relatively small: they go to zero in the SU(n) gauge theory when n goes to infinity.

In the pion case, PCAC tell us that the pion is a quasi-Goldstone boson whose scattering length should go to zero, when m_π goes to zero, as m_π^2. The fluctuations in $E_\pi - m_\pi$ must become proportional to $1/L^3 m_\pi$ when m_π goes to zero, note however that this result may be not true on the lattice when chiral symmetry is broken explicitly.

In the rho and delta cases let us assume that there are no exotic four quark bound states. In this situation the only danger may come from the continuum below: two pions for the rho, two pions or two rhos for the delta. Similar arguments suggest that:

$$\begin{aligned} \langle G_\rho^2 \rangle - \langle G_\rho \rangle^2 &\simeq \frac{A_{\rho\pi}}{L^3} \langle G_\pi^2 \rangle \\ \langle G_\Delta^2 \rangle - \langle G_\Delta \rangle^2 &\simeq \frac{A_{\Delta\pi}}{L^3} \langle G_\pi^2 \rangle + \frac{A_{\Delta\rho}}{L^3} \langle G_\rho^2 \rangle \end{aligned} \tag{8}$$

where the A's are some kind of scattering amplitudes. It is evident that in this case the fluctuations will become very large at large t: the only relief is that the prefactor of the most dangerous term ($\langle G_\pi^2 \rangle$) should become small with the pion mass.

It is now clear that a careful analysis of the statistical errors gives important informations on the Zweig violating scattering amplitudes: a full computation of the scattering length in exotic S-channel states (like $\pi^+\pi^+$, pp, K^+p, ...) should not present serious difficulties, because in this case only quark interchange diagrams are present (quarks annihilation diagrams are absent).

For practical purposes it is interesting to note that the final estimate of the error on the meson propagators is essentially:

$$(A/(NL^3)^{1/2}) \exp(Bt) \tag{9}$$

where the prefactor should be rather small. Eq. (9) tells us that the statistical error is controlled by NL^3: in a regime where only CPU time matters and not memory, the same amount of computer time would be needed on a small and on a large lattice, the second solution being clearly the best because it minimizes the finite volume systematical effects. It is also evident that unless the amount of CPU time is really large the data at a too large value of t will contain too much noise and would be useless. It would be much better to extract masses working at a very small value of t; unfortunately systematic effects are present in this region as we will see in section 4.

If we restrict to the study of masses in the pure gauge sector (i.e. glueballs) the situation is strongly dependent on the method used.

The conventional measurement of correlations is very painful and requires a too large amount of time [9]. Masses can also be extracted by considering not the correlation function but the response function, as it was pointed out in ref. [10]. In these two cases the analysis of the statistical errors is straightforward and no interesting phenomena are present. The most promising method seems to be the measurement of the linear response function by studying the difference in the signal in two different runs where the same set of random numbers has been used [10, 11]; this method does not work if the computer simulation is done using the Monte Carlo method, it works reasonably well for a Langevin-type simulation [12]. Unfortunately a careful analysis of the physical meaning and of the origin of the systematic effects in this case have not been done at the present moment.

3. Finite lattice spacing

In computer simulations a mesh must be introduced in the Euclidean space-time. It is usual to work with an hypercubic lattice, the lattice spacing being denoted by a. In an asymptotically-free theory without mass parameter like QCD with only zero mass quarks, only dimensionless quantities are relevant and all masses may be measured in units of a parameter Λ, which can be extracted from the experimental data on the scaling violations at high energy (in physical units Λ should stay in the range 70–200 MeV); while no parameter exists in the continuum, on the lattice we can change the value of the bare coupling constant $g_B^2 \sim 1/\beta$.

The dimensionless quantity $a\Lambda$ is a function of β, which in the relevant region of large β becomes:

$$a\Lambda = \pi(\beta)^{\omega} \exp[-\beta C_1 + C_0 + C_{-1}/\beta + \cdots] \tag{10}$$

where the constants ω and C_1 can be computed in the continuum while a computation on the lattice is needed to obtain C_0. The presence of corrections proportional to high powers of β^{-1} is rather annoying: a safe estimate would consider that $a\Lambda$ is known in the relevant region inside a factor two of incertitude until the next order correction has been computed: at the present moment it is rather useless to compare the value of Λ extracted from different forms of the lattice action.

The situation changes if we consider quantities which have a direct physical meaning, e.g. the ratio $R(\beta)$ of the masses of two glueballs with different quantum numbers. Here for large β we have that:

$$R(\beta) = R(\infty) + O(a^2\Lambda^2) = R(\infty) + O(\exp(-C_1\beta)) . \tag{11}$$

The corrections to the continuum result are exponentially small in β. Better results can be obtained if, as stressed by Symanzyk [13], we start to change the form of the lattice action; roughly speaking we have:

$$R(\beta) = R(\infty) + a^2\Lambda^2 (s_0 + s_1/\beta + s_2/\beta^2 + \cdots) . \tag{12}$$

The quantities s_k depend on the lattice action, there is a form of the action such that all the s_k' are zero; moreover it is possible by a K loop computation on the lattice to extract the form of the action for which all the s_k for $k < K$ are zero.

It is therefore possible to find a form of the lattice action such that the effects of finite lattice spacing should be very small: Monte Carlo experiments for the O(3) two-dimensional sigma model [14] and analytic results for the O(∞) [15] model show that a definite improvement is obtained also if one considers only a simple form of the action in which only tree diagrams have been used to find the coefficients of the lattice action.

It is reasonable to think that after the improvement of the action the spectrum of the low-lying states will be slightly modified for a lattice spacing of 0.1 F (and may be 0.2 F) with respect to the continuum values.

4. Finite time

We have seen in section 2 that the statistical errors in the propagator increase exponentially with the time so that it would be much better to extract the masses of the states from measurements done at small values of t. Generally speaking the comparison of Euclidean and Minkowski field theories tell us that the

$$G_A(t) = \langle A(t)\,A(0)\rangle = \sum_{n=1}^{\infty} \exp(-E_n|t|)\,|\langle 0|A|n\rangle|^2/E_n\,. \tag{13}$$

Only in the region where

$$\exp((E_2 - E_1)|t|) \gg |\langle 0|A|2\rangle|^2/|\langle 0|A|1\rangle|^2 \tag{14}$$

we can safely approximate $G_A(t)$ with $\exp(-E_1|t|)|\langle 0|A|1\rangle|^2/E_1$.

If $|\langle 0|A|2\rangle|^2/|\langle 0|A|1\rangle|^2$ would be a number of order 1 the effective mass

$$m(t) = -\frac{\mathrm{d}}{\mathrm{d}t}\ln[G_A(t)] \tag{15}$$

should be near to the true mass for not too large t.

Unfortunately if A is a local operator in space (it must be a local operator in time if eq. (13) is correct) in 4 dimensions it has at least dimension 3 for mesons and dimension $4\frac{1}{2}$ for barions and the transition matrix elements $|\langle 0|A|n\rangle|^2$ strongly increase with n. In other words a high dimension operator is much more coupled to high mass states than to low mass states. In this situation we need to study the behaviour of G_A at somewhat large t. We face an impasse: if t is small the estimated mass is not the true mass and if t is large the statistical error may be so large that nothing may be measured.

As stressed by Wilson [16] the situation may be improved if we consider a set of operators A_i and we study the correlation matrix [17]

$$G_{ik}(t) = \langle A_i(t)\,A_k(0)\rangle = \sum_{n=1}^{\infty} \exp(-E_n|t|)\,\frac{\langle 0|A_i|n\rangle\langle n|A_k|0\rangle}{E_n}\,. \tag{16}$$

Up to now this method has been mainly applied to the study of the glueball spectrum with only somewhat marginal effects [17] (in most of the cases the estimated mass was only slightly smaller than the one estimated in the naive approach). One of the possible reasons of this marginal improvement is the following: this method is mostly efficient if the operators A_i are really different, one from the other, so that they have different matrix elements with different states. It is quite possible that the operators used in numerical simulations where too similar (e.g. a plaquette of lattice side 1 and lattice side 2) and the method would have been much more successful if also quite different operators (e.g. plaquettes of much higher side) would be included.

It is rather evident that the natural thing to do would be to construct block variables by averaging the gauge fields and the quark field in space and later to construct the meson and the glueball fields by using the block variable fields. Some minor problems are posed by gauge invariance: fields at different points of the space transform in a different way; this difficulty may be bypassed, or by averaging after that the appropriate parallel transport has been done or by averaging operators at different points after a gauge fixing.

Although the question of the best choice of the set of operators is not a very deep question, it is related to the physical structure of the hadrons in a very straightforward way and its solution is a crucial step if we want to have a good measurement of the hadronic spectrum beyond the π, ρ, p and J. Careful and systematic investigations in this direction would be highly welcome.

5. Finite volume

Finite volume effects are very important: their study is not so straightforward especially in gauge theories and this has caused a certain amount of misunderstanding, e.g. it has been claimed that finite volume effects may be estimated in the framework of the potential theory (in this case they would be terribly high [18]). Fortunately enough potential theory is not a good representation for QCD and finite volume effects are much smaller than was estimated in this way.

For simplicity of arguments it may be much better to consider the case in which only one spatial direction is finite while all the other directions are infinite; of course this is not a realistic situation, however as soon as finite volume effects are small, it may be reasonable to suppose that the total finite volume effects are the sum of those coming from each direction independently. If only one direction is finite and periodic boundary conditions have been used general arguments tell us that finite volume effects go to zero like $\exp(-mL)$, m being the mass gap.

Let us consider in detail the case of pure gauge theories in the limit $N \to \infty$. The same analysis of ref. [19] tells us that we have two regimes: a confined phase in which the thermal loops are disordered and a deconfined phase where the average of thermal loops is different from zero. If one direction is finite (of length L), the system stays at a temperature T equal to $1/L$. In the confined low temperature phase the free energy of the system can be written as the zero temperature free energy plus a temperature dependent term which has a simple expression in terms of the spectrum and the S matrix of the system. Now when N goes to infinity, the spectrum and the S matrix have a finite limit, while the zero temperature free energy is proportional to N: the relative variation of the free energy with L disappears in the low temperature phase when N goes to infinity; this phenomenon is the translation in physical terms of the Eguchi Kawai [20] effect for pure gauge theory. Of course in the high temperature phase, where confinement does not hold anymore, the free energy will have a non-trivial dependence on L.

This analysis implies that for the physical case (N finite) finite volume corrections will be rather small (i.e. proportional to $1/N$) as soon as thermal and space loops are disordered. For the SU(3) case the confinement–deconfinement transition is rather sharp (first order) [21] and happens at a temperature of about 350 MeV which corresponds to a box side of about 0.6 F [22] (the value of L at which the space and thermal loops are disordered may be slightly different in the case of only one direction finite or of 4 directions finite).

If we want to have small finite volume effects we need to stay in a situation where the loops winding through the lattice are disordered: it is therefore necessary in computer simulations with small lattices to monitor them.

The same analysis may be extended to the case where quarks are present: if we restrict ourselves to the quenched approximation we notice that also in the high temperature phase finite volume effects for fermions are depressed by the presence of the gauge fields: the existence of different gauge configurations having the same action but a value of the space loops which differs from an element of the center of the group (Z_3), implies that the momenta of the quark can take the following values:

$$p_n = \frac{2\pi}{3L} n , \qquad n = 0, \ldots, 3L - 1 \tag{17}$$

not $(2\pi/L)n$ as happens in the free theory without gauge fields. In other words the quarks feel a volume which is three times larger than the true one [23]. This fact explains why finite volume effects cannot be estimated from potential theory. Of course the hadrons are singlets of SU(3) colour and they do not feel Z_3 transformations: finite volume effects on the masses of the hadrons can be related to the physical static potential between hadrons: a naive estimate for large L is [1]

$$F(L) = F(\infty) + 6 V_h(L) \tag{18}$$

where $V_h(r)$ is the static hadronic potential. A more careful analysis [8] shows that eq. (18) is essentially correct for the barions, while for the mesons only the Zweig violating part of the potential is relevant. The experimental fact that strong forces between quarks are essentially saturated and that the residual hadron–hadron force is relative weak (it disappears when N goes to infinity) indicates that finite volume effects should be much smaller that the estimates which can be done by assuming a non-relativistic potential model for the quark–quark interaction.

At any rate it is clear that it is better to work on boxes which are not too small than on the diameter of the hadron. A naive estimate of the hadron diameter can be taken to be the square root of the cross section at intermediate energies (i.e. 10 GeV).

If we use the experimental data and the SU(3) flavour symmetry we find that the proton diameter is 1.8 F while the pion diameter is 1.2 F; if we substitute strange quarks to the light quarks the hadronic radius should go down a factor of 2.

A qualitative estimate of the finite volume effects for the proton may be obtained by considering a box such that the system has the same density of the nuclear matter: this happens for a box of about 1.8 F; in this situation we expect to have a shift in the proton energy of the order of the nuclear matter binding energy, i.e. 14 MeV.

The reader interested in planning a computer experiment may use this information as he likes: personally I would recommend a box of at least 0.8 F for any kind of measurement: on such a box the masses of strange mesons may be reasonably estimated, the masses of the strange barions likely need a larger box (1.0 F) while the nucleon needs a larger box (1.5–2.0 F).

It may sound strange that I am suggesting to use boxes just of the same size as the lightest physical particles (i.e. the pion whose mass is about 1.5 F). Fortunately enough the pion is a quasi-Goldstone boson who decouples (as we have already seen) in the zero mass limit: everything must be finite when the pion acquires zero mass also in a finite box. It is useful to remark that this last statement is true if the chiral symmetry is spontaneously broken; if the chiral symmetry is broken on the lattice explicitly, the pion acquires a residual interaction which does not vanish in the zero mass limit. At the present moment it is unclear how much this effect is really annoying: it shows up in the fluctuations of the critical value of K at which the pion becomes massless in the Wilson theory. If it is needed, a drastic decision can be taken, e.g. averaging the quark propagator (not the hadronic propagator) over configurations which differs for Z_3 transformations.

6. Extrapolation in the quark masses

As we have seen from the previous sections both systematic and most unfortunately statistical errors increase strongly by decreasing the quark masses (this last point has been completely lost in the analysis of ref. [24]). This difficulty may be bypassed by studying theories where the up and down quarks have a mass higher than the physical one, only at the end we extrapolate at nearly zero mass. This procedure is imperative for the lattices of the present dimensions and it is likely that the computation of the hadronic masses at the physical point (i.e. pion mass equal to 140 MeV) will be a useless and time consuming effort also in the near future. Indeed quantities like the width of the ρ or the Δ, which are very sensitive to the pion mass, cannot be computed directly (they can be estimated from the knowledge of the $\rho\pi\pi$ or $\Delta p\pi$ couplings which can be obtained or from vector meson dominance or PCAC) while other quantities are reasonable smooth functions of the quark mass.

It is a general rule that reasonable extrapolations are possible only if something is known on the function we need to extrapolate; in the present case the dependence of physical quantities on the quark masses has been widely studied in the contest of chiral perturbation theory [25]: non-analytic terms like $m_q^{3/2}$ or $m_q^2 \ln m_q$ (m_q being the quark mass) are present whose coefficients are known as functions of the chiral parameters (e.g. f_π or g_A). It is quite reasonable (also if we consider the success of the Gellman Okubo sum rule), that a careful computation of the hadronic spectrum and of the low energy coupling constants in the region where the pion mass is not smaller than 300–400 MeV will be more than enough to obtain quite reliable extrapolations if the results of chiral perturbation theory are used.

It is a pleasure for me to thank the people I have worked with on these problems, who have strongly contributed to my present understanding: M. Falcioni, F. Fucito, H. Hamber, E. Marinari, G. Martinelli, M. Paciello, R. Petronzio, F. Rapuano, C. Rebbi, B. Taglienti, Wu Yong-shi and Zhang Yi-cheng.

References

[1] A representative set of this first generation papers is:
E. Marinari, G. Parisi and C. Rebbi, Phys. Rev. Lett. 47 (1981) 1978;
H. Hamber and G. Parisi, Phys. Rev. 23 (1983) 247;
D. Weingarten, Nucl. Phys. B215 [FS7] (1983) 1;

F. Fucito, G. Martinelli, C. Omero, G. Parisi, R. Petronzio and F. Rapuano, Nucl. Phys. B210 [FS6] (1982) 407.
G. Martinelli, C. Omero, G. Parisi and R. Petronzio, Phys. Lett. 117B (1982) 434.
A. Hasenfratz, P. Hasenfratz, Z. Kunszt and C.B. Lang, Phys. Lett. 117B (1982) 81
[2] See the contribution to this workshop of R. Petronzio and D. Wallace
[3] F. Fucito, E. Marinari, G. Parisi and C. Rebbi, Nucl. Phys. B180 [FS2] (1981) 369;
E. Marinari, G. Parisi and C. Rebbi, Nucl. Phys. B190 [FS3] (1981) 734.
H.W. Hamber, E. Marinari, G. Parisi and C. Rebbi, BNL preprint (1982)
[4] H.W. Hamber, E. Marinari, G. Parisi and C. Rebbi, Saclay preprint (1983)
[5] H.B. Callen, Phys. Lett. 4 (1963) 161
[6] R.L. Dobrushin, Theory Prob. Appl. 13 (1969) 197;
O.E. Lanford III and D. Ruelle, Comm. Math. Phys. 13 (1969) 194
[7] G. Parisi, R. Petronzio and F. Rapuano, Phys. Lett. 128B (1983) 418.
[8] G. Parisi, talk given at the Trieste Workshop (1982)
[9] At least in the region of large values of β, reasonable results have been obtained for intermediate values of β.
[10] G. Parisi, Nucl. Phys. B180 [FS2] (1981) 378, B205 [FS5] (1982) 337;
see also K.H. Mutter and K. Shilling, Cern preprint TH3246 and (1982).
C. Michael and I. Teasdale, Nucl. Phys. B215 [FS7] (1983) 433
[11] F. Falcioni, E. Marinari, M.L. Paciello, G. Parisi, B. Taglienti and Zhang Yi chen, Nucl. Phys. B215 [FS7] (1983) 289.
[12] G. Parisi and Wu Yong-shi, Scientia Sinica 24 (1981) 483
[13] See Symanzik's contribution to this workshop
[14] M. Falcioni, G. Martinelli, M.L. Paciello, G. Parisi and B. Taglienti, Nucl. Phys. B225 [FS9] (1983) in press
[15] R. Musto, F. Nicodemi and R. Pettorino, Napoli preprint IFTN 500/83 (1983)
[16] K. Wilson, talk given at the Abingdon meeting (1981)
[17] M. Falcioni, E. Marinari, M.L. Paciello, G. Parisi, F. Rapuano, B. Taglienti and Zhang Yi cheng, Phys. Lett. 110B (1982) 295.
B. Berg, A. Billoire and C. Rebbi, Annal. Phys. (NY) 142 (1982) 185
[18] P. Hasenfratz and I. Montvay, Desy preprint (1982)
[19] G. Bhanot, U. Heller and H. Neuberger, Phys. Lett. 113B (1982) 47
[20] T. Eguchi and H. Kawai, Phys. Rev. Lett. 48 (1982) 1063
[21] J. Kogut, M. Stone, H.W. Wyld, W.R. Gibbs, J. Shigemitsu, S.H. Shenker and D.K. Sinclair, Univ. of Illinois preprint, Ill-TH 82-39 (1982)
[22] We use as a reference the rho mass of (2), no reliable measurement of the string tension exists in the scaling regime
[23] G. Martinelli, G. Parisi, R. Petronzio and F. Rapuano, Phys. Lett. 117B (1982) 56;
for related considerations see also G. Parisi and Zhang Yi cheng, Nucl. Phys. FS, to be published
[24] I.M. Barbour, J.P. Gilchrist, H. Schneider, G. Schierholz and M. Teper, Desy preprint 83-012 (1983)
[25] H. Pagels, Phys. Reports 16C (1975) 219

PROLEGOMENA TO ANY FUTURE COMPUTER EVALUATION OF THE QCD MASS SPECTRUM

Giorgio Parisi

Universita' di Roma II "Tor Vergata" 00173 Roma (Italy)
and
INFN-Laboratori Nazionali Frascati 00040 Frascati (Italy)

I. INTRODUCTION

In recent years we have seen many computer based evaluations of the QCD mass spectrum[1]. At the present moment a reliable control of the systematic errors is not yet achieved; as far as the main sources of systematic errors are the non zero values of the lattice spacing and the finite size of the box, in which the hadrons are confined, we need to do extensive computations on lattices of different shapes in order to be able to extrapolate to zero lattice spacing and to infinite box. While it is necessary to go to larger lattices, we also need efficient algorithms in order to minimize the statistical and systematic errors and to decrease the CPU time (and the memory) used in the computation.

In these lectures the reader will find a review of the most common algorithms (with the exclusion of the application to gauge theories of the hopping parameter expansion in the form I have proposed[2]: it can be found in Montvay's contribution to this school); the weak points of the various algorithms are discussed and, when possible, the way to improve them is suggested.

For reader convenience the basic formulae are recalled in the second section; in section third we find a discussion of finite volume effects, while the effects of a finite lattice spacing are discussed in section fourth; some techniques for fighting against the statistical errors and the critical slowing down are found in section fifth and sixth respectively. Finally the conclusions are in sections seventh.

Although these last four years have seen a great improvement of the techniques, there is still a lot to do: it would be very nice if our collegues, which have never used a computer, would start to study these problems, not with the aim of doing simulations themselves but for finding theoretically the best way for performing simulations: the need of better algorithms cannot be overstimated.

II. BASIC FORMULAE

In this Section the basic formulae of Euclidean field theory are recalled. Let us consider a four dimensional box of sides L^3xT (with periodic boundary conditions or any other kind of homogeneous boundary condition); in most of the cases we suppose $L \gg T$ and T so large that it can be practically considered to be infinite).

If only bosons are present, there is a probability measure $d\mu[A]$ (A being the generic Bosonic field) which is proportional to $\exp\{-S[A]\} d[A]$, $S[A]$ being the Euclidean action: it can be written as the integral of a local function.

If our Euclidean theory satisfies the Osterwalder Schrader axioms (which imply the existence of a corresponding Wightman type field theory in Minkowski space), we have that:

$$O_i(t)\, O_j(o) = \Sigma_n\, C_i^{(n)}\, C_j^{(n)}\, \exp(-E^{(n)}t) \qquad (1)$$

where the operators $O_i(t)$ are functionals depending only on the A field at time t, the E_n's are the energies of the states at rest (which are supposed to be discreate) in the Minkowski space in a box of side L^3. For most of the physical application we are interested to compute the E_n's in the limit $L \to \infty$, although interesting informations on the low energy hadron-hadron intcraction may be obtained if we study the L dependence.

In presence of fermionic field ($S_F = \int d^Dx\, \bar{\psi} \Delta[A] \psi$) if C invariance is not violated, after the elimination of the Fermionic field by Gaussian integration, we remain with an effective probability distribution for the Bosonic field:

$$d\mu_F[A] = d\mu[A] \det[\Delta] \equiv d[A] \det[\Delta] \exp\{-S[A]\} \qquad (2)$$

The correlation functions of the fermionic field can be evaluated easyly using formulae like[3]:

$$\begin{aligned} \langle \bar{\psi}(x)\gamma_5\psi(x)\, \bar{\psi}(o)\gamma_5\psi(o)\rangle &= \int d\mu_F[A]\, |G(x,o|A)|^2 \\ \Delta[A]\, G(x,y|A) &= \delta(x-y) \end{aligned} \qquad (3)$$

Eq. 3 holds only if the bilinear operator $\bar{\psi}\gamma_5\psi$ is not a flavour singlet, otherwise a slightly more complex formula holds.

Although the "natural" formulation of fermions is done using anticommuting variables, only commuting quantities enter in Eqs. (2) and (3): it is possible to generate the bosonic field according to the probability distribution (2) by using a modified Monte Carlo method[4] and the Green function $G(x,y|A)$ can be analytically computed using a fast method for solving elliptic differential equations. In the so called quenched approximation the determinant in (3) is removed: this correspond to neglect virtual quarks loops.

This program may be implemented only by introducing in the space time a mesh of size a (i.e. we consider lattice field theory): on the top of the statistical errors common to any probability based computation we have systematic errors due to the non vanishing of L^{-1} and a. Although at the end we need to do an extrapolation by considering different values of a and L, it is convenient to use algorithms which have the smallest possible systematic errors; they will be described in the next section.

III. FINITE VOLUME EFFECTS

Le us discuss firstly finite volume effects in an SU(N) theory in the limit $N \to \infty$. In this case in a box of size $L^3 \times \infty$ two phases are possible:

a) $\langle P \rangle = 0$
b) $\langle P \rangle \neq 0$

where P is the trace of the Wilson loop of lenght L winding the box. It has been argued that for $L > L_c$ $\langle P \rangle = 0$ and in this phase no physical quantity depends on L[5]. This result is confirmed by the explicite formulae for finite volume correction written in terms of the S matrix, if we use the conjecture that the S-matrix is the identity for an SU(∞) theory.

For infinite N, $L > L_c$ would be enough for killing all the finite volume corrections. For finite N we cannot have phase transitions in a finite box and L_c is not sharply defined, however we can speak of two different regimes a) and b). In regime a) the effect of finite volume corrections may be systematically evaluated by considering the effect of virtual particles winding through the box[6,7]; these effects are rather small for all the virtual particles but the pion: they are proportional (roughly speaking) to[6]

$$\exp(-mL) \tag{4}$$

Moreover the corrections to the meson spectrum are Zweig suppressed and have a small prefactor, unfortunately the corrections to the barion spectrum are not Zweig suppressed and may be relatively high.

For boxes of reasonable size (i.e. 2 Fermi) the only effects on the masses may come from virtual pion exchange; fortunately these exchange is suppresed due to the Goldstone nature of the pion (Adler zeros); the decoupling of the pion in this limit may be checked by computing the zero energy scattering lenght following Ref. 4.

At the present moment the most fashionable method for decreasing finite volume effects due to meson exchange consists in changing the masses of the quarks and to perform extrapolations to small masses[FI]. A more fancy possibility for reducing finite volume effects consists in playing with the boundary conditions.

In the rich men version we introduce an addictional U(9) smell group: quarks transform under the fundamental representation of the group and are fermion of parastatistic 1/9; physical objects are singlect under this group; we can now impose twisted boundary conditions in the U(9) directions[9], strongly reducing the finite volume effects. Obviously in the infinite volume limit we recover the original theory.

In the poor men version we introduce only an U(3) smell group and the boundary conditions are imposed using the following matrix diagonal in smell space[10]:

$$\begin{vmatrix} 1 & 0 & 0 \\ 1 & \exp(i2\pi/3) & 0 \\ 0 & 0 & \exp(i4\pi/3) \end{vmatrix} \tag{5}$$

Different quarks with different smell get different phases at the boundary: for an SU(2) theory the same prescription correspond of imposing periodic and antiperiodic boundary conditions for the two smells respectively.

The poor men version kills only the leading terms ($\exp(-m_M L)$) leaving subleading terms $\exp(-m_M \sqrt{2}L)$, while the rich version kills all terms up to $\exp(-3m_M L)$ but costs more in CPU time and memory.

The poor man version is recommended in all cases where memory and not CPU is the limiting factor while the rich man version is compulsory for studing more subtle effects like the ϱ width.

VI. THE LATTICE SPACING

In order to perform a simulation it is necessary to introduce a lattice spacing (let us call it a).

In a pure gauge theory it was proven by Symanzik[11] that the finite lattice spacing corrections are proportional to a^2 and it is possible to find out an action on the lattice such that these corrections are absent: in an asymptotically free theory the action may be computed in perturbation theory in the bare coupling constant. When fermions are present the corrections are proportional to a and are much more serious.

The advantages of using an improved action have been carefully investigated by Monte Carlo in the case of the two dimensional σ-models[12] and start to be investigated in the case of the lattice gauge theories.

In my opinion the effects of the improvement would be small for pure gauge theories (provided that we stay far away from the critical point in the β-fundamental β-adjoint plane[12]) while the improvement seem to be absolutely necessary for fermions where theeffects are much strongers; careful studied in this direction would be very interesting: the field is rapidly developing and it is difficult to provided an updated list of references.

A related subject which is not yet studied is how to improve the results of a simulation done in the Langevin approach[13] by trying to compensate the finite time step effects (in this context time is the fith dimensional computer time) with the finite lattice spacing effects. Let us consider a trivial case: massless free field theory in one dimension. In the continuum the Langevin equation is

$$\dot{\varphi}= \Delta\varphi + \eta \quad \overline{\eta(x_1 t)\,\eta(y_1 t)}=2\ \delta(x-y)\ \delta(t-t') \tag{6}$$

When we introduce a lattice spacing a and a time spacing ε the Langevin equation can be written as:

$$\begin{aligned} \tilde{\varphi}_n(i) &= \varphi_n(i)+R_n(i)\,\varepsilon^{\frac{1}{2}} \qquad \overline{R_n(i)R_m(j)}= \frac{2}{a}\ \delta_{i,j}\ \delta_{n,m} \\ \varphi_{n+1}(i) &= \tilde{\varphi}_n(i)+\varepsilon\left[\tilde{\varphi}_n(i+1)+ \tilde{\varphi}_n(i-1)-2\ \tilde{\varphi}_n(i)\right]/(2\ a^2) \end{aligned} \tag{7}$$

An easy computation in momentum space tell us that:

$$G(P)=\varphi(P)\varphi(-P) = \frac{1}{P^2+(\frac{\varepsilon}{2} - \frac{a^2}{12})P^4+ O(P^6)} \tag{8}$$

at the magic value $\varepsilon=a^2/6$ the effects of order a^2 cancels with those of order ε. Independently of the possibility of cancelling errors of different origines, it is clear that if the continuum limit is done at a^2/ε constant, the final errors remain of order a^2. A more careful investigation of these problems would be welcome, also given the relevance of the Langevin equation for introducing Fermion loops.

V. FIGHTING AGAINST STATISTICAL ERRORS

Everyone would agree that if we want to measure something (e.g. $\langle 0 \rangle$ near the continuum limit) it is better to consider a quantity which has a definite probability distribution in the continuum limit, i.e. if (P(z) is the probability that O=Z, we would like that in the continuum limit the limiting probability $P_c(Z)$ exists and it is such that:

$$\int_{-\infty}^{+\infty} P_c(z)\, dz = 1 \tag{9}$$

As far as the statistical errors are proportional to $< 0^2 > - < 0 >^2$ we would like that the noise to signal ratio

$$R= \frac{N^2}{S^2} = \frac{< 0^2>}{< 0 >^2} - 1 \tag{10}$$

remains finite in the continuum limit.

Unfortunately it is well known that if 0 is a local operator in more than one dimensions:

$$P_c(z) = o \tag{11}$$

Fields are not functions but distributions: only observable constructed with smeared fields have a well defined probability distribution.

If we use local operators to compute the masses the ratio noise to signal diverges in the continuum limit; e.g. if we measure the glueball mass by looking to the plaquette plaquette correlation function the noise to signal ratio diverges like a^{-16}, making the computation impossible for small a.

Generally speaking in order to compute masses it is convenient to consider observables which are functional of the field smeared in space and not in the time in order to preserve Eq. (1). This can be trivially done in a non gauge theory. In a gauge theory we have two options:

a) we fix the gauge (i.e. the Coulomb gauge[14]) and we smear the gauge variables in this gauge;

b) we use a gauge invariant construction for the block gauge field(e.g. the one proposed by Wilson restricted on a space slice). A very simple and efficient procedure is discussed in Ref. 15, as far as the computation of the glueball is concerned.

We notice en passant that all the informations on the low energy spectrum are conteined in the block fields and going from the original lattice to a new lattice with twice lattice spacing we loose only informations on the high energy part of the spectrum, which is not so important; however the number of variables will decrease of a factor 8-16. The variational approach for computing the eigenvalus can be done by starting directly from the block field configurations (reducing the amount of work needed if we consider many operators); moreover we could save on a tape only the block field configurations and the block quarks propagator, strongly reducing the input-output problem.

It is also possible to decrease the statistical error[16] by a careful use of the DLR equations in order to find a new observable O' such that $\langle O' \rangle = \langle O \rangle$ and $\langle (O')^2 \rangle \ll \langle O^2 \rangle$.

This procedure has been used for computing the string tension with rather good results. The most efficient way I can think to measure the sting tension using the DLR identities is the following: we consider a lattice $L^3 x T$ where all gauge fields at t=0 are equal to zero (U=1). The expectation value of the Poliakoff loop P in the x direction decrease for large L as

$$\langle P(t) \rangle \simeq \exp(-KLt) \tag{12}$$

The appropriate DLR identity is:

$$\langle \Pi\, U_x \rangle = \overline{\Pi \ \langle U_x \rangle_x} \tag{13}$$

where $\langle U_x \rangle_x$ denotes the average over the links in the x direction and the bar the average over the other links. The implementation of the identity is rather simple: we start from independent gauge field configurations and we upgrade only the field in the x direction with the other fields quenched; in this way we compute an approximate expectation value of each link U_x and the Poliakoff loop is computed as the product of all the approximated expectation values. At the end we average the configurations of the gauge fields. This procedure may be also used to compute the correlation function of two Poliakoff loops at distance larger than 1 if the slice at t=0 is not cold (i.e. A=0) but it is at thermal equilibium and we quench also the values of the U_x's on the two slices (i.e. at t=0 and t=T/2). This procedure should be used only if one is interested to compute the potential, not only the string tension; indeed it is well known that unless special techniques are used the best way for computing the exponential decay of correlation functions consists in measuring the responce function[17].

VI. FIGHTING AGAINST THE CRITICAL SLOWING DOWN

From the lattice point view the continuum limit is a second order phase transition at which the coherence lengh becomes infinite. For a theory whose Lagrangian is quadratic in the momentum it is well known that the dynamics of the low frequency modes slows down of a factor a^2, i.e. the number of Monte Carlo interaction needed to produce independent configurations increase as a^{-2}, neglecting logarithms. The same result follows in the Langevin approach where the time step must be of order a^2. Pictorial we could say that the information diffuses on the lattice (or makes a random walk) so that we need a time proportional to ℓ^2 in order to change the block variable associated to a region, of size ℓ, in lattice spacing units.

This slowing down should be avoided as far as possible: at my knowledge there are two methods which solve this problem, the FFT preconditioned Langevin equation and the Multi Grid Monte Carlo.

Let us first describe the Langevin approach for a scalar field theory with action

$$S[\varphi]=\int d^dx\ \frac{1}{2}\left[(\partial_\mu\ \varphi)^2+V(\varphi)\right] \tag{14}$$

As we have already mentioned it is well known that:

$$<0[\varphi]>=\overline{0[\varphi]} \tag{15}$$

where as usual the bar denotes the average over the noise and φ is the solution of the Langevin equation:

$$\dot{\varphi}=\Delta\varphi-V'(\varphi)+\eta\quad \eta(x\ t)\ \eta(y\ t')=2\delta(t-t')\delta(x-y) \tag{16}$$

Now we would like to increase the speed of the slow variables at low momenta; a very interesting theorem tell us that equation (15) still holds if we consider the generalized Langevin equation:

$$\begin{aligned}\dot{\varphi}(x)&=\int dy\ Q(x-y)\ \left[\Delta(y)-V'(\varphi(y))\right]+\eta_Q(x)\\ &\eta_Q(x\ t)\ \eta_Q(y\ t')=2\delta\ (t-t')\ Q(x-y)\end{aligned} \tag{17}$$

where the kernel Q has been chosen in such a way to have a fast speed up the low momentum region: e.g. $Q(x)=dk\int\exp(ikx)\ (k^2+m^2)^{-1}$.

The computation of the convolution can be done in momentum space and the Fourier transform can be done using the Fast Fourier Transform algorithm which is quite efficient (the slowing down is now proportional to ln a). It looks like that the best choice of Q(x) consists in taking

$$Q(x)\simeq<\varphi(x)\varphi(0)> \tag{18}$$

however this question should be investigated in a more detailed way.

The same technique may be used for gauge theories (in the Landau gauge it is likely that Q(x) should behaves like $\exp(-x^2)$ at large distance) if we remove the constraint that the U variables belong to the gauge group and we replace it with an appropriate weight in the action; on the other hand this method could be well used for Fermions both in the pseudofermions approach and in the computation of the Green function, using the Gauss Seidel or the conjugate gradient methods.

The basic idea of the multigrad Monte Carlo approach consists in introducing a varaible for each field and also for each block field.[18] In lattice of size $L=2^n$ we can write

$$\varphi(i) = \sum_{0}^{n}{}_{k} \quad \sum_{j} c^{(k)}_{i,j} \quad \varphi^{(k)}\ (j),$$

where each component of the the index j runs from 1 to 2^{n-k} and the $c^{(k)}_{i}$ are appropriate constant.

We can now perform a Monte Carlo simulation on all the variables together; also in absence of tricks for fastening the computation, the total time for a multigrad Monte Carlo cycle will be at worse proportional to n, i.e. the algorithms is slow only of a factor ln a. The benefit of the multigrad algorithm is that the efficiency for changing the low momentum variables should be quite high and no critical slowing down should be present. This may happens if the constants C(k) are chosen in the appropriate way: simple minded arguments suggest that for an action with two derivatives the c's must be linear in j while for the action with only one derivative the c's can be constant inside each block.

The method should be particularity efficient for quadratic actions: (the CPU time for a multigrid sweep is similar to the one for an usual sweep) expecially for the computation of the Fermionic propagator which is a first order differential equation. On the contrary the application of the multigrid method to the gauge field sector seems to be particularily paintfull but it may be rewarding.

It could be very interesting to see how these methods work in a concrete case.

VII.CONCLUSIONS

I believe that in the future the methods described in the two last sections will lead to a strong reduction of the CPU time needed to perform a computation. The use of block fields for constructing the observable decreases the statistical errors; moreover as far as the block fields should excite from the vacuum only low energy particles Eq. (1) will be dominated by the lowest energy state also at moderate values of t, allowing therefore an unbiased determination of the masses of the lowing lying states. On the other hand the multigrid method should strongly fasten the slow part of the computation, i.e. the treatment of the Fermions.

It seems that we have at our disposal all ingredients for performing a successful computation of the hadronic spectrum keeping under control the systematic errors.

At the present moment, on most of the super computers the main limitation seems to be lack of memory space; the trend is reversing: new generation supercomputers will have a reasonable amount of memory (may be not enough). With the advent of 256 kb chips also dedicated computers may be equipped with a sufficient large amount of memory. A serious problem that slows down the progress in this field is the difficulty in writing down the computer code which implements the various algorithms, i.e. the software problem. This implements the various algorithms, i.e. the software problem. This problem becomes stronger if we want to write down efficient codes for pipelined or parallel machines. A possible way to overcome these difficulties is discussed in Wilson's contribution to this school.

Of course the final solution for decreasing the memory space (may be not the CPU time) is to push Symansik's improvement program or/and Wilson's renormalization group approch. This may be crucial expecially for the properties of the pion in a realistic simulation in order of not upsetting the cancellation of virtual pion exchange due to the quasi Goldstone nature of the pion.

ACKNOWLEDGMENTS

It is a pleasure to thank M. Falcioni, E. Marinari, G. Martinelli, M. Paciello, R. Petronzio, F. Rapuano, B. Taglienti, Y.C. Zhang for many interesting discussions.

I am also grateful to H. Hamber and K.G. Wilson for useful discussions and suggestions concerning the multigrid method.

REFERENCES

1. See for example H. Hamber, Proc. of the Intern. Conf. on "Matematical Physics", Boulder (1983), to be published; G. Parisi, Proc. Summer School Cargese (1983), to be published.
2. G. Parisi, Nuclear Phys. B205, /FS 5/, 337 (1982).
3. H. Hamber and G. Parisi, Phys. Rev. Letters 47, 1972 (1981); E. Marinari, G. Parisi and C. Rebbi, Phys. Rev. Letters 47, 1975 (1981).
4. F. Fucito, E. Marinari, G. Parisi and C. Rebbi, Nuclear Phys. B180, /FS 2/, 369 (1981); H. Hamber, E. Marinari, G. Parisi and C. Rebbi, Nuclear Phys. to be published and references therein.
5. T. Eguchi and H. Kawai, Phys. Rev. Letters 48, 1063 (1982); G. Bhanot, U. Heller and H. Neuberger, Phys. Letters 113B, 47 (1982); G. Parisi, Phys. Letters 112B, 463 (1982); G. Parisi and Y.C. Zhang, Nuclear Phys. 215, /FS 4/, 182 (1983).
6. H. Hamber and G. Parisi, Phys. Rev. D27, 3215 (1982).
7. See H. Luscher contribution to this school.
8. For a recent review see J. Gasser and H. Leutvyler, Phys. Reports, in press.
9. T. Eguchi, J. Jurkiewicz and C.P. Korthals Altes, in Proc. Workshop Word Scientific (1982).
10. G. Martinelli, G. Parisi, R. Petronzio and F. Rapuano, Phys. Letters 122B, 283 (1983).
11. K. Symanzik in "Methematical Problems in Theoretical Physics" eds. R. Schrader et al. Lecture Notes in Physics 153, (Springer, Berlin, 1982).
12. M. Falcioni, G. Martinelli, M.L. Paciello, G. Parisi, B. Taglienti, Nuclear Phys. B225, /FS 9/, 313 (1983); B. Berg, S. Meyer, I. Montvay and K. Symanzik, Phys. Letters 126B, 467 (1983).

13. G. Parisi and Y.S. Wu, Scientia Sinica 24, 483 (1981); M. Falcioni, E. Marinari, M.L. Paciello, G. Parisi, B. Taglienti and Y.S. Zhang, Nuclear Phys. B215, /FS 7/, 265 (1983).
14. G. Parisi, R. Petronzio, R. Rapuano, in preparation.
15. M. Falcioni, G. Parisi, M. Paciello and B. Taglienti, in preparation.
16. G. Parisi, R. Petronzio and F. Rapuano, Phys. Letters 128B, 418 (1983).
17. For a discussion of this point see G. Parisi in the Proc. of the Workshop "Word Scientific", Trieste (1982).
18. For a review of the multigrid method see A. Brandt and N. Dinar in "Numerical Methods for Partial Differential Equations" ed. by S.V. Parter, (Academic Press 1979).

FOOTNOTES

F1. It is well know that in reality the masses of the barions and the mesons contain non analytic terms like $m_q^{3/2}$ and $m_q^2 \ln m_q$ (m_q being the quark mass) with computable coefficients; moreover there are indications from the experimental value of the sigma term[8] in pion proton scattering that strong non linearities in the barionic masses are present when the quark mass changes from the up to the strange mass. These terms should be practically absent in the quenched approximation: they are Zweigh suppressed: similar arguments suggest that the nucleon mass may be about 1100 MeV in the quenched approximation, i.e. it should be similar to the purified mass of Ref. 8.

F2. The observed "scaling" behaviour of the gluball mass near 5.6 implies that these data are useless to conclude something on the continuum limit: indeed both the string tension and the deconfinement critical temperature do not scale in this region and the scalar glueball mass is the most sensitive quantity to the existence of the nearby critical point: it seems likely the observed "scaling" of the glueball is simple a coincidence without any deep meaning.

COURSE 2

A SHORT INTRODUCTION TO NUMERICAL SIMULATIONS OF LATTICE GAUGE THEORIES

Giorgio PARISI

Dipartimento di Fisica
Università Roma II "Tor Vergata"
Rome 00173, Italy

and

INFN, sezione di Roma

K. Osterwalder and R. Stora, eds.
Les Houches, Session XLIII, 1984
Phénomènes critiques, systèmes aléatoires, théories de jauge/
Critical phenomena, random systems, gauge theories

Contents

1. Introduction

It is widely believed that strong interactions are described by a field theory of spin-1/2 particles coupled to non-Abelian gauge particles (quarks and gluons).

Present experience strongly suggests that a computation of the hadronic spectrum of such a theory can be done with small errors (statistical and systematic) on computers available in the near future by means of numerical simulations.

In these lecture notes I will show how such a program may be accomplished and which are the principles at the basis of numerical simulations of gauge theories. I would like to stress that numerical simulations (applied also to other models) have a very strong heuristic power: they may be used to establish which results must be proved by analytic considerations.

These notes are organized as follows: in sect. 2, we find some general considerations on numerical simulations for Euclidean field theory; in sect. 3, the approach based on the Langevin equation is described; in sect. 4 we show how to deal with fermionic degrees of freedom; in sect. 5 we introduce the Monte Carlo algorithm; in sect. 6 we mention some of the various problems connected with the analysis of Monte Carlo data; in sect. 7 we remind the reader of some of the basic concepts in continuum gauge theories; in sect. 8 we introduce lattice gauge theories and we review some of the analytic results; in sect. 9 we present some of the most important results for Abelian gauge theories; in sect. 10 we show some numerical results for non-Abelian gauge theories (only SU(3) for reasons of space) also in presence of fermions; finally in the last section we discuss some possible techniques which may be useful to decrease the statistical errors by a large factor.

2. Numerical simulations

In Euclidean field theory [1] (i.e. in the statistical mechanics of a continuous medium) we are mainly interested in the evaluation of the statistical expectation values defined by

$$\langle f[\varphi]\rangle = Z^{-1}\int \mathrm{d}[\varphi]\exp\{-S[\varphi]\}f[\varphi]\,, \tag{2.1}$$

where Z is a normalization factor such that $\langle 1\rangle = 1$, φ is a function (a set of functions) of the space variables x, $S[\varphi]$ is the Euclidean action which often can be written as

$$S[\varphi] = \int \mathrm{d}^D x\, \mathscr{L}(x)\,, \tag{2.2}$$

where $\mathscr{L}(x)$ depends only on the field φ and its derivatives at the point x.

Under certain conditions, we can associate to the Euclidean field theory a quantum field theory in Minkowski space, characterized by a vacuum state $|0\rangle$ and by a field $\hat{\varphi}$, which is an operator acting on the Hilbert space to which the vacuum belongs; the correspondence is, roughly speaking, given by

$$\langle f[\varphi]\rangle = \langle 0|f[\hat{\varphi}]|0\rangle\,. \tag{2.3}$$

More precisely for the correlation function of two local operators $A(x)$ and $B(y)$ one finds

$$\begin{aligned}\langle A(x)B(y)\rangle_c &\equiv \langle A(x)B(y)\rangle - \langle A(x)\rangle\langle B(y)\rangle \\ &= \sum_n a_n b_n \exp\{-E_n|x-y|\}\,,\end{aligned} \tag{2.4}$$

where

$$a_n = \langle 0|\hat{A}|n\rangle\,, \qquad b_n = \langle n|\hat{B}|0\rangle \tag{2.5}$$

and E_n is the energy of state $|n\rangle$ (in most cases we should substitute an integral for the sum of eq. (2.4)).

Eqs. (2.4) and (2.5) show that the lowest energy levels control the asymptotic decay of the correlation functions. In particular if

$$\langle A(x)A(0)\rangle \sim \exp(-|x|/\xi_A) \quad \text{for } x\to\infty\,, \tag{2.6}$$

$\xi_A^{-1} \equiv m_A$ is the mass of the particle at the lowest energy level having the same quantum numbers of A. The minimum over A of $m_A(m)$ is called the mass gap and is the inverse of the so-called coherence length ξ (the maximum over A of ξ_A).

The main idea at the basis of numerical simulations is to evaluate numerically the integral in eq. (2.1) and to extract the mass spectrum of the theory from a careful analysis of the exponential decay of the correlation functions.

It is evident that this program can be done only when the number of degrees of freedom of φ is finite. To this end it is convenient to consider only the case in which the field φ is defined in a box of size L^D (D being the dimension of the space); it is known that if periodic (or antiperiodic) boundary conditions are used we have

$$\langle \ \rangle_L = \langle \ \rangle_\infty + O(\exp(-L/\xi)) , \tag{2.7}$$

so that, if $L \gg \xi$, practically no corrections are present (L must be larger than ξ also if we want to apply eq. (2.6) to compute m_A). A mesh must be also introduced in space: if a is the lattice spacing (let us consider cubic lattices) the field φ will be defined only at the points of the lattice; the total number of lattice points (V) is $(L/a)^D$ which is finite.

Let us suppose that $S_a[\varphi]$ is an action for the lattice field such that

$$\frac{\int d[\varphi]_a \exp\{-S_a[\varphi]\} f[\varphi]}{Z_a} \underset{a\to 0}{\longrightarrow} \frac{\int d[\varphi] \exp\{-S[\varphi]\} f[\varphi]}{Z} . \tag{2.8}$$

Very often (if $S_a[\varphi]$ is well chosen) the difference among the r.h.s. and the l.h.s. may be of $O(a^2)$, although better results may be obtained by a careful choice of S_a [2]. We only need to compute the l.h.s. of eq. (2.8) for different values of a and L and to extrapolate the results up to $a = 0$ and $L = \infty$. The errors connected with this extrapolation are sometimes called systematic errors; in these lecture notes we will mainly address the problem of computing expectation values for a system with a finite, but large, number of degrees of freedom, and neglect the theory of systematic errors [3].

The integrals in eq. (2.8) cannot be computed in a practical way using conventional methods, i.e. evaluating the integrals in a given set of points: if n points are taken for each variable, the total number of points needed is n^V which is a gigantic number.

The solution to this difficulty consists in undoing what Boltzmann, Gibbs and Einstein did: We consider a system $\varphi(t)$, with a given law of evolution in time (e.g. $(\mathrm{d}/\mathrm{d}t)^2\varphi_i = -\partial S/\partial\varphi_i$) and we compute

$$\bar{f} = \lim_{T\to\infty} \frac{1}{T}\int_0^T \mathrm{d}t'\, f[\varphi(t')] . \tag{2.9}$$

If ergodic theorems are valid, we have

$$\bar{f} = \langle f \rangle , \tag{2.10}$$

provided that the initial energy (i.e. $\Sigma_{i=1}^{V} \frac{1}{2}\dot{\varphi}_i + S[\varphi]$) has been chosen in such a way that

$$\overline{\left(\frac{\mathrm{d}\varphi_i}{\mathrm{d}t}\right)^2} = \frac{1}{2} . \tag{2.11}$$

If the time T which must be used in eq. (2.9) is not terribly large, i.e. if it does not increase faster than a power of V, the CPU time (i.e. the number of operations) needed to evaluate the l.h.s. of eq. (2.10) is a polynomial in V while the CPU time for the r.h.s. is exponential in V. From the narrow point of view of a computer the work of the founding fathers of statistical mechanics has been to transform a simple problem into a terrible mess. Although we do not share this point of view, it is clear that it is better to compute numerically the l.h.s. of eq. (2.10). Different algorithms will correspond to different equations of motion (we could for example use $\mathrm{d}\varphi_i/\mathrm{d}t = -\mathrm{sign}(\partial S/\partial\varphi_i)$). Generally speaking deterministic equations of motion may be dangerous because the appropriate ergodic theorems have not been established; it is more convenient to consider stochastic equations of motions, in which possibly the time is a discrete variable. These stochastic methods will be described in the following sections; they rely on the availability of sequences of random numbers. Although no finite state machine like a computer can produce a sequence of really random numbers (in the same way that no generic real number can be stored in the memory of such machine), it is possible to produce sequences of numbers which for all practical purposes are random.

3. The Langevin equation

The prototype for a stochastic equation for which eq. (2.10) is satisfied, is the Langevin equation [1, 4]

$$\frac{\mathrm{d}\varphi_i}{\mathrm{d}t} = -\frac{\partial S}{\partial \varphi_i} + \eta_i(t)\,, \tag{3.1}$$

where the η's are Gaussian distributed random numbers with zero mean and covariance

$$\overline{\eta_i(t)\eta_j(t')} = 2\delta(t-t')\delta_{ij}\,. \tag{3.2}$$

For the purist who does not like this formulation we can rewrite eq. (3.1) as (for $t>0$)

$$\varphi_i(t) - \varphi_i(0) = -\int_0^t \mathrm{d}t' \, \frac{\partial S}{\partial \varphi_i}(t') + W_i(t)\,, \tag{3.3}$$

where the W's are still Gaussian distributed variables with zero mean and covariance

$$\overline{W_i(t)W_j(t')} = 2\delta_{ij}\min(t, t') \quad (t>0, t'>0)\,. \tag{3.4}$$

It is easy to show that if $\int \mathrm{d}[\varphi]\exp\{-S[\varphi]\}$ is finite and $S[\varphi]$ is a sufficient smooth function (and the space of the φ's is a connected manifold) eq. (2.10) holds. We denote by $P_{\varphi_0}(\varphi \mid t)$ the probability that $\varphi(t) = \varphi$ if $\varphi(0) = \varphi_0$; eq. (2.10) is satisfied if

$$P_{\varphi_0}(\varphi \mid t) \xrightarrow[t\to\infty]{} \exp\{-S[\varphi]\}\,, \tag{3.5}$$

independently of φ_0.

If we introduce a time mesh of size ε, eqs. (3.1) and (3.2) become

$$\varphi_i(n\varepsilon + \varepsilon) - \varphi_i(n\varepsilon) = -\varepsilon \frac{\partial S}{\partial \varphi_i}[\varphi(n\varepsilon)] + \varepsilon^{1/2} R_i^{(n)}\,,$$

$$\overline{R_i^{(n)} R_j^{(m)}} = 2\delta_{nm}\delta_{ij}\,. \tag{3.6}$$

By studying carefully the limit $\varepsilon \to 0$, we find that $P(\varphi \mid t)$ satisfies the Fokker–Planck equation

$$\frac{\partial P}{\partial t} = \sum_{i=1}^{V} \left\{ \frac{\partial}{\partial \varphi_i} \left[P \frac{\partial S}{\partial \varphi_i} \right] + \frac{\partial^2}{\partial \varphi_i^2} P \right\}, \tag{3.7}$$

where we have not indicated the dependence of $P(\varphi \mid t)$ on φ_0.

In order to study the solutions of eq. (3.7) it is convenient to introduce the "reduced probability" ρ defined by

$$\rho(\varphi \mid t) = \exp\{\tfrac{1}{2}S[\varphi]\} P(\varphi \mid t) . \tag{3.8}$$

Eq. (3.8) implies that ρ satisfies the following equation:

$$\frac{\partial \rho}{\partial t} = \left\{ \sum_{i=1}^{V} \left(\frac{\partial}{\partial \varphi_i} \right)^2 - U[\varphi] \right\} \equiv -\hat{\mathcal{H}}\rho ,$$

$$U[\varphi] \equiv \sum_{i=1}^{V} \left[\frac{1}{4} \left(\frac{\partial S}{\partial \varphi_i} \right)^2 - \frac{1}{2} \frac{\partial^2 S}{\partial \varphi_i^2} \right]. \tag{3.9}$$

$\hat{\mathcal{H}}$ is a self-adjoint operator and $\psi_0[\varphi] \propto \exp\{\frac{1}{2}S[\varphi]\}$ is an eigenstate of $\hat{\mathcal{H}}$ with zero eigenvalue. The well-known variational theorems imply that $\psi_0[\varphi]$, being always positive, is the unique ground state of $\hat{\mathcal{H}}$.

We finally find:

$$P_{\varphi_0}(\varphi \mid t) = \sum_n \psi_n[\varphi]\psi_n[\varphi_0] \exp[-E_n t]\psi_0[\varphi]/\psi_0[\varphi_0] ,$$

$$\hat{\mathcal{H}}\psi_n[\varphi] = E_n \psi_n[\varphi] . \tag{3.10}$$

Eq. (3.10) reduces to eq. (3.5) in the limit $t \to \infty$.

The time needed to forget the initial conditions, i.e. for the system to go to equilibrum, is of the order of $(E_1)^{-1}$ $(E_0 = 0)$.

A more careful analysis shows that for a general trajectory we have at large times

$$\frac{1}{t} \int_0^t f[\varphi(t')] \, dt' = \langle f \rangle + e(t) , \tag{3.11}$$

where the error $e(t)$ is a Gaussian distributed variable, with zero mean and variance

$$\overline{e^2(t)} = \frac{2}{t} \sum_{n=1}^{\infty} |\langle \psi_0 | f | \psi_n \rangle|^2 / E_n . \tag{3.12}$$

The statistical error is going to zero as $t^{-1/2}$, as could be argued a priori.

Let us consider a simple example, the free-field case. We can work directly in the continuum limit; the Langevin equation becomes

$$\frac{\partial\varphi}{\partial t} = -(-\Delta + M^2)\varphi + \eta(x,t) ,$$
$$\overline{\eta(x,t)\eta(y,t')} = 2\delta(x-y)\delta(t-t') . \tag{3.13}$$

This equation can be easily solved; for example, if $\varphi(x,0) = 0$ we have

$$G(p,t) \equiv \int d^D x \exp(ipx)\overline{\varphi(x,t)\varphi(0,t)}$$
$$= \frac{1}{p^2 + M^2}[1 - \exp(-2(p^2 + M^2)t)] . \tag{3.14}$$

We see that the time needed to forget the initial condition is (in the small-momenta region) proportional to $1/M^2$, as can be guessed by dimensional analysis; also the statistical errors in such a region are proportional to $(tM^2)^{-1/2}$. This phenomenon is called critical slowing down and it is very important because it slows down numerical simulations when M goes to zero.

For a φ^4 interaction the Langevin equation becomes

$$\frac{\partial\varphi}{\partial t} = -(-\Delta + M^2)\varphi + g\varphi^3 + \eta(x,t) . \tag{3.15}$$

The usual techniques can be applied to expand the solutions of eq. (3.15) in powers of g. A diagrammatic approach can be used and, after a small tour de force, one recovers the usual Feynman diagrams for the equal-time correlations at equilibrium. Using arguments based on the renormalization group one finds that for $g \neq 0$

$$t_{eq} \equiv (E_1)^{-1} \sim \xi^z \equiv m^{-z} ,$$

where $m = \xi^{-1}$ is the physical mass and z is a critical exponent which is equal to 2 for $D \geq 4$ and can be computed using the $4 - \varepsilon$ expansion for $D < 4$. For most systems z is not very far from 2 also for $D < 4$.

This slowing down implies that if we want to go to the continuum limit at fixed ratio L/ξ, L being the size of the box, the CPU time increases essentially as $(\xi/a)^{D+2}$, which is much worse than the naive estimate $(\xi/a)^{D}$. Some proposals for overcoming this difficulty are briefly discussed in the last section.

4. Fermions

When anticommuting fields are present, the previous approach is not so straightforward; the simplest thing to do is to integrate out the fermionic variables using the Mattews–Salam formula

$$\int \mathrm{d}[\psi]\exp[-\bar{\psi}D\psi] = \det[D]. \tag{4.1}$$

For example if our action is

$$S[\psi, A] = \sum_{f=1}^{n} \bar{\psi}_f D[A]\psi_f + S[A], \tag{4.2}$$

we have

$$\langle f[A]\rangle \propto \int \mathrm{d}[A]\exp[-S_{\mathrm{eff}}[A]],$$

$$S_{\mathrm{eff}}[A] = S[A] - n_{\mathrm{F}}\,\mathrm{Tr}\ln[D[A]], \tag{4.3}$$

where we have assumed that $\det[D[A]]$ is real and non-negative.

Eq. (4.3) would also hold for bosons if we substitute n_{F} with $-n_{\mathrm{B}}$ (n_{B} being the number of bosonic flavours). The possibility of computing $\langle f\rangle$ for negative n_{F}, using the techniques of the previous sections with an action

$$S[\varphi, A] = \sum_{f=1}^{-n_{\mathrm{F}}} \bar{\varphi}_f D[A]\varphi_f + S[A], \tag{4.4}$$

and later extrapolating to positive h_{F}, is certainly interesting, but fails for theories like QCD with nearly-zero- (zero-)mass quarks, where a phase transition is present for n_{F} nearly zero (zero).

If we look directly to eq. (4.3), we can associate to it the Langevin equation

$$\dot{A}(x,t) = -\frac{\delta S}{\delta A(x,t)} + n_F \operatorname{Tr}\left[\frac{\delta D}{\delta A(x,t)} D^{-1}\right] + \eta(x,t),$$

$$\overline{\eta(x,t)\eta(y,t')} = 2\delta(x-y)\delta(t-t'). \tag{4.5}$$

By introducing the Green function $G(x, y \mid A)$, which satisfies the equation

$$D_x[A]G(x,y \mid A) = \delta(x-y), \tag{4.6}$$

eq. (4.5) can be written as

$$\dot{A} = -\frac{\delta S}{\delta A} + n_F \frac{\delta D}{\delta A} G(x,x \mid A) + \eta. \tag{4.7}$$

Eq. (4.7) is rather difficult to evaluate numerically because the computation of $G(x, x \mid A)$ is a nonlocal operation which may take a lot of time.

A possible method to deal with fermions is the so-called pseudo-fermion method (also other methods are available). One introduces a field $\varphi(x, t)$ (commuting) and writes down the following equations [5]:

$$\dot{A} = -\frac{\delta S}{\delta A} + n_F \bar{\varphi} \frac{\delta D}{\delta A} \varphi + \eta, \qquad \dot{\varphi} = -\tau D(A)\varphi + \tau^{1/2}\zeta,$$

$$\overline{\zeta(x,t)\zeta(y,t')} = 2\delta(x-y)\delta(t-t'), \qquad \overline{\eta\zeta} = 0. \tag{4.8}$$

When $\tau \to \infty$ the evolution of φ is much faster than the evolution of A, so that for small Δt with probability

$$\frac{1}{\Delta t}\int_t^{t+\Delta t} dt' \, \bar{\varphi}(x,t')\, \varphi(y,t') \to G(x,y \mid A(t)) \quad (\tau \to \infty). \tag{4.9}$$

Explicit computations for a system with a finite number of degrees of freedom show that the solutions of eq. (4.8) go to those of eq. (4.7), with an error of order τ^{-1}. Therefore, by taking τ sufficiently large (the CPU time increases like τ) the error may become as small as we want.

The same results can also be proved in an elegant way in perturbation theory by using a diagrammatical approach.

The pseudo-fermion method gives us a way to compute the properties of interacting fermions with controlled errors; the only difficulty may be the extrapolation to $\tau = \infty$. A simple formula can be written for $\langle \bar{\psi}\psi \rangle$:

$$\langle \bar{\psi}\psi \rangle = -n_{\mathrm{F}} \langle \bar{\phi}\phi \rangle . \tag{4.10}$$

The situation is more complex if one looks for the correlation fucntions of four fermions. Indeed, generally speaking

$$\begin{aligned} &\langle \bar{\psi}_a(x)\, \psi_b(x)\, \bar{\psi}_b(0)\, \psi_a(0) \rangle \\ &\quad = \int \mathrm{d}\mu_{\mathrm{eff}}[A] \{ G(x,0 \mid A) G(0,x \mid A) \\ &\qquad\qquad - \delta_{ab} G(x,x \mid A) G(0,0 \mid A) \} , \end{aligned} \tag{4.11}$$

where $\mathrm{d}\mu_{\mathrm{eff}}[A] \propto \mathrm{d}[A] \exp\{-S_{\mathrm{eff}}[A]\}$, while

$$\begin{aligned} &\langle \bar{\varphi}(x)\varphi(x)\bar{\varphi}(0)\varphi(0) \rangle \\ &\quad = \int \mathrm{d}\mu_{\mathrm{eff}}[A] \{ G(x,0 \mid A) G(0,x \mid A) + G(x,x \mid A) G(0,0 \mid A) \} . \end{aligned} \tag{4.12}$$

One solution consists in doubling the number of pseudo-fermions while the other solution consists in adding a term proportional to $\bar{\psi}(0)\psi(0)$ to the action and observe the variation of $\bar{\psi}(x)\psi(x)$. If, however, we are interested in computing eq. (4.12) with $a \neq b$, a very efficient method consists in computing $G(x,0 \mid A)$ analytically by solving eq. (4.6) for each field A. The last method is very efficient for computing the large-distance decay of the correlation functions; rather accurate and fast converging methods for computing $G(x,0 \mid A)$ are the Gauss–Seidel method [6] or the conjugate gradient method; a more efficient way is the FFT preconditioned Gauss–Seidel method as discussed in ref. [3].

5. The Monte Carlo method

For practical purposes the Langevin equation can be integrated only after the introduction of a time mesh as in eq. (3.6). It is clear that the results will depend on ε. Different strategies are possible:

(a) We take ε very small and we extrapolate to $\varepsilon = 0$; for eq. (3.6) the deviations are proportional to ε.

(b) We use a better discretization in time (as in the Runge–Kutta method) and we reduce the deviations to $O(\varepsilon^2)$ or $O(\varepsilon^3)$; the algorithm is more complex but ε does not need to be very small.

(c) We take the continuum viewpoint and we try to cancel effects of order $O(\varepsilon)$ with those of $O(a^2)$ [3].

(d) We discretize the theory in such a way that the statistical expectation values are not changed [1, 7].

We concentrate our attention on the last strategy. Our time-discretized algorithm will be characterized by the transition probability $TP[\varphi, \psi]$, which is the probability of having the state φ at "time" $n+1$ if ψ was the state at "time" n. Conservation of probability implies that

$$\int d[\varphi]\, TP[\varphi, \psi] = 1 \quad \forall\psi . \tag{5.1}$$

Eq. (3.12) suggests that we must have

$$TP[\varphi, \psi] = \exp[-\tfrac{1}{2}S(\varphi) + \tfrac{1}{2}S[\psi]]K[\varphi, \psi] , \tag{5.2}$$

where K is a symmetric kernel.

Eq. (5.2) can also be written as

$$\exp[-S[\psi]]\, TP[\varphi, \psi] = \exp[-S[\varphi]]\, TP[\psi, \varphi] , \tag{5.3}$$

the so-called detailed balance condition.

The transition probability for going from ψ to φ in n steps is given by

$$TP^{(n)}[\varphi, \psi] = \exp[-\tfrac{1}{2}S[\varphi] + \tfrac{1}{2}S[\psi]]\, K^{(n)}[\varphi, \psi] , \tag{5.4}$$

where $K^{(n)}[\varphi, \psi]$ is the kernel of the nth power of the operator $\hat{K}$. Under mild hypotheses $\hat{K}$ has only one eigenvalue equal to 1; if we call $\omega[\varphi]$ the corresponding eigenvalue, we obtain:

$$\lim_{n\to\infty} TP^{(n)}[\varphi, \psi] = \exp[-\tfrac{1}{2}S[\varphi] + \tfrac{1}{2}S[\varphi]]\omega[\varphi]\omega[\psi] . \tag{5.5}$$

Conservation of probability (i.e. $\int d[\varphi]\, TP^{(n)}[\varphi, \psi] = 1$) implies

$$\omega(\psi) \propto \exp[-\tfrac{1}{2}S[\psi]] ,$$

$$TP^{(n)}[\varphi, \psi] \underset{n\to\infty}{\propto} \exp[S[\varphi]] . \tag{5.6}$$

More generally, if we know that when $n \to \infty$

$$\mathrm{TP}^{(n)}[\varphi, \psi] \to P_{\mathrm{eq}}[\varphi] , \tag{5.7}$$

the distribution $P_{\mathrm{eq}}[\varphi]$ must satisfy the condition

$$\int \mathrm{d}[\psi]\, \mathrm{TP}[\varphi, \psi]\, P_{\mathrm{eq}}[\psi] = P_{\mathrm{eq}}[\varphi] . \tag{5.8}$$

Therefore we may have $P_{\mathrm{eq}}[\varphi] \propto \exp[-S[\varphi]]$ only if the "balance condition"

$$\int \mathrm{d}[\psi]\, \mathrm{TP}[\varphi, \psi] \exp[-S[\psi]] = \exp[-S[\varphi]] \tag{5.9}$$

is satisfied (by integrating eq. (5.3) it is easy to see that the detailed balance condition implies the balance condition).

Summarizing, if the balance condition is satisfied and eq. (5.7) holds or if the detailed balance condition is satisfied we have

$$\lim_{n\to\infty} \frac{1}{n} \sum_{n'=1}^{n} f[\varphi^{(n')}] = \langle f \rangle . \tag{5.10}$$

It is important to note that in the previous discussion there was no need for having φ near to ψ so that we are not restricted to small ε. We can also forget the Langevin equation and to try to implement directly eq. (5.3) or eq. (5.9).

A rather simple method for implementing these conditions is the following. The algorithm breaks down into two steps: given a configuration $\varphi^{(n)}$ a configuration $\bar{\varphi}^{(n)}$ is suggested according to a given algorithm; afterward a decision is taken: either $\varphi^{(n+1)} = \bar{\varphi}^{(n)}$ or $\varphi^{(n+1)} = \varphi^{(n)}$. If the kernel $S[\varphi, \psi]$ of the suggestor algorithm is symmetric (i.e. $S[\varphi, \psi] = S[\psi, \varphi]$) and if the suggestion is accepted (i.e. $\varphi^{(n+1)} = \bar{\varphi}^{(n)}$) with probability $g(\Delta S)$, $\Delta S \equiv S[\bar{\varphi}^{(n)}] - S[\varphi^{(n)}]$, where the function $g(y)$ satisfies the relation $g(y) = \exp(-y) g(-y)$ (e.g. $g(y) = 1$ for $y < 0$, $g(y) = \exp(-y)$ for $y > 0$), the detailed balance condition is satisfied; indeed one finds

$$\mathrm{TP}[\varphi, \psi] = g(S[\varphi] - S[\psi]) S[\varphi, \psi] + A[\psi] \delta[\psi - \varphi] , \tag{5.11}$$

where $A[\psi]$ is fixed by the condition of conservation of probability.

The decision algorithm is very simple to implement: one takes a random number (r) uniformly distributed in the interval [0, 1] and if $g(\Delta S) \geq r$ the suggestion is accepted. A practical way to construct the suggestor algorithm is to pick randomly one site i of the system and put

$$\bar{\varphi}_i^{(n)} = \varphi_i^{(n)} + r_\lambda \,, \qquad \bar{\varphi}_j^{(n)} = \varphi_j^{(n)} \quad \text{for } i \neq j \,, \tag{5.12}$$

where r_λ is a random number uniformly distributed in the interval $[-\lambda, \lambda]$.

The efficiency of the algorithm, i.e. the number of times we need to apply it in order to obtain a statistically uncorrelated distribution, strongly depends on λ. If as usual we take $g(x) = \exp(-x)$ for $x \geq 0$, and if λ is very small, the suggestion is nearly always accepted, but the new configuration will be very similar to the old one. If λ is very large, the suggestion will likely be rejected and the algorithm is slow. It is often stated than an acceptance rate of the order of 50% is optimal; the precise value, however, of the optimal acceptance rate depends on the system.

In many cases it is more convenient to update the configuration in a way similar to eq. (5.12), but choosing the sites in a preassigned order (not randomly), for example sequentially. In this last case a Monte Carlo sweep consists in applying the algorithm to all the sites of the system, one after the other; a simple analysis shows that the algorithm satisfies only the balance condition, which is enough if only one equilibrium probability exists.

It is interesting to note that if we set

$$\lambda = \varepsilon^{1/2} \,, \qquad t = N\varepsilon \,, \tag{5.13}$$

in the limit $\varepsilon \to 0$, $\varphi(t)$ satisfies the Langevin equation (3.1), or better the evolution in time of the probability satisfies the Fokker–Planck equation (3.4). It can be argued that the considerations on the critical slowing down of sect. 3 apply here as well, the number of sweeps playing the role of time.

Up to now we have considered continuous systems without constraints: if the variables are discrete like in the Ising model or satisfy a constraint like in SU(3) gauge theories, the form of the suggestor must be different.

In these two models we can set respectively

$$\bar{\sigma}_i^{(n)} = -\sigma_i^{(n)} \,, \qquad U_i^{(n)} = g_r U_i^{(n)} \,. \tag{5.14}$$

where g_r is an SU(3) matrix chosen randomly with the condition that g_r^* has the same probability to be chosen as g_r; this can be done by constructing a list of matrices from which g_r is chosen.

It is easy to write down a computer code for doing Monte Carlo simulations; an example for the two-dimensional Ising model on a square lattice of size $L \times L$ ($L \leqslant 100$) with periodic boundary conditions can be found in appendix 1; the program is rather slow: much faster codes may be obtained by using a table look-up for exp(effective_force * current_spin), avoiding the evaluation of the exponential at each updating.

The CPU time for a sweep is in a first approximation proportional to the number of variables that must be updated. On a minicomputer like the Vax 11/780 it may range from 25 μs per spin for the three-dimensional Ising model to 15 ms per link for the SU(3) gauge theory in four dimensions with Wilson action. These times are typical for a well-written Fortran Code; it is certainly possible to do better by writing a part or all of the code in machine language. On special purpose computers the time may be much smaller and can reach 40 ns for a spin update in the Ising model.

6. Some general considerations

Let us suppose that we have a Monte Carlo program which is working and that after N_0 sweeps the new configuration is statistically independent (in a first approximation) from the previous one. What can we do?

If we decide to measure an observable A we have

$$\frac{1}{N} \sum_{N'=1}^{N} A^{(N')} = \langle A \rangle + e_N , \tag{6.1}$$

where e_N is a Gaussian random number with variance

$$\overline{e_N^2} \simeq \frac{\langle A^2 \rangle_c}{N-1} , \qquad \langle A^2 \rangle_c \equiv \langle A^2 \rangle - \langle A \rangle^2 , \tag{6.2}$$

and for simplicity we have assumed $N_0 = 1$. We can thus define a noise-to-signal ratio

$$R_A^2 \equiv \frac{\langle A^2 \rangle_c}{\langle A \rangle^2} = \frac{\langle A^2 \rangle}{\langle A \rangle^2} - 1 . \tag{6.3}$$

To measure $\langle A \rangle$ with good relative accuracy is an easy task if R_A is small; on the contrary, if R_A is large, the situation is very bad. If we stay far from a phase transition and A is a local variable, taking care of the transitional invariance of the system ($A = V^{-1} \Sigma_{i=1}^{V} A_i$, V being the total number of points) we have (if $\xi = O(1)$)

$$R_A \simeq \sqrt{C/V}\,, \qquad \overline{e_N^2} \simeq C/VN\,, \tag{6.4}$$

C being a constant for large V.

For a large system only a few Monte Carlo sweeps are enough to get a very good accuracy. On the contrary, if we need to measure a correlation function like $\langle O(0)O(x)\rangle_c$ which decays at large x like $\exp(-|x|/\xi_0)$, the noise-to-signal ratio increases exponentially at large x:

$$\begin{aligned} &\langle (O(x)O(0))^2\rangle_c \simeq \langle O^2(0)\rangle^2\,, \quad x \to \infty\,, \\ &R \sim \exp(|x|/\xi_0)\,. \end{aligned} \tag{6.5}$$

In this last case the measurement is very difficult because, unfortunately, the spectrum of the theory is determined by the expontential tail of the correlation functions. We will see in sect. 11 some of the techniques which can be used to overcome this unpleasant situation.

Summarizing, the internal energy and similar thermodynamics-related quantities are easily measured while the spectrum always gives some problems.

When we go near phase transitions new problems are present. Near a first-order phase transition we may be locked in a metastable phase for an incredibly long time (in this case a precise determination of the critical temperature can be achieved by doing mixed-phase runs). Near second-order phase transitions we are interested in computing the critical exponents, for which the two most efficient methods are the finite size scaling method and the Monte Carlo renormalization group method. Both methods have been used recently to produce rather accurate measurements of the critical exponents for the three-dimensional Ising model.

Extra difficulties are present in many random systems (random magnetic field or spin glasses) for which the time to equilibrate may be very, very large.

Apart from the determination of the phase diagram and other related thermodynamic quantities in nonrandom system, special purpose techniques must be invented to be efficient measurements of other quantities.

Before presenting the results of the application of these techniques to lattice gauge theories let me recall to the reader some simple facts on continuum gauge theories.

7. Gauge theories

We start by defining some geometrical concepts for gauge fields [8]. Let G be a group and let g denote an element of the group. Like in general relativity we can define a "parallel transport", i.e., if C_{xy} is a path going from x to y we can associate to it a group element $t(C_{xy})$.

If we transport through the space an object, which belongs to a representation of the group (in the same way as we transport vectors in general relativity), $\langle\psi|$ transported from x to y along the path C_{xy} becomes

$$\langle\psi'| = \langle\psi|t(C_{xy}) . \tag{7.1}$$

It is natural to suppose that, when the path C_{xy} goes from x to z along C_{xz} and from z to y along C_{zy}, and $(C_{xy})^{-1}$ is the reverse of the path C_{xy},

$$\begin{aligned} t(C_{xy}) &= t(C_{xz})t(C_{zy}) , \\ t(C_{xy}^{-1}) &= [t(C_{xy})]^{-1} . \end{aligned} \tag{7.2}$$

By definition a gauge transformation consists in applying a group rotation in each point of space; i.e., if $\langle\psi|$ is at the point x, after a gauge transformation it becomes

$$\langle\omega_{\mathrm{R}}| \equiv \langle\psi|g(x) , \tag{7.3}$$

where $g(x)$ is a group-valued function which generates the gauge transform. After a gauge rotation the new parallel transport is related to the old one by the relation

$$t_{\mathrm{R}}(C_{xy}) = g^{-1}(x)t(C_{xy})g(x) . \tag{7.4}$$

In the same way as in general relativity the interesting objects are invariant under general coordinate transformations, it is assumed here that the theory is invariant under gauge transformations. The prototype of a gauge-invariant object is the Wilson loop: We consider a closed circuit L and let C_{xx} be the path going from x to x along L; if χ is the character of the group (if the group is represented as matrices it is the trace, normalized to $\chi(1) = 1$), we have, by definition,

$$W_\chi(L) = \chi(t(C_{xx})) . \tag{7.5}$$

It is easy to check (using the property $\chi(g^{-1}Ag) = \chi(A)$) that $W_\chi(L)$ is gauge invariant and does not depend on x.

If $t(C_{xy})$ is a continuous function of y, G must be a continuous group. For definiteness let us assume that G is a compact finite-dimensional Lie group and let us call $\mathcal{R}$ the associated Lie algebra.

If we assume that $t(C_{xy})$ is a smooth function of y (apart from isolated singularities) we can define the gauge field A_μ as

$$\mathrm{i}A_\mu(y) = [t(C_{xy})]^{-1} \frac{\partial}{\partial y_\mu} t(C_{xy}) . \tag{7.6}$$

It is evident that A_μ belongs to $\mathcal{R}$.

Using the conditions (7.2) it is easy to see that $A_\mu(y)$ does not depend on C_{xy}; moreover if C_y^δ is a straight path going from y to $y + \delta$ we have formally;

$$t(C_y^\delta) = \exp[\mathrm{i}\delta_\mu A^\mu + O(\delta^2)] . \tag{7.7}$$

A covariant derivative D_μ can be defined as

$$D_\mu|\psi\rangle = (\partial_\mu + \mathrm{i}A_\mu)|\psi\rangle , \tag{7.8}$$

in such a way that under a gauge transformation $D_\mu|\psi\rangle$ transforms in the same way as $|\psi\rangle$.

The variation of the gauge field A_μ under a gauge transformation can be readily obtained from eqs. (7.4) and (7.7):

$$A_\mu \to g^{-1}(x) A_\mu g(x) + \mathrm{i}g^{-1}(x)\partial_\mu g(x) . \tag{7.9}$$

We can introduce the equivalent of the curvature tensor of general relativity, i.e.

$$F_{\mu\nu} = \mathrm{i}[D_\mu, D_\nu] = \partial_\mu A_\nu - \partial_\nu A_\mu + \mathrm{i}[A_\mu, A_\nu] . \tag{7.10}$$

If C_x^σ is a small circular loop characterized by a surface tensor $\sigma_{\mu\nu}$, we find

$$t(C_x^\sigma) = \exp[\mathrm{i}F_{\mu\nu}\sigma^{\mu\nu} + \mathrm{O}(\sigma^2)] . \tag{7.11}$$

Under a gauge transformation we have

$$F_{\mu\nu}(x) \to g^{-1}(x) F_{\mu\nu}(x) g(x) . \tag{7.12}$$

The simplest gauge-invariant form of the action is

$$S[A] = \frac{1}{e^2} \int \mathrm{d}^D x \, \tfrac{1}{4} \Big[\sum_{\mu,\nu} \mathrm{Tr}(F_{\mu\nu}^2) \Big] . \tag{7.13}$$

In Minkowski space the classical equations of motions are:

$$D_\mu F_{\mu\nu} = 0 . \tag{7.14}$$

It has been recently discovered that these equations can also be written in terms of purely gauge-invariant quantities, i.e.

$$\frac{\partial}{\partial x_\mu} \frac{\delta}{\delta \sigma_{\mu\nu}} W_x(C) = 0 . \tag{7.15}$$

In ref. [9] the reader can find a precise definition of the symbols used in eq. (7.15).

Eq. (7.15) can be generalized to the quantum case [10], but its consequences are not fully understood. The Euclidean quantum field theory is obtained by assuming a probability distribution for the A field given by

$$\mathrm{d}\mu[A] = \mathrm{d}[A] \exp\{-S[A]\} . \tag{7.16}$$

A perturbation expansion is possible when the coupling constant e goes to zero. In this limit we can see $e\tilde{A}_\mu = A_\mu$ and obtain a Gaussian probability distribution for $\tilde{A}_\mu$.

One finds for the expectation value of a smooth Wilson loop

$$\langle W_x(C) \rangle = \exp\Big[-\frac{e^2}{4\pi} c_x \oint \mathrm{d}x_\mu \oint \mathrm{d}y_\mu \frac{1}{(x-y)_+^2} - \mathrm{O}(e^4) \Big] , \tag{7.17}$$

where the +-sign indicates that the integral is formally linearly divergent when x goes to y; therefore, we must subtract from it the infinite part which is proportional to the length of C; c_χ is a constant which depends on the representation.

It is remarkable that rather complex objects can be written in a simple way by using the expectation values of the Wilson loop; for example, we have formally

$$\langle \mathrm{Tr} \ln(-D_\mu D^\mu + M^2) \rangle = \int_0^\infty \frac{\mathrm{d}t}{t} \int \mathrm{d}^D x \int \mathrm{d}P'_{xx}[\omega] \exp(-tM^2) \langle W_\chi(\omega) \rangle \,,$$

where $\mathrm{d}P'_{xx}[\omega]$ is the Wiener measure for a trajectory to go from x to x in a time t.

This kind of formula shows that the static potential between two massive charged particles (from now on we will only consider χ to be the fundamental representation) is given by

$$V(x) = \lim_{t\to\infty} \frac{1}{t} \ln\langle W(C_{xt}) \rangle \,, \tag{7.19}$$

where C_{xt} is a rectangular loop with sides x and t. In the limit $e^2 \to 0$ one finds

$$V(x) = \alpha[1/x - \text{const.}] \,, \tag{7.20}$$

where $\alpha = e^2/4\pi$ and the constant is a linearly divergent quantity as the classical self-energy of the electron. No divergencies are present if we consider the force defined as

$$F(x) = |\nabla_x V| = \alpha/x^2 \,. \tag{7.21}$$

If the theory is Abelian, eqs. (7.17), (7.20) and (7.21) are exact. If the theory is non-Abelian, higher-order corrections are present. In order to compute these corrections we must introduce an ultraviolet cut-off.

An explicit computation shows that in SU(N) gauge theories [11]

$$\alpha_{\mathrm{R}}(x) = x^2 F(x) = \alpha + b\alpha^2(\log x\Lambda + h) \,,$$
$$x\Lambda \gg 1 \,, \qquad b = 11N/6\pi \,, \tag{7.22}$$

where h depends on the definition of the ultraviolet cut-off.

The infinite cut-off must be done by setting

$$\alpha = \alpha(\lambda^{-1}), \tag{7.23}$$

where $\alpha(\lambda^{-1})$ is an appropriate function of λ which behaves as $1/(b \ln \lambda)$ when $\lambda \to \infty$. The running coupling constant $\alpha_R(x)$ remains finite when λ goes to ∞ and simultaneously α, the bare coupling constant, goes to zero at short distance; it can be parametrized (in a first approximation) as

$$\alpha_R(x) = \frac{1}{(b \ln(x\Lambda))}, \tag{7.24}$$

where Λ is a scale parameter which has the dimension of mass (in the physical world it is about 100 MeV).

If α is the coupling constant appearing in the action, one finds that for $\lambda \to \infty$:

$$\alpha(\lambda^{-1}) = \alpha_R(\lambda^{-1}) + H\alpha_R^2(\lambda^{-1}) + O(\alpha_R^3), \tag{7.25}$$

where the constant H depends on the precise way in which the ultraviolet cut-off λ has been defined. Eqs. (7.24) and (7.25) imply

$$\Lambda = C\lambda \exp(-1/b\alpha), \qquad C \equiv \exp(H/b). \tag{7.26}$$

In other words, the continuum limit $\Lambda/\lambda = \infty$ is reached when $\alpha \to 0$.

From this analysis it should be clear that the small-x behaviour of $F(x)$ can be studied in perturbation theory, but nothing can be said on the large-x behaviour.

It is often conjectured that in non-Abelian gauge theories

$$\lim_{x \to \infty} F(x) \equiv K \neq 0, \tag{7.27}$$

i.e., "charged" particles are confined. K/Λ^2 is a pure number that should be computed by using nonperturbative techniques.

Similarly it is assumed that no massless particles are present and the correlation functions of closed Wilson loops decay like $\exp(-|x|m_g)$ (in Abelian theories we have $1/x^8$), m_g being often referred to as the "glueball" mass. The ratio $m_g/K^{1/2}$ is a very interesting pure number.

Another interesting quantity is related to the temperature for the transition from confinement to deconfinement. A finite-temperature (T)

Minkowski field theory corresponds to a Euclidean field theory defined on a strip of side $l = T^{-1}$ with periodic boundary conditions.

If we consider the Polyakov loops $P(x)$, i.e. the straight Wilson loops winding the strip going from x to x, we have two regime for $x \to \infty$:

$$\text{(a)} \quad \langle \mathrm{Tr}\, P(x)\, \mathrm{Tr}\, P^*(0) \rangle \sim \exp(-lK(l)|x|)\,;$$

$$\text{(b)} \quad \langle \mathrm{Tr}\, P(x)\, \mathrm{Tr}\, P^*(0) \rangle \sim \text{const.}\,, \tag{7.28}$$

where $K(\infty) = K$. For large l (a) holds, while (b) is realized for small l. The critical temperature $T_c \equiv l_c^{-1}$ is fixed by the value of l which separates the two regimes.

It is rather interesting to comute T_c and, in the case of a first-order transition, the latent heat. We will see in the following sections how such computations can be done by using the Monte Carlo method for lattice gauge theories.

8. Lattice gauge theories

Let us consider a hypercubic lattice, a being the lattice spacing [12]. In the simplest approach the objects are situated at points of the lattice and we need to specify the parallel transport from one point to the nearest one, i.e. $U(\boldsymbol{x}, \boldsymbol{x} \pm \boldsymbol{\mu})$ where $\boldsymbol{\mu}$ is a vector in the direction μ of length a. Eq. (7.2) implies

$$U(\boldsymbol{x}, \boldsymbol{x} + \boldsymbol{\mu})U(\boldsymbol{x} + \boldsymbol{\mu}, \boldsymbol{x}) = 1\,. \tag{8.1}$$

Roughly speaking,

$$U(\boldsymbol{x}, \boldsymbol{x} + \boldsymbol{\mu}) = \exp(\mathrm{i}aA_\mu(x))\,. \tag{8.2}$$

It is natural to take the U's (which belong to the group) as natural integration variables. The action (7.13) may be transcribed on the lattice by using eq. (7.11) for the smallest possible circuits we can have on the lattice, i.e. a square; in a first approximation we have

$$\begin{aligned} &U(x, x + \mu)U(x + \mu, x + \mu + \nu)U(x + \mu + \nu, x + \nu)U(x + \nu, x) \\ &\quad \simeq \exp(\mathrm{i}F_{\mu\nu}(x)a^2)\,. \end{aligned} \tag{8.3}$$

Let us denote the product of the four U's by $U_{\square}$, where the square (plaquette) denotes the circuits going through x, $x+\mu$, $x+\mu+\nu$, $x+\nu$, x. Naively we would set

$$\frac{1}{4}\int d^D x\, \mathrm{Tr}(F_{\mu\nu})^2 = \tfrac{1}{2}a^{D-4}\sum_{\Box} \mathrm{Re}\,\mathrm{Tr}(1-U_{\Box}) \equiv S_L[U]\,. \qquad (8.4)$$

The probability distribution of the U's will be proportional to

$$d[U]\exp(-\beta S_L[U])\,, \qquad \beta \equiv 1/e^2\,, \qquad (8.5)$$

$d[U]$ being the product of the Haar measure for each U.

It is possible to prove that in the limit $a \to 0$ we recover the perturbative expansion where a^{-1} plays the role of the ultraviolet cut-off λ. Moreover, for all dimensionless quantities like m_g/Λ or $T_c/K^{1/2}$ the error induced by the lattice spacing is of order a^2. Eq. (7.27) becomes:

$$\Lambda = \bar{\Lambda}[1+O(1/\beta)]\,, \qquad \bar{\Lambda} = (\pi a)^{-1}C\exp(-4\pi\beta/b)\,, \qquad (8.6)$$

where for SU(N) theories

$$C \simeq 2.4\exp\left[\frac{3\pi^2}{11}\,\frac{N^2-1}{N^2}\right]. \qquad (8.7)$$

Unfortunately $\Lambda(\beta)/\bar{\Lambda}(\beta)$ is not a known function; it can be computed in perturbation theory but it has not been computed yet, so only the ratio $m_g/\bar{\Lambda}(\beta)$ can be computed. It is clear that we expect for this ratio corrections to the continuum limit proportional to $1/\beta$, i.e. $1/\ln(a^2)$, which are much larger than for other quantities (e.g. m_g/T_c). It has been suggested [13] that it may be possible that the corrections are smaller if we consider the ratio $m_g/\bar{\bar{\Lambda}}$, where

$$\bar{\bar{\Lambda}} = (\pi a)^{-1}\bar{C}\exp\left[-\frac{4\pi\beta}{b}\,\mathrm{Re}\langle\mathrm{Tr}(U_{\Box})\rangle\right], \qquad \bar{C} \simeq 2.4 \qquad (8.8)$$

[we still have $\Lambda(\beta)/\bar{\bar{\Lambda}}(\beta) = 1 + O(1/\beta)$].

Lattice gauge theories can be defined also for discrete groups like Z_2, because the Lie algebra does not play any role in the main definitions. For reasons of space we will skip all discussions concerning discrete groups.

Just after the introduction of lattice gauge theories it has been found that in the region $\beta \simeq 0$ (the continuum theory for non-Abelian theories is around $\beta \simeq \infty$) the theory can be solved exactly. In this case one finds that for small β all gauge theories are confined, independently of the group, the string tension and the glueball mass being for small β

$$K = a^{-2}[\ln(1/\beta) + O(1)] ,$$

$$m_g = a^{-1}[4 \ln(1/\beta) + O(1)] . \tag{8.9}$$

In principle one would expand K and m_g in powers of β up to an arbitrary order and use these series to compute the large-β behaviour; unfortunately the series are rather short and the results obtained are not fully satisfactory. Our most safe results for not too small β (in the region where the perturbative $1/\beta$ expansion cannot be applied) come from the Monte Carlo method as the reader will see in the following sections.

9. Abelian gauge theories

In this section we will mainly consider the U(1) gauge theory in four dimensions. We have already seen that this theory is very simple in the continuum: the static potential $V(x)$ is equal to α/x and all positive values of α are allowed.

The situation is more complex for the lattice. The gauge variables U's belong to U(1) and they can be written as $\exp(i\theta)$, θ being defined modulo 2π. The contribution of a single plaquette to the action is

$$\frac{1}{2\alpha}(1 - \cos\theta_p) = \frac{1}{\alpha}\sin^2(\theta_p/2) , \tag{9.1}$$

where θ_p is the sum (with the appropriate signs) of all the θ's corresponding to the links belonging to the plaquette.

The lattice theory is not exactly soluble because the energy is not a quadratic form in the θ_p's. If the gauge group would be R (and not U(1)), the action would be simply:

$$\frac{1}{4\alpha}\theta_p^2 , \tag{9.2}$$

and the theory would be soluble. Indeed in the U(1) case the action must be a periodic function of θ_p.

When α is very large or very small the theory is well understood. In the small-α region a perturbative computation can be done: one finds that for $x/a \gg 1$

$$V(x) \sim \alpha_R/x \,, \tag{9.3}$$

where α_R is a function of α (i.e. $\alpha_R = \alpha + C_2\alpha^2 + C_3\alpha^3 + \cdots$).
In the large-α region we have

$$V(x) \sim Kx \,, \tag{9.4}$$

where K is of order $1/a^2$.

For different values of α the system stays in two different phases: the "Coulomb" phase and the confined phase. A simple analysis shows that a massless photon is present at small α (the plaquette–plaquette correlation function decreases like $1/x^8$ at large distances), while at large α the lightest particle is a glueball of mass of order $1/a$. The theory exists in the continuum limit only in the small-α region.

A transition must be present to separate the two different phases (the so-called deconfinement transition). The nature of this transition may be partially elucidated by interpreting it as a transition from a disordered state at high temperature (large α) into an ordered state at low α. It is tempting to associate the α/x term in the potential with the exchange of a Goldstone boson. This possibility has been suggested a long time ago by Ferrari and Picasso [14] and I will present here my personal version of this argument.

In a box of size L^4 the theory is invariant under the group G^{L^4} (i.e., the product of the group G with itself L^4 times), because we can do an independent gauge transformation at each point. It is well known that this group is too large to be broken; we can, however, fix the gauge and reduce the symmetry.

Let us consider the Coulomb gauge: For each configuration of the U's, the corresponding configuration in the Coulomb gauge is the configuration which (still being a gauge transform of the original configuration) maximizes:

$$\sum_i \sum_{\mu=1}^{3} \mathrm{Re}\{\mathrm{Tr}[U_\mu(i)]\} \,, \tag{9.5}$$

i.e. the configuration which has the spacelike U's as near to the identity as possible. Although according to Gribov many local maxima can be present (maximizing expression (9.5) is equivalent to finding the ground state of a spin glass), in the general case only one global maximum will be present (apart from symmetry considerations).

It is well known that if we transform the U's with a gauge transform that is time dependent but space independent, expression (9.5) will not change. If the average in the Coulomb gauge is defined as the average over all gauges that maximize expression (9.5), this average will be invariant under the group G^{L}. We suggest that in the infinite-volume limit this symmetry is broken in the Coulomb phase and that it is not broken in the confined phase. More precisely, we may have two possibilities:

$$\text{(a)}\quad \langle \mathrm{Tr}[U_4(\boldsymbol{x},0)U^{+}(\mathbf{0},0)]\rangle \xrightarrow[x\to\infty]{} 0\,,$$

$$\text{(b)}\quad \langle \mathrm{Tr}[U_4(\boldsymbol{x},0)U^{+}(\mathbf{0},0)]\rangle \underset{x\to\infty}{\sim} \exp(-A|\boldsymbol{x}|)\,, \tag{9.6}$$

where the correlation functions are computed at the same time.

As far as $\langle U_4\rangle$ is zero by symmetry considerations, in case (a) the clustering property is violated and the G^{L^3} symmetry is broken spontaneously: a Goldstone boson is expected to be present, producing a connected correlation function proportional to $1/x$. In case (b) the symmetry is not broken and A is approximately related to the string tension (we expect that $aK \sim A$).

Let me notice en passant that the study of the correlations of products of links in the time direction in the Coulomb gauge is likely to be the best method for obtaining information on the static potential.

The argument is suggestive and properties (a) and (b) can be verified in the small- and large-α regions, respectively; however, it is not clear to me if the argument is always correct, in particular for non-Abelian gauge theories and/or in the presence of matter fields.

Another point which is unclear is what happens when the U(1) group is replaced by the group Z_N, for large values of N. The argument here presented would suggest that, as far as the symmetry is discrete, the Goldstone boson (and by consequence also the Coulomb force) disappears. This result is in contrast with the accepted folklore; it may be interesting to analyze carefully this point.

Having more or less understood the deconfinement transition from a symmetry point of view, we can now discuss the nature of the transition. Two possibilities are open:

(a) The transition is second order: At the transition point (coming from the confined phase) the glueball mass (m_g) becomes zero in lattice spacing units; the ratio k/m_g^2 should go to a finite value (probably different from zero). If the value of the renormalized charge at the critical point is finite (coming from the deconfined phase) its value α_T^R should not depend on the way the action on the lattice has been chosen and it should be the maximum value allowed for the charge. In this way we could define a nontrivial confining U(1) theory: it is likely that magnetic monopoles are present in such a theory and the confinement is an effect of their condensation.

(b) The transition is first order: The internal energy is discontinuous and the mass of the glueball remains finite at the transition point. In this case no new U(1) theory is obtained.

It may be discussed at length which of the two cases is realized. In the absence of a clear-cut theoretical argument (in the Ising model the situation is much better), only experiment (i.e. computer simulations) can settle the question.

The first simulations on really small lattices (5^4) [15] indicated a second-order phase transition. More careful experiments with larger lattices indicate a "small" first-order transition, i.e. a small discontinuity in the expectation value of the action. This negative result does not imply that the deconfinement transition for U(1) is never second order. For example, in the three-dimensional three-state Potts model, when the action depends on two parameters, we have a line of first-order transitions which ends and becomes a line of second-order transitions [16]. A similar phenomenon seems to be present here.

The action

$$\frac{1}{2\alpha}\{f\sin(\theta_p^2/2)+\tfrac{1}{4}g\sin^2(\theta_p)\}\,,\quad f+g=1\,, \tag{9.7}$$

has been studied in ref. [17]. Apparently for $g<g_c<0$ ($|g|$ not too large), the transition is second order, opening again the possibility of having a nontrivial U(1) in the continuum. A still more careful analysis would be welcome.

It seems that at the present moment analytic studies are excellent in predicting the various possibilities which are open, while only numerical work (guided by the theoretical understanding obtained from the analytic studies) may discriminate among the various possibilities and pick the correct one.

10. SU(3) gauge theories

In the real world fermions are present; however, it is rather difficult to do reliable computations in presence of fermions; a lot of work has been done to understand pure gauge theories in the absence of fermions. These efforts are not useless; one gains a lot of experience in finding the different sources of systematic errors; moreover, it is known that the effects of fermionic loops are proportional to $1/N$ in a SU(N) gauge theory and it is hoped that they are not large for $N = 3$.

The simplest quantities to be computed are the glueball mass and the static potential. Very large scale evaluations of the static potential have been done recently (see for example ref. [18]) and others are running at the present moment. The introduction of the DLR identities (see next section) has been very useful and has led to a gain in CPU time of a factor of up to 10^4. It is likely that by the end of 1985 the situation should be rather good. For the glueball mass only shorter simulations are available and the situation is more confused (see for example ref. [24]).

A fast way to study the properties of hadrons containing fermions (i.e. quarks) is the quenched approximation, first introduced in refs. [19, 20]. It consists of using eq. (4.11) with the standard measure $\mathrm{d}\mu[A]$ for the gauge fields instead of $\mathrm{d}\mu_{\mathrm{eff}}[A]$. The computation is essentially correct in the limit $N \to \infty$ and it is likely that the final results are at least in qualitative agreement with the true ones.

I will not discuss here the various results obtained within this approximation (these can be found in ref. [21]); I will only mention some interesting points.

In the continuum, if the quark mass is zero, the action is invariant under chiral transformations (γ_5 symmetry); this symmetry is believed to be spontaneously broken in the real world in the limit of zero mass; the pion is a quasi-Goldstone boson and its mass is proportional to the square root of the quark mass. It is possible to quantize the theory on the lattice, preserving part ot this symmetry. The numerical simulations show that this symmetry is spontaneously broken in the limit of zero quark mass, as was expected.

Various quantities have been computed (e.g. the masses of the rho, nucleon and delta, the decay constant of the pion, the magnetic moment of the baryons, the weak decay constant of the kaons, etc.). The pattern is essentially correct with a discrepancy with respect to the real world of 20–30% on average. Unfortunately, a careful analysis of systematic errors, in order to make a reliable comparison of the experimental data with the numerical computations, has not been done. It is clear that one should work in the range $a \ll \xi \ll L$, a, ξ and L being the lattice spacing, the range of the strong interactions ($\xi \simeq 1/m_p = 3$ fm) and the size of box, respectively, and carefully measure the dependence of the observables on ξ/a at fixed L/ξ and on L/ξ at fixed ξ/a. This will be done.

At the present moment there are some computations that have been done in the presence of a fermionic loop. The results are very preliminary because the implementation of the pseudo-fermionic algorithm depends on many parameters [21]; the interesting papers deal with the optimization of the algorithm. Work is in progress.

11. Statistical errors

Generally speaking the statistical errors are large either because of the disappearance of the signal (e.g. correlation functions at large distance) or because of the critical slowing down.

First I will present some techniques to decrease the noise-to-signal ratio. The main idea is the following: as far as the noise for measuring $\langle A \rangle$ is proportional to $\langle A^2 \rangle_c$, if one finds a quantity B such that

$$\langle A \rangle = \langle B \rangle, \qquad \langle A^2 \rangle_c \gg \langle B^2 \rangle_c, \tag{11.1}$$

it is much more convenient to measure B than A. A simple example of this approach can be constructed using the DLR equations: In an Ising model with nearest-neighbour interaction we have

$$\begin{aligned} \langle \sigma_i \rangle &= \left\langle \tanh\left[\beta \sum_k J_{ik}\sigma_k\right]\right\rangle, \\ \langle \sigma_i\sigma_j \rangle &= \left\langle \tanh\left[\beta \sum_k J_{ik}\sigma_k\right] \tanh\left[\beta \sum_{k'} J_{jk'}\sigma_{k'}\right]\right\rangle, \quad |i-j|>1, \end{aligned} \tag{11.2}$$

where J_{ik} is different from zero and equal to 1 only if i and k are nearest neighbours. It is evident that

$$\left\langle \tanh^2\left(\beta \sum_k J_{ik}\sigma_k\right)\right\rangle < \langle \sigma_i^2 \rangle = 1. \tag{11.3}$$

The quantities on the r.h.s. of (11.2) fluctuate much less than those on the l.h.s. For the correlation functions it is possible to decrease the error squared by a factor of O(10), and therefore to decrease the CPU time by a factor of O(10); the precise value of this factor depends on the value of β. A similar technique can be applied to gauge theories for measuring large Wilson loops; here the gain in CPU time may be of the order of a factor 100–1000 [22].

Another approach which may be quite useful is based on the response theory: If we consider the action

$$S_\varepsilon = S_0 + \varepsilon A(0) , \tag{11.4}$$

we have (apart from possible cancellations)

$$\begin{aligned} &\langle A(x)\rangle_\varepsilon - \langle A(x)\rangle_0 \cong \exp(-|x|/\xi_A) , \quad x \to \infty , \\ &\langle A(x)A(0)\rangle_0^c \sim \exp(-|x|/\xi_A) , \quad x \to \infty . \end{aligned} \tag{11.5}$$

More precisely, for small ε we have

$$\langle A(x)\rangle_\varepsilon - \langle A(x)\rangle_0 = -\varepsilon\langle A(x)A(0)\rangle_0^c + O(\varepsilon^2) . \tag{11.6}$$

We thus face two strategies; either compute eq. (11.6) for small ε or for large ε. In this second case the signal is of O(1) and can be measured without any difficulty, at least for not too large x.

Let us see a simple example: We have a D-dimensional Ising model in a box $L^{D-1} \times T$, with periodic boundary conditions; if we see $\sigma = 1$ on the time slice at $x_D \equiv t = 0$ we have

$$m(t) = \frac{1}{L^{D-1}} \sum_{x_D = t} \sigma(x) \sim \exp(-t/\xi) , \tag{11.7}$$

for $1 \ll t \ll T/2$, at least in the high-temperature phase. An elementary estimate shows that the increase in efficiency due to this method is proportional to L^{D-2} for large L, so that another factor 10–100 is easily obtained in $D = 4$ [23].

At first sight we cannot use eq. (11.6) for small ε because the error in $\langle A\rangle$ is of O(1); this can be done only if $\langle A\rangle_\varepsilon - \langle A\rangle_0$ is computed as average of quantities which are of $O(\varepsilon)$. A possible framework is provided by the Langevin equation [24]: We consider for simplicity the case in which $A = \varphi$. We have

$$\dot{\varphi}_\varepsilon(x) = -\frac{\delta S}{\delta\varphi(x)} + \varepsilon\delta(x) + \eta(x, t) . \tag{11.8}$$

Therefore

$$\dot{C}(x) = \int \mathrm{d}y \, \frac{\delta^2 S}{\delta\varphi_0(x)\, \delta\varphi_0(y)} \, C(y) + \delta(x) , \tag{11.9}$$

where

$$C(x, t) = \frac{\partial}{\partial\varepsilon} \varphi_\varepsilon(x, t)\bigg|_{\varepsilon=0} , \qquad \overline{C(x)} = \langle \varphi(x)\varphi(0)\rangle . \tag{11.10}$$

For example, in the φ^4 case

$$\dot{\varphi}_0 = -(-\Delta + m^2)\varphi_0 - g\varphi_0^3 + \eta ,$$

$$\dot{C} = -(-\Delta + m^2 + 3g\varphi_0^2)C + \delta(x) . \tag{11.11}$$

It easy to check that the method is very efficient if $g = 0$ and $m^2 > 0$, while it is less adequate when $K(x, y) = \delta^2 S/\delta\varphi(x)\, \delta\varphi(y)$ is no longer a positive operator, e.g. $m^2 < 0$, $g > 0$. The efficiency of the method in practical applications strongly differs from case to case. It is evident also that a prerequisite for applying this method is the finiteness of $\overline{C^2(x)}$, which is not evident a priori.

An easy computation [25] for only one degree of freedom shows that $\overline{C^2}$ is finite for the equation

$$\dot{\varphi} = -\frac{\mathrm{d}V}{\mathrm{d}\varphi} + \eta , \tag{11.12}$$

only if the Hamiltonian

$$\hat{H}_2 = -\frac{\mathrm{d}^2}{\mathrm{d}\varphi^2} + \frac{1}{4}\left(\frac{\mathrm{d}V}{\mathrm{d}\varphi}\right)^2 + \frac{3}{2}\frac{\mathrm{d}^2V}{\mathrm{d}\varphi^2} = \hat{H} + 2\frac{\mathrm{d}^2V}{\mathrm{d}\varphi^2} \tag{11.13}$$

has no negative eigenvalues. Indeed, let us define the response function $R(t, t')$ as

$$\frac{\delta}{\delta\nu(t')} \varphi_\nu(t) \equiv R(t, t') , \tag{11.14}$$

where $\varphi_\nu(t)$ satisfies the equation

$$\dot{\varphi}_\nu = -\frac{\mathrm{d}V}{\mathrm{d}\varphi} + \eta + \nu(t)\,. \tag{11.15}$$

Following ref. [25] the generalized Lyapunov exponent $\lambda^{(n)}$ can defined as

$$\overline{|R(t,t')|^n} \underset{|t-t'|\to\infty}{\longrightarrow} \exp(+\lambda^{(n)}|t-t'|)\,. \tag{11.16}$$

The usual Lyapunov exponent λ is given by

$$\lambda = \left.\frac{\mathrm{d}\lambda^{(n)}}{\mathrm{d}n}\right|_{n=0}. \tag{11.17}$$

It is evident that in the linear approximation

$$C(t) = \int_{\infty}^{t} \mathrm{d}t'\, R(t,t')\,. \tag{11.18}$$

On the other hand

$$R(t,t') = \exp\left\{-\int_{t'}^{t} \frac{\mathrm{d}^2V}{\mathrm{d}\varphi^2}(t'')\,\mathrm{d}t''\right\}, \tag{11.19}$$

and therefore

$$\lambda^{(n)} = -E_0^{(n)}\,, \tag{11.20}$$

where $E_0^{(n)}$ is the ground-state energy of the Hamiltonian

$$\hat{H}_n = \hat{H} + n\,\frac{\mathrm{d}^2V}{\mathrm{d}\varphi^2}\,. \tag{11.21}$$

The conclusion is that regions where $\mathrm{d}^2V/\mathrm{d}\varphi^2$ is negative are very dangerous and these regions are unavoidable if the potential is periodic. The most dangerous situation would be a sharp, easily crossed barrier between two potential wells, as could be present between sectors with different topological charges.

In the same spirit, i.e. writing the correlation functions as solutions of differential equations which depend on the time-dependent field configurations, we could try to write down similar equations with no reference to the time evolution.

We consider the case of the φ^4 theory in the continuum limit; of course, our considerations can and must be extended to lattice theories.

Following Symanzik we have [26]

$$\begin{aligned}\langle \varphi(x)\varphi(0)\rangle &= \frac{1}{Z}\int \mathrm{d}[\varphi]\exp\{-S[\varphi]\}\varphi(x)\varphi(0)\\ &= \frac{1}{Z}\int \mathrm{d}[\sigma]\,\mathrm{d}[\varphi]\exp\{-S[\varphi,\sigma]\}\varphi(x)\varphi(0)\,,\end{aligned}\tag{11.22}$$

where

$$\begin{aligned}S[\varphi] &= \int \mathrm{d}^D x\,[\tfrac{1}{2}(\partial_\mu\varphi)^2 + \tfrac{1}{2}m^2\varphi^2 + \tfrac{1}{2}g\varphi^4]\,,\\ S[\varphi,\sigma] &= \int \mathrm{d}^D x\,[\tfrac{1}{2}(\partial_\mu\varphi)^2 + \tfrac{1}{2}m^2\varphi^2 + \tfrac{1}{2}\sigma^2 + \mathrm{i}\sqrt{g}\,\varphi^2\sigma]\,.\end{aligned}\tag{11.23}$$

Using the properties of Gaussian integrals we obtain

$$\langle \varphi(x)\varphi(0)\rangle = \frac{1}{2}\int \mathrm{d}[\varphi]\,\mathrm{d}[\sigma]\exp\{-S[\varphi,\sigma]\}G(x,0\,|\,\mathrm{i}\sqrt{g}\,\sigma)\,,\tag{11.24}$$

where G is the φ propagator in the presence of the external field σ, i.e., G satisfies the following differential equation:

$$[-\Delta_x + m^2 + A(x)]G(x,0\,|\,A) = \delta(x)\,.\tag{11.25}$$

By shifting the integration path over σ in the complex plane we obtain [24]

$$\begin{aligned}\langle \varphi(x)\varphi(0)\rangle = \int \mathrm{d}[\varphi]\,\mathrm{d}[\sigma]\exp&\Big\{-\frac{1}{2}\int \mathrm{d}^D x\,\sigma^2(x) - S[\varphi]\Big\}\\ &\times G(x,0\,|\,g\varphi^2 + \mathrm{i}\sqrt{g}\,\sigma)\,.\end{aligned}\tag{11.26}$$

This last representation is the equivalent of eq. (4.11) for fermions. It can be practically implemented by performing a usual Monte Carlo calculation over the φ field (σ is white noise that can be easily generated), and computing G for each configuration of φ and σ. This method is rather efficient because the noise-to-signal ratio does not increase dramatically with the distance. Indeed, we should have

$$\langle G(x,0 \mid g\varphi^2 + \mathrm{i}\sqrt{g}\,\sigma)\rangle \sim \exp - |x|\xi_1 ,$$

$$\langle |G(x,0 \mid g\varphi^2 + \mathrm{i}\sqrt{g}\,\sigma)|^2\rangle \sim \exp - |x|\xi_2 , \tag{11.27}$$

where for not too large g, ξ_2 should not be very different from $2\xi_1$. In this case also the noise decreases to zero exponentially.

The best results can be obtained if we integrate over σ analytically. This can be done by using the random-walk representation for G:

$$G(x,0 \mid A) = \int_0^\infty \mathrm{d}t \int \mathrm{d}P'_{[0,x]}[\omega] \exp\left\{ -tm^2 - \int_0^t \mathrm{d}t'\; A[\omega(t')]\right\} , \tag{11.28}$$

where $\mathrm{d}P'_{[0,x]}[\omega]$ is the usual Wiener measure on the ω trajectories going from 0 to x in time t. We finally find [27]

$$\langle \varphi(x)\varphi(0)\rangle = \frac{\int \mathrm{d}[\varphi] \int_0^\infty \mathrm{d}t \int \mathrm{d}P'_{[0,x]}[\omega] \exp\{S[\omega,\varphi]\}}{\int \mathrm{d}[\varphi] \exp\{-S[\varphi]\}} ,$$

$$S[\varepsilon,\varphi] = S[\varphi] + g\int_0^t \mathrm{d}t'\; \varphi^2(\omega(t'))$$

$$+ \frac{g}{2}\int_0^t \mathrm{d}t' \int_0^t \mathrm{d}t''\; \delta(\omega(t') - \omega(t'')) . \tag{11.29}$$

If we leave the end point x free to be in the region $R < x < R + \delta$ with R very large, we can do a simultaneous Monte Carlo calculation on φ, ω and x. The correlation length can be easily extracted from the dependence of the correlation functions. In this way we have reached our goal of measuring the decay rate of the correlation functions at large distance with errors which do not increase with the distance. It would be very instructive to perform Monte Carlo simulations using this approach.

In non-Abelian gauge theories like SU(2), representations similar to eq. (11.29) exist for the expectation values of the Wilson loop,

$$\langle W(C)\rangle = \frac{1}{Z}\int \mathrm{d}[U]\,\mathrm{d}[\Sigma]\,\rho(\Sigma, U)\,, \tag{11.30}$$

where the integral over Σ runs over all the surfaces having C as boundary. Unfortunately, the function ρ is not positive; it is not clear if the representation (11.30) may be very useful in numerical simulations.

No very satisfactory technique is available for gauge theories, although the representation (4.11) has produced very nice results for quark–antiquark bound states.

I hope to have convinced also the constructivist reader that the search for efficient numerical techniques has many points of contact with proving rigorous theorems; in both cases we must use our best knowledge of field theory in order to find the representation for the correlation functions which is best suited to our aims.

The last difficulty is the critical slowing down; the results of sect. 3 show that, roughly speaking, when the lattice spacing a goes to zero, in order to change the average value of φ on a region with side proportional to the coherence length ξ, we need to update each local variable φ a number of times proportional to $(\xi/a)^2$.

The natural suggestion would be to divide the phase space of the field φ in many regions corresponding to different characteristic lengths and to perform different but coupled Monte Carlo simulations for each scale. The practical way (if any) to implement this suggestion is unclear; there have been two proposals [2], i.e. the multi-grid Monte Carlo and the FFT preconditioned Langevin equation, whose theoretical and practical soundness have not yet been proven. It is evident that the Monte Carlo renormalization group is a possible solution when the block-spin Hamiltonian is not too complex.

Appendix 1. Monte Carlo simulation; program listing

```
C  Auxiliary vectors forward and backward are used to impose periodic
C  boundary conditions.
C  Each spin may take only values plus or minus 1.
        PARAMETER maximal_side=100
        INTEGER forward(maximal_side),backward(maximal_side)
        COMMON /boundaries/ forward,backward
        INTEGER side,spin(maximal_side,maximal_side)
        INTEGER random_seed
        REAL magnetic_field,magnetization_density
```

```
      CALL read_input(side,number_of_iterations,beta,magnetic_field)
      CALL get_random_seed(random_seed)
      CALL compute_backward_and_forward(side)
      CALL set_spin_to_1(spin,side)
      DO 1 iteration=1,number_of_iterations
        CALL one_Monte_Carlo_cycle(spin,side,random_seed,
     1  beta,magnetic_field,energy_density,magnetization_density)
        CALL write_output(iteration,energy_density,magnetization_
     1  density)
1     CONTINUE
      END

C *****************************************************************

      SUBROUTINE compute_backward_and_forward(side)
      PARAMETER maximal_side=100
      INTEGER forward(maximal_side),backward(maximal_side)
      COMMON /boundaries/ forward,backward
      INTEGER position,side

      DO 1 position=1,side
        forward(position)=mod(position,side)+1
        backward(position)=mod((position-2+side),side)+1
1     CONTINUE
      RETURN
      END

C *****************************************************************

      SUBROUTINE set_spin_to_1(spin,side)
      PARAMETER maximal_side=100
      INTEGER side,spin(maximal_side,maximal_side)
      INTEGER x_coordinate,y_position

      DO 1 x_position=1,side
        DO 2 iy_position=1,side
          spin(x_position,iy_position)=1
2       CONTINUE
1     CONTINUE
      RETURN
      END

C *****************************************************************

      SUBROUTINE one_Monte_Carlo_cycle(spin,side,random_seed,
     1 beta,magnetic_field,energy_density,magnetization_density)
      PARAMETER maximal_side=100
      INTEGER forward(maximal_side),backward(maximal_side)
      COMMON /boundaries/ forward,backward
      INTEGER side,random_seed,spin(maximal_side,maximal_side)
      REAL magnetic_field,magnetization_density
      INTEGER current_spin,sum_of_the_neighbours
```

```
        total_magnetization=0
        total_energy=0
        DO 5 x_position=1,side
        DO 6 y_position=1,side
          current_spin=spin(x_position,y_position)
          sum_of_the_neighbours=
     1    spin(x_position,forward (y_position))+
     2    spin(x_position,backward(y_position))+
     3    spin(forward (x_position),y_position)+
     4    spin(backward(x_position),y_position)
          effective_force=sum_of_the_neighbours+magnetic_field
          IF ( EXP(-beta*effective_force*current_spin*2.)
     1    .GT. RAN(random_seed) ) THEN
              new_spin=-current_spin
          ELSE
              new_spin=current_spin
          END IF
          spin(x_position,y_position)=new_spin
          total_magnetization=total_magnetization+new_spin
          total_energy=total_energy+
     1    (0.5*sum_of_the_neighbours+magnetic_field)*new_spin
C  please note the factor 0.5 multiplying sum_of_the_neighbours
6         CONTINUE
5       CONTINUE
        magnetization_density=total_magnetization/FLOAT(side**2)
        energy_density=total_energy/FLOAT(side**2)
        RETURN
        END


C   **********************************************************************


        SUBROUTINE read_input(side,number_of_iterations,beta,magnetic_field)
        REAL magnetic_field
        INTEGER side
        PARAMETER maximal_side=100

1       WRITE(6,*) 'Which is the length of the side ?
     1     (Please less than',maximal_side,')'
        READ(5,*) side
        IF (side .GT. maximal_side .OR. side .LT. 1) THEN
             WRITE(6,*) 'The value of side is not good'
             go to 1
        END IF
        WRITE(6,*) 'How many iterations?'
        READ(5,*) number_of_iterations
        WRITE(6,*) 'Which is the value of beta ?'
        READ(5,*) beta
        WRITE(6,*) 'Which is the value of the magnetic field?'
        READ(5,*) magnetic_field
        RETURN
        END
```

```
C   ******************************************************************

        SUBROUTINE get_random_seed(random_seed)
        INTEGER random_seed

        WRITE(6,*)'Which is the random seed?'
        WRITE (6,*)'Please insert a positive odd number of 7-8 digits'
        READ(5,*) random_seed
        RETURN
        END

C   ******************************************************************

        SUBROUTINE write_output(iteration,energy_density,magnetization_
     1                 density)
        REAL magnetization_density

        WRITE (6,*) 'Iteration=       ',iteration
        WRITE (6,*) 'Energy density =       ', energy_density
        WRITE (6,*) 'Magnetization density=       ',magnetization_density
        RETURN
        END
```

References

[1] For a review see, for example, G. Parisi, Statistical Field Theory (Benjamin, New York, 1985) and references therein.

[2] This has been suggested by K. Symanzik, in: New Developments in Gauge Theories, eds. G. 't Hooft, A. Jaffe, H. Lehman, P.K. Mitter, K. Symanzik and R. Stora (Plenum, New York, 1980); for an historical review see G. Parisi, Nucl. Phys., to be published.

[3] A discussion of this point can be found in G. Parisi, Phys. Rep. 103 (1984) 203; and in: Cargèse Summer School 1983, eds. G. 't Hooft, A. Jaffe, H. Lehman, P.K. Mitter, I.M. Singer and R. Stora (Plenum, New York).

[4] C. DeWitt-Morette and K.D. Elworthy, eds., Phys. Rep. 77 (1981) 123 and references therein.

[5] F. Fucito, E. Marinari, G. Parisi and C. Rebbi, Nucl. Phys. B180 [FS2] (1980) 369; F. Fucito and E. Marinari, Nucl. Phys. B190 [FS3] (1981) 237.

[6] D.N. Petcher and D.H. Weingarten, Phys. Lett. 100B (1980) 571.

[7] For a review of the Monte Carlo approach see C. Rebbi, ed., Lattice Gauge Theories and Monte Carlo Simulations (World Scientific, Singapore, 1983).

[8] For a introduction to gauge theories see R. Stora, in: New Developments in Quantum Field Theory and Statistical Mechanics, eds. M. Levy and P. Mitter (Plenum, New York, 1977); and C. Itzykson and J.B. Zuber, Quantum Field Theory

[9] A. Neveu, in: Recent Advances in Field Theory and Statistical Mechanics, Les Houches 39, eds. J.B. Zuber and R. Stora (North-Holland, Amsterdam, 1984).

[10] Yu.M. Makeenko and A.A. Migdal, Phys. Lett. 88B (1979) 135.

[11] H.D. Politzer, Phys. Rev. Lett. 30 (1973) 1346; G. 't Hooft, unpublished remarks at the Marseille Conference on Gauge Theories.

[12] For a review of lattice gauge theories (introduced by K. Wilson, Phys. Rev. D14 (1974) 2455) see J.C. Drouffe and J.B. Zuber, Phys. Rep. 102 (1984) 1; and J. Kogut, in: Recent Advances in Field Theory and Statistical Mechanics, Les Houches 39, eds. J.B. Zuber and R. Stora (North-Holland, Amsterdam, 1984).

[13] G. Parisi, in: High Energy Physics—1980, eds. L. Durand and L.G. Pondrom (American Institute of Physics, New York, 1981).

[14] R. Ferrari and R. Picasso, Nucl. Phys. B31 (1971) 316.

[15] B. Lautrup and M. Nauenberger, Phys. Lett. 95B (1980) 63.

[16] F. Fucito and A. Vulpioni, Phys. Lett. 89A (1982) 33.

[17] H. G. Evertz, J. Jersák, T. Neuhaus and P.M. Zerwas, Nucl. Phys. B251 [FS13] (1985) 279.

[18] D. Barkai, K.J.M. Moriarty and C. Rebbi, Phys. Rev. D30 (1984) 1293.

[19] G. Parisi, E. Marinari and C. Rebbi, Nucl. Phys. B190 [FS3] (1981) 734.

[20] H. Hamber and G. Parisi, Phys. Rev. Lett. 47 (1981) 1972;
E. Marinari, G. Parisi and C. Rebbi, Phys. Rev. Lett. 47 (1981) 1975.

[21] For a review of the results obtained in the quenched approximation see H. Hamber, in: Mathematical Physics VII, eds. W.E. Brittin, K.E. Gustafson and W. Wyss (North-Holland, Amsterdam, 1984); a discussion of the strategy for including the effects of fermionic loops can be found in H. Hamber, E. Marinari, G. Parisi and C. Rebbi, Nucl. Phys. B225 [FS9] (1983) 475.

[22] G. Parisi, R. Petronzio and F. Rapuano, Phys. Lett. B113 (1983) 112.

[23] This estimate can be found in G. Parisi, Non-perturbative Field Theory and QCD, eds. R. Jengo, A. Neveu, P. Olesen and G. Parisi (World Scientific, Singapore, 1983).

[24] G. Parisi, Nucl. Phys. B205 [FS5] (1982) 337;
M. Falcioni, E. Marinari, M. Paciello, G. Parisi, R. Taglienti and Zhang Yi-Cheng, Nucl. Phys. B215 [FS7] (1983) 256.

[25] R. Benzi, G. Paladin, G. Parisi and A. Vulpiani, submitted to J. Physique, and references therein.

[26] K. Symanzik, in: Local Quantum Theory, ed. R. Joos (Academic Press, New York, 1969).

[27] See for example C. Aragao de Carvalho, S. Caracciolo and J. Fröhlich, Nucl. Phys. B215 [FS7] (1983) 209.

COURSE 9

PRINCIPLES OF NUMERICAL SIMULATIONS

G. PARISI

Dipartimento di Fisica
II Università Roma "Tor Vergata", Roma, Italy

E. Brézin and J. Zinn-Justin, eds.
Les Houches, Session XLIX, 1988
Champs, Cordes et Phénomènes Critiques
/Fields, Strings and Critical Phenomena

Contents

1. Introduction

After the advent of fast computers, numerical simulations have been very useful in many domains of physics. There are many reasons for which numerical simulations may be extremely useful:

a) We can check the conclusions of doubtful analytic proofs,

b) We can use them as heuristic for later analytic studies.

c) They can play the role of an interface between theory and experiments (e.g. a strong argument for the correctness of a short range Ising Hamiltonian for metallic spin glasses comes from the successful comparison of the numerical simulations with the experimental data).

d) Numerical simulation can do hard computations which cannot be done analytically. Indeed, also if the theory is simple at an elementary level, most of the experiments are not done at this level and there are many quantities (e.g. the spectrum of Bismuth) which quite likely will never be computed analytically.

In general, a brute force approach to numerical simulations is not paying and a lot of ingenuity is needed in order to obtain the wanted results. Numerical simulations for statistical systems or quantum field theories have their own problems, whose solution cannot be found in the books of numerical analysis. The aim of these lectures is to present the principle of these computations and to describe in details some results obtained recently for quantum chromodynamics. Indeed it is extremely likely that quantum chromodynamics will not be solved analytically in the next future and that the only sources of unbiased information on its properties are numerical simulations.

2. The floating numbers

A strong difference among analytic computations and the usual computations on a computer is that real numbers are used in the first case while floating point numbers are used in the second case [1]. Generally speaking a floating point number can be written as

$$f = am2^b, \tag{2.1}$$

where a and b are integers in a given range, e.g. $2^{-23} \leq a < 2^{23}$, $-2^7 \leq (b+24) < 2^7$. In this case a floating point number may span the interval 10^{-38} - 10^{38} (with both signs) plus the zero and may approximate any real number in this interval with a relative precision of less than 10^{-7}.

The fundamental operations that are done on a computer are additions and multiplications, (these operations are called floating point operations) while divisions and powers can be expressed in terms of these more basic operations. The problems arise from the rounding, i.e. the fact that the sum (or the product) of two floating point numbers cannot be represented exactly as a floating point number. The usual solution consists in doing the computation practically exactly and to approximate the results with the nearest floating point number. In this way the relative error of a single computation is smaller than 10^{-7}. If we do a long computation of n steps, we could argue that at each step we introduce an error of order 10^{-7} which is random in sign so that the final error should be of order $\sqrt{n}10^{-7}$.

This argument is quite sensible, however there are many cases in which it fails. For example, if we compute $a - (a - b)$ where $a \sim 10^9$ and $b \sim O(1)$ the result is identically zero while if we compute $(a-a)+b$ the result is identically b. Another case is the computation of the sequence of complex numbers $\{C\}$ where

$$|C_0| = 1,$$
$$C_n = (C_{n-1})^2. \tag{2.2}$$

It is obvious that $|C_n| = 1$, however if $|C_0| = 1 + \varepsilon, \varepsilon \sim O(10^{-7})$, we expect that $|C_n| \sim \exp(\varepsilon 2^n)$. The error becomes of $O(1)$ after a number of steps $n \simeq -\ln_2 \varepsilon \simeq 24$. On the contrary, if we compute the solution of the equation

$$x = F(x) \tag{2.3}$$

by a convergent Newton recursion (i.e. $x_{n+1} = F(x_n)$), the relative error on x_∞ does not depend on the number of steps needed to reach the fixed point.

In other words, when we study the time evolution of a system, if the Lyapunov exponent is positive, we can follow the evolution only for a very short time, because the errors become too large; on the contrary, if the Lyapunov exponent is negative, the errors are always small; indeed, $\delta C(t) \sim \delta C(0) m \exp(\lambda t)$ for large t, where λ is the Lyapunov exponent and $\delta C(t)$ is the error at time t; if C is an N dimensional vector space, we have consider the matrix $\partial C_i(t)/\partial C_k(0)$ whose eigenvalues are proportional to $\exp(\lambda_i t)$, where the $N \lambda's$ are the Lyapunov exponents, the largest Lyapunov exponent is sometimes called the Lyapunov exponent.

These considerations seems to imply that the large time behavior of a system with positive Lyapunov exponent cannot be studied numerically. More precisely it seems that it would be impossible to compute the large time probability distribution $P(C)$ defined as

$$\lim_{t\to\infty} \frac{1}{t} \int_0^t f(C(t'))\,\mathrm{d}t' = \int \mathrm{d}C P(C) f(C), \tag{2.4}$$

where f is an arbitrary function and $P(C)$ depends on $C(0)$.

However there is a way out. If the time evolution is ergodic in the sense that there is only one $P(C)$ for every (or almost every) $C(0)$, it is often argued that the effect of rounding would make the system jump from one trajectory to another, and the equilibrium probability $P(C)$ would not be seriously affected if the time evolution is computed in presence of rounding. The argument is rather sound, although exceptions may be present.

3. Random numbers

Obviously no computer can generate a sequence of random numbers according to a deterministic algorithm, it is possible however to generate numbers that are random for all practical purposes [2] (it is also possible to use a tape where is written a sequence of random numbers generated by a physical process like radioactive decay, which should be the ultimate random process).

The concept of random sequence becomes very sharp if we consider an infinite (or nearly infinite) sequence r_i, where the $r's$ are often integer numbers in the range $0-1$, or $0-(2^{32}-1)$ (although we can consider also floating point numbers in the same range). The sequence is random if all connected correlations functions at different points, i.e. $\langle r_i r_{i+k}\rangle - \langle r_i\rangle^2$, are zero, where

$$\langle r_i\rangle = \lim_{n\to\infty} \frac{1}{n}\sum_{i=1}^{n} r_i$$

$$\langle r_i r_{i+k}\rangle = \lim_{n\to\infty} \frac{1}{n}\sum_{i=1}^{n} r_i r_{i+k}. \tag{3.1}$$

It is usual to consider random numbers which are equidistributed in there range of variation ($\langle r\rangle = 1/2$).

A random sequence is obviously aperiodic, while a finite state machine (like a computer) can generate only sequences shorter than number of states; this limitation is not relevant because it gives a bound of order 10^{10^6} on usual computers. The general theory of random number generations is very poor; as far as I know, there are no theoretical bounds on the non randomness as function of the amount of memory used by the program. The most known results describe some algorithms which are known to have small correlation.

A well know method for generate sequence of zero and one's consist in assigning the first p elements of sequence (they play the role of the random seed and are needed to initialize the sequence; it is crucial that at least one is different from zero) and to generate the rest as using the rule

$$r_i = r_{i-p} + r_{i-p+q} \pmod 2 \tag{3.2}$$

with $0 < q < p$.

This rule can be understood as the following: we can consider the finite field (Galois field) A_p of elements of the form

$$\sum_{k=1}^{p} C_k X^k, \tag{3.3}$$

where $C_k \in Z_2$ and $X^P = 1$.

We can assign a sequence of $n - p$ elements of A_p to the sequence of random numbers $\langle r\rangle$ ($A_l = \sum_{k=1}^{p} r_{l+k} X^k$) in such a way that eq. (3.2) becomes

$$A_l = gA_{l-1} = g^{l-1}A_1, \tag{3.4}$$

where

$$g = 1 + X^q. \tag{3.5}$$

If p and q are suitable chosen, g is a primitive element and all the non zero elements of A can be written as g^k for a suitable k ($1 \leq k < 2^p$). Therefore the sequence A_l has a very large period 2^p ; a typical choice is $p = 55 \quad q = 33$ ($p = 3 \quad q = 2$ produces a much shorter sequence).

It is evident that the connected correlation functions of random numbers are zeros (if averaged over the whole period) when the distance is smaller than p. On the other hand it is evident that

$$\langle r_i(r_{i-24} + r_{i-57})\rangle_C = 1 - \frac{1}{4} = \frac{3}{4}. \tag{3.6}$$

In this method short range correlations are small while medium range correlation are large.

A quite different method for generate integer in the range $0 - (2^{32} - 1)$ uses the relation

$$r_n = Ar_{n-1} + B \pmod{2^{32}}, \tag{3.7}$$

where A and B are chosen in such a way that A is large (e.g. $A \approx 2^{16}$) and the period is 2^{32}.

In this case the Lyapunov exponent of the process is of order A: a small short range correlation may be present, but long range correlations are supposed to be small.

A way to produce very small correlations is the following. If we already constructed $r^{(1)}$ and $r^{(2)}$, which are two approximatively random sequences of 0 and 1 with average 1/2, we can construct the new random sequence

$$r_i = r_i^{(1)} + r_i^{(2)} \quad \mathrm{mod}\, 2. \tag{3.8}$$

The connected correlations of r are given by

$$\langle r_i r_{i+j}\rangle_C = \left\langle r_i^{(1)} r_{i+j}^{(1)} \right\rangle_C \left\langle r_i^{(2)} r_{i+j}^{(2)} \right\rangle_C. \tag{3.9}$$

If the sequences $r^{(1)}$ and $r^{(2)}$ are produced using techniques such that these two sequences have non negligible correlations in different regions, the final sequence of the r's will have practically zero correlations.

The apparent conclusion (which is also confirmed by the approximate independence of the results of many numerical simulations on the random number generator) is that it is possible to construct sequences of random numbers which have arbitrary small correlations (by mixing many different random numbers). It should be stressed however that for particularly sensitive computations, a special care should be done in the choice of the random number generator and the usual machine random numbers may give some bad surprises, as shown for example in [3].

4. Solving differential equations

There exists a large literature on how solve numerical differential equations [1], here we sketch some of the points which will be relevant in study of numerical simulations of Quantum Field Theory.

We concentrate our attention on the time evolution of a system. We consider firstly the very simple equation

$$\dot{x} = F(x). \tag{4.1}$$

In our presentation x is a number, however the extension to the case where x is a vector is quite straightforward.

Our aim is to compute $x(t)$ for a given $x(0)$, under the hypotheses that F is a smooth function and the behavior of F at large x is such that $x(t)$ does not escape to infinity.

We will introduce a time step ε and compute the sequence $x_n \equiv x(n\varepsilon)$.

In the simplest approximation we have

$$x_{n+1} = x_n + \varepsilon F(x_n). \tag{4.2}$$

It easy to check that the error on $x(t)$ is of order ε when $\varepsilon \to 0$ at fixed t. Better algorithms can be found, e.g.

$$\begin{aligned} \tilde{x}_{n+1} &= x_n + \varepsilon F(x_n) \\ x_{n+1} &= x_n + \varepsilon[F(x_n) + F(\tilde{x}_n]/2, \end{aligned} \tag{4.3}$$

where it is easy to check that the error at fixed t is of order ε^2.

In general, there are algorithms which reduce the error to $O(\varepsilon^k)$ (e.g. Runge–Kutta) with arbitrary large k. It must be noticed that ε^k for large k is a quite small number, if ε is small, but it is quite large as soon ε is greater than one. In other words high order methods perform magnificently for $\varepsilon < \varepsilon_c$, but they are a disaster for $\varepsilon > \varepsilon_c$.

First order Hamiltonian equations are often studied; in the simplest case

$$\begin{aligned} \dot{x} &= p \\ \dot{p} &= -\frac{\partial H}{\partial x}. \end{aligned} \tag{4.4}$$

It is well known that the time evolution of an Hamiltonian flow is area preserving; it would be quite useful if this property still holds after the time discretization. This can be achieved by using the leap frog algorithm defined as

$$\begin{aligned} p_{n+\frac{1}{2}} &= p_n - \frac{1}{2}\varepsilon\frac{\partial H}{\partial x}(x_n) \\ x_{n+1} &= x_n + \varepsilon p_{n+\frac{1}{2}} \\ p_n &= p_{n+\frac{1}{2}} - \frac{1}{2}\varepsilon\frac{\partial H}{\partial x}(x_{n+1}). \end{aligned} \tag{4.5}$$

It is easy to check that the equations (4.5) are time reversible: if we go from $(x_0,\ p_0)$ to $(x_n\ ,p_n)$ and at this point we change the sign of the momenta (i.e. we set $p_n = -p_n$), we eventually find $x_{2n} = x_0$ and $p_{2n} = p_0$. Mostly importantly the transformation (4.5) is area preserving: the Jacobian

$$J = \begin{pmatrix} \frac{\partial x_{n+1}}{\partial x_n} & \frac{\partial x_{n+1}}{\partial p_n} \\ \frac{\partial p_{n+1}}{\partial x_n} & \frac{\partial p_{n+1}}{\partial p_n} \end{pmatrix} \tag{4.6}$$

is identically equal to 1.

If we want to compute the solution of a partial differential equations like

$$\dot{\varphi}(x,t) = \Delta\varphi(x,t) + F(\varphi(x,t)), \tag{4.7}$$

we must introduce a mesh in both space and time: a is the lattice spacing and ε is the time step.

If we use a method like in eqs. (4.2) or (4.3) for the time evolution, we face the problem of approximating the Laplacian on the lattice. In the simplest case, where x is one dimensional, we set $\varphi_{i,0} = \varphi(ai,0)$ and an approximation to the Laplacian at $x = a\ i$ is given by

$$\left(\varphi_{i+1,0} + \varphi_{i-1,0} - 2\varphi_{i,0}\right)/a^2 . \tag{4.8}$$

The error in (4.8) is of order a^2.

Higher order approximations to the Laplacian with error $O(a^{2k})$ can be easily written, however, if the function φ is analytic, the best approximation is obtained by going in Fourier space. Indeed the analyticity of $\varphi(x,0)$ implies that its Fourier transform $\tilde{\varphi}(k,0)$ decreases as $\exp(k/\lambda)$ for large k.

The introduction of a lattice modifies $\tilde{\varphi}(k,0)$ only for $|k| \sim 1/a$ where $\tilde{\varphi}$ is exponentially small. Therefore, if we go from φ to $\tilde{\varphi}$, multiply $\tilde{\varphi}$ by k^2 and Fourier transform again, the error on the Laplacian is exponentially small. The practical implementation of this approach is linked to the existence of Fast Fourier Transforms [4], i.e. algorithms which compute the Fourier transform of a sequence of N numbers in about N ln N operations (not N^2, as one could naively think).

If the time evolution is such that analyticity of $\varphi(x,t)$ is lost, as it happens for the Navier–Stokes equation in the zero viscosity limit, the effects of the grid are quite large and, if they go to zero, they generally vanishes as a power of a. A similar situation happens in field theory (cf. Lüscher's lectures at this school) where φ is far from being a smooth function.

5. Simulating statistical mechanics

In statistical mechanics we have to compute integrals of the type

$$\int \mathrm{d}CP(C)f(C), \tag{5.1}$$

where

$$P(C) = 1/Z\exp(-H(C)), \qquad \int \mathrm{d}CP(C) = 1. \tag{5.2}$$

If the configuration space C has a very large dimensions, the direct evaluation of (5.1) is practically impossible; the only way out consists in finding a sequence C_n of configurations such that

$$\lim_{n\to\infty} \frac{1}{n}\sum_{i=1}^{n} f(C_i) = \int \mathrm{d}CP(C)f(C). \tag{5.3}$$

There are many methods for constructing a sequence of C's, such that (5.3) is satisfied.

A possibility consists in using a Markov chain which is characterized by the transition probability $T(C,C')$ of going from the configuration C at step n to C' at step $n+1$ [5]. If the detailed balance condition,

$$P(C)T(C,C') = T(C,C')P(C), \tag{5.4}$$

is satisfied, then eq. (5.3) certainly holds.

On the other hand, if the evolution is ergodic and only one equilibrium state is present, independently from the initial configuration, the simple balance condition,

$$\sum_{C} P(C)T(C,C') = P(C'), \tag{5.5}$$

implies eq. (5.3). The simple balance condition is less strong the detailed balance condition; the latter implies the former.

In usual applications C is an highly dimensional space and we change only one component (e.g. in the Ising case we flip only one spin) at each step. In the Monte Carlo approach we have a suggestor algorithm with symmetric transition probability ($S(C,C') = S(C',C)$) which, starting from the configuration C_n, suggest a new configuration $\tilde{C}_n$. The ratio

$$r = P(\tilde{C}_n)/P(C_n) \tag{5.6}$$

is computed and we set

$$\begin{aligned} C_{n+1} &= \tilde{C}_n \quad \text{with probability} \quad p(r) \\ C_{n+1} &= C_n \quad \text{with probability} \quad 1 - p(r) \end{aligned} \tag{5.7}$$

where the probability $p(r)$ must satisfy the constraint

$$p(r)/p(1/r) = r. \tag{5.9}$$

It is easy to prove that the detailed balance condition is satisfied by this algorithm. Different choices of the suggestor may lead to more or less efficient algorithms.

In the heath bath method the algorithm is constructed in such a way that

$$T(C,C') = P(C'). \tag{5.10}$$

In this case the balance condition is obviously satisfied.

There are other methods are based on continuous equations: for example, if the configuration space is $\{\psi_i,\ i = 1,\dots,N\}$, we could write the Langevin equation [6]

$$\dot{\psi}(t)_i = -\frac{\partial H}{\partial \psi_i} + \eta_i(t), \tag{5.11}$$

where $\eta_i(t)$ is a white noise, normalized to

$$\eta_i(t)\eta_j(t') = 2\delta_{ij}\delta(t-t') \tag{5.12}$$

In this case it may be proved that

$$\lim_{t\to\infty} \frac{1}{t}\int_0^t dt' f(\psi(t')) = \int d\psi P(\psi) f(\psi) \tag{5.13}$$

Alternatively we could introduce auxiliary momenta p and consider the microcanonical equations

$$\begin{aligned} \dot{\psi}_i(t) &= p_i(t) \\ \dot{p}_i(t) &= -\frac{\partial H}{\partial \psi_i} \end{aligned} \tag{5.14}$$

If the energy is well chosen and the time evolution (5.12) is ergodic, we should have for large volume that the $p's$ distribution is Gaussian and eq. (5.12) is satisfied.

It is quite interesting to note that at thermal equilibrium, if we compute the transition probability $P(\psi,\psi';t)$ (averaged over the noise or over the momenta) we find for small time t the surprising result

$$P_{\mathrm{Lang}}(\psi,\psi';t) = P_{\mu\mathrm{can}}(\psi,\psi';t^{1/2}). \tag{5.15}$$

This remark is at the base of hybrid methods in which some microcanonical steps are done intermixed with Langevin steps [7]. Equation (5.14) is also important because it is much easier to write down algorithms with very small error associated to the time step in the microcanonical approach than in the Langevin approach.

In real simulations, if we approach $\langle f \rangle$ with the left hand side of eq. (5.3) evaluated after a finite number of steps n, we find a statistical error, which is essentially proportional to $(\tau/n)^{1/2}$, where τ is the decorrelation time, i.e. the number of steps needed to obtain a statistically independent configuration.

If the Hamiltonian (and consequently the laws which control the time evolution) is local, we readily obtain that τ of $O(1)$ in the region where the correlation length ξ is small, otherwise τ diverges as

$$\tau \simeq \xi^z \tag{5.14}$$

when ξ is large; z is a critical exponent equal or near to 2 in many non disordered systems.

The divergence of τ with ξ is called critical slowing down and it is quite annoying, because it strongly affect the value of the errors in the critical region where "ξ" is large.

Some time ago I have realized [8] that the use of non local algorithms may strongly reduce (or eliminate) the divergence of τ. This proposal of mine has become popular only after it has been further elaborated in [9]. These non local algorithms are actually under investigation and the question of their efficiency is not completely settled in the general case, although it is quite reasonable that the can be extremely efficient for asymptotically free theories.

6. Preliminaries on field theories

In order to compute the spectrum of a (Bosonic) quantum theory it is convenient to use the Euclidean formulation of the theory, to introduce a lattice spacing a and a box $L^3 \times L_t$ (for simplicity we assume $L^t >> L$). In this way, if the Hamiltonian is real we have a system of statistical mechanics with $L^3 \times L^t$ degrees of freedom and the techniques of the previous section can be used.

If we stay in the region $L >> \xi >> a$, the decay of correlations in the time direction is related to the energy spectrum of the theory, i.e. if

$$C(t) \equiv \langle O(t)O(0)\rangle \simeq \exp(-m_O t) \qquad (t \to \infty), \tag{6.1}$$

then m_O is the mass of the lowest created by the action of O on the vacuum.

An estimation of the mass m_O can be obtained by computing

$$m(t) = -\ln[C(t)/C(t-1)]. \tag{6.2}$$

Obviously $m(t)$ depends on L and a and it approaches m_O when t goes to infinity.

The results continuum theory in the infinite volume limit are obtained in the triple limit $t \to \infty$, $L \to \infty$, $a \to 0$.

As discussed in Lüscher's lessons, the error induced by finite value of L decrease as $\exp(-mL)$, m being a suitable mass; the errors coming from non zero a are usual proportional to a^2 (a for fermions).

Finite time effects appears as consequence of

$$m(t) \to m_O + A\exp(-\Delta mt) + \cdots, \tag{6.3}$$

where Δm is the mass gap between the first and the second excited state in the channel having the quantum numbers of the operator O ; we assume that O is at zero spacial momentum (as in eq. (6.4)); in this way the energy coincides with the mass. If we change the operator O, but we keep the same quantum numbers, the mass m does not change, but A changes and it may become zero. The technique of comparing the correlation functions of different operators is crucial in getting the control over the errors due to finite time.

The reader should notice that if we have

$$O(t) = \int d^3x A(x,t), \tag{6.4}$$

where $A(x,t)$ is a local operator, $\langle O(0)^2\rangle$ is a divergent quantity in the limit $a \to 0$ as soon as the dimension of A is greater than 3/2.

This result is unpleasant: for small t we have just the perturbative result

$$C(t) \sim \frac{1}{t^{2d_A-3}}$$
$$m(t) = \frac{2d_A - 3}{t} \qquad (6.5)$$

and the interesting informations on $m(t)$ are confined in the region of large t where the statistical noise is large.

This problem becomes particularly acute in gauge theories, where all local gauge invariant operators have high dimensions because they are composite operators. This difficulty may be bypassed by introducing a smeared local field

$$\varphi_f(x,t) = \int \mathrm{d}^3 y f(x-y)\varphi(y,t) \qquad (6.6)$$

and constructing the composite operators as powers of φ_f. If the function f is sufficiently smooth the correlations of the composite operators are well defined also at time zero. Obviously in gauge theories we must be very careful with gauge invariance (for a gauge invariant procedure see [10].

There is no way (apart from doing the computation with different boundary conditions) to control the systematic effect without changing the actual value of L. Sometimes the use of twisted boundary conditions may helps in decreasing these systematic errors.

How to change a is more tricky.

If the theory is asymptotically free and contains no massive paramceter, the continuum theory will depend only by the scale Λ. On the lattice the bare coupling constant $g_0^2 \equiv 1/\beta$, is not zero but it is related to a and Λ by a formula of the kind

$$\Lambda^2 a^2 = C\ (\beta)^B \exp(-A\beta)[1 + O(\frac{1}{\beta})], \qquad (6.7)$$

where A, B, C are computable (and computed) constants. Unfortunately the $1/\beta$ corrections have practically never being computed: as far as β^{-1} goes to zero very slowly as $1/\ln(\Lambda^2 a^2)$, it is very difficult to extract the precise value of a from eq. (6.7) also if we know Λ. However eq. (6.7) is important as far as it tell us qualitatively how to change a by changing β.

The size of the errors proportional to a^2 are estimated by computing the same dimensionless quantity, i.e. a mass ratio, at different values of a^2.

7. Pure gauge QCD: the spectrum

We consider in this section pure gauge Quantum Chromodynamics (i.e. we neglect quarks).

The gauge group is SU(3) and the action in the continuum limit is

$$\frac{1}{g^2}\int \mathrm{d}x\, F^2(x). \tag{7.1}$$

On the lattice the gauge fields belong to the group SU(3), they are defined on the links of a lattice (which we can suppose to be hypercubic) and the action can be written as

$$-\beta \sum_{\text{plaquettes}} \mathrm{Tr}\, UUUU, \tag{7.2}$$

where we sum over all plaquettes (elementary squares) the trace of the Wilson loop along the plaquette.

Which this definition of β we have

$$\Lambda_L \equiv \beta^{-51/101} \exp(-\frac{11}{\pi^2}\beta) \tag{7.3}$$

and the usual Λ's, e.g. Λ_{Mom} is related to Λ_L by

$$\Lambda_{\mathrm{Mom}} = 33.5\Lambda_L[1 + O(\frac{1}{\beta})] \tag{7.4}$$

In the real words Λ_{Mom} is supposed to be in the 100–200 MeV range, but it is a quantity which is very difficult to extract from the experimental data.

Unfortunately, the absence of a computation of the orders $1/\beta$ makes the application of eq. (7.4) quite doubtful ; indeed $1/\beta$ goes to zero quite slowly, (as $1/(-\ln a^2)$); in absence of the $1/\beta$ (and $1/\beta^2$) corrections, it is better to concentrate the analysis on quantities, like mass ratios, whose error goes to zero like a^2.

If we want to do a simulation of QCD, there are some basic numbers which may be important. For each link there is an SU(3) matrix, which can be represented by 9 complex numbers; there are 4 links per site: the total memory requirement is $72L^4$ words on a L^4 lattice; however, using tricks the memory requirement may go down to approximately $36L^4$ words. A lattice of 35^4 is the maximum possible on a memory of $64MW$ (a typical upper bound with present computers)

The numbers of floating point operations is dependent on the algorithm, in a typical case we need to do about 2.10^3 floating point operation for link updating. For the updating of the whole lattice (with say

$L = 32$) we need about 8.10^9 floating point operations ; if the computer is performing with a speed of .5 10^9 floating point operations (we need a Cray II or APE [11]) it takes 16 seconds to do one step. A long computation may take 10^5 steps and the results may be available in less than one month.

Although such a computation is at the boundary of the present technology the situation will change in the future: *APE* 100 (a new computer which is planned to be constructed in Italy) should be about 100 times faster of *APE* and it should have a memory of 2 *GW*. On such a machine the update of a 64^4 lattice should take only 2 seconds.

The general strategy to estimate the systematic errors connected to the extrapolation in L and a consists in computing the masses with very small systematic error at given L and a, in changing L at fixed a to control the finite volume effects in changing a at fixed L to study the effects of the lattice spacing.

In these lectures I will mainly describe the results obtained by the *APE* group, mainly because they are the most accurate, with better control of systematic error and also because I know them very well.

I will concentrate my attention on three quantities, the glueball mass, the string tension and the order deconfinement transition.

Let us firstly consider the string tension and the glueball mass. In the continuum limit we have

$$\langle O \mid F^2 \mid G \rangle \neq 0 \tag{7.5}$$

where $\mid G\rangle$ is the glueball and F is the gluonic equivalent of the electromagnetic tensor (we have already seen it in eq. (7.1)). The lowest glueball is the lightest state satisfying eq. (7.5).

Consequently the glueball can be found on the lattice by looking to the plaquette - plaquette correlations.

The string tension is defined in term of the static potential between charges in the fundamental representation:

$$V(r) \sim \sigma r \quad \text{when } r \to \infty \tag{7.7}$$

On an $L^3 \times \infty$ lattice we can consider the trace of the Wilson loops winding in the space direction (the Polyakov loops). If we consider the correlations of these operators we have

$$\begin{aligned} \langle P(t)P(0)\rangle &\sim \exp(-m_P t). \\ m_P &\sim \sigma L - \frac{\pi}{3L} + \cdots \end{aligned} \tag{7.8}$$

The factor $\pi/3L$ is derived by assuming the presence of flux tube (which behaves like a string) between two confined charge, but the presence of this factor is not absolutely sure (long range excitations along the string can modify this factor).

More abstractly we could remark that the theory is symmetric under the global $(Z_3)^3$ group, each Z_3 group acts only on the Polyakov loops in one given direction, leaving invariant the local gauge invariant quantities.

The spectrum cane divided in two pieces, one piece corresponds to states which transform as singlets under the group $(Z_3)^3$ and the other piece to states which transform as a non trivial representation of the same group.

We have an energy gap in both piece of the spectrum; they are respectively the glueball mass (for the singlet) and m_P for the non singlet. In the work of the APE collaboration [10] simulations have been done on lattices of size $9^3\times32$ at $\beta = 5.7$, $9^3\times32$, $12^3\times32$ and $16^3\times32$ at $\beta = 5.9$ and $16^3 \times 32$ at $\beta = 6.1$. In all cases the time direction is much larger than the space and finite time length effects may be neglected.

The comparison of the three lattices at $\beta = 5.9$ gives us an idea on the finite volume effects. The three lattices 9^3 at $\beta = 5,7$, 12^3 at $\beta = 5.9$ and 16^3 at $\beta = 6.1$, have approximatively the same physical dimensions, i.e. in all the three cases we find $L\sigma^{1/2} \simeq 2.5$ (if we assume that $\sigma^{1/2}$ is 420 MeV, the box is about 1.2 Fermi). In this way we can compare the results obtained at the three lattice spacing (which range from .14 to .075 Fermi) independently of the finite volume effects. This procedure (changing a at fixed volume) is quite useful to disentangle finite volume effects for non zero lattice spacing effects.

Instead of measuring correlations we have used the alternative method of changing the Hamiltonian at time $t = 0$ and to measure the expectation values of the operators as function of time.

For example, if

$$\tilde{H} = H - \epsilon O(0), \tag{7.9}$$

usual perturbation theory tell us that for small ϵ we have that

$$\langle O(t)\rangle_\epsilon = \langle O(t)\rangle_{\epsilon=0} + \epsilon\langle O(t)O(0)\rangle_{\epsilon=0} + O(\epsilon^2). \tag{7.10}$$

If ϵ is not small we still have that

$$\langle O(t)\rangle = A + B\exp(-m_O t) + C\ \exp(-m'_O t) + \cdots, \tag{7.11}$$

where m'_O is the maximum of the first exited state having the same quantum number of O or O^2. A measurement of $\langle O(t)\rangle$ when the Hamiltonian is modified at time 0, allow us to extract the masses in the same way as the correlation function. The advantage of this method is that statistical errors are much smaller here than for the correlations: in the limit of large L one gains a factor $O(L^{-3/2})$ in the size of the error.

Actually we put all the links at $t = 0$ in the x and y direction equal to one; in this way we could monitor the string tension and the 0^{++} and 2^{++} glueballs: the string tension is obtained by looking the trace of the Polyakov loops while the 0^{++} glueball is obtained by looking to the sum of the trace of the plaquettes in the xy, yz, and zx plane.

The masses can be estimated by studying the effective mass $m(t)$ defined as

$$m(t) = -\ln[(\langle O(t) - O(\infty)\rangle)/(\langle O(t-1) - O(\infty)\rangle) \tag{7.12}$$

where $O(\infty)$ is obtained as the average in the large t region.

Unfortunately in a typical simulation for the glueball the effective mass $m(t)$ has small errors only for $t \leq 4$: in this region $m(t)$ is not t independent if we construct the plaquettes using local operators, as it can be expected from the discussion in previous chapter.

We could cross the fingers and hope that

$$m(4) \simeq m(\infty), \tag{7.13}$$

but it would be hard to assign any systematic error to the approximation (7.13).

Our approach consists in introducing the operators $U^n(x)$ which are smeared in a region of size $O(n^{1/2})$ around x. If only two masses are important (and this should happens for not to large t) we can write

$$m^n(t) = m_\infty + C_n \exp(-\Delta mt)t + \cdots \tag{7.14}$$

where $m^n(t)$ is the effective mass computed using the U^n operator and $\Delta m = (m_O - m_O')$.

Assuming that we stay at a sufficient large value of n such that eq. (7.14) is well satisfied without considering still higher states, if there is a value of n such that

$$m^{\bar{n}}(t) = m^{\bar{n}}(t-1), \tag{7.15}$$

$m^{\bar{n}}(t)$ is a good estimator of m_∞. This observation is sometimes useful to find the best value of n for which the effective mass is a better indicator of the true mass.

In the continuum a possible smoothing could be done by using a Gaussian kernel in eq. (6.6). Equivalently we could introduce a family of fields $\varphi_\tau(x,t)$ which satisfies the differential equation

$$\frac{\partial}{\partial \tau}\varphi_\tau(x,t) = \Delta_x \varphi_\tau(x,t) \tag{7.16}$$

with boundary condition

$$\varphi_0(x,t) = \varphi(x,t), \tag{7.17}$$

whose solution is given by

$$\varphi_\tau(x,t) = \frac{C}{\tau^{3/2}} \int dy \exp -\frac{(x-y)^2}{4\tau}\varphi(y,t). \tag{7.18}$$

In other words φ_τ is a smooth (analytic) field averaged on the region $(x-y) = O(\tau^{1/2})$.

The generalization of eq. (7.16) to gauge theories is

$$\frac{\partial}{\partial \tau}A^\mu_\tau(x,t) = -\mathcal{D}^\tau_\nu F^{\mu\nu}_\tau(x,t), \tag{7.19}$$

where F_τ and $\mathcal{D}_\tau$ (the covariant derivative) are computed using the field A_τ.

In the limit $\tau \to \infty$ A_τ is a solution of the classical three dimensional Yang Mills equation $\mathcal{D}_\nu F^{\mu\nu} = 0$.

Equation (7.19) has been implemented on the lattice in a gauge invariant way, the integration step is ϵ and the operators U^n are related to the solution of the lattice equivalent of eq. (7.19) at time $\tau = n\epsilon$ (i.e. after n steps).

The results for the effective masses at $\beta = 5.9$, $L = 12$ are shown in fig. 1 for different n and t.

As can be seen form fig. 1(1), the effective masses are strongly n dependent for $t = 1, 2$; the n dependence becomes weaker by increasing n and the time dependence is very small in the region $n \sim 15$. An overall fit to the data allow is to extract the mass of the glueball with very systematical error.

A similar analysis is done on for the glueball; the final results are given in the table.

Table

σ	m	β	L
0.122	0.84	5.7	9
0.047	0.65	5.9	10
0.056	0.75	5.9	12
0.053	0.76	5.9	16
0.026	0.54	6.1	16

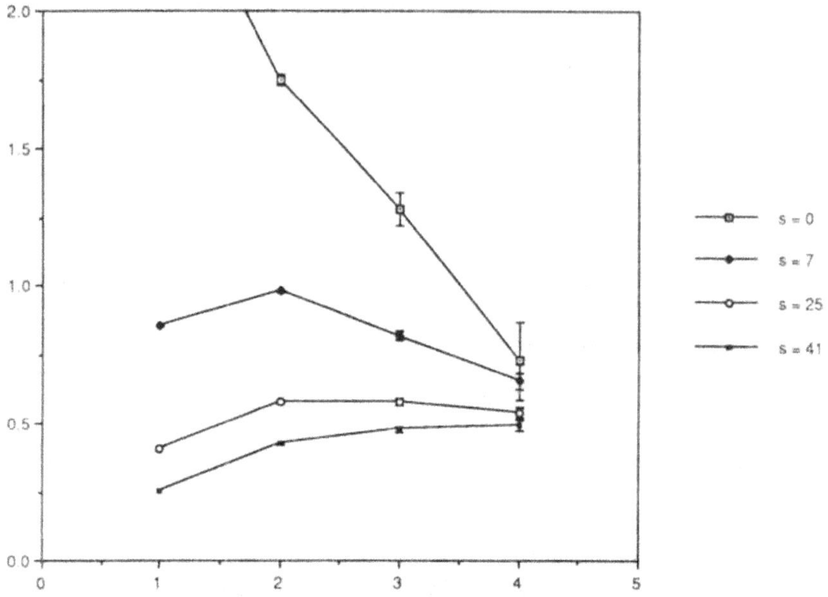

Fig. 1. $m(t)$ as function of t computed for different values of the smearing

The errors that are around two three percent.

A few comments are in order. The L dependence of σ is not in strong disagreement with eq. (7.8) as can be seen from the data at $\beta = 5.9$. The L dependence of m_g at $\beta = 5.9$ is not expected. If we use the Lüscher formula for the L dependence of m_g

$$m(L) = m_\infty - \frac{G^2}{16\pi m_\infty} \frac{3}{m_\infty L} \exp -(\mu L)$$
$$\mu = \frac{\sqrt{3}}{2} m_\infty, \tag{7.20}$$

we obtain an extremely high value for G^2, i.e. the three glueballs coupling constant:

$$G^2 \simeq 250 - 1000\ m_g^2. \tag{7.21}$$

The lower value having being obtained by using in the computation of μ $m(L)$ at the place of m_∞.

On the other hand physical arguments suggest that we should have

$$m(L) = m_\infty - \frac{C^2}{16\pi m_\infty} \frac{3}{m_\infty(L-\mathcal{D})} \exp -\mu(L-\mathcal{D}), \qquad (7.22)$$

where $\mathcal{D}$ is the diameter of the glueball and C is a quantity of $O(1)$.

A comparison of eq. (7.22) and (7.20) tell us that

$$G^2 = C^2 \exp(\mathcal{D}\mu). \qquad (7.23)$$

If the diameter $\mathcal{D}$ is much greater than $1/\mu$, G^2 may be extremely large. If we assume a simple minded model in which the mass of the glueball is equal to the energy of a string of length $\pi\mathcal{D}$, we have that

$$m_g = \sigma\pi\mathcal{D} \qquad (7.24)$$

Using the numerical value of m_g and σ we find $\mathcal{D} \simeq 3m_g^{-1}$: in other words glueballs may be rather large objects. The naive estimate may be easily modified by a factor 2 ; in this case an extremely large value of G^2 may be not surprising: in other terms is the glueball diameter could be $0.8F$ with a mass of 1.5 GeV, $m_g\mathcal{D} \simeq 6$.

In fig. 2 we show the product

$$\sigma\Lambda_L^2. \qquad (7.25)$$

It is difficult to extract the asymptotic value of this quantities. We can only say that it is possible that the quantity in eq. (7.25) goes to a finite non zero limit when $\beta \to \infty$, but it would be very hard to extract the asymptotic value from the data.

More interesting results arise if we plot

$$R(a) = m_g/\sigma^{1/2} \qquad (7.26)$$

as function of σa^2 (see fig. 3).

In this case we see a reasonable dependence of R on the lattice spacing (we recall that we should have $R(a) = R_{\text{cont}} + O(a^2)$, where R_{cont} is the value of R in the continuum limit $a \to 0$). The data may already be extrapolated at $a = 0$ giving

$$R_{\text{cont}} = 3.4 \pm 0.3, \qquad (7.27)$$

which, using the assumption that the string tension is (420 MeV) $^{-2}$, gives a glueball mass of 1500 $\pm$ 150 MeV.

The same value of the string tension has been used to estimate the size of the lattice in Fermi. The number 420 MeV comes from the string model for strong interactions, but, as you will see, it is not a bad approximation to reality.

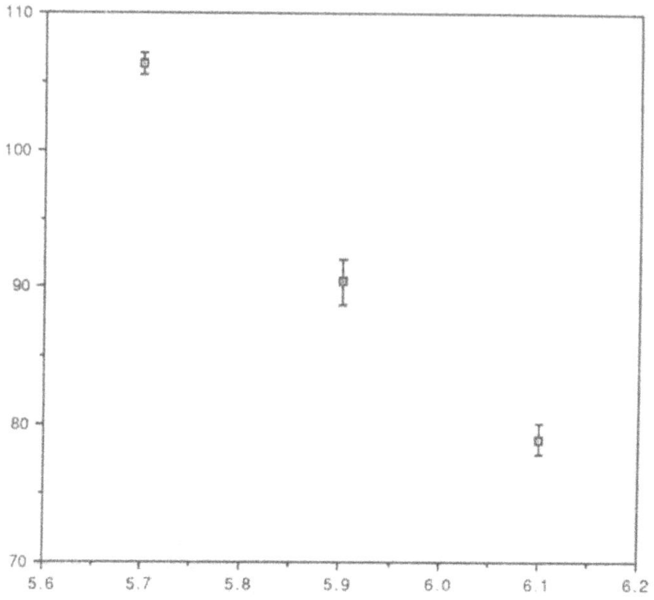

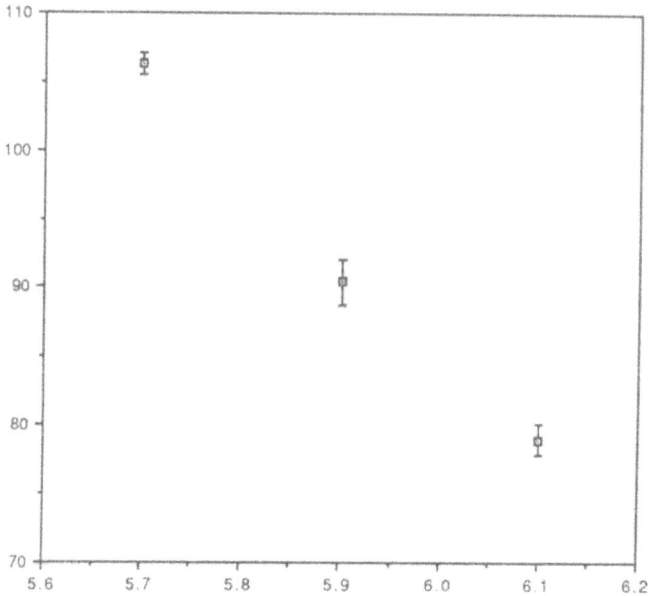

Fig. 2. Plot of $\sigma\Lambda_L^2$ for three values of β

8. The deconfinement transition in pure gauge QCD

Also if color is confined at zero temperature, a perturbative analysis shows that color becomes deconfined when we increase the temperature: there must be a transition temperature (T_c), which separate the confined from the deconfined phase. We would like to compute T_c and to investigate the nature of the transition.

In the Euclidean formulation the behavior at finite temperature can be obtained if we study a strip of width $L_t = T^{-1}$ in one direction and infinite in the others three directions. In this case the theory is invariant under a global Z_3 group and the trace of Polyakov loops in the time direction are not invariant under this transformation.

General principles suggest the existence of two phases:

$$\langle P(x)P(0)\rangle \sim \exp(-|x|/\xi), \qquad T < T_c;$$
$$\langle P(x)P(0)\rangle \sim |\langle P\rangle|^2 + \exp(-|x|/\xi), \qquad T > T_c. \tag{8.1}$$

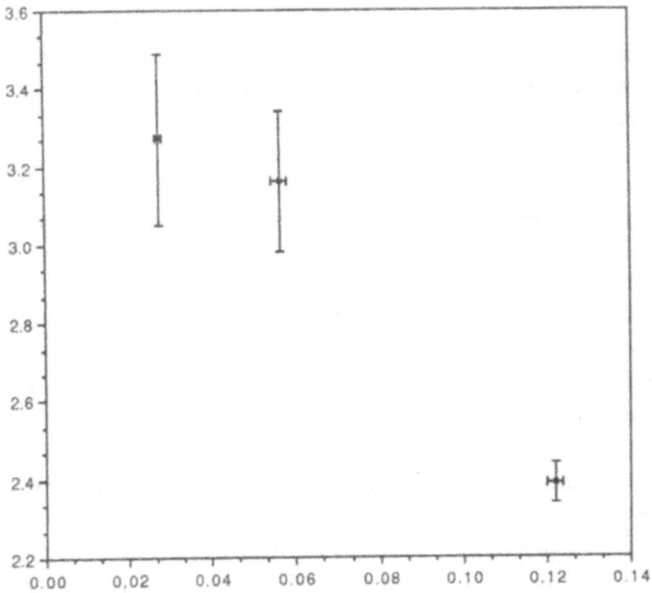

Fig. 3. Plot of $R(a) = m_g/\sigma^{1/2}$ as function of σa^2.

The quantity $-1/L_t \ln|\langle P\rangle|$ has the meaning of the energy of a static isolated quark in the deconfined phase.

We may ask if the transition is first order or second order. In the first case ξ remains finite at the transition and $|\langle P\rangle|$ jumps ; in the second case ξ becomes infinite at the transition and $\langle P\rangle$ vanishes continuously.

It has been argued that the transition should be first order: the Z_3 model on the lattice with nearest neighbor interaction has a first order transition (at least there are numerical indications) and universality has been used to conclude that the transition must be first order, in agreement with the Landau criterion.

This hasty conclusion has demotivated many people from carefully investigating the order of the transition.

However it well known that in two dimensions, the Z_3 invariant spin model (i.e. the three state Potts model) has a second order phase transition with nearest neighbor coupling, while if the range of the interaction is sufficiently large, but finite, the transition is first order because mean field theory is a very good approximation.

Generally speaking it is well know that the same kind of transition may be first or second order as function of a parameter: usually a tricritical point separates the two regimes. Indeed it has been suggested in the past that three dimensional Z_3 model may have (for some values of the parameters) a second order phase transition [12]: this question is under careful numerical investigation in Rome.

Numerical simulations must be done on lattices $L_x \times L_y \times L_z \times L_t$, where the $L's$ in the space directions are much larger than in the time direction in order to remove finite lattice effects which would smooth the transition. Unfortunately many long simulations are done on lattices which practically have $L_x = L_y = L_z \simeq L_t$ (e.g. $16^3 \times 8$ or $17^3 \times 13$) [13]. In this case the conclusions on the order of the transition depend only on the author's mood.

Also if more reasonable lattices are used (e. g. $16^3 \times 6$) most of the literature has mainly considered the dependence on β of $|\langle P \rangle|$; here β is not T^{-1}, but it is defined in eq. (7.2)): to change β is equivalent to change a and consequently to change the temperature in physical units if the number of steps in the time direction is kept fixed: therefore for any fixed time size N_t there is a value β, β_c for which the transition happens.

However $\langle | P | \rangle$ jumps at a first order transition, while it vanishes as

$$\langle | P | \rangle \sim |\beta - \beta_c|^b \tag{8.2}$$

at a second order transition, with b is likely not very far from 1/3. The distinction between $b = 0$ (i.e. first order) and $b \neq 0$, (i. e. second order) may be not clear cut, especially if finite volume rounding are present.

In order to have a clear-cut result we have measured the correlation function ξ in a geometry where $L_x = L_y = L$ and $L_2 >> L$. In this situation we should have that

$$\begin{aligned} \xi &= |T - T_c|^{-\nu} f((T - T_c)/L^{1/\nu}) \\ \xi &= CL \quad \text{at } T = T_c. \end{aligned} \tag{8.3}$$

The divergence of the correlation length near T_c is a clear signal of a second order phase transition: finite volume effects should produce a correlation length ξ proportional to L.

We have studied the theory with $L_t/a = 4$, where the transition was known to be near $\beta = 5.70$.

We [14] have done simulations on lattices of size $8^2 \times 24 \times 4$, $12^2 \times 36 \times 4$ and $16^2 \times 48 \times 4$ and we have found the results shown in fig. 4.

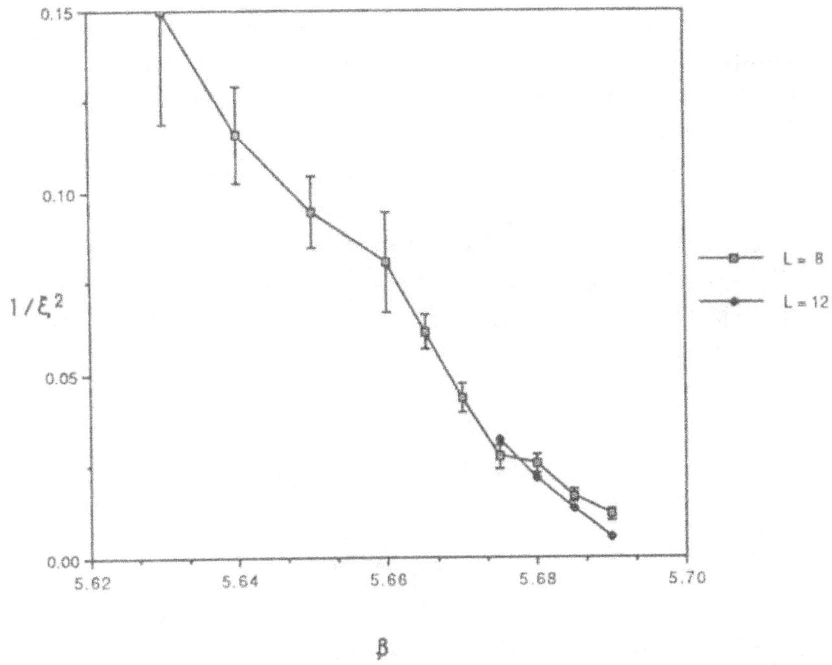

Fig. 4. The correlation length as function of β near the phase transition.

The value of the correlation length at $\beta = 5.69$ (which is our estimation for the critical point) inside the experimental errors is strictly proportional to L. The value of $\langle P \rangle$ has also been computed and its behavior is consistent with eq. (8.2) with b in the 1/3 - 1/4 region, although it is difficult to reach definite conclusions.

All the results are consistent with a second order transition and no signals of a first order transition are observed: also if a first order phase transition is present, it should be a weak first order transition, characterized by a very small latent heat. Of course such a possibility can never be excluded by any numerical computation on a finite lattice: numerical simulations, as real experimental data, neither say nor hide, they hint.

9. Quenched fermions

Up to now we have considered pure gauge QCD. In the real world fermions (i.e. quarks) are present and all the observed hadrons contain quarks. If n flavour quarks of the same mass are present, the effective action for gauge fields is

$$S = \frac{1}{g^2}\int d^4x F^2 - n\,\mathrm{tr}[\ln(\mathcal{D}+m)]. \tag{9.1}$$

Numerical simulations with eq. (9.1) are possible, but they are quite difficult, so most of the computations for fermions have been done in the quenched approximation (i. e. $n = 0$), where everything becomes simple.
These computations are based on the following exact formula

$$\begin{aligned}&\langle\bar{\psi}^a_\alpha(x)\Gamma_{\alpha\beta}\psi^b_\beta(x)\psi^a_\gamma(0)\Gamma_{\gamma\delta}\psi^b_\delta(0)\rangle\\ &\quad= \langle\mathrm{tr}(G(x,A)\Gamma\gamma_5 G^*(x,A)\gamma_5)\rangle,\end{aligned} \tag{9.2}$$

where the expression in r.h.s. of eq. (9.2) depends only on the gauge fields A, the indices a and b are different and

$$(\mathcal{D}+m)G(x,A) = \delta(x). \tag{9.3}$$

Equation (9.2) and its generalization allow us to compute the masses of hadrons for $n = 0$ (and also $n \neq 0$).
Generally speaking, if we study equation (9.3) on a lattice, its solution can be found in a number of steps which is proportional to $1/(ma)$, although it may be possible to find a technique which needs only $-\ln(ma)$ steps.
When we quantize fermions on a lattice, in the usual Wilson approach the chiral symmetry is explicitly broken: the lattice equivalent of $\mathcal{D}+m$ depends on a parameter k such that

$$m \propto (k_c - k). \tag{9.4}$$

If we consider for simplicity the case of only two quarts (up and down) of equal mass, we can compute the mass of the pion (the lowest mass pseudo-scalar meson) and rho (the lowest mass vector meson); the physical value of k is fixed by the condition that the ratio m^2_π/m^2_ρ takes the physical value (about .04).
It is usually believed that the chiral symmetry is spontaneously broken in the continuum limit and the pion is the Goldstone boson associated to this symmetry; on the lattice the situation is more tricky because the

chiral symmetry is not exact. The quasi Goldstone nature of the pion implies that $m_\pi^2 \propto (k_c - k)$; if chiral symmetry is approximatively exact at low energy; in this case we expect that all the masses are smooth functions of m_π^2, with some possible terms $m_\pi^4 \ln m_\pi^2$ for mesons or m_π^3 for baryons, although these terms should be depressed in the quenched approximation.

As usual a smoothing procedure strongly improve the quality of the data: in the case of fermions we found more efficient not to use the same procedure we have followed for gauge fields; the simpler solution is to perform a gauge transform on the gauge fields in order to go in a Coulomb gauge: after the gauge transform we [15] define

$$\psi_s(x,t) = \sum_{y \in B_x} \psi(y,t) \tag{9.5}$$

where B_x is a box centered at the point x. We show in fig. 5 the effective mass for the pion on a $9^3 \times 24$ lattice at $\beta = 5.7$ for the rho propagator; the effect of smearing is very strong and inpressive [15].

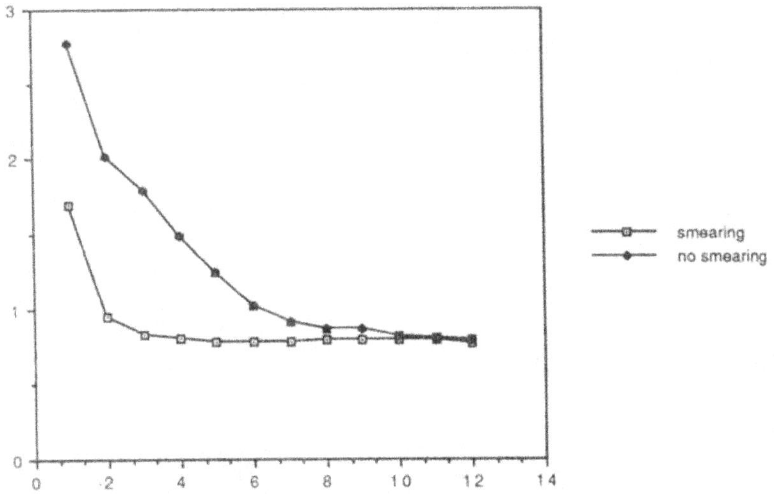

Fig. 5. The effective mass for the rho propagator with and without smearing.

In figs.(6) and (7) we show the data for the masses squares of pion, rho, proton and delta as function of k (they are obtained as the average of 289 propagators on a $12^3 \times 32$ lattice at $\beta = 5.7$).

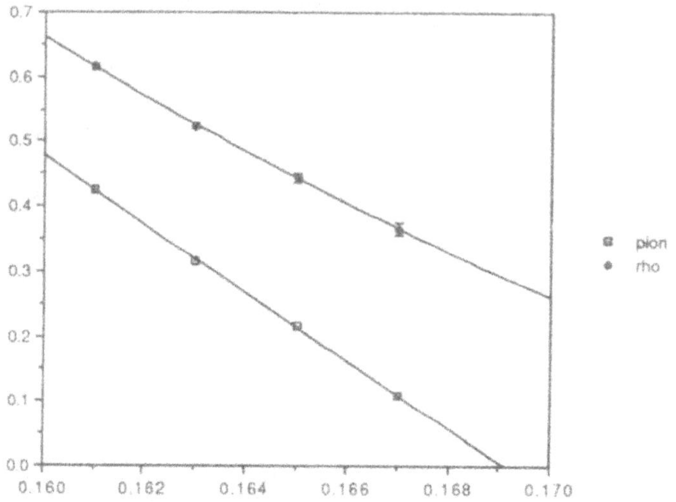

Fig. 6. The masses squared of the pion and of the rho as function of k.

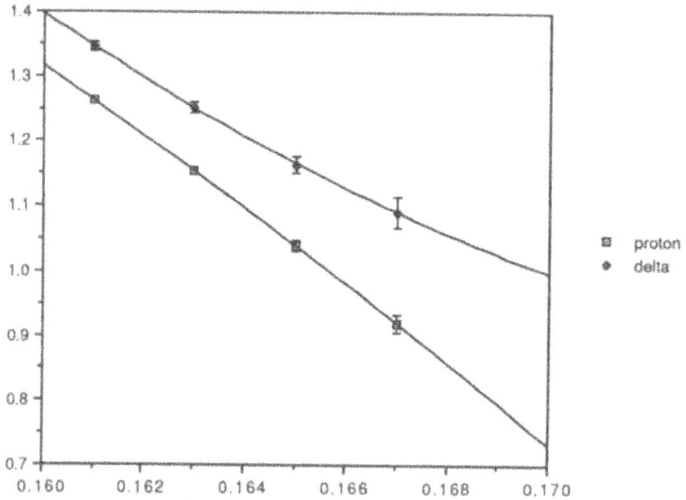

Fig. 7. The mass of the proton and of the delta as function of k

The data are smooth, the mass of the pion squared seems to vanish linear at k_c (which may be estimated to be at $k = .1696$) and no difficulty seems to be involved in the extrapolation at k very near to k_c.

In fig. 8 we plot the data for the ratio m_p/m_ρ as function of m_π^2/m_ρ^2, computed at $\beta = 5.7$ and at $\beta = 6.0$ (on 114 propagators on a $18^3 \times 32$ lattice).

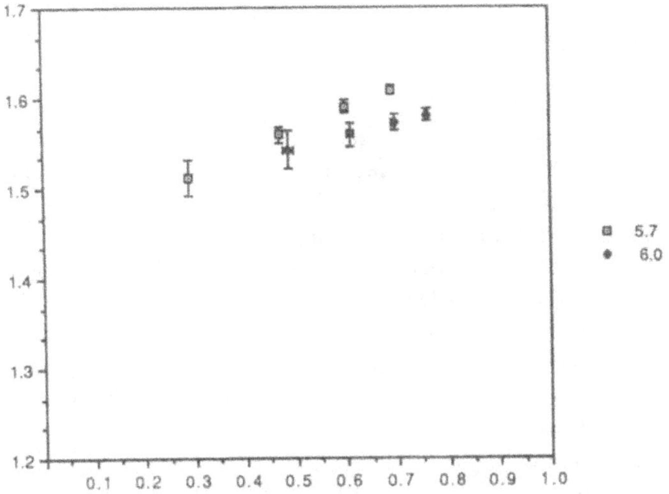

Fig. 8. The data for the ratio m_p/m_ρ versus m_π^2/m_ρ^2, at $\beta = 5.7$ and 6.0.

There is an apparent shift of the ratio of a few per cent going from $\beta = 5.7$ to $\beta = 6$. Unless something strange happens to the data in the region of smaller m_π or higher β, we could conclude that in the quenched approximation the ratio m_p/m_ρ is about 1.35-1.40, i.e. 10% higher than the experimental value 1.22. We plan to do a more detailed study of this point in the next future). It would be not surprising if the quarks loops shift the ratio m_p/m_ρ of 10% .

Many others results with fermions have been obtained and we invite the interested reader to look to the original literature [15].

We remark that if we compare the pure gauge computations described in the previous section with these results, we can compute the absolute value of string tension using as input the mass of the rho: we get

$$\begin{aligned} \sigma^{-1/2} &= 505 \pm\ 15 \text{ MeV}, \qquad \beta = 5.7 \\ \sigma^{-1/2} &= 470 \pm\ 20 \text{ MeV}, \qquad \beta = 6. \end{aligned} \tag{9.6}$$

The value of σ has a very small variation from 5.7 to 6.0 and it is surprisingly consistent with the value of 420 MeV in the continuum limit as suggested by the naive application of string models.

We are confident that we are not far from the goal of getting the mass spectrum of quenched QCD in the continuum limit with an accuracy of 5–10%.

10. Open problems

It is discussed in many papers how fermions loop may be introduced: there are many effective ways of doing it, which however produce a slowing down of the computation of a factor which may range from 10 to 100 (for a review see [16]). Although the situation is not fully satisfactory, such computations start to be in the range of existing machines. There is still work to be done in order to improve the algorithm, especially in the part in which the lattice Dirac equation is numerically solved.

The real unsolved problem is how to formulate a numerical simulations for a supersymmetric theory in such a way to preserve supersymmetry (also if we forget the problem connected with the lattice). In other words we would like to have a pure bosonic formulation of supersymmetric theory in which anticommuting quantities are not present.

In the two dimensional case it has been show that the Wess-Zumino model is equivalent (at least perturbatively) to the stochastic differential equation [17]

$$\frac{\partial}{\partial z}\Phi + \frac{\partial \bar{W}}{\partial \Phi} = \eta(z) \qquad (10.1)$$

where

$$\langle \eta(z)\bar{\eta}(z)\rangle = \delta^2(z). \qquad (10.2)$$

Unfortunately these kind of stochastic differential equations cannot easily extend to the generic case. It is also not clear how to implement these equations in a practical way. In principle it is reasonable to believe that supersymmetry could be used to simplify the numerical simulation, but this has yet to be done.

An other problem, which is unsolved at the present moment. is to find an algorithm for doing efficient simulations for a complex action; a complex action appears in many interesting cases as in presence of a topological term or of a chemical potential; a complex action appears also if we formulate quantum mechanics for real time, (not for Euclidean time).

The key idea is that it may be possible to write

$$\int_{-\infty}^{+\infty} \mathrm{d}x P(x) f(x) = \int \mathrm{d}z \tilde{P}(z) f(z), \tag{10.3}$$

where f is a polynomial (may be an entire function) of z, $P(x)$ is an analytic non real function of x and $\tilde{P}(z)$ is a non-negative function of the complex variable z.

Although some suggestion have already been made [18,19] (they are based on the proposal to use the Langevin equation with a complex drift, but a real noise), an efficient and reliable approach to computations with complex action has not yet been obtained. A breakthrough in this direction would strongly extend our limited possibilities of large scale numerical simulations.

References

[1] For a more careful discussion of this point see O. Lanford, Les Houches, Session XLII, 1984, Critical Phenomena, Random System, Gauge Theories, ed. by K. Osterwalder and R. Stora, Elsevier Science Publisher 1986.

[2] D. E. Knuth, The Art of Computer Programming, Addison-Wesley, Reading, Mass. 1973.

[3] G. Parisi and F. Rapuano, Phys. Lett. 157B (1985) 301.

[4] For a very efficient Fast Fourier Transform, see A. Nobile and V. Roberto, Comp. Phys. Comm. 42 (1986) 233.

[5] See for a review method: Monte Carlo Methods in Statistical Mechanics, ed. by K. Binder, Springer-Verlag, New York (1979); for a short theoretical discussion see G. Parisi, Statistical Field Theory, Addison Wesley New York 1988 or G. Parisi, Les Houches, Session XLII, 1984, Critical phenomena, random system, gauge theories, ed. by K. Osterwalder and R. Stora, Elsevier Science Publisher 1986.

[6] G. Parisi Nucl. Phys B205 [FS5] (1982) 337.

[7] S. Duane and J. Kogut, Nucl. Phys B275 (1986) 338.

[8] G. Parisi, Progress in Gauge Field Theory, ed. by 't Hooft et al., Plenum, New York 1984.

[9] G. G. Batrouni, G. R. Katz, A. S. Kronfeld, G. P. Lepage, B. Svetisky and K. G. Wilson, Phys. Rev. D 10 (1986) 398.

[10] The Ape Collab., P. Bacilieri et al., Phys. Lett. 205 B (1988) 535.

[11] The Ape Collab., P. Bacilieri et al., Conf. Computing in High Energy Physics, ed. by L. O. Hertsberg and W. Hoogland, North-Holland, Amsterdam 1985. See also G. Parisi et al., Lattice Gauge Theory Using Parallel processors, ed. by Li Xiaoyuan et al., Gordon and Breach, London 1987.

[12] F. Fucito A. Vulpiani Phys. Lett. 89A (1982) 33.

[13] For a review see M. Fukugita, Lattice Gauge Theory using parallel processors, ed. by Li Xiaoyuan et al., Gordon and Breach, London 1987.

[14] The Ape Collab., P. Bacilieri et al., Phys. Rev. Lett. 61 (1988) 1545.

[15] The Ape Collab., P. Bacilieri et al., Phys. Lett. 209B (1988) 145 and Nucl. Phys. [FS] (in press).

[16] For a review see the articles of J. B. Kogut, M. Fukugita, O. Martin and R. Gupta in Lattice Gauge Theory using parallel processors, ed. by Li Xiaoyuan et al., Gordon and Breach, London 1987.

[17] G. Parisi, N. Sourlas, Nucl. Phys. 217B (1983) 203.

[18] G. Parisi Phys. Letters 101A (1982) 333.

[19] B. Söderberg, Nucl. Phys. 245B (1988) 765.

Computer Physics Communications 45 (1987) 345-353
North-Holland, Amsterdam

THE APE COMPUTER: AN ARRAY PROCESSOR OPTIMIZED FOR LATTICE GAUGE THEORY SIMULATIONS

M. ALBANESE [d], P. BACILIERI [a], S. CABASINO [b], N. CABIBBO [c], F. COSTANTINI [d], G. FIORENTINI [d], F. FLORE [d], L. FONTI [a], A. FUCCI [e], M.P. LOMBARDO [d], S. GALEOTTI [d], P. GIACOMELLI [h], P. MARCHESINI [e], E. MARINARI [c], F. MARZANO [b], A. MIOTTO [f], P. PAOLUCCI [b], G. PARISI [c], D. PASCOLI [f], D. PASSUELLO [d], S. PETRARCA [b], F. RAPUANO [b], E. REMIDDI [a,g], R. RUSACK [h], G. SALINA [b] and R. TRIPICCIONE [d]

[a] INFN-CNAF, Bologna, Italy
[b] Dipartimento di Fisica, I Universita' di Roma "La Sapienza" and INFN-Sez. di Roma, Italy
[c] Dipartimento di Fisica, II Universita' di Roma "Tor Vergata" and INFN-Sez. di Roma, Italy
[d] Dipartimento di Fisica, Universita' di Pisa and INFN-Sez. di Pisa, Italy
[e] CERN, Geneva, Switzerland
[f] Dipartimento di Fisica, Universita' di Padova and INFN-Sez. di Padova, Italy
[g] Dipartimento di Fisica, Universita' di Bologna and INFN-Sez. di Bologna, Italy
[h] The Rockefeller University, New York, USA

The APE computer is a high performance processor designed to provide massive computational power for intrinsically parallel and homogeneous applications. APE is a linear array of processing elements and memory boards that execute in parallel in SIMD mode under the control of a CERN/SLAC 3081/E. Processing elements and memory boards are connected by a 'circular' switchnet. The hardware and software architecture of APE, as well as its implementation are discussed in this paper. Some physics results obtained in the simulation of lattice gauge theories are also presented.

1. Introduction

APE (Array Processor with Emulator) is a high performance computer designed to provide massive number-crunching capabilities for applications that are intrinsically parallel and homogeneous. So far, APE has been used for lattice gauge theory (LGT) simulations.

The full scale machine consists of a linear array of 16 cells, each consisting of a floating point processor and a memory-board. Floating point processors and memories are connected through a 'circular' switchnet. The number of cells is in principle arbitrary and can eventually be enlarged. The array runs in SIMD (Single Instruction Multiple Data) mode under the control of a CERN/SLAC 3081/E (the Controller), integrated in a general purpose host environment (presently a VAX/VMS system).

The theoretical speed of the machine is 1 Gflops while the total memory size is 256 Mbytes.

As an example of performance, a prototype consisting of 4 cells updates one link in an SU(3) pure gauge theory in 40 μs, running a program written in a high-level specially developed language. The efficiency for such a program, whose source code is larger than 1500 lines, is 70% of the theoretical speed. This figure can be compared with 35 μs obtained on a CRAY-XMP1 code in which the most time consuming routines were written in assembly language.

APE can efficiently perform many primitive computations such as matrix manipulation, fast Fourier transform, multidimensional convolution and, in general, algorithms that can be parallelized while requiring extensive communication among the processing elements.

While achieving a high throughput, typical of special-purpose processors, APE has a high degree of programmability. A high level FORTRAN-like language has been developed to take full advantage of the features of the machine. Although the user must keep in mind the parallel structure of the machine while writing application programs, no deep understanding of the structure of APE is required to achieve good efficiency. An optimizing step is in fact included in the APE compiler.

Two prototypes consisting of 4 cells each have been fully developed and are operational since september 1986. They have been used for SU(3) LGT simulations. Production of the full scale machine is underway (partially contracted to industrial firms). A full scale operational prototype is expected for spring 1987.

This paper describes the hardware and software architecture of APE and its implementation. We first give an overview of the machine and then give a more detailed description of the hardware structure of APE. Software aspects are then considered while some physics results and concluding remarks are presented in the last section.

2. Overview

APE, as shown in fig. 1, is a linear array of SIMD processors. A linear array is easy to implement, and can be expanded to larger size in a straightforward way. Furthermore a more complex structure (e.g. a planar array) can be efficiently simulated on a linear array, the reverse being not always true.

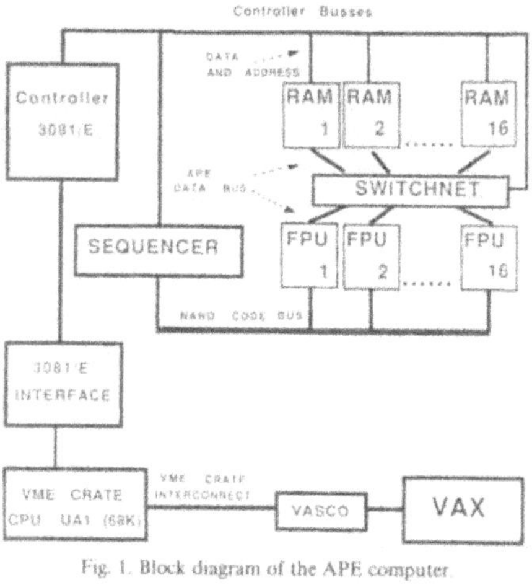

Fig. 1. Block diagram of the APE computer

A SIMD architecture is simple from the point of view of implementation, since it requires only one sequencing unit, which controls all the processing elements via a broadcast micro-code. It also has conceptual simplicity, in that all problems of synchronization between processors are avoided.

The range of problems that can be solved on a SIMD machine clearly depends on the communication capabilities of the switchnet. As far as LGT simulation is concerned communication between first neighbor processors is adequate. This solution would however over-specialize our SIMD machine, preventing its use in even moderately non-local problems.

These considerations lead us to consider a 'circular' switchnet, capable of connecting in a 'rigid' way each FPU to each memory. Specifically the switchnet can establish a bi-directional data-flow between FPU(k) and memory($k+l$), $0<k,l<15$, periodic boundary conditions being used to wrap around the edge of the array.

This structure is adequate to handle non-local problems, provided that they can be solved by parallel algorithms. Still it is relatively simple and requires a limited number of control signals. Its simplicity can be exploited to attain a speed sufficient to sustain the peak performance of the processing elements. APE is supported by a specially developed software. The APE software is essentially a two level package, designed to adress two main issues. At the user level, a FORTRAN-like language naturally reflects the parallel structure of the machine. At a level closer to the hardware architecture of the machine, maximization of the performance of the pipelines of the machine is the key issue. This fact has favoured the decision to build a fully microcoded machine, completely avoiding any assembly-language level of programming. From the point of view of the high level user, APE is controlled by a host computer (currently a VAX). The Host resident APE Kernel Software (Hack) provides a VMS environment for the high level user. Hack consists of the APE compiler, the symbolic I/O manager, the

Debugger, the Program Loader and the Backup software.

3. Hardware

The APE computer has 4 elements: an array of 16 'cells', a switchnet, a sequencer and a controller running synchronously with a clock cycle of 120 ns.

Each cell comprises one Floating Point Unit (FPU) and one memory board. The switchnet provides a data-path between memories and FPU's. The connection is usually between the FPU and memory belonging to the same cell, but the data-path can be re-directed under program control to bi-directionally connect FPU's and memories of different cells. A sequencer broadcasts the same microcode (labelled nano-code in APE jargon) to all the FPU's. Nano-code sequences are stored in the (writable) sequencer memory and are broadcast to the FPU's under control of the sequencer itself.

The controller executes the integer-arithmetic and logic sections of the application program that are not mapped on the FPU's. It also generates appropriately sinchronized addresses and controls for the memory banks and the switchnet. Finally it controls the logic flow of the FPU program supplying branch adresses to the sequencer. The controller is connected to a VME bus via dedicated boards. The VME QBUS/UNIBUS connec-

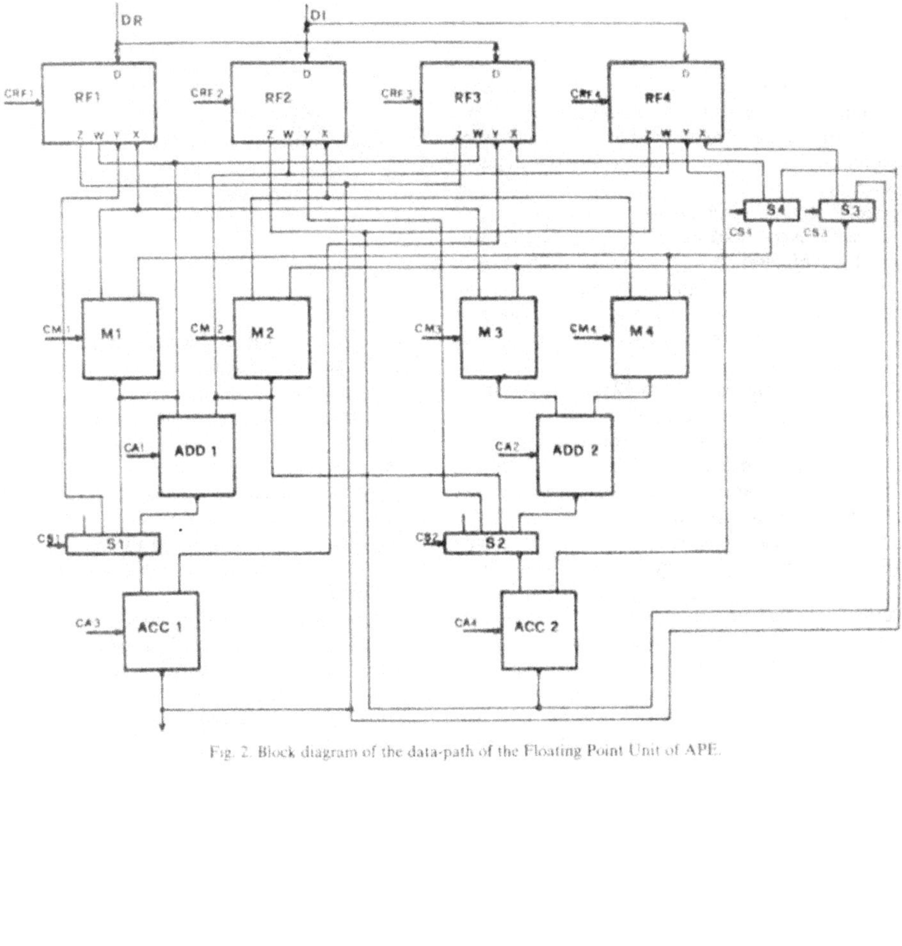

Fig. 2. Block diagram of the data-path of the Floating Point Unit of APE.

tion is finally accomplished through a dedicated interface. These hardware items are now described in more detail.

3.1. FPU

The Floating Point Unit is the processing element of the elementary APE cell. It has been designed to efficiently operate on complex floating point numbers with 32 bit accuracy. The floating point format conforms to the IEEE standard. A block diagram is presented in fig. 2. Communication to/from the memory is accomplished through two complex register files (each containing 32 registers). These are implemented with 4 Weitek WTL1066 register file chips, connected to the memory port through their bi-directional ports.

The FPU is optimized to perform the operation

$$A = B * C + D. \qquad (1)$$

A, B, C, D being complex numbers. The FPU uses four floating point multipliers and four floating point ALU's, respectively Weitek WTL1032 and WTL1033 chips. They can start a new operation every clock cycle in pipeline mode, with a latency of 5 cycles. A suitable configuration of these devices is capable of obtaining a new result of operation (1) every clock cycle, after pipeline startup. Such a configuration is shown in fig. 3. The process of computing eq. (1) can be logically split in three steps. All products of the four real numbers in $B = (b_1, b_2)$ and $C = (c_1, c_2)$ are evaluated in step 1. The complex number $B * C = (b_1c_1 - b_2c_2, b_1c_2 + b_2c_1)$ is evaluated in step 2 by appropriately adding/subtracting the result of step 1. Finally $B * C$ and D are accumulated onto A in step 3 and the result is saved onto any of the two register files for later use or transmission to the memory. The latency of the full operation is 18 cycle.

The data-path can be re-arranged for greater flexibility on a cycle by cycle basis. For instance pure real arithmetic can be performed efficiently using the real and imaginary parts of each data word as two independent real numbers.

Special hardware is used to implement IF...THEN...ELSE structures supported by the APE language. On our SIMD machine IF structures can be implemented at a local and global level. At the local level writing of data onto memory is performed or inhibited in each cell according the value of the condition code generated on the cell itself by the (local) IF instruction. The local IF structure cannot alter the logical flow of the program. The latter can be modified by the global IF instruction also implemented in hardware. In this mode, the logical AND of the local condition codes is passed to the controller which will act consequently. In both cases four levels of nesting are hardwired.

Heavily used special functions like exp and log are evaluated in the FPU's using a first approximation provided by hardwired logic. This is also true for the sqrt and inverse functions, with the help of an 8 bit look-up table built in the register files.

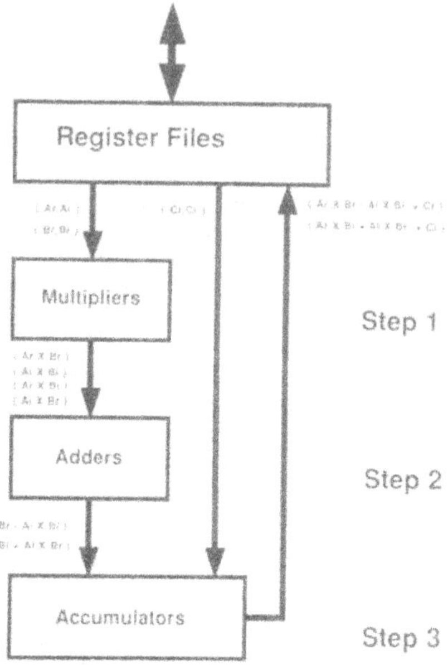

Fig. 3. Simplified diagram of the data-path of the Floating Point Unit of APE, showing the pipeline structure optimizing the evaluation of complex scalar products.

3.2. Memory

Typical LGT applications require a large data base (our long standing goal is a memory size of 1 Gbytes). This requirement dictates the use of dynamical memory. On the other hand there is a requirement of a large memory band-width, sufficient to keep the FPU's busy. As an example, the calculation of Tr(A∗B∗C∗D), A B C and D being SU(3) matrices (a typical kernel of LGT codes) requires about 200 floating point operations on 24 complex numbers (or 200 bytes). Hence a band-width of 8 bytes per cycle is required. The conflicting requirements of large size and high band-width are met by organizing the memory in 8 interleaved banks, accessed in sequence. Memory access time is one word every three clock cycles for sparse access.

Higher speed is achieved in FAST mode. In this mode a block of memory up to 32 kbytes can be transferred at a rate of one word (8 bytes) per cycle after a start-up of 3 cycles. The 64 bit wide word is organized as 2 banks of 32 bits each. Seven check bits are used for each half-word to provide single error correction and double error detection, adding to the reliability.

8 Mbyte boards have been built using standard 256 kbit DRAM chips. The two working APE prototypes use such boards. The final version will use SIP packaging of 256 kbit chips to squeeze 16 Mbytes of memory into one board, giving a total memory of 256 Mbytes. This figure will be expanded to 1 Gbytes when denser chips are available at reasonable prices.

Each memory board can be accessed by the controller and the FPU on independent busses, addressing and control being provided by the controller in both cases. A common bus is shared by all memories for communication with the controller, while separate busses connect each memory to an FPU via the switchnet. Either a data transfer between one of the memories and the controller is active or a parallel transfer between each memory and the connected FPU is in progress at each clock cycle.

3.3. Switchnet

The switchnet provides a re-configurable data-path connecting FPU's and memories.
As already mentioned, the switchnet can connect FPU(k) and memory[mod($k+j$, 16)] $0 < k, j < 16$. The offset j is specified by an appropriate number of bits in the memory address.

The switchnet is a separate board housed between the crate containing the memories and the FPU crate. Flat cables ensure electrical continuity. Physically each data-path is 32 bits wide. The transfer rate is 1 64-bit word per clock cycle, multiplexed as 2 half words of 32 bits each.

Two prototypes connecting the FPU's and memories of 4 cells have been built using wire-wrap technology. Extensive use of programmable logic (PALs) has been made. The switchnet is transparent, in that the transfer time of one data word through the switchnet is shorter than 1 clock cycle.

The full scale version will follow a pipelined scheme. The transfer of 1 data word will take one full clock cycle. This brings the effective pipeline latency of the memory from 3 to 4 cycles, negligibly affecting the overall performance of the system. This choice has been made to ensure appropriate reliability to a device expected to deliver a band-width of 1 Gbyte/s and to have sufficient design margins to allow experimentation of more complex switchnet architecture in the future.

Using off-the-shelf components for the switchnet would result in an oversized board of high power consumption. To overcome these problems, we have designed a semicustom chip based on 2 μm CMOS gate array technology. This device, built by Plessey Ltd., connects one bit in the data-path across the sixteen FPU's and memories. Pipeline flip-flops are contained inside the chip. The switchnet requires 32 such chips, packaged in standard 40 pin DIP's. A prototype of the board is being tested at present.

3.4. Sequencer

The APE computer has one dedicated sequencer board, broadcasting the same nano-code to all the FPU's on a dedicated bus. The nano-code, controlling the cycle by cycle status of the machine, is 64 bits wide.

APE is a microcoded machine, completely bypassing any assembly language level of control. As a consequence, the size of the control store is large, nano-code sequences being usually very long (typically of order of 100–500 machine cycles). At present, the size of the writable control store of the APE sequencer is 16 Kwords of 96 bits each, to be expanded to 64 kwords in the near future. A 64 bit pattern is broadcasted to the FPU's, while 32 bits are used within the sequencer to control the flow of the nano-code stream. Sequencing is accomplished using a standard architecture based on the AMD-2909 μsequencer chips. Four AMD2909 are used to provide 16 bit addressing. The writable control store uses static CMOS RAM chips (16 k × 1 bit, to be replaced by 64 k × 1 bit) with access time of 45 ns. Parity bits are also stored (on a byte basis) and are checked at every clock cycle.

3.5. The controller

The controller performs all integer arithmetic and logical operations that are not mapped onto the FPU array. It also controls the operation of the FPU's through the sequencer and generates addresses for the memories and controls for the switchnet.

The controller is based on a CERN/SLAC 3081/E processor, described in details elsewhere [1]. This processor is well suited as the APE controller, since it is a synchronous machine with a clock period of 120 ns, well matching the FPU clock. It also has integer processing power adequate to fulfill the addressing requirements of the memory array. Finally, the 3081/E is interfaced to the VME bus via dedicated boards.

3.6. Host interface hardware

The host system of the APE machine is a VAX/VMS computer. The connection is done in two steps. First, the 3081/E is interfaced to the VME bus via dedicated boards. Programs running on a VME CPU (at present a Data Sud CPUA-1) control the 3081/E at the hardware level. Communication between VME and the VAX is accomplished via a specially developed VME-QBUS/UNIBUS controller, having a peak transfer rate of 500 kbytes/s. The main advantage of this configuration is that of keeping the user level and the hardware level well separated, allowing a relatively easy migration to different host systems. In fact, we are considering at present the possibility of using a VME system running UNIX as the host of APE.

The current APE implementation uses conservative design principles. TTL-compatible parts are used throughout and no custom-made components are used. Custom components have been developed and will be used in the switchnet for the full scale machine however. Fig. 4 is a picture of one 4 cells APE prototype. A single 19" rack houses the 3081/E the memory banks and the FPU array. The two first prototypes use wire wrap technology. The full scale prototype is being built using PC cards.

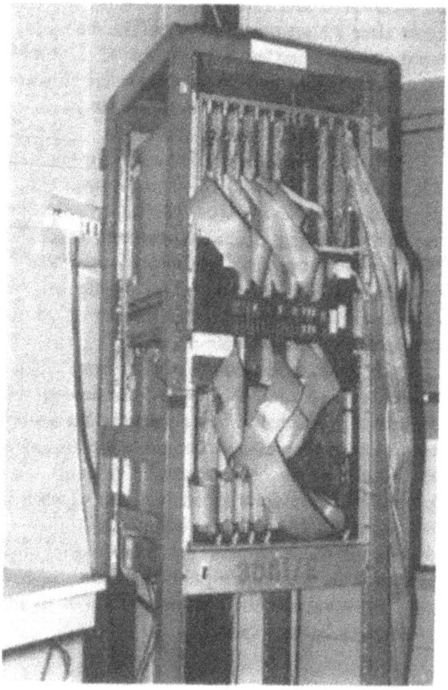

Fig. 4. Picture of the APE prototype #2.

4. Software

High level users can fully control APE capabilities while logged onto the host computer.

A typical APE working session can be described as follows:
The user logs onto the host computer and edits an APE language source program. He then invokes the APE language compiler to obtain an APE executable file and subsequently loads the executable code on APE, runs and debugs it, examines computed results and backups them using Hack: the Host resident APE Control Kernel.

4.1. The APE language and its compiler

APE language is a structured language quite similar to Fortran, even though it faithfully reflects the APE SIMD architecture.
The flow control is managed by language instructions directed to the controller: a branch is generated if data on the controller or cells satisfy the programmed condition.
Control statements are divided into two classes:
The first group consists of:

- if(condition)then . . . else . . . endif,
- for . . . endfor,
- repeat . . . until(condition),
- while(condition) endwhile.

These statements are used when the tested data reside on the controller memory: typically an address or loop counter.
The second group refers to parallel processing:

- where(condition) . . . endwhere,
- ifall(condition)then . . . elseall . . . endifall,
- repeatall . . . untilall(condition),
- whileall(condition) . . . endwhile.

The "where(condition) . . . endwhere" executes the nested instruction block only on the FPUs where data satisfy the required condition.

Using control statements such as "ifall(condition)then . . . endifall", the condition must be referred to data residing on FPUs. The global condition will be statisied only if it is true on all cells.

Data storage is allocated by explicit declaration of "integer" (32 bits), "real" (32 bits) or "complex" (32 + 32 bits) variables and multi-dimensional arrays.

However the instruction set allows closer control on data storage. When scalar variables are declared, data storage may be forced by specifiers:

- register: data will reside on registers (FPUs or controller).
- static: data will reside on memories (cells or controller).

Arrays of FPU registers may be declared: they are named "fast" arrays because a DMA is performed when loading or storing them. "fast in" and "fast out" statements permit data transfer between static and "fast" arrays. When needed the "allocate . . . endallocate" statements actually allocate controller registers for local variables.

The compiler automatically optimizes sequencer code by rearranging it. A good APE programming experience will improve code efficiency, relieving part of the burden of the optimizing step of the compiler. An expert APE programmer will minimize controller branches during inner loops to reduce pipeline breaking. He/she will obtain maximum performance using, as much as possible, controller and FPUs register variables and "fast" arrays when vector processing is required. Source optimization is facilitated by optimization symbolic data produced by the compiler. These tables allow easy identification of time critical source segments.

A preprocessor supporting macro expansion, token definition and source text inclusion is available. Its performance is comparable to that of a C language-preprocessor, while APE oriented improvements are implemented.

The compiler saves the symbols tables for the symbolic Hack debugger. Time synchronization between controller and sequencer micro-codes is performed by a time-linker.

4.2. The system software

The structure of the system software of APE present two levels (fig. 5)

- Hack is the higher level software running on the host computer.
- The lower level Ac&Bc (APE software Channel & Backup software Channel) is hardware dependent. It runs distributed over the APE ↔

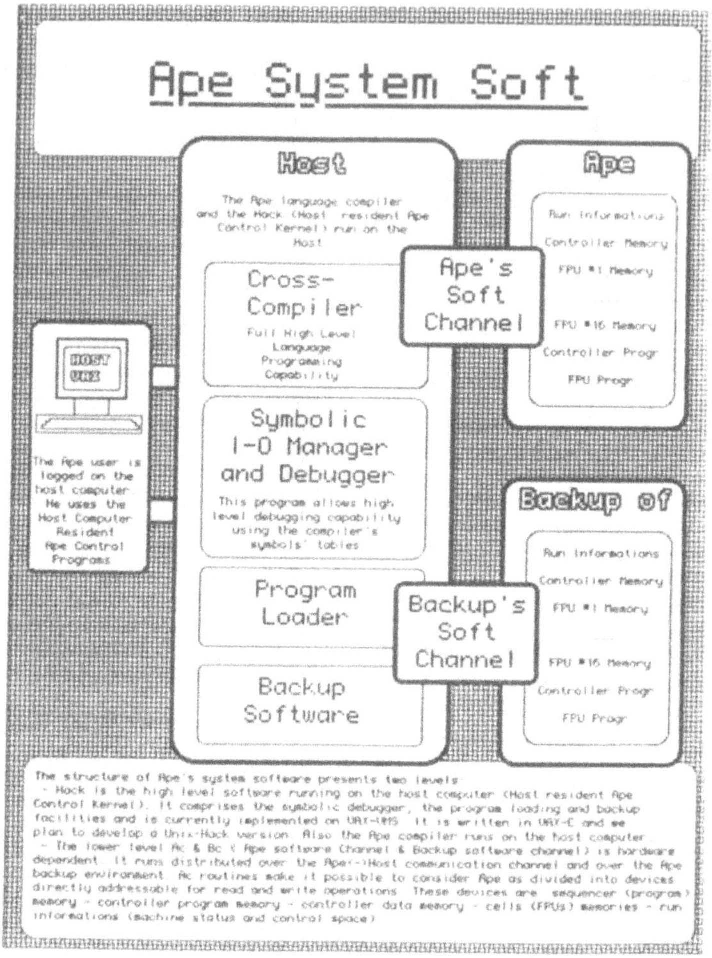

Fig. 5. The system software of APE.

host communication channel and over the APE ↔ backup environment.

Hack comprises the symbolic debugger, the program loading and backup facilities and is currently implemented on Vax-Vms. It is totally Vax C written and we plan to develop a Unix-Hack version.

The user interactive interface is provided by the Hack command interpreter supporting the Hack command language. The user can also easily incorporate calls to Hack routines in host programs. For instance a Vax-Fortran or Vax-Dcl program can call Hack's routines to run a program on APE and wait for APE run completion, and symbolically access APE data for further use.

The symbolic debugger can access variables and arrays either by name or by address. It can also start programs at given labels. Comparison of symbolically selected data on APE memories and corresponding data in backup save-sets is also possible.

Monitoring and machine status display routines are also available. The Hack command language is easily extendable, as new features are needed.

APE is network accessible, since Hack is integrated in the host computer environment.

Hack is independent of the channels' hardware as it works calling the lower level Ac&Bc.

Ac routines make it possible to consider APE as divided into devices directly addressable for read and write operations. These devices are:

sequencer (program) memory,
controller program memory,
controller data memory,
cells' memories,
machine status and control space.

Ac allows monitoring, starting and stopping of APE by reading and writing status registers and control space locations of the machine.

APE backups are structured by a similar architecture and are managed by Bc. This similarity allows Hack to access data resident on APE or on backup save sets in the same fashion.

Hack commands reconfigure Ac&Bc calls to match the actual APE hardware configuration when memory size and number of cells vary.

5. Physics applications and conclusions

The four cell prototypes have been used since september 1986 to perform lattice simulations of pure gauge systems. In particular, the glueball spectrum has been analysed, yielding results for the masses of the 0^{++} and 2^{++} states, published elsewhere [3]. The typical source code used in the simulation is about 1500 lines long, while the size of the executable codes is 180 Kbytes (or 11 Kwords). The time required to update one link is 40 μsec. This corresponds to an average speed of 70% of the theoretical peak speed. Simulations have been performed on lattices of size $10^3 * 32$, $12^3 * 32$ and $16^3 * 32$, the last lattice size almost saturating the present memory size of 32 Mbytes. More than 20000 lattice iterations have been performed in the three cases. This has required a total CPU time of about 450 hours. For comparison, a similar calculation would require about 500 CPU hours on a CRAY 1XMP supercomputer.

Overall system reliability has been very high. No system crashes have ever occurred during production runs that have lasted about 20 days.

In summary, a special purpose processor has been built and is being used for theoretical physics simulation, while application in other fields (e.g. signal processing) are being considered.

Performance of the system is comparable to that obtainable on state of the art supercomputers, for the specific application for which APE has been conceived, while development costs are roughly one order of magnitude lower than the tag price of a supercomputer.

Key features of the machine are the development of an optimized (but not over-specialized) architecture, the use of recently made available floating point chips and, possibly most important, the development of a high-level programming environment reflecting the structure of the machine in an user friendly fashion.

References

[1] P.M. Farran et al., Proc. of the Conf. on Computing in High Energy Physics, eds. L.O. Hertzberg and W. Hoogland (North-Holland, Amsterdam, 1986).

[2] The APE collaboration, Phys. Lett. to be published.

THE APE-100 COMPUTER: (I) THE ARCHITECTURE

CLAUDIA BATTISTA, SIMONE CABASINO, FRANCESCO MARZANO, PIER S. PAOLUCCI, JARDA PECH*, FEDERICO RAPUANO, RENATA SARNO, GIAN MARCO TODESCO, MARIO TORELLI, WALTER TROSS and PIERO VICINI

Infn Sezione di Roma
and Dipartimento di Fisica
Università di Roma La Sapienza
P. Aldo Moro, 00187 Roma (Italy)

NICOLA CABIBBO, ENZO MARINARI†, GIORGIO PARISI and GAETANO SALINA

Infn Sezione di Roma Tor Vergata
and Dipartimento di Fisica
Università di Roma Tor Vergata
V. della Ricerca Scientifica, 00173 Roma (Italy)

FILIPPO DEL PRETE, ADRIANO LAI, MARIA PAOLA LOMBARDO‡ and RAFFAELE TRIPICCIONE

Infn Sezione di Pisa
56100 Pisa (Italy)

ADOLFO FUCCI

Cern
Geneva (Switzerland)

---

* on leave from Institute of Physics, Czechoslovak Academy of Sciences, 18040 Prague 8, Czechoslovakia

† and NPAC and Physics Department, Syracuse University, Syracuse, N.Y. 13244, USA

‡ and Physics Department, University of Ilinois at Urbana-Champaign, Urbana, IL 61801, USA

ABSTRACT

We describe APE-100, a SIMD, modular and fine-grained parallel processor architecture for large scale scientific computations. The largest configuration that will be implemented in the present design delivers a peak speed of 100 Gflops. This performance is, for instance, required for high precision computations in Quantum Chromo Dynamics, for which APE-100 is very well suited.

Keywords: Parallelism, architectures, floating point, VLSI.

1. Overview. In the years 1985–1987, the APE collaboration has been involved in a major effort to design and build a parallel computer in the 1 Gflops range. APE [1] has been one among several projects [2] that have built floating point engines mainly tailored to the requirements of numerical simulations of Lattice Gauge Theories (LGT) and especially of Lattice Quantum Chromo Dynamics [3] (QCD), the gauge theory which, in the continuum limit, is supposed to describe the strong interactions between elementary particles. Three APE units, featuring floating point performances between 256 Mflops and 1 Gflops, have been completed and have been heavily used for LGT simulations. A large variety of physics results have been obtained on the APE computer. We have been studying, for example, the pure gauge and the hadronic mass spectrum in quenched QCD, and the deconfinement phase transition of quarks and the hadron spectrum in lattice QCD [4].

We will describe here the architecture of the next step, APE-100, a design that, by including all of the positive features of APE, has been able to push the peak speed to 100 Gflops [5]. The details of the hardware and software implementation of this architecture can be found elsewhere[6]. In Tables 1 and 2 we give the basics technical informations about the Ape-100 computers.

We recall that although APE was used mainly for Lattice Gauge Theory simulations, both APE and APE-100 are general purpose SIMD computers. The architecture of APE-100 is more flexible than that of APE and we plan to use APE-100 for many physical applications beyond LGT (e.g. fluidodynamics).

The SIMD architecture of APE-100 is based on a three dimensional cubic mesh of nodes with periodic boundary conditions, each node being connected to its 6 neighbors. APE-100 has a modular architecture. The building block is a $2\times2\times2$ cube, while a 2048-node APE-100 can be configured as a $8\times8\times32$ lattice. The 6 Gflops version is implemented with a simple array of 16 cubes. Such connection grids are well suited for the simulation of homogeneous physical systems (including, of course, Lattice Gauge Theories and Lattice QCD, and Statistical Lattice Systems [7], [8]). Discussing various aspects of the APE-100 design we will stress here the new ideas and implementations which allow the feasibility of a 2048-node machine, with a floating point performance of 100 Gflops peak speed. In principle the design is such that

TABLE 1
Ape-100 at a glance - 1

Three dimensional topology.
Fine grained modular SIMD architecture (from 8 to 2048 nodes).
4 Mbyte, 128 registers and 50 Mflops per node.
Hardware support for local (non-SIMD) conditional structures.
Centralized address and integer computations.
25 MHz master clock.
8 nodes per Processing Board.
1 to 16 crates, each with 16 Processing Boards and 1 Controller.

TABLE 2
Ape-100 at a glance - 2

	1 PB	1 Crate	Max Conf
Floating Point	400 Mflops	6.4 Gflops	100 Gflops
Data Memory Size	32 Mbyte	512 Mbyte	8 Gbyte
RAM to reg rate	400 Mbyte/sec	6.4 Gbyte/sec	100 Gbyte/sec

one could also build a 32768 node machine on a $8 \times 64 \times 64$ lattice, with a peak speed of 1.6 Tflops, but the construction of such a machine is not at present planned. The viability of this project has been specially due to the large use of custom VLSI components[9] and to the large effort devoted to the software development[10].

2. The APE-100 architecture. The architecture of APE-100, fig. 1, can be seen as composed of two layers: a synchronous kernel and an asynchronous shell. The software environment handles all the interactions, which are completely transparent to the user.

2.1. The synchronous kernel. The synchronous kernel is a Single Instruction Multiple Data (SIMD) 3*D* cubic mesh of nodes. Each node consists of a floating point unit (the *Multiplier and Adder Device,* MAD, a Communication and Control Unit (the *Commuter*) and a Memory Bank. The nodes are driven by a synchronous computer: the S-CPU[11]. All the elements of the kernel are controlled by a 25 MHz clock.

The heart of the APE-100 computing power is the floating point processor MAD [9], a custom VLSI device. MAD is a floating point 32 bit real arithmetics processor, and contains one adder and one multiplier. It has a peak floating point performance of 50 Mflops. The MAD architecture, which

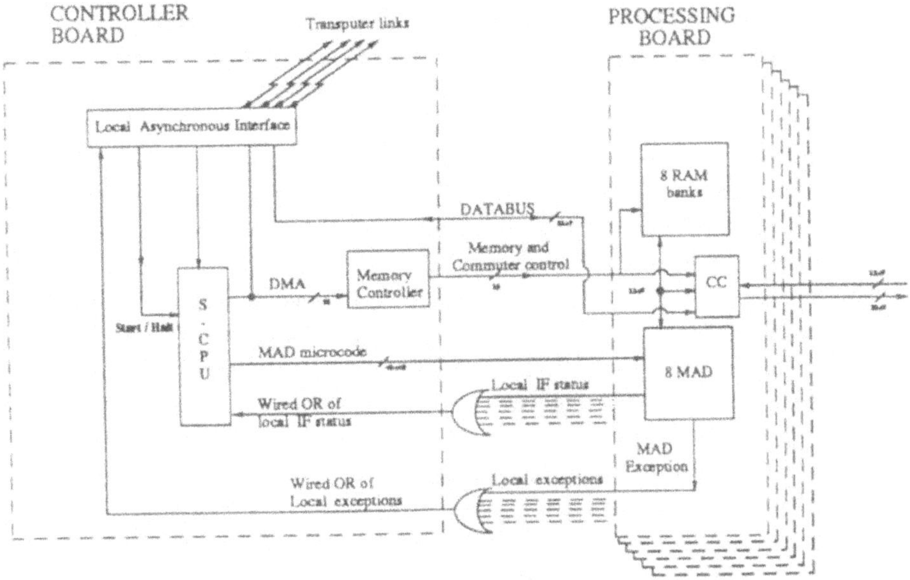

FIG. 1. *Schematic diagram of the controller and of the processing boards.*

we will describe in the following, guarantees that a big fraction of this theoretical peak speed can be attained for an important variety of numerical algorithms. MAD has built in a large amount of support circuitry for error detection and correction. It contains specialized hardware for performing conditional write instructions and look up tables.

The communication is handled by a second custom VLSI device, the Commuter, which takes care of the node-to-node data transfer as well as of the communications with the asynchronous interface.

A processing node contains one MAD and its 4 Mbyte local memory. Eight such nodes are assembled on a Processing Board (PB), together with a Commuter. A 2048 node model contains 256 PB's (with a total of 8 Gbyte of memory) assembled in 16 crates (housed in 4 racks). The 128 node 6 Gflops version is built with 16 PB's which are housed in a single crate.

The S-CPU is housed in the Controller Board, and controls the nodes. The Controller is replicated, and there is one S-CPU in each crate. This replication gives, among others, the advantage that APE-100 can be segmented into smaller independent units. The 2048-node version, for example, can be configured as four 512 node machines (or sixteen 128 node machines). Segmented configurations are supported by the Transputer network (which we will describe in the following), and are made possible by re-arranging subsections of the global mesh of processors in smaller blocks with periodic boundary connections. It is obvious that in this configuration communication among different machines is only supported by the Transputer network.

2.2. The asynchronous shell. The asynchronous shell can be seen as composed by two main elements: the host computer and the asynchronous interface. The host computer is the user interface: it runs the cross-compiler and the host resident part of the monitor/debugger. These services are available through local terminals or via network. The host is connected to the synchronous kernel by the asynchronous interface. Via this interface the host can load programs and data onto the synchronous kernel, define its configuration, start and stop the execution of synchronous programs, examine the result of the computation. The asynchronous interface is also responsible for the management of the APE-100 mass memories. It can save/load the content of the PB's memories to disks. The asynchronous interface is based on a network of Transputer which can be programmed to execute these tasks. The I/O facilities can be easily used from the user applications programs running on APE-100.

The asynchronous interface is distributed. A Transputer is the heart of the *Local Asynchronous Interface* (LAI), associated to each S-CPU and housed in the controller board. Another Transputer, the root Transputer, forms the *Root Asynchronous Interface* (RAI), connected to the host computer. In the processing board the Commuters have an asynchronous section accessible by the LAI. In single controller configuration (i.e. up to 6 Gflops) there is no need for the RAI, as the LAI is directly connected to the host computer.

2.3. Data types and program flow. Our architecture naturally distinguishes two types of data: *global data* (on the S-CPU) and *local data* (on node). In general, we regard S-CPU data as integer and local data as floating point (IEEE-754 standard). Local data reside in the node memories or on MAD registers, while global data reside in the S-CPU memory and registers.

We have implemented 3 kinds of conditional structures: IF, IF-ALL and WHERE. The IF structure is the classical non parallel Fortran-like instruction, which controls program flow according to a condition which is a Boolean function of global (in the above sense) variables. The IF-ALL instruction is a global structure, executed by the S-CPU on the basis of the logical AND of local conditions evaluated by all nodes which take part in the computation. The WHERE structure is local: it causes a block of instructions to be enabled only in those nodes where a certain local condition (the Boolean result of computations done by the local MAD) is fulfilled. Synchronization is preserved by executing the conditioned code on all nodes but disabling write operations on the nodes which do not fulfill the logical condition. This construct allows an effective breaking of the SIMD sequence.

When multiple S-CPU are active in a given program, they will execute

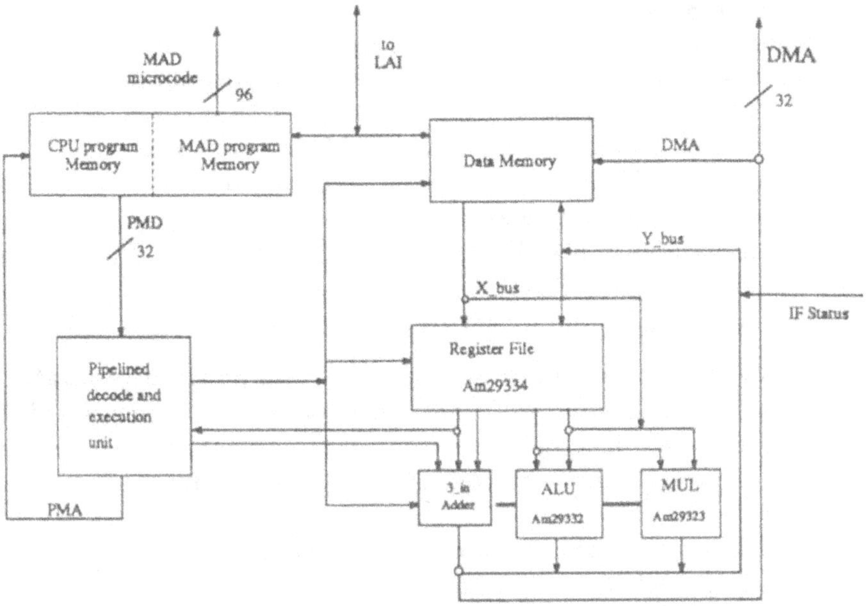

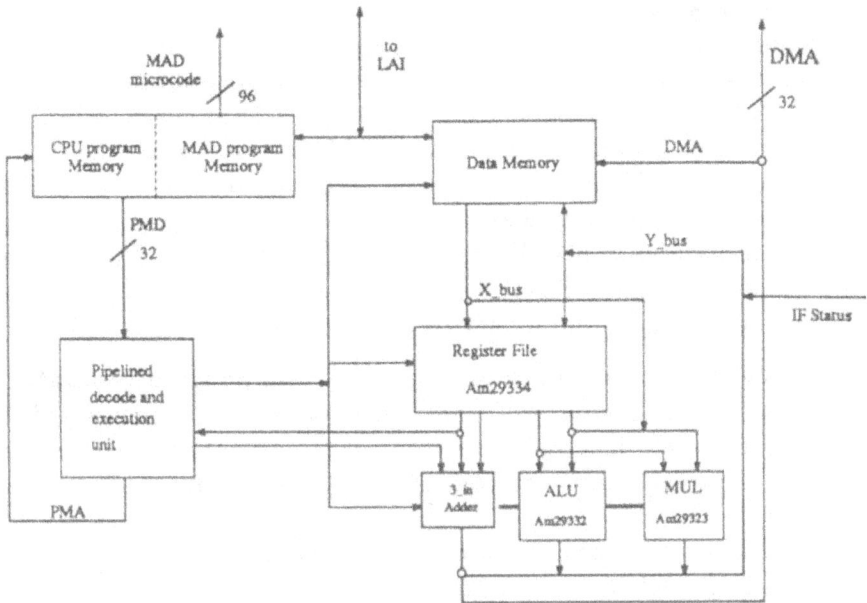

FIG. 2. *Controller block diagram.*

the same branch, be it an IF (they all have identical copies of the global variables), or an IF-ALL (they all share the conditions returned by the nodes which they control).

These basic structures can be used to implement complex ones, such as the typical Fortran-like DO loops, or new parallel ones, such as REPEAT-UNTIL-ALL or REPEAT-WHILE-ALL.

3. The hardware structure. We will discuss in the following the controller and the processing boards of the computer (with the floating point MAD chip and the communication network).

3.1. The controller. We show the controller sub-system in fig. 2. It includes the Synchronous CPU (S-CPU), program and data memories, the memory controller and a LAI for the communication with the external world. The whole sub-system is housed in a single board.

3.1.1. The Scalar CPU. The first implementation of the S-CPU [11] has been a synchronous integer processor, operating at a cycle time of 80 ns (similar to a processor developed at CERN as an improvement of the original 3081 – E[12]).

The S-CPU runs at a speed 2 times lower than the MAD: it executes one instruction for every four MAD floating point instructions (since MAD contains one adder and one multiplier). The S-CPU controls the program flow. The control word issued to the nodes can be regarded as an extension of the instruction word controlling the S-CPU itself: the same Program

Memory Address (PMA) points at both control sections. In principle the nodes can be regarded as special purpose devices of just one processor: a controller branch directs all nodes. A single program instruction contains one S-CPU instruction and two MAD instructions (since, as we said, there is a difference in the S-CPU and the MAD clock speeds).

The S-CPU also executes global conditional structures: IF and IF-ALL. The IF instruction executes a test and branch on an internal controller logical condition. The IF-ALL executes a logical AND of the conditional expressions (local IF status) on all the nodes which run the program and then a conditional branch set according to the result of the AND.

3.1.2. Node distributed memory and address generation. Data Memory Addresses (DMA), identical for all nodes, are generated by the S-CPU or by the LAI and are converted into memory and commuter control signals. All the memory control circuitry is centralized in the controller board.

There are 40 Mbyte of dynamic memory on each processing board. Each node uses 4 Mbyte to store data and 1 Mbyte to store the EDAC code used to correct single error, detect all double errors and some multiple errors.

The memory uses 4 Mbit memory chips organized as 1 Mega $\times$ 4 bits: each chip contains one bit of data belonging to four different nodes, so that after a complete failure of one chip (4 bits wrong) 4 nodes would detect a single error which would be corrected, and the machine would continue working properly.

The management of the remote data transfers are handled by the same logical circuitry as the address generation. From the programmer's point of view the first neighbor nodes can be seen as an extension of the local address space. The hardware takes care of respecting the appropriate timings.

The transfer of contiguous data from node memory to MAD proceeds at a rate of 1 word per S-CPU cycle. The data access to the memory of the first neighbor is 4 times slower. In the time interval we need to transfer a single word, four floating point operations may be executed on MAD.

3.1.3. The asynchronous interface. The LAI processor is a Transputer placed on the controller board. When the S-CPU is halted (this condition is always under LAI control) the LAI is able to assert all external busses of the controller and to access node memories. When the crate controlled by the LAI is in local mode, the LAI has complete control on it, including exception handling. Finally, the LAI monitors environmental conditions, like temperature or supply voltage, and issues alarms as appropriate.

The LAI handles exceptions produced by all nodes or by the controller. It manages the clock distributed to the synchronous sections, the global

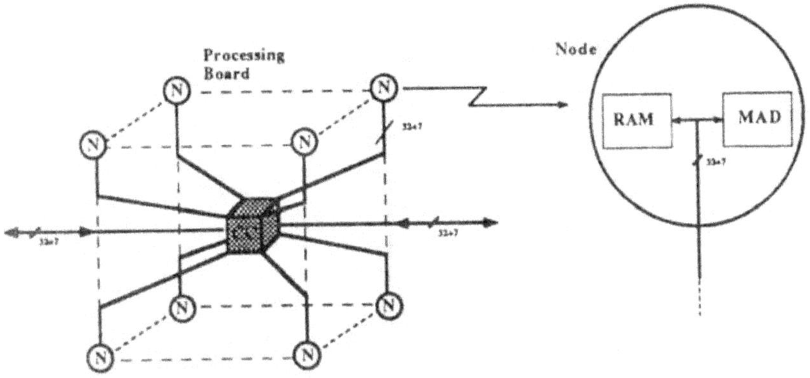

FIG. 3. *Arrangement of nodes on a Processor Board.*

start/stop signal and it controls the run/halt condition of the S-CPU.

Exceptions are collected and delivered to the RAI which controls the CDU (Clock Distribution Unit). After an exception the RAI sends a global stop signal. If the machine is divided in crate wide segments, the exceptions are treated locally by the controller of the crate.

Each controller is synchronized via a common clock routed to all crates, and they all run the same program with the same data. The S-CPU are synchronized, under the control of the LAI, via a global start-stop signal generated in the CDU. If the machine is segmented (e.g into four racks), there is a separate start-stop signal for each rack.

3.2. The nodes and the processing boards. As discussed above the nodes of APE-100 lie on a 3 dimensional simple cubic mesh. On a given PB there are 8 nodes, forming an elementary 2^3 cube. The PB also contains 10 Commuters (see fig. 3), which provide mutual inter-connection between nodes either locally (i.e. on the same PB) or remotely (i.e on neighboring boards). The cubic lattice topology is thus implemented through a distributed inter-connection system. Every PB also contains some support circuitry which allows connection of the PB's to the controller.

The PB's receive the MAD code from the Controller and deliver it to the 8 MAD's. While the MAD executes an instruction every clock cycle, the PB's are fed with a double instruction every two cycles. The instructions are unpacked locally and delivered every cycle. All the elements of the SIMD kernel are driven by a single clock which synchronizes their activities cycle

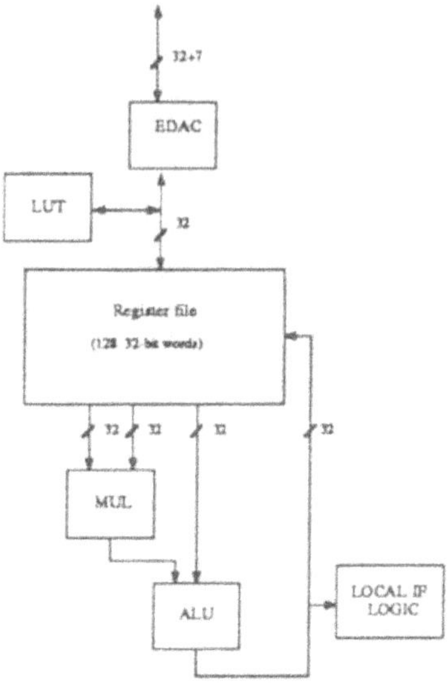

FIG. 4. *Simplified block diagram of MAD.*

by cycle.

3.2.1. The MAD. Here we present a short description of MAD; more detailed information can be found in [9]. MAD (see fig. 4) is a 40 ns pipelined VLSI custom device controlled by a 48 bit wide control word supplied on a cycle by cycle basis.

The main components of the MAD are the register file, the floating point multiplier, the logic and arithmetic floating point unit, the look-up tables, the error detection and correction unit and a status register.

The register file contains 128 32-bit registers and supports 5 simultaneous accesses (3 read, 1 write and 1 read or write). Thus on each cycle it is possible to start one floating point addition, one floating point multiplication and one I/O operation.

Input/output MAD operations use a bidirectional port of 32 data and 7 check bits, which can transfer one data word every clock cycle. When MAD inputs a word, it uses the check bits to correct single-bit errors and detect double-bit (and some multiple bit) errors. In an output operations MAD generates a 7 bit modified Hamming code. Error correction is useful both against memory errors and inter-node transmission errors. Multiple errors are fatal and stop the machine. Single errors are not fatal, but on each occurrence MAD updates a counter which can be read when the synchronous kernel is idle. The I/O bandwidth of the MAD is twice the bandwidth that can be actually used in APE-100, due to the memory access time and the absence of interleaving.

MAD contains Look-Up Tables (LUT's). LUT's are used to obtain approximated terms used in the computation of some frequently used functions like: $\frac{1}{x}$, $\log(x)$, $\frac{1}{\sqrt{(x)}}$, $\exp(x)$.

The arithmetic part of MAD consists of a multiplier and an adder hard wired for the **normal** operation: $D = (A \times (\pm B)) \pm C$. At each clock cycle one MAD can produce the results of one **normal** operation. Two of the 128 registers are permanently loaded with the values 0 and 1 so that simpler operations can be executed (without breaking the homogeneity of the code, which helps in producing well optimized codes):

- Simple sum: $D = (1 \times (\pm B)) \pm C$.
- Simple multiplication: $D = (A \times (\pm B)) \pm 0$.
- No-Operation: $0 = (0 \times (\pm 0)) \pm 0$.

The adder's result is used cycle by cycle to produce Boolean values for elementary conditions (i.e. $== 0$, > 0, $>= 0$, $!= 0$, using C language notations) which are made available to the IF circuitry. This is a 1-bit wide stack- based machine, which allows the evaluation of complex conditions by building on the Boolean operation AND, OR, NOT, to generate the *Local IF Status*. The stack allows nesting of up to eight conditional structures.

When the execution of a code section is conditioned (inside a WHERE block) and the *Local IF Status* is **false** all MAD operations are converted to NOP's (No-Operations), forbidding write operations to registers and memory. The AND of *Local IF Status* of all MAD's is delivered to the controller to obtain the global condition used by IF-ALL.

MAD is able to detect floating point exceptions (adder or multiplier overflow and LUT exceptions), non recoverable errors in the data bus and parity errors in the code bus. All the exceptions are maskable.

3.2.2. The communication network. Let us first discuss inter-node communication from the point of view of the application programmer, and then move on to hardware considerations. APE-100 is simple from the point of view of the programmer, since he can concentrate his attention on the behavior of a single node. The parallelism manifests itself through the existence of replicas of node data structures which are logically placed along the six possible directions in space. These replicas can be accessed by using predefined constant displacements. For instance, if $V(i)$ is an element of an array belonging to a given node, $V(i + Left)$ is the corresponding element on the closest left-side replica of the array. Inter-node communication can thus be controlled by an extension of the address field.

At the hardware level this model is implemented by splitting the address into two fields: a memory address, which is sent to the RAM chips, and a communication field, which is sent to the Commuters.

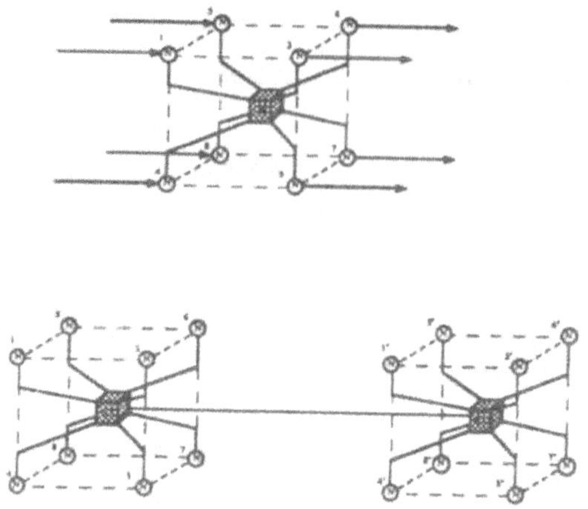

FIG. 5. *Connections between adjacent nodes.*

As we show in fig. 5, for full communication speed during an inter-node transaction the PB should receive and transmit four data words. In practice, a 4 : 1 time multiplexing in inter-node communications is implemented, reducing the amount of cabling required for inter-crate links, while commutation within a node is supported at full speed.

In a fully connected first neighbor $3D$ machine each PB would be provided with 6 links to neighboring boards. However, in our design only one pair of links is active on each memory access: the left/right pair, as in fig. 5, the up/down pair or the front/back pair. For each given PB only two (not six) links are thus sufficient. We have left to a separate board, the Connection Board (CB) (which we will describe in the following) the further routing of these links to other boards.

In the full size, 100 Gflops drive, the 2048 nodes are arranged on a $8 \times 8 \times 32$ lattice. In this section it is convenient to view this pattern as a $4 \times 4 \times 16$ lattice of PB's, each holding a $2 \times 2 \times 2$ cube of nodes. The PB's are distributed among crates and racks as shown in fig. 6. Boards with the same value of Y and Z are in the same crate, and each crate holds four X-rows, displaced along the Y direction. A rack houses a full $X - Y$ plane of PB's.

There is one CB for every PB. The CB shares the same connector of the PB and is mounted on the opposite side of the backplane. Fig. 7 schematically represents the links between CB's and PB's for a fraction of the $3D$ lattice. A link in the Y direction is highlighted at the bottom left side of the figure. The connections in the X direction run on the crate backplane, those along the Y and Z directions run over bidirectional differential lines

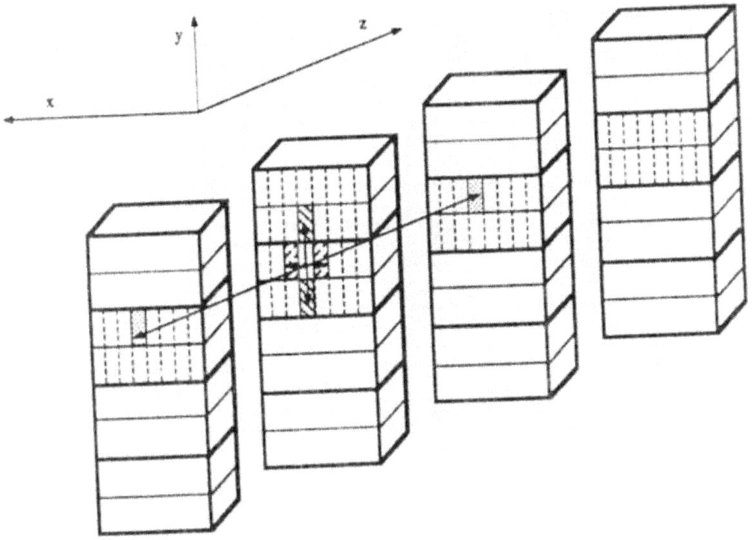

FIG. 6. 256 *PB's in* 16 *crates in* 4 *racks.*

on twisted pair ribbon cables.

To understand the way in which the topology is implemented, note that each CB is assigned the (x, y, z) coordinates of the PB which precedes it in the direction of increasing X. Assume the range of the coordinates to be $0...N_x$, $0...N_y$, $0...N_z$. The Y and Z cables are laid according to the following rule: given a CB of coordinates (x, y, z) its *top* (as one can see in fig. 7) is linked to the *bottom* of the CB $(x-1, y, z+1)$, while its *back* is linked to the *front* of CB $(x-1, y+1, z)$, all coordinates being taken modulo $N_x + 1$, $N_y + 1$, $N_z + 1$.

In conclusion, communication tasks are partitioned between PB's and CB's. This solution has several advantages with respect to the alternative (the PB's take care of all communication tasks, i.e. non CB's):

1. There are no cables directly plugged into the PB's, making them easily replaceable.
2. A smaller number of backplane pins is used for outboard connections and the complexity and pin count of the Commuter is also reduced.
3. The *hot* bipolar chips needed for buffering and level translation to/from differential lines are mounted far away from the *cool*, (mostly) CMOS chips on the PB's.

4. The software structure. In the following we will discuss the operating system (OS), the compiler, the *Apese* language, the optimizer and the assembler language. A more detailed description can be found in [10].

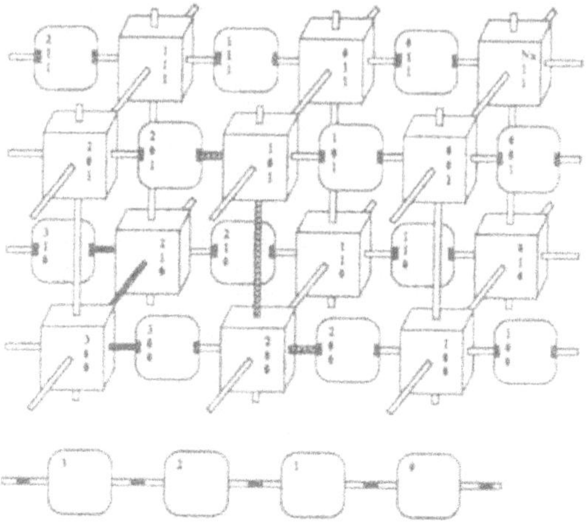

FIG. 7. *3D connections of PB's and CB's. The cubes represent CB's, the rounded rectangles PB's. The link between PB's with* $(x, y, z) = (3, 0, 0)$ *and* $(3, 1, 0)$ *is highlighted in black.*

4.1. The operating system. As we have seen before the user controls APE-100 by means of the host computer, that provides a conventional file system and application development environment. The communication of APE-100 with the external world (host computer and mass storage devices) is asynchronous and is controlled by a network of Transputer.

To describe the functionality of the APE-100 OS it is useful to distinguish two different operating modes: user and system. The user writes programs in a high level or assembly language, from which code for the S-CPU and FPU's is generated. When an user program needs an asynchronous service (e.g. an I/O operation), it halts the S-CPU and switches to system mode under Transputer control.

The Transputer is able to gain control over the system also when abnormal conditions occur. Typical examples are the raising of an exception, or an operator request, or de-scheduling of an user program that exceeded some resource quota (for example the assigned maximum CPU time).

The code runs on the host computer, which contains the user interface called HACK (Host-resident Ape Control Kernel). HACK provides the user with an interface to access APE-100 devices, issue I/O operations, load and run programs, monitor the machine status, etc.

4.2. The compiling system. The compiler chain runs on the host computer. The main steps in the chain are the compiler (or the assembler) and the optimizer. The optimizer has some characteristics of generality which has allowed its porting to from APE to APE-100 with only minor changes. In the following, we describe the organization of the compiler chain.

4.2.1. The *Apese* language. Here is an overview of the *Apese* language:

- there is a minimal subset of *Apese* that can be easily learned by every physics application oriented user. This subset is a structured language inspired by Fortran and C and is similar to the language used with the first generation APE machines;
- the language faithfully matches the APE-100 hardware architecture;

Even if the *Apese* language helps to write effective programs, an APE-100 user should be well aware of some architectural characteristics. The programmer should consider APE-100 computers as a three dimensional mesh of nodes (from 8 up to 2048) with periodic boundary connections.

Data words on nodes are single precision 32 bits IEEE-754 standard floating point numbers. All the nodes execute the same code, typically and hopefully acting on different data. Parallel processing on APE-100 is limited to operations on floating point numbers. Every time the *Apese* programmer declares a floating point data structure memory will be reserved on all node distributed memories. The allocated memory will be placed on every node at the same local address, that will be associated to the name of data structure. Therefore each operation written in *Apese* language, acting on that name, will actually activate the same operation on every node of the mesh, choosing the data stored at the associated local address.

The programmer should also consider that integer arithmetic is performed by the Controller (which is in charge of the instruction flow and of addressing). As we have already discussed at length there are three type of conditioning that can modify the program execution: the first two IF and IF-ALL are managed by the Controller and may cause a true branch in the program flow, while the third one WHERE is locally managed by each node and may result in a temporary suspension of the effects of the program execution on that node, allowing a synchronous form of local program conditioning.

4.2.2. The optimization. During the first step in the compilation the compiler optimizes the register usage and the I/O to and from memory. The second step of the compilation chain produces code optimization. The optimizer reads the intermediate code produced by the compiler and generates the executable code.

Let us now describe the principles of operation of the optimizer, which mainly acts on the floating point part of the code. Its structure is quite similar to that of the optimizer used in APE. In the first phase the optimizer tries to combine multiplications and additions into so-called **normal** operations for which the architecture of MAD is optimized. In a second phase (pipeline and device usage optimization) the optimizer rearranges and packs the code, attempting to find the position of each instruction that minimizes the total number of machine cycles. Two kinds of constraints are taken into account: the operands of an instruction must be calculated before being used, and two instructions cannot use the same hardware device at the same time. The cycle number at which each device of the computer (busses, registers, I/O ports, floating point adder and multiplier) is reserved by each instruction is defined by optimizer tables. Apart from these constraints, the optimizer is free to re-organize the code with the goal of filling the pipeline. The last phase consists in the allocation of the registers. Finally the executable code is produced.

Acknowledgements. It is a pleasure for us to thank Ettore Remiddi and Pedro Mato for useful discussions and suggestions.

REFERENCES

[1] P. Bacilieri, S. Cabasino, F. Marzano, P. Paolucci, S. Petrarca, G. Salina, N. Cabibbo, C. Giovannella, E. Marinari, G. Parisi, F. Costantini, G. Fiorentini, S. Galeotti, D. Passuello, R. Tripiccione, A. Fucci, R. Petronzio, F. Rapuano, D. Pascoli, P. Rossi, E. Remiddi and R. Rusack, *The Ape Project: a 1 Gflops Parallel Processor for Lattice Calculations*, in *Computing in High Energy Physics*, edited by L. O. Hertzberger and W. Hoogland, North-Holland, Amsterdam 1986;

M. Albanese, P. Bacilieri, S. Cabasino, N. Cabibbo, F. Costantini, G. Fiorentini, F. Flore, L. Fonti, A. Fucci, M. P. Lombardo, S. Galeotti, P. Giacomelli, P. Marchesini, E. Marinari, F. Marzano, A. Miotto, P. Paolucci, G. Parisi, D. Pascoli, D. Passuello, S. Petrarca, F. Rapuano, E. Remiddi, R. Rusack, G. Salina and R. Tripiccione, *The Ape Computer: an Array Processor Optimized for Lattice Gauge Theory Simulations*, Comp. Phys. Comm., 45 (1987), pp. 345-353; see also:

G. Parisi, F. Rapuano and E. Remiddi, *The APE Computer and First Physics Results*, in *Lattice Gauge Theories Using Parallel Processors*, edited by Li Xiaoyuan et al., Gordon and Breach, London 1987);

[2] For detailed reviews, see

N. H. Christ, *QCD Machines*, Nucl. Phys. B (Proc. Suppl.) 9 (1989) pp. 549-556;

R. Tripiccione, *Dedicated Computers for Lattice Gauge Theories*, Nucl. Phys. B (Proc. Suppl.) 17 (1990) pp. 137-145;

N. H. Christ *QCD Machines - Present and Future*, Nucl. Phys. B (Proc. Suppl.), 20 (1991) pp. 129-137;

N. Cabibbo, talk given at the LAT91 Conference on Lattice Field Theory, Tsukuba, Japan, November 1991, to be published in Nucl. Phys. B (Proc. Suppl.);

[3] K. G. Wilson, *Confinement of Quarks*, Phys. Rev. D10 (1974) pp. 2445-2459;

[4] See for example:

S. CABASINO, F. MARZANO, J. PECH, F. RAPUANO, R. SARNO, W. TROSS, N. CABIBBO, E. MARINARI, P. PAOLUCCI, G. PARISI, G. SALINA, G. M. TODESCO, M. P. LOMBARDO, R. TRIPICCIONE AND E. REMIDDI, *The Ape with a Small Jump*, Nucl. Phys. B (Proc. Suppl.), 17 (1990) pp. 218-222;
S. CABASINO, F. MARZANO, J. PECH, F. RAPUANO, R. SARNO, W. TROSS, N. CABIBBO, A. L. FERNANDEZ, E. MARINARI, P. PAOLUCCI, G. PARISI, G. SALINA, A. TARANCON, G. M. TODESCO, M. P. LOMBARDO, R. TRIPICCIONE AND E. REMIDDI, *The Ape with a Small Mass*, Nucl. Phys. B (Proc. Suppl.), 17 (1990) pp. 431-435;
S. CABASINO, F. MARZANO, J. PECH, F. RAPUANO, R. SARNO, G. M. TODESCO, W. TROSS, N. CABIBBO, M. GUAGNELLI, E. MARINARI, P. PAOLUCCI, G. PARISI, G. SALINA, M. P. LOMBARDO, R. TRIPICCIONE AND E. REMIDDI, *Ape Quenched Spectrum*, Nucl. Phys. B (Proc. Suppl.), 20 (1991) pp. 399-405;

[5] N. AVICO, P. BACILIERI, S. CABASINO, N. CABIBBO, L. A. FERNANDEZ, G. FIORENTINI, A. LAI, M. P. LOMBARDO, E. MARINARI, F. MARZANO, P. S. PAOLUCCI, G. PARISI, J. PECH, F. RAPUANO, E. REMIDDI, R. SARNO, G. SALINA, A. TARANCON, G. M. TODESCO, M. TORELLI, R. TRIPICCIONE, W. TROSS, *From APE to APE-100: From 1 to 100 Gflops in Lattice Gauge Theory Simulations*, Comp. Phys. Comm. 57 (1989) pp.285-289;
N. AVICO, C. BATTISTA, S. CABASINO, N. CABIBBO, F. DEL PRETE, A. FUCCI, A. LAI, M. P. LOMBARDO, E. MARINARI, F. MARZANO, P. S. PAOLUCCI, G. PARISI, J. PECH, F. RAPUANO, R. SARNO, G. SALINA, G. M. TODESCO, M. TORELLI, R. TRIPICCIONE, W. TROSS, P. VICINI, *A 100 Gflops Parallel Computer*, Preprint n.737, Dipartimento di Fisica, Università di Roma *La Sapienza* (1990);

[6] A. BARTOLONI, C. BATTISTA, S. CABASINO, F. MARZANO, P. PAOLUCCI, J. PECH, F. RAPUANO, R. SARNO, G. TODESCO, M. TORELLI, W. TROSS, P. VICINI, R. BORGOGNONI, F. DEL PRETE, A. LAI, R. TRIPICCIONE, N. CABIBBO, A. FUCCI, Preprint n. 838, Dipartimento di Fisica, Università di Roma *La Sapienza* (1990);

[7] S. K. MA, *Modern Theory of Critical Phenomena*, (Benjamin, New York, USA 1976);

[8] G. PARISI, *Statistical Field Theory*, (Addison-Wesley, Redwood City, USA 1988);

[9] A. BARTOLONI, C. BATTISTA, S. CABASINO, N. CABIBBO, F. DEL PRETE, F. MARZANO, P. S. PAOLUCCI, R. SARNO, G. SALINA, G. M. TODESCO, M. TORELLI, R. TRIPICCIONE, W. TROSS, P. VICINI, E. ZANETTI, Particle World 2 (1991) 65;

[10] THE APE COLLABORATION, in preparation;

[11] THE APE COLLABORATION, in preparation;

[12] P. M. FARRAN ET AL., *The 3081 – E Emulator*, in *Computing in High Energy Physics*, edited by L. O. Hertzberger and W. Hoogland, North-Holland, Amsterdam 1986;

www.ingramcontent.com/pod-product-compliance
Lightning Source LLC
Chambersburg PA
CBHW070741220326
41598CB00026B/3719
* 9 7 8 9 8 1 0 2 1 3 5 6 5 *